Advances in Intelligent Systems and Computing

Volume 979

The series "Advances in Intelligent Systems and Computing" contains publications on theory, applications, and design methods of Intelligent Systems and Intelligent Computing. Virtually all disciplines such as engineering, natural sciences, computer and information science, ICT, economics, business, e-commerce, environment, healthcare, life science are covered. The list of topics spans all the areas of modern intelligent systems and computing such as: computational intelligence, soft computing including neural networks, fuzzy systems, evolutionary computing and the fusion of these paradigms, social intelligence, ambient intelligence, computational neuroscience, artificial life, virtual worlds and society, cognitive science and systems, Perception and Vision, DNA and immune based systems, self-organizing and adaptive systems, e-Learning and teaching, human-centered and human-centric computing, recommender systems, intelligent control, robotics and mechatronics including human-machine teaming, knowledge-based paradigms, learning paradigms, machine ethics, intelligent data analysis, knowledge management, intelligent agents, intelligent decision making and support, intelligent network security, trust management, interactive entertainment, Web intelligence and multimedia.

The publications within "Advances in Intelligent Systems and Computing" are primarily proceedings of important conferences, symposia and congresses. They cover significant recent developments in the field, both of a foundational and applicable character. An important characteristic feature of the series is the short publication time and world-wide distribution. This permits a rapid and broad dissemination of research results.

**** Indexing: The books of this series are submitted to ISI Proceedings, EI-Compendex, DBLP, SCOPUS, Google Scholar and Springerlink ****

More information about this series at http://www.springer.com/series/11156

Debashis Dutta · Biswajit Mahanty
Editors

Numerical Optimization in Engineering and Sciences

Select Proceedings of NOIEAS 2019

Springer

Editors
Debashis Dutta
National Institute of Technology Warangal
Warangal, Telangana, India

Biswajit Mahanty
Indian Institute of Technology Kharagpur
Kharagpur, India

ISSN 2194-5357 ISSN 2194-5365 (electronic)
Advances in Intelligent Systems and Computing
ISBN 978-981-15-3214-6 ISBN 978-981-15-3215-3 (eBook)
https://doi.org/10.1007/978-981-15-3215-3

This Springer imprint is published by the registered company Springer Nature Singapore Pte Ltd.
The registered company address is: 152 Beach Road, #21-01/04 Gateway East, Singapore 189721, Singapore

Contents

About the Editors

Dr. Debashis Dutta is a Professor in the Department of Mathematics, National Institute of Technology, Warangal, India. He obtained his M.Sc in Mathematics in 1988, and Ph.D. in Operations Research from IIT Kharagpur in 1994. He has teaching experience of 25 years and research experience of 29 years. He has supervised five Ph.D students and 49 post-graduate dissertations. He has successfully completed 2 sponsored research projects by MHRD. He is a member of ISTE, APSMS, and SDSI. Dr. Dutta is a reviewer of 9 journals and is on the editorial board of 2 journals. He has published 35 articles in international journals, 6 in national journals, and has also authored 4 books. He has organized two short term training programs in Statistics and Optimization Techniques.

Dr. Biswajit Mahanty is a Professor at the Department of Industrial and Systems Engineering at the Indian Institute of Technology (IIT) Kharagpur, India. In the recent past, he was Dean (Planning and Coordination) at IIT Kharagpur. He obtained his B.Tech (Hons) in Mechanical Engineering, and his M.Tech and Ph.D. in Industrial Engineering and Management, all from IIT Kharagpur. He has had a varied professional career with over six years of industrial experience and 28 years of teaching, research, and industrial consulting experience. His areas of interest include supply chain management, e-commerce, transportation science, technology management, software project management, and system dynamics. He has guided 15 doctoral and more than 150 undergraduate and post-graduate level dissertations. Dr. Mahanty has also carried out more than 20 industrial consulting projects and 10 sponsored research projects. He has more than 100 publications in national and international journals and conferences of repute. He has authored a book titled "Responsive Supply Chain" by CRC press. He has also taught at the School of Management at AIT, Bangkok as a visiting faculty member. He is on the editorial board of the journal Opsearch published by the Operational Research Society of India.

Hydro-Chemistry for the Analysis of Sub-surface Water Quality in North-Eastern Haryana: A Fast-Urbanizing Region

Sandeep Ravish, Baldev Setia and Surinder Deswal

Abstract Hydro-geochemical characteristics of sub-surface water in the concerned region comprising Yamunanagar and Ambala districts of Haryana, India, were estimated. The hydro-geochemical feature of sub-surface water in the concerned region was investigated by collecting 30 sub-surface water samples. Groundwater samples from specific deposits were analysed for physico-chemical elements, i.e. pH, TDS and prime ion contents, i.e. potassium, sodium, magnesium, calcium, bicarbonate, chloride and sulphate. These ions were as abundance of sodium > calcium > magnesium > potassium, and bicarbonate > chloride > sulphate, respectively. The principal component analysis (PCA) and hydro-geochemical diagrams have been found to be in good agreement in optimizing the significant elements. Analysis of chemical dataset represented that the predominant hydro-chemical facies in the area of study were Na^+–HCO_3–Cl^- and 'Ca^{2+}–Mg^{2+}–HCO_3–Cl^-' types. Sub-surface water in the area of study is normally very hard, moderately hard and slightly saline in most of the region. Chloro-alkaline indices (CAI) revealed that the most of the water samples showed positive magnitude showing reverse ion exchange process in sub-surface water. Scholler assortment of water pointed out that there was longer residence period of aqua with more prominent base exchange. The outcomes of the appraisals were explained with hydro-geology, and the chemical contents in the sub-surface water vary temporally and spatially. As per the observations of year 2017, 13.33, 66.67% of water samples in TDS, TH, respectively, of the study area are suffering from non-suitability of aqua for drinking and irrigation purposes.

Keywords Sub-surface water · Hydro-geochemistry · Aquifers · Quality

S. Ravish (✉) · B. Setia · S. Deswal
Civil Engineering Department, National Institute of Technology, Kurukshetra, Haryana 136119, India
e-mail: mmec11081224@gmail.com

B. Setia
e-mail: setia_b@rediffmail.com

S. Deswal
e-mail: deswal.leo@gmail.com

D. Dutta and B. Mahanty (eds.), *Numerical Optimization in Engineering and Sciences*, Advances in Intelligent Systems and Computing 979,
https://doi.org/10.1007/978-981-15-3215-3_1

1 Introduction

During the last few decades, there has been a rapid increase in aqua shortage and need for fresh aqua in semi-arid and arid areas due to intense irrigation practices, industrialization, urbanization and population increase in various parts of the globe. In India, most of the people are relying principally on sub-surface water resources for agricultural, industrial, domestic and drinking activities, due to insufficient supply of surface aqua. In India, many cities and myriad large towns derive aqua supply from sub-surface water for discontiguous purposes via large number of private dug wells and also from municipality network. Therefore, knowledge on hydro-geochemistry of fresh aqua is significant to evaluate the pre-eminence of sub-surface water in any rural/urban region or/and basin that impacts the suitability of aqua for industrial, irrigation and domestic purposes. Significant hydro-geologic elements, i.e. biological activity, topographic relief, mineral weathering and precipitation in a concerned study area are significant for hydro-geochemical reactions and controlling recharge responsible for hydro-chemical constituents polluting the sub-surface aqua. Because of the significance of sub-surface water in domestic and in other purposes, its environmental aspects, i.e. pollution transport, have been significantly investigated. Various investigators have studied on pollution of sub-surface water and hydro-geochemical signatures in discontiguous rural/urban as well as in basins that attributed due to human activities principally by domestic and industrial waste aqua and irrigation activities [1]. The aqua pre-eminence may yield knowledge about the inner hydro-geologic atmospheres via which the aqua has circulated. The hydro-chemical variations of rainfall aqua rely on various elements, i.e. human activities, mineral species dissolution and water–soil interaction [2]. The over-exploitation of sub-surface aqua has hazardously influenced its quantity and quality.

In Yamunanagar and Ambala districts (Haryana) environs, the source of about 90% of domestic and agricultural aqua is from sub-surface aqua resources. However, this aqua resource is facing issues including pre-eminence danger in several regions where the exposure to contamination from irrigation and other ion contamination in deep/shallow sub-surface aqua aquifers makes the aqua unsuitable for consumption of human. Land use for agricultural and urbanization activities in the Yamunanagar and Ambala districts vicinity has enhanced at an alarming rate in last few times. In sophisticated multi-layered alluvium deposits, the shallowest Pheratic aquifer is often most susceptible to saline intrusion and most permeable to human contamination. A number of investigations on sub-surface aqua pre-eminence with respect to domestic and agricultural uses have been recorded in discontiguous parts of India [1]. The investigation region is predominantly an irrigation zone with dense irrigation practices and also situated near the hill-cum-plain area. The majority of the inhabitants in this area rely on irrigation (such as agricultural and cultivators workers). Both for domestic and agricultural practices, substantial quantity of sub-surface aqua is consumed in this region. The investigation of sub-surface water samples from a concerned region offers clues to several hydro-geochemical variations that the meteoric sub-surface aqua undergoes before acquiring different hydro-geochemical signatures. Therefore, this investigation formed the baseline attempt on the hydro-geochemistry or geochemical process of groundwater, suitability of sub-surface aqua resources and aptness for agricultural and domestic in the Yamunanagar and Ambala districts, Haryana region.

2 Materials and Methods

2.1 Region of Study

The Yamuna is the main river of Haryana running from northeast to south-west direction. The study region is principally drained by one perennial river Yamuna in the North-Eastern part of the Yamunanagar district and 03 non-perennial rivers such as Tangri (Dangri), Markanda, Ghagghar and their tributaries. Yamunanagar, Ambala districts and their vicinity are situated at an average elevation of about 255–300 m above the MSL. The study region covers an about 3330 km^2 region in and around Yamunanagar and Ambala districts and located between the 76° 30′ to 77° 28′ E longitude and 30° 06′ to 31° 35′ N latitude (Fig. 1). Geomorphologically, the study region is situated in the north-west part of the Indian subcontinent. Its climate is sub-humid, hot summer, dry and mild winter and subtropical monsoon with a marked seasonal influence. The districts receive about 81% of its annual average precipitation of around 2183 mm from the south-west monsoon during July to September month. In winter, the minimal temperature drops to 6.8 °C, and in summer, the maximal temperature rises to 48.8 °C with the annual average temperature being 24.1 °C. The area of study spreading over 3330 km^2 is part of the Indo Gangetic plain that comprises sedimentary rocks of tertiary to quaternary alluvium deposits, which occupy the southern and north part of the region [3]. In the region investigated, however, in alluvium formations, the permeable granular zones consist of fine to medium grained sand and occasionally coarse sand and gravel. The aquifers form highly potential aquifers and comprise sand, silt, gravels and kankar associated with clay. The formation of the kankar may be the rainfall of the $CaCO_3$ from the sub-surface water, and origin of clays, sand and silt is from alluvium deposits. The sand beds without or with kankar in the region form zones of the principal aquifer of the multi tier aquifer network. The deeper -aquifers are in conditions of confined to semi-confined, as benchmarked to shallow sub-surface aqua under Pheratic condition. As stated sooner, the sedimentary deposits occur in almost the overall area and are presented by tertiary and quaternary deposits. Sub-surface water occurs in these deposits under confined conditions as well as under aqua table and is extracted by means of bore wells, bore-cum-dug wells and hand pumps. Both hand pumps and tube wells are used for sub-surface aqua abstraction for discontiguous purposes in the study region. The hand pumps diameter varies from two to eight metre and varies in depth from 19 to 64 m. The usual depth of tube wells varies from 21 to 77 m below sub-surface level. The intensive extraction of aqua due to urbanization elements and

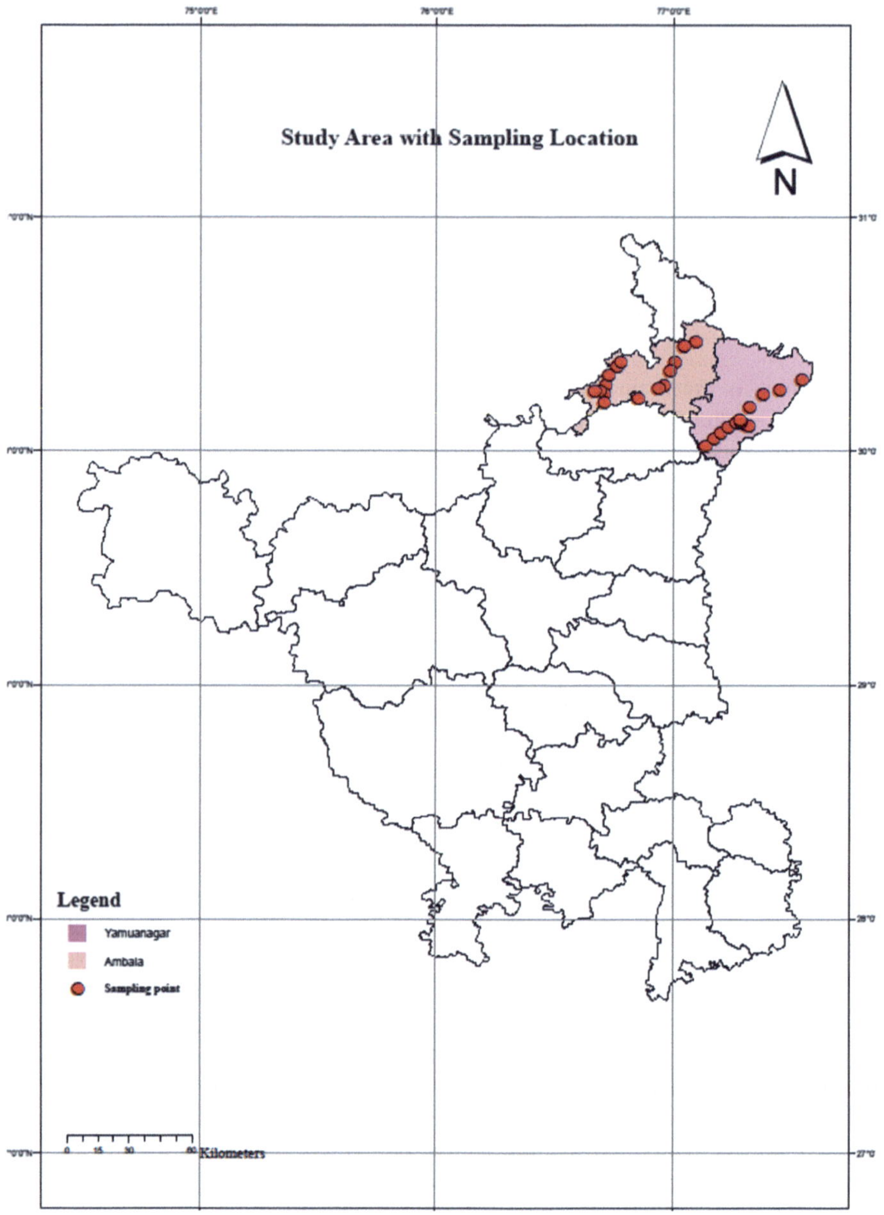

Fig. 1 Groundwater sample location map of the study area with sampling sites

populace increase in Yamunanagar and Ambala presents a decreasing trend of the aqua level in various parts of the perusal region. The availability of sub-surface aqua and nature of occurrence in the perusal region have been monitored by conducting hydro-geologic studies.

2.2 *Methodology*

Totally 30 sub-surface water samples have been identified and collected in order to represent the whole study area (Fig. 1) from hand pumps and tube wells during April 2017 and appraised to understand the hydro-chemical alterations of groundwater pre-eminence constituents applying standards procedures [4]. Acid-washed (precleaned) polyethylene bottles of one-one litre capacity were applied for the collection of sub-surface aqua samples. Entire sub-surface aqua samples were appraised for total dissolved solids (TDS), pH, prime anions and cations. TDS (HACH, HQ40d) and pH (EUTECH Instruments pH meter pH-700) were appraised applying portable meters. Magnesium (Mg^{2+}) and calcium (Ca^{2+}) were measured by ethylenediaminetetraacetic acid (EDTA) titrimetric method. Bicarbonate (HCO_3^-) and also chloride (Cl^-) were measured by titration methods. Potassium (K^+) and sodium (Na^+) were determined by flame photometer EI-380. SO_4^{2-} (Sulphate) determinations were measured by the spectrophotometer HACH DR-2800. The hydro-chemical appraisal accuracy was tested by computing ion balance error percentage (IBEP) where the errors in the groundwater samples were usually within 5% [5, 6]. Further, principal component analysis (PCA) using statistical package for social sciences (SPSS) V20.0 and geochemical plots was used to optimize the significant parameters, which are mainly responsible for regulating the hydro-geochemistry of sub-surface water in the investigation region. In this study, 16 elements (pH, TDS, TA (Total Alkalinity), TH, Cl, SO_4, CO_3, HCO_3, Na, K, Ca, Mg, F, NO_3, Fe, Cr) were determined but after applying PCA nine elements were resulted to be significant (i.e. eigen value > 1.0 or more and having strong positive loading > 0.50). The PCA analysis yielded nine principal components (pH, TDS, Ca, Mg, Na, K, HCO_3, SO_4, Cl) with higher eigen values, accounting for 100% of the total variance. Hence, majority of the hydro-geochemical elements (100%) loaded under pH, TDS, TA (Total Alkalinity), TH and Cl were having strong positive loading (0.75), and these were principally responsible for regulating the water chemistry of sub-surface aqua in the investigation region.

3 Results and Discussion

Minimum and maximum levels for the hydro-chemical constituents of sub-surface water are presented in Table 1, benchmarked with BIS guidelines. The profusion of metals is in the following sequence: sodium > calcium > magnesium > potassium and bicarbonate > chloride > sulphate, respectively. Hydro-geochemical diagrams

Table 1 Maximum and minimum levels of the hydro-chemical composition of sub-surface water samples

Elements	Range	BIS 2003
pH	7.07–8.12	6.5–8.5
TDS	220–2770	500–2000
Ca^{2+}	36–188	75–200
Mg^{2+}	4.80–88.80	30–100
Na^{+}	8.5–521	*
K^{+}	0.0–20.9	*
HCO_3^{-}	244–1061.40	*
SO_4^{2-}	18–460	200–400
Cl^{-}	56.80–766.80	250–1000

All the concentrations are in milligram per litre except pH; *Well-being-based standards concentrations have not been established

were also applied to optimize the prime reactions/processes that exert a control over the hydro-geochemical composition of the sub-surface water.

3.1 Hydro-Chemical Indices

Piper plots [7] are employed by delineating the ratios (in meq) of the prime anions (SO_4, Cl, HCO_3, CO_3) on one triangular plot, the ratios of the prime cations (K, Na, Mg, Ca) on another and superimposing the dataset from the two triangles on a quadrilateral. The location of this delineating points out the relative composition of sub-surface aqua in terms of the cation–anion groups that correspond to 04 vertices of the zone. The hydro-geochemical assessment can be explained from the Piper diagram (Fig. 2). The sub-surface water samples were taken from Yamunanagar and Ambala districts of N-E Haryana, India.

Hydro-geochemical constituent of aqua varies in space and period due to hydro-geochemical processes between the porous medium and the aqua and due to variations in patterns of flow and composition of recharge. Such alterations in hydro-chemical signature are applied to subdivide a hydrosome into 'hydro-chemical indices' or 'characteristic fields' [8]. In the investigation region, the majority of the sub-surface aqua samples were concentrated in the 'calcium-magnesium-bicarbonate', 'sodium bicarbonate' and 'sodium chloride' type (Fig. 2), pointing the mixed water type, hard water and slightly saline nature of the sub-surface water. In usual, a gradual rise of the sub-surface aqua mineralization and shift from the predominant anion bicarbonate through sulphate to chloride are found in aqua flowing from shallow to greater depth, due to increasing rock–water interaction and decreasing sub-surface aqua circulation.

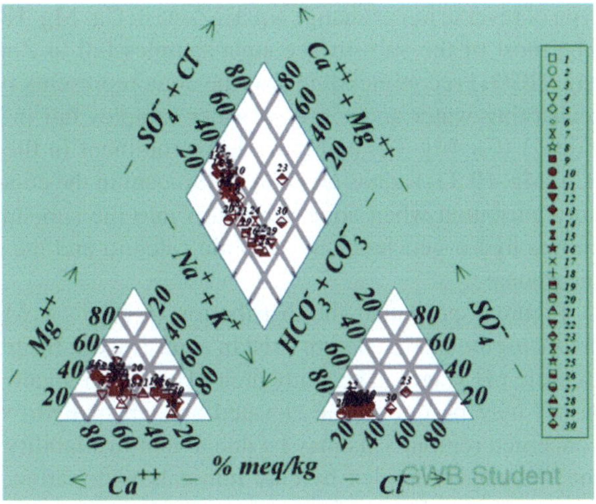

Fig. 2 Piper facies plot for sub-surface water samples

3.2 *Hydro-Chemical Process Assessment*

A hydro-geochemical plot suggested by Chadha's [9] has been employed in this investigation to evaluate/optimize the hydro-geochemical processes occurring in the concerned region (Fig. 3).

Datasets were modified to % reaction concentrations (meq %) and shown as the difference between strong acidic anions ($SO_4^{2-} + Cl^-$) and weak acidic anions ($CO_3^{2-} + HCO_3^-$) and the difference between alkali ($K^+ + Na^+$) and alkaline earths ($Mg^{2+} + Ca^+$) metals for cations. The hydro-geochemical processes recommended by Chadha's [9] are pointed in each of the four zones of the diagram. These are extensively grouped as: Zone 1: sodium bicarbonate type of base ion exchange aquas, Zone 2: sodium chloride type of end-member aquas (sea aqua), Zone 3: Ca^{2+}–Mg^+–Cl^-

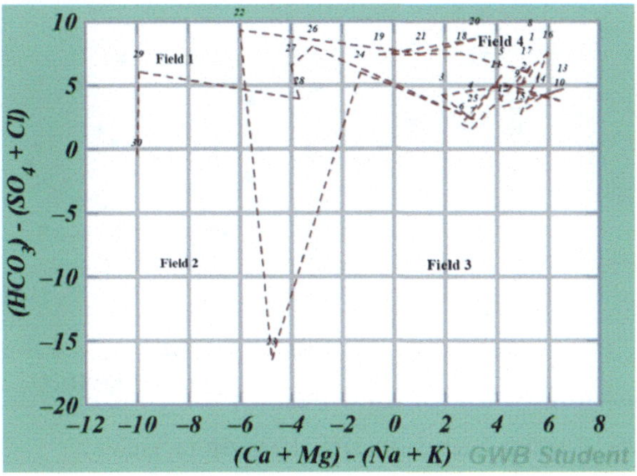

Fig. 3 Chadda's indices hydro-geochemical process evaluation plot (all ions are in meq/l)

type of reverse ion exchange aquas, Zone 4: Ca–Mg–HCO_3 type of recharging aquas. The most of the sub-surface aqua samples fall in Zones 1 (Na–HCO_3) and 4(Ca–Mg–HCO_3) recommending that the aqua represents type of base ion exchange and recharging water and a few of water samples fall in Zone 2 (end-member water). Zone 3 (Ca–Mg–Cl) aquas are less prominent in the investigation region. Zone 4 (Ca–Mg–HCO_3) aquas are more significant in the concerned region. It may be probably attributed when aqua percolates into the sub-surface from the surface, and it carries hydro-geochemically mobile calcium and the dissolved CO_3 in the form of bicarbonate.

Figure 4 presented the distribution of $(Ca^{2+} + Mg^{2+})/HCO_3^-$ ratio to pH. pH estimates the CO_3^{2-} nature exist in aqua as CO_3, bicarbonate and $H_2CO_3^-$, in acidic to basic pH stages. The study area pH showed alkaline condition. The gradual rise of pH may be due to the elevated content of hydroxyl ions (H^+) presence in the concerned region. This may be due to non-availability of neutralizing ions or strong ion exchange complex by clay minerals with cation. This process also aids us to determine circulation of the hydroxyl ions in the sub-surface aqua. Elevated content of H^+ ions present in alluvium aquifers is neutralized by the process of dissolution and weathering. The sub-surface water from the investigation region mostly showed the (Ca + M)/HCO_3 proportions below 1.0 in all the water samples. Water samples with lower proportions indicated additional bicarbonate input from albite mineral weathering, rather than from calcium and magnesium formation processes alone.

The mK + mNa − mCI versus mMg + mCa − mSO_4 relationship gives information on the hydro-geological sources of Mg and Ca in the groundwater. To pose for meteoric calcium from the dissolution of $CaSO_4$, an amount of calcium equal to the content of sulphate is subtracted from the sum of alkaline earths metals (magnesium + calcium). Computation of sodium levels depletion posed by exchange of cation was done by supposing that all meteoric sodium inputs were from sodium

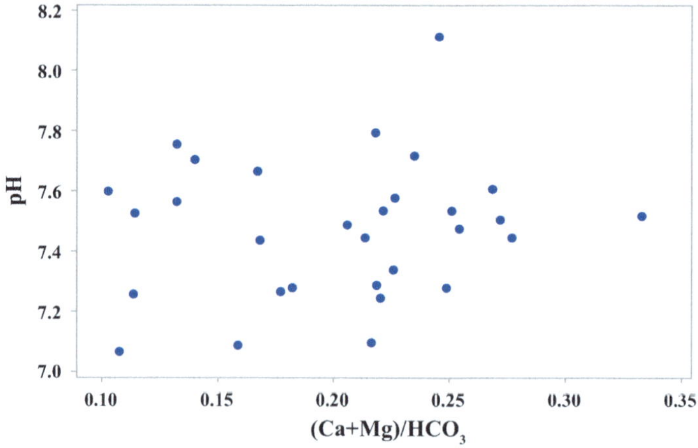

Fig. 4 Distribution of pH to (Ca + Mg)/HCO_3 ratio

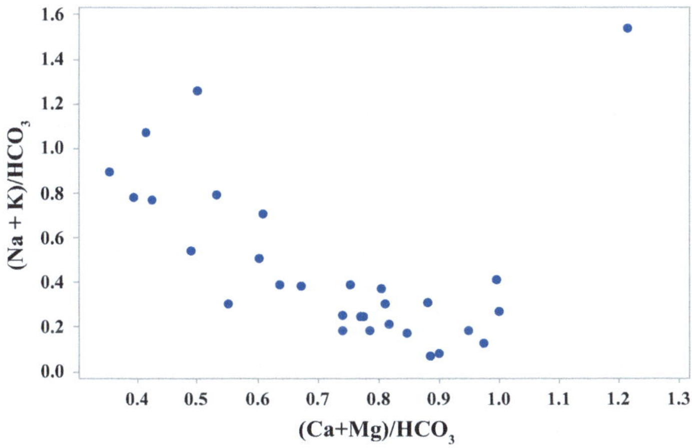

Fig. 5 Ratio between the (Na + K)/HCO_3 and (Ca + Mg)/HCO_3

chloride. Because all chloride metals are meteoric in origin, subtracting chloride from the total sodium estimates the meteoric sodium concentration. Figure 5 showed the distribution of (Ca + Mg)/HCO_3 and (Na + K)/HCO_3. It is to be recorded that the '*X*-axis' traversed the '*Y*-axis' at 0.90, i.e. the line along which (sodium + magnesium)/bicarbonate is equal to 0.90. The plot showed higher ratio of (Ca + Mg)/HCO_3 and (sodium + potassium)/bicarbonate, with calcium + magnesium – bicarbonate – sodium + potassium aqua-type (Fig. 5). The figure displayed the predominance of excess calcium + magnesium and higher sodium + potassium. This displayed that the region has predominance of excess calcium + magnesium and sodium + potassium, with no important impacting anions indications due to pollution.

Most of the sub-surface water samples in the concerned region, showing Ca–Mg–Na–K–HCO_3 water type (Fig. 5). This recommends that silicates minerals weathering is the important contributors to the hydro-chemistry of the study area.

4 Quality of Water

4.1 Quality of Domestic Water

The hydro-chemical characteristics related to the water salinity can be evaluated/optimized by the assessments of the following elements: Cl content and TDS. Desjardin [10] grouped water typology pursuance to the total dissolved solids concentration. Table 2 clearly showed that the higher presentations of the sub-surface aqua samples in the area of study were observed in freshwater and moderately fresh-brackish water types. Salinity occurs in sub-surface water due to anthropogenic sources, leaching from topsoil, weathering of rocks and along with minor climate

Table 2 Water typology according to their content of TDS [10]

Type of water	Limit (mg/l)	Number of water samples	Percentage (%)
Slightly brackish aqua	1000–5000	04	13.33
Moderately fresh-brackish aqua	500–1000	15	50
Freshwater	>500	11	36.67

impact [11]. The concentration of sodium and bicarbonate in agriculture sub-surface water influences the area drainage and the soil permeability [12, 13].

4.2 *Total Hardness (TH)*

TH is expressed as the sum of content of their metals defined in milligram per litre of calcium carbonate. TH increases from metals mixed in aqua. TH is applied as the scale formation rate indicator in hot aqua heaters in low-pressure boilers. The United States of Geological Survey TH [14] showed four classes of hardness: very hard, moderately hard, slightly hard and soft. Depositions and scaling issues in air-conditioning plants are accounted with the TH of aqua. The TH of greater than one hundred eighty milli gram calcium carbonate per litre can be categorized as 'very hard' aqua and can pose to scaling issues in air-conditioning plants [15]. In the concerned region, 30% of water samples were 'moderately hard group', 3.33% fall under 'slightly hard group' and 66.67% are in the 'very hard group' (Table 3).

Table 3 Hydro-geochemical classification summary

Class	Range (mg/l)	Number of samples (30)	%
USGS hardness [14]			
Very hard	>300	20	66.67
Moderately hard	150–300	09	30.00
Slightly hard	75–150	01	03.33
Classification of chloride [18]			
Brackish	8.46–28.20	02	06.67
Fresh brackish	4.23–8.46	03	10.00
Fresh	0.84–4.23	25	83.33
Very fresh	0.14–0.84	00	00.00
Base-exchange indices (BEI) Scholler [16]			
(sodium + potassium) g.w. → (calcium/magnesium) rock		15	50.00
(sodium + potassium) rock → (calcium/magnesium) g.w.		10	50.00

4.3 Base-Exchange Index (BEI)

Scholler [16] suggested a term known as 'Base-Exchange Index' (BEI) to interpret the hydro-geochemical processes taking place in sub-surface water. There are substances which exchange and absorb their ions with ions existing in sub-surface water. Those substances are known as 'permutolites', e.g. organic substances and clay minerals like zeolites, glauconite, halloysite, chlorite, illite and kaolinite. Halloysite, chlorite, illite and kaolinite are the clay minerals in which metals capacity of ionic exchange is low and is present at edges. This condition is reverse in vermiculite and montmorillonite. However, when the number of metals present on the surface is more, the exchange capacity is higher. Indices of chloro-alkaline, CAI1 and CAI2 are applied to estimate the extent of base exchange during interaction of water–rock using Eqs. 1 and 2 [16].

$$\text{Chloro-alkaline-indices 1} = [\text{chloride} - (\text{sodium} + \text{potassium})]/(\text{chloride}) \quad (1)$$

$$\text{Chloro-alkaline-indices 2} = [\text{chloride} - (\text{sodium} + \text{potassium}]/(\text{sulphate} + \text{bicarbonate} + \text{carbonate} + \text{nitrate}) \quad (2)$$

(All metallic content is represented in milliequivalent per litre.)

Where there is no exchange of potassium (K^+) and sodium (Na^+) in sub-surface aqua with calcium (Ca^{2+}) or magnesium (Mg^{2+}) in alluvium/rock, both the indices are positive and vice versa. In most of the samples, reverse ion exchange is the predominant process in the area of study. The base-exchange indices point that there is a prominent exchange of sodium + potassium in sub-surface water into the calcium + magnesium in alluvium from the matrix, whereas the vice versa is less important and the exchange of sodium + potassium in alluvium to the calcium + magnesium in sub-surface water is less observed [17]. More than 50% of water samples fall in (sodium + potassium) sub-surface water → magnesium/calcium alluvium and 50% of the water samples are in (sodium + potassium) alluvium → magnesium/calcium. Chloride classification by Styfzands [18] showed that 19% of water samples were in the 'brackish category', 10% were under the 'fresh-brackish category', and 83.33% were fresh in nature in rock (Table 3).

5 Conclusions

Dominance of the anion is in the following sequence: sodium > calcium > magnesium > potassium and that of cations is bicarbonate > chloride > sulphate. The concerned region forms a part of the inland aquifers, and the elevated level of TDS extended from the north-eastern to south-western part influenced by anthropogenic sources, leaching from topsoil, weathering of rocks and along with minor climate impact. Elevated content of total dissolved solids was also found in the, central and south-western, western, north-eastern, north-western, parts due to the metals leaching. Majority of the sub-surface aqua samples varied from 'slightly hard' to 'very hard' group. Cl classification by Styfzands showed that all the groundwater samples fall in 'brackish' to 'fresh' range in the study area. Piper plot showed that 'calcium-magnesium-bicarbonate' type is the predominant indices in the sub-surface aqua with few presentations of Na–Cl indicating the recharge and end-members water and showed the mixed water type, hard water and slightly saline nature of the sub-surface water. Chadda's diagram showed that base ion exchange, recharging water and end-member water are more common in rock, which has more presentations of recharging water polluted samples. The PCA analysis yielded nine principal components (pH, TDS, Ca, Mg, Na, K, HCO_3, SO_4, Cl) with higher eigen values, accounting for 100% of the total variance. Hence, majority of the hydro-geochemical elements (100%) loaded under pH, TDS, TA (Total Alkalinity), TH and Cl were having strong positive loading (0.75), and these were principally responsible for regulating the water chemistry of sub-surface aqua in the investigation region. TDS is higher and exceeded the guideline ranges for domestic purpose in few of the aqua samples in Yamunanagar and Ambala districts. Few sub-surface aqua samples of the investigation region were not suitable for domestic and drinking purposes, and only few sites needs some kind of treatment for better water quality for the human consumption. Hydro-geochemical diagrams and principal component analysis (PCA) may be helpful to optimize the prime reactions/processes that exert control over the hydro-geochemical composition of the sub-surface water. The present study may be helpful for the execution and planning for quality of water, protection of environment and formulation of policies.

References

1. Ravish, S., Setia, B., Deswal, S.: Groundwater quality in urban and rural areas of north-eastern Haryana (India): a review. ISH J. Hydraul. Eng. (2018). https://doi.org/10.1080/09715010.2018.1531070
2. Faure, G.: Principles and Applications of Geochemistry, 2nd edn. Prentice Hall, Englewood Cliffs (1998)
3. CGWB.: Groundwater Year Book-India. Central Ground Water Board, Ministry of Water Resources Government of India, Faridabad (2012)
4. APHA: Standard Methods for the Examination of Water and Wastewater, 19th edn. APHA, Washington, DC (1998)
5. Freeze, A.R., Cherry, J.A.: Groundwater. Prentice-Hall Inc, Englewood cliffs, p. 604 (1979)
6. Domenico, P.A., Schwartz, W.: Physical and Chemical Hydrogeology, 2nd edn. Wiley, New York, p. 506 (1998)
7. Piper, A.M.: A graphical procedure in the geochemical interpretation of water analysis. Trans. Am. Geophys. Union **25**, 914–923 (1944)
8. Back, W.: Origin of hydrochemical facies of groundwater in the Atlantic Coastal plain. In: 21st International Geological Congress, Copenhagen 1960, Rept pt. 1, pp. 87–95 (1960)
9. Chadha, D.K.: A proposed new diagram for geochemical classification of natural waters and interpretation of chemical data. Hydrogeol. J. **7**(5), 431–439 (1999)
10. Desjardins, R.: Le traitement des eaux. Edition II revue. Edition de l'Ecole Polytechnique de Montre´al, Montre´al (1988)

11. Prasanna, M.V., Chidambaram, S., Gireesh, T.V., Jabir Ali, T.V.: A study on hydrochemical characteristics of surface and subsurface water in and around Perumal Lake, Cuddalore District, Tamil Nadu, South India. Environ. Earth Sci. **64**(5), 1419–1431 (2011)
12. Tijani, J.: Hydrochemical assessment of groundwater in Moro area, Kwara State. Nigeria. Environ. Geol. **24**, 194–202 (1994)
13. Kelly, W.E.: Geoelectric sounding for delineating groundwater contamination. Ground Water **14**(1), 6–11 (1976)
14. Handa, B.K.: Modified classification procedure for rating irrigation waters. Soil Sci. **98**(2), 264–269 (1964)
15. Hem, J.D.: Study and Interpretation of the Chemical Characteristics of Natural Water, 2nd edn. USGS Water Supply, 1473, p. 363 (1970)
16. Scholler, H.: Hydrodynamic Dam Lekar Collogue Doboronik 1, pp. 3–20 (1965)
17. Chidambaram, S.: Hydrogeochemical studies of groundwater in Periyar District, Tamilnadu, India. In: Ph.D. Thesis, Annamalai University (2000)
18. Stuyfzand, P.J.: Non point sources of trace elements in potable groundwater in the Netherlands. In: Proceedings 18th TWSA Water Workings. Testing and Research Institute, KIWA (1989)

Numerical Optimization of Pile Foundation in Non-liquefiable and Liquefiable Soils

M. K. Pradhan, Shuvodeep Chakroborty, G. R. Reddy and K. Srinivas

Abstract Numerical optimization techniques are used widely for different engineering fields. But there are limited applications of this method in geotechnical engineering. However in this study, topology optimization of pile foundation for different site conditions and loading conditions is obtained through a finite element (FE) analysis study. The suitable topology of piles in foundation system offering minimum internal energy, i.e. maximum stiffness for a given fraction of material is studied. The study is also enhanced to the piles which are located in the soils prone to liquefaction. In the present study, the design methodology for cost optimization of construction of a pile group with a raft foundation is also presented through a case study. In the optimization algorithm, the raft dimensions, no of piles, pile diameter, pile length are taken as the design variables.

Keywords Pile foundation · Topology optimization · Liquefaction · Cost optimization

1 Introduction

Pile foundation is used for large number of purposes in geotechnical fields like to cater the heavy load of the structure to harder strata, to support the structure where the uplift force is high, to resist lateral loads, supporting retaining walls, bridge piers, abutments, etc. In general, there are two types of pile, namely driven piles and bored piles. Once the decision for pile foundation has been taken, the engineer must choose the type, topology and size of the pile which is most suitable for a particular soil and loading condition. Using the numerical optimization technique, it will be much easier to obtain the topology of pile foundation which is most suitable and will optimize the material use.

M. K. Pradhan (✉)
Bhabha Atomic Research Centre & HBNI, Mumbai, India
e-mail: mpradhan@barc.gov.in

S. Chakroborty · G. R. Reddy · K. Srinivas
Bhabha Atomic Research Centre, Mumbai, India

D. Dutta and B. Mahanty (eds.), *Numerical Optimization in Engineering and Sciences*,
Advances in Intelligent Systems and Computing 979,
https://doi.org/10.1007/978-981-15-3215-3_2

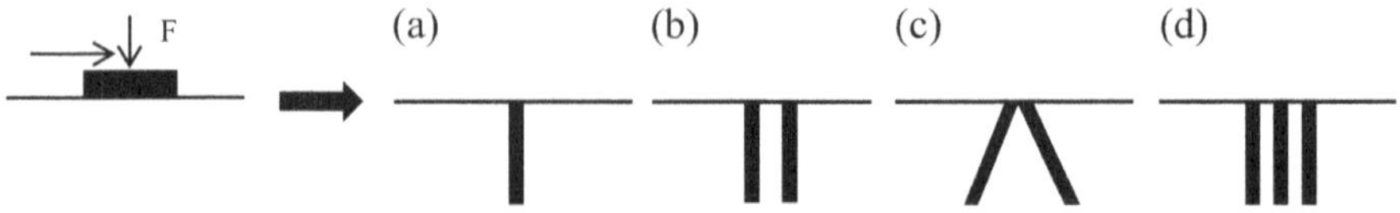

Fig. 1 Topology optimization of pile foundation: four different topologies

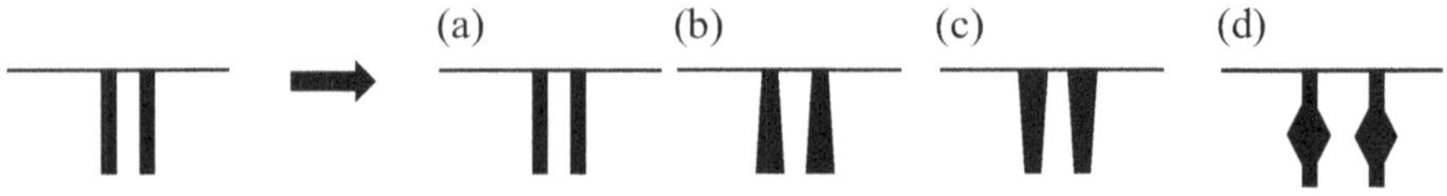

Fig. 2 Shape optimization: four different shapes for foundation

1.1 Topology Optimization of Pile Foundation

When a structure is carrying heavy load and near ground, soil strata are comparatively week, it is often inevitable to go for pile foundation. At this stage, topology optimization can be conducted which will improve deformational behaviour of structure along with cost-saving, economy in material use. Possible topology of piles under a strip footing may be any one of the followings as depicted in Fig. 1a–d.

Depending on the magnitude and direction of the forces acting on the strip footing and the soil characteristics existing in site, the best suitable pile foundation topology can be established. Further using shape optimization, more knowledge about the suitable shape of that chosen topology can be obtained (Fig. 2a–d).

Further, size optimization may be used to optimize size of pile (Fig. 2a) or the dimensions of varied parts in piles (Fig. 2b–d).

Topology optimization was first used in geotechnical problem for the underground excavation in linear elastic rock material by Ren et al. [1]. Pucker and Garbe [2] presented topology optimization under a strip footing in granular hypoplastic material. In this paper, the application of topology optimization is presented by FE analysis for both liquefiable and non-liquefiable soils under different loading conditions. The FE analysis of topology optimization is based on solid isotropic material with penalization (SIMP) method which is illustrated briefly in Sect. 2.

1.2 Cost Optimization of Pile Foundation with a Raft

In general, the foundation cost of real-world structures can vary from 5 to 20% of the construction cost of the superstructure [3]. For a conservative design, much attention is not paid to cost of the construction of such foundation. Hence, sometime it may happen that the number of pile used in foundation for design is much more than actual requirement and this unnecessary increases cost of the project. Hence, an optimum design methodology fulfilling all structural criteria should be studied to minimize cost prior to the detail designing of the proposed structure. The main objective is to minimize the pile foundation cost with the consideration of all the constraints like bearing capacity of the soil beneath the raft, pile load-bearing capacity, settlement criteria as in IS: 2911 Part IV [4]. The optimization formulation in the present study is carried out through a case study and is based on the evolutionary algorithm by Lagaros et al. [5].

2 Topology Optimization of Pile Foundation

In FE analysis, the topology optimization is carried out by using a numerical optimization algorithm, i.e. SIMP by Sigmund [6] through an iterative procedure. SIMP method assumes that material is uniformly placed in the design domain. The material property depends on the relative density ρ. The relative density varies over the design domain. It is considered that material is concentrated at highly loaded regions. The relative density should be either zero or one in the design domain after optimization completed. Zero relative density signifies no material and unit value signifies material.

The aim of the optimization procedure is to minimize the compliance, i.e. the internal energy of the structure in the design domain so that stiffness of the structure can be maximized for a given fraction of material in the design domain.

Optimization task is basically,

$$\text{Minimise: internal energy } c(x) = U^T KU \text{ i.e also the objective function} \quad (1)$$

Subject to:

$$KU = F \quad (2)$$

$$V_\delta = V_0.\delta \text{ These are the constraints.} \quad (3)$$

where U is the global deformation tensor, K is the global stiffness matrix and x is tensor of design parameters. Where F is the external forces and V_0 is the initial volume in the design domain, V_δ is the volume after optimization and δ is fraction of initial volume. The relative density ρ of the element changes as iteration progresses. The material change over of material with a relative density ρ and a chosen penalty term p from elastic modulus E_1 to E_2 is given by Eq. 4.

$$E_2 = E_1.\rho^p \quad (4)$$

2.1 FE Modelling

Optimized topology of piles is obtained using this method in a 2D-FE analysis with a strip footing for both non-liquefiable and liquefiable conditions of soil under different types of loadings.

2.1.1 In Non-liquefiable Soil

The existing strip footing of 5 m wide and 1 m depth carries both vertical and horizontal loads is resting on a soil of elastic modulus of 70 MPa and poison ratio 0.3. The elastic modulus of concrete is 25,000 MPa. The soil mass continuum of 80 m wide and 60 m deep under the strip footing is considered for developing the numerical model. The design domain is of 20 m wide and 15 m deep as shown in Fig. 3a. The *x* and *y* displacement of the soil continuum is restricted in both the *y* and *x* plane, respectively. FE analysis has been carried out for both horizontal and vertical loading.

In the present study, the soil and concrete have been analyzed using linear elastic material property. For meshing free mesh control for both soil and strip footing with plane stress, linear element of reduced integration and hourglass control has been chosen. As the design domain is our area of concern, the size of meshing is reduced to capture the results efficiently and is shown in Fig. 3b. For modelling the strip footing and soil, it is assumed that no relative movements occurred between them. Hence they are modelled integrally. The main objective of this modelling is to minimize the stain energy of the whole model by using 10% of the material that is in the design domain. Thus, the optimum topology for pile foundation can be obtained.

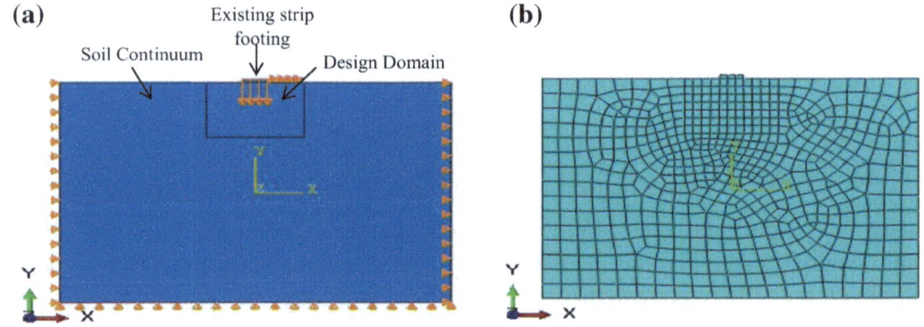

Fig. 3 **a** Boundary and loading conditions of the soil mass under the strip footing tube optimized, **b** meshing of the soil continuum, design domain and the strip footing

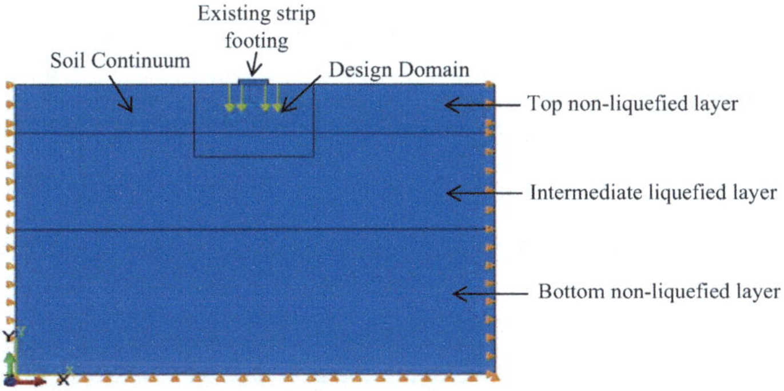

Fig. 4 Boundary and loading conditions of the soil mass prone to liquefaction under the strip footing

2.1.2 Liquefiable Soil

In this analysis, it is considered that the soil is susceptible to liquefaction under seismic condition. Hence to optimize the topology for the pile foundation for such soils, the post-liquefaction modulus of elasticity has been considered, i.e. generally, one-tenth of initial elastic modulus. As it can be observed in Fig. 4, the soil has been divided into three layers of 10, 20 and 30 m from top to bottom. Considering post-liquefied state in the middle layer, elastic modulus is taken as 7 MPa. Concrete property and other soil layers property has been considered same as described in the earlier section for non-liquefiable soil condition. The geometric condition, boundary condition, element type are identical to the previous section. The analysis carried out for a vertical loading of 500 KN.

2.2 Results and Discussion

Figure 5 depicts the progress of FE analysis in successive iterations by minimizing the strain energy and keeping material fraction 10% of original volume.

It can be observed from Fig. 6a that when there is no eccentricity in loading, a simple vertical profile of pile is obtained. From Fig. 6b, it can be seen that a vertical pile along with an inclined pile be the efficient topology for pile for a vertical and horizontal loading condition. Figure 6c shows the same topology but with a thicker inclined pile is required than the previous one as the horizontal load increases in this case. Nowadays, any profile of piles can be constructed efficiently by concrete jet grouting method. Hence, this method realizes pile construction with any optimized topology. In the case of the liquefiable soil, it remarked from Fig. 6d that the battered pile or inclined pile will be most suitable to provide a very stiff foundation within minimum concrete use. But due to being stiff, the design for induced seismic forces and ductility check under seismic condition is very important to consider.

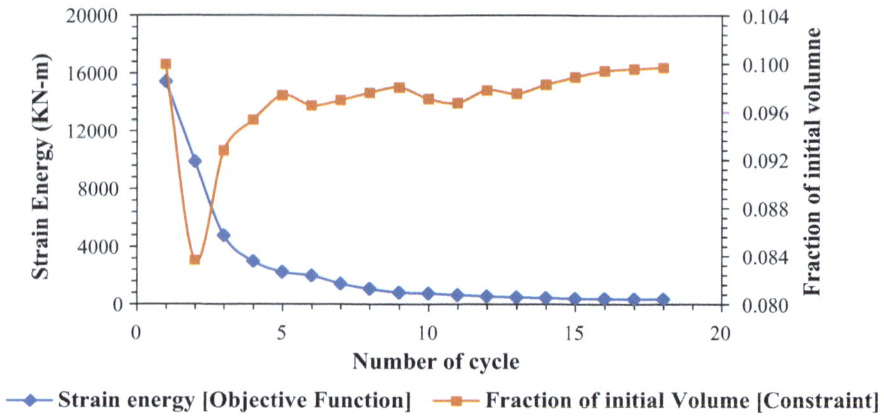

Fig. 5 Objective function minimization and material fraction versus number of cycles

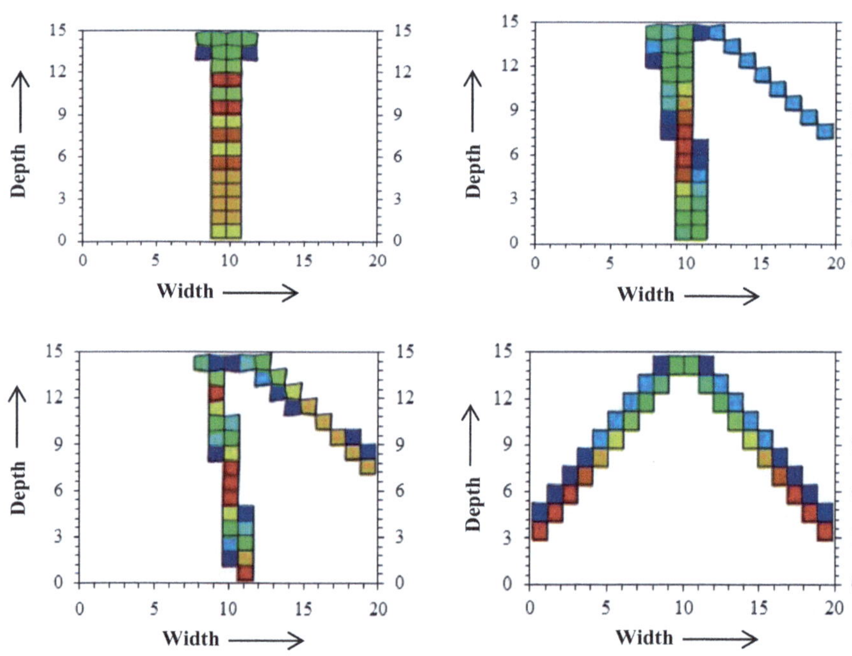

Fig. 6 Topology optimization under strip footing for **a** vertical loading, $P = 500$ KN, horizontal loading, $H = 0$ KN in non-liquefiable soil; **b** $P = 500$ KN, $H = 100$ KN in non-liquefiable soil; **c** $P = 500$ KN, $H = 250$ KN in non-liquefiable soil; **d** $P = 500$ KN, $H = 0$ KN in liquefiable soil

It is also observed in the present study that as material volume percentage increases, the settlement of the foundation gets reduced.

3 Cost Optimization of a Pile Foundation with Raft

Cost optimization design methodology is described for the structure with the column arrangements as shown in Fig. 7. It is assumed that foundation rests on soil strata of medium stiff clayey soil up to 20 m below foundation level. The undrained compressive strength of soil is 100 kPa. The factored loads in the columns in x-direction are 2200 KN, 2200 KN, 2200 KN, 2400 KN, 2200 KN, 2000 KN and 2200 KN, respectively, on the edge strips, and in the middle strip, columns loads are twice that of the edge strip columns.

3.1 Objective Function and Constraints of the Optimization Algorithm

The main objective of this study is to observe how the design methodology is to be applied so that after satisfying all the design criteria, the foundation can be built with optimum cost. The overall cost of raft including excavation, reinforcement, casting of concrete is 0.5 lakh per cum and for pile is 2.6 lakh per cum in Mumbai region. The constraints are bearing pressure under raft, the pressure induced in each pile and pile group settlement. Pile group settlement is calculated by using formula given by Randolf and Wroth [8]. The length of the raft has been assumed as 1.16 B. The solution of the optimization problem is performed with the evolutionary algorithms [5].

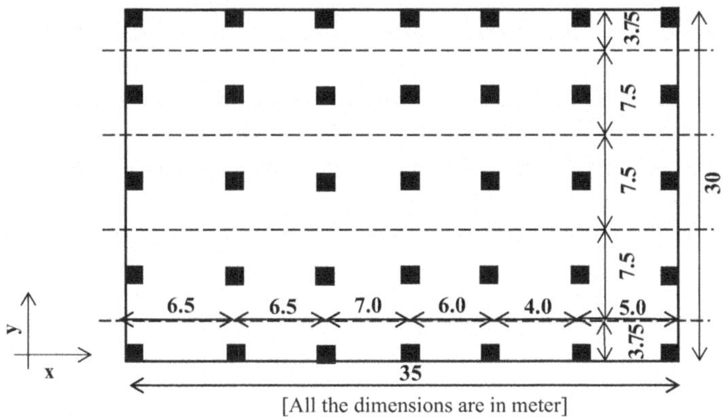

Fig. 7 Column arrangement for the case study [7]

Table 1 Variables range and the optimum values of variable in the algorithm

S. No.	Variable	Range	Design values
1	Width of raft (m)	35–45	38
2	Thickness of raft (m)	0.35–0.5	0.43
3	Pile dia (m)	0.6–2.5	2.1
4	Pile length (m)	10–40	15
5	No. of pile	50–350	72

3.2 Results and Discussion

The variables and the result obtained are discussed in Table 1.

Total optimized cost of the pile foundation along with raft is 46.5 crore considering the structure to be in Mumbai region. The soil pressure as obtained from the numerical formulation under raft is 82 kPa. As the soil is medium stiff clayey soil, the observed pressure is less than allowable bearing capacity of soil. It is considered in this analysis that 60% load is shared by pile and 40% load is shared by raft. Each pile has to carry a load of 810 KN which is within safe load carrying capacity of each pile. The group settlement of the pile is 10 mm which is lower than the value mentioned in IS: 2911 Part (IV) (1979).

4 Conclusion

From the present Numerical Optimization Study, it can be concluded that vertical piles are best suited for loading without eccentricity. If there is eccentric loading, vertical pile along with inclined pile is most suitable. In the case of liquefiable soil, battered pile provides the stiffer solution than vertical pile. But the associated cost for realization of these topologies of piles in site must be investigated. Nevertheless, optimization of topology of piles provides significant opportunities in the design process of pile foundation. In the present cost optimization, study approximate costs are considered for pile and raft construction in clayey soil. For other types of soil as the cost of pile and raft construction changes, the objective function also changes. So, a new function should be developed with the presented constraints to minimize the cost. Hence, both topology and cost optimization of pile foundation are very much required to be considered in design from both structural and economic perspective.

Acknowledgements The academic and research support provided by institution HBNI, BARC and ECIL are gratefully acknowledged by the authors.

References

1. Ren, G., Smith, J.V., Tang, J.W., Xie, Y.M.: Underground excavation shape using an evolutionary procedure. Comput. Geotech. **32**, 122–132 (2005)
2. Pucker, T., Grabe, J.: Structural optimization in geotechnical engineering—basics and application. Acta Geotech. **6**, 41–49 (2011)
3. Letsios, C., Lagaros, N.D., Papadrakakis, M.: Optimum design methodologies for pile foundations in London. Case Stud. Struct. Eng. **2**, 24–32 (2014)
4. IS: 2911 Part IV: Code of Practice for Design and Construction of Pile Foundations (1985)
5. Lagaros, N.D., Papadrakakis, M., Kokossalakis, G.: Structural optimization using evolutionary algorithms. Comput. Struct. **80**(7–8), 571–587 (2002)
6. Sigmund, O.: A 99 line topology optimization code written in Matlab. Struct. Multi. Optim. **21**, 120–127 (2001)
7. Beaks, G.K., Stravroulakis, G.E.: Cost optimization of a raft foundation including pile group design optimization and soil improvement considerations. In: 11th HSTAM International Congress on Mechanics. Athens, Greece (2016)
8. Randolph, M.F., Wroth, C.P.: An analysis of the vertical deformation of pile groups. Géotechnique **29**, 423–439 (1979)

Nonlinear Regression for Identifying the Optimal Soil Hydraulic Model Parameters

Navsal Kumar, Arunava Poddar and Vijay Shankar

Abstract This study is focussed on the determination of soil hydraulic parameters and the analysis of the moisture retention function for different soil textures. Experiments utilizing the pressure plate apparatus were conducted to estimate actual soil moisture characteristics. Optimal values of soil hydraulic parameters of Brooks and Corey (Hydraulic Properties of Porous Media. Civil Engineering Department, Colorado State University, Fort Collins, Colorado, 1964) [14], Van Genuchten (Soil Science Society America Journal 44:349–386, 1980) [16] and modified Van Genuchten (2006) [18] models were estimated using nonlinear regression in SPSS. These parameters were used as input in HYDRUS-1D forward simulation to yield analytical soil moisture retention curve. The analytically obtained moisture retention curve is compared with the actual SMC curve to assess the performance of nonlinear regression-based approach for identifying optimal soil hydraulic parameter values.

Keywords Soil moisture · Matric potential · Iteration · Unsaturated zone · Optimization

1 Introduction

Water transport in saturated soils takes place as liquid flow and in unsaturated soils as liquid–vapour flow [1]. The soil water flow is generally specified by solving the proper governing partial differential equations for different initial and boundary conditions [2, 3]. The soil hydraulic parameters at the soil surface and in the unsaturated zone are key variables, because they are essential inputs to large-scale hydro-climatic and hydrologic processes [4]. An important problem commonly encountered in computer-based numerical solutions is obtaining the primary input parameters, that is, soil matric potential and hydraulic conductivity as a function of soil water content [5]. Soil water retention functions describe the ability of soils in storing and releasing water and form an essential parameter in the unsaturated flow and infiltration models

N. Kumar (✉) · A. Poddar · V. Shankar
Civil Engineering Department, National Institute of Technology, Hamirpur, Himachal Pradesh 177005, India
e-mail: navsal.happy@gmail.com

D. Dutta and B. Mahanty (eds.), *Numerical Optimization in Engineering and Sciences*, Advances in Intelligent Systems and Computing 979,
https://doi.org/10.1007/978-981-15-3215-3_3

[6, 7]. Water retention function is dependent on particle size distribution, organic matter content, mineralogy of clay and hysteresis.

The concept of SMC was initially proposed by [8]. For many years, SMC has been subjected to substantial research and numerous laboratory, field and theoretical techniques are developed for its determination [9–11]. The soil moisture characteristic (SMC) expresses the functional relationship between the suction head (ψ) and volumetric moisture content (θ) in an unsaturated soil [12, 13].

Modelling SMC is generally based on the solution of governing mathematical equations containing various parameters describing the system. The most popular analytical models available in the literature are Brooks and Corey [14], Campbell [15], Van Genuchten [16], Kosugi [17] and modified Schaap and Van Genuchten [18]. In order to have a reliable model, the estimated value of the parameters should fit their actual ones. The parameters are obtained either experimentally or analytically. In the present study, optimal values of parameters of three analytical models [14, 16, 18] are estimated using nonlinear regression. Following objectives have been set:

i. Estimation of field parameters of SMC through laboratory experiments and the development of experimental SMC using the pressure plate apparatus.
ii. Calculation of optimal soil hydraulic model parameters through nonlinear regression using SPSS.
iii. Comparison between experimental SMC and model predicted SMC using HYDRUS-1D.

2 Materials and Methods

2.1 Analytical Models

Three analytical models are considered in the study as given below:

Brooks and Corey (BC):

$$\frac{\theta - \theta_r}{\theta_S - \theta_r} = \left(\frac{\psi_b}{\psi}\right)^{\lambda} \quad \text{for } \psi \leq \psi_b \tag{1}$$

Van Genuchten (VG):

$$\frac{\theta - \theta_r}{\theta_S - \theta_r} = \left[\frac{1}{1 + \|\alpha_v \psi\|^{n_v}}\right]^m \quad \text{for } \psi \leq 0 \tag{2}$$

Modified Van Genuchten (MVG)

$$\frac{\theta_m - \theta_r}{\theta_S - \theta_r} = \left[\frac{1}{1 + \|\alpha_{mv} \psi\|^{n_{mv}}}\right]^m \quad \text{for } \psi \leq 0 \tag{3}$$

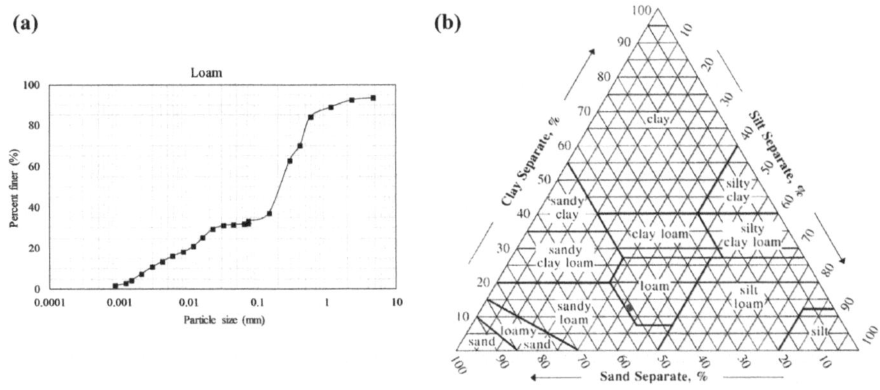

Fig. 1 **a** Particle size distribution, **b** soil textural classification chart for loam soil

where $\theta_m = \theta_r + (\theta_s - \theta_r)(1 + ||\alpha h_s||^n)^m$

In the above models, ψ is the soil matric potential, ψ_b is the bubbling pressure, λ is the pore size index, α_v and n_v are unsaturated soil parameters with $m = 1 - (1/n_v)$, θ is the actual volumetric moisture content ($cm^3\ cm^{-3}$) and the subscripts s and r represent saturation and residual values of the moisture content.

2.2 *Experimental Data*

Laboratory experiments were utilized to estimate relevant soil textural and hydraulic parameters. Three different sites were selected, and 30 samples from each site were subjected to sieve and hydrometer analysis [19]. The soil texture comes out to be loam, sandy loam and loamy sand from sites 1, 2 and 3, respectively. Figure 1 shows the particle size distribution and soil textural classification chart (USDA) for loam soil. Experimental SMC was estimated using the pressure plate apparatus (PPA), which involved concurrent measurements of soil moisture and soil suction. The duration of experiments varied for each sample and for different pressure readings. θ_s is assumed to be equal to soil porosity, and θ_r is estimated from PPA.

2.3 *Nonlinear Regression*

The optimal values of model parameters are determined through nonlinear regression using IBM SPSS (Statistical Package for Social Sciences) software. In nonlinear regression, a function which is nonlinear combination of model parameters is used for modelling the data. The function is dependent on one or more independent variables. In the present study, nonlinear regression is applied through nonlinear statistical

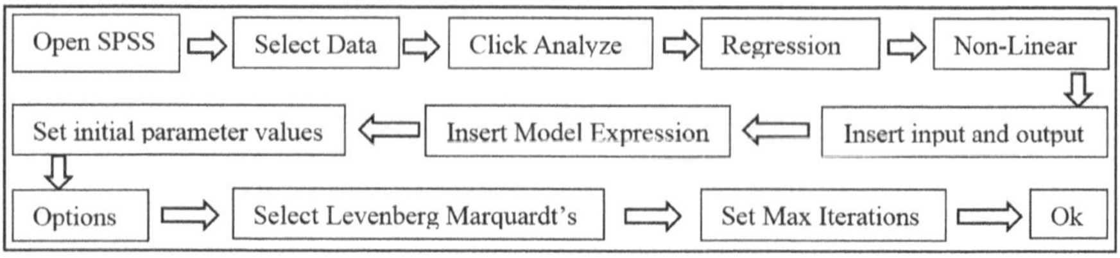

Fig. 2 Step-by-step procedure of nonlinear regression in SPSS

modelling, in which parameter estimates are obtained by "method of least squares". The approximate solution of nonlinear equations is obtained through iterative procedures. Levenberg–Marquardt's method is used for estimating parameters through certain number of iterations. The method is better than others (linearization or steepest descent) because it always converges and does not slow down in the later stages of iterative process.

The experimentally obtained input (ψ, θ_s, θ_r) and output (θ) variables were used. An initial guess of parameters is provided, and the iterations are performed. Each of the models has two unknown parameters whose value needs to be determined. Initial guess for the model parameters h_b, λ, α_v, n_v, α_{mv} and n_{mv} were 1, 0.1, 0.01, 1, 0.01 and 1, respectively, for all the soil textures considered in the study. After a certain number of model and derivative assessments, the iteration stops resulting in model fit parameter values. The procedure for nonlinear regression in SPSS is outlined in Fig. 2.

2.4 *HYDRUS*

The parameters of the analytical models calculated from SPSS and variable obtained from laboratory experiments are used as input data for forward simulation in HYDRUS-1D. HYDRUS is based on the soil moisture flow equation [20]. The simulated results in the development of analytical SMC from each model considered in the study.

3 Results and Discussion

3.1 *Soil Moisture Characteristics*

Numerous experiments were conducted using PPA for different soil textures. Thirty sets of sample observations were taken for each soil texture to accurately determine its SMC curve. The SMC curves obtained for different soil textures considered in the

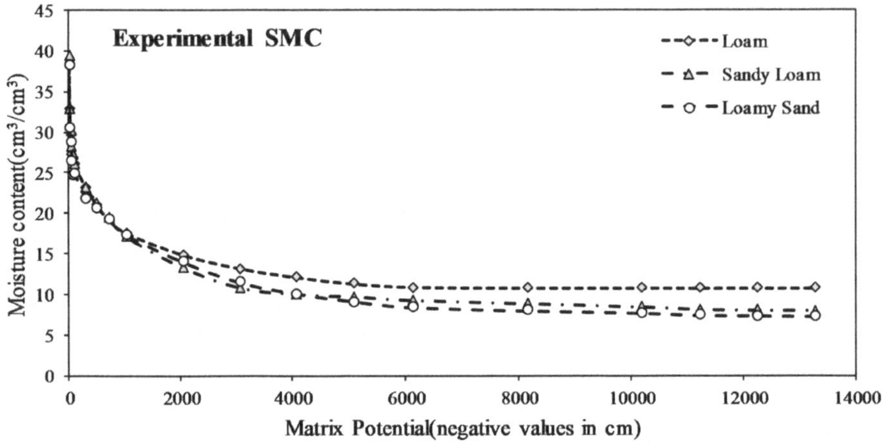

Fig. 3 Experimental soil moisture characteristics curve for different soils

study are shown in Fig. 3. There is no clear difference between the moisture retention near the saturation end; however at the drier end, the loam exhibits higher moisture retention in comparison with sandy loam and loamy sand soil.

3.2 *Soil Hydraulic Parameter Estimation*

The soil hydraulic parameters of three analytical models are estimated through nonlinear regression using Levenberg–Marquardt's iterative optimization. The model fit optimal values of these parameters for loam, sandy loam and loamy sand are given in Table 1. The parameters are found to be different for each soil texture. The statistical results for nonlinear regression are shown in Table 2. For each model, a number of model assessments (MA) and derivative assessments (DA) are run which converges to a final value as mentioned in Table 2. The *R*-squared (R^2) value of analysis of variance is also given in Table 2. R^2 is found to be above 0.95 for each model; in case of all soil textures considered in the study which suggests that the estimated values are reliable. Considering brevity in paper, a detailed analysis of one of the cases

Table 1 Values of soil hydraulic model parameters

Soil texture	SMC models					
	Brooks and Corey		Van Genuchten		Modified Van Genuchten	
	ψ_b (cm)	λ	α_v (cm^{-1})	n_v	α_{mv} (cm^{-1})	n_{mv}
Loam	8.952	0.28	0.032	1.405	0.053	1.402
Sandy loam	16.522	0.467	0.048	1.682	0.051	1.663
Loamy sand	11.831	0.278	0.017	1.457	0.024	1.362

Table 2 Statistical results for nonlinear regression

Soil texture	SMC models								
	Brooks and Corey			Van Genuchten			Modified Van Genuchten		
	MA	DA	R^2	MA	DA	R^2	MA	DA	R^2
Loam	12	6	0.97	16	8	0.95	16	8	0.95
Sandy loam	12	6	0.96	16	8	0.95	14	7	0.96
Loamy sand	14	7	0.96	16	8	0.96	14	7	0.96

Table 3 Results of statistical comparison between experimental and analytical SMC

Soil texture	SMC models					
	Brooks and Corey		Van Genuchten		Modified Van Genuchten	
	COD	COV	COD	COV	COD	COV
Loam	0.812	0.19	0.924	0.06	0.824	0.22
Sandy loam	0.884	0.05	0.902	0.04	0.801	0.12
Loamy sand	0.786	0.21	0.967	0.02	0.795	0.20

is provided as Appendix. The description includes the results of analysis of variance (ANOVA), correlation of parameters and 95% confidence interval (CI) values (Table 3).

3.3 *SMC Comparison*

The analytical SMC is computed using HYDRUS-1D employing the equations mentioned in Sect. 2.1. Each soil texture exhibited its peculiar SMC curve as shown in Figs. 4, 5 and 6. The graphical comparison between experimental SMC and analytical SMC obtained from three models is shown in Figs. 4, 5 and 6. From the graphical comparison, it is found that Van Genuchten presented precise SMC to experimental one. In case of sandy loam soil, all the three models were found to present close agreement with the experimental SMC curve. These observations are further substantiated by the results of statistical analysis using the coefficient of determination

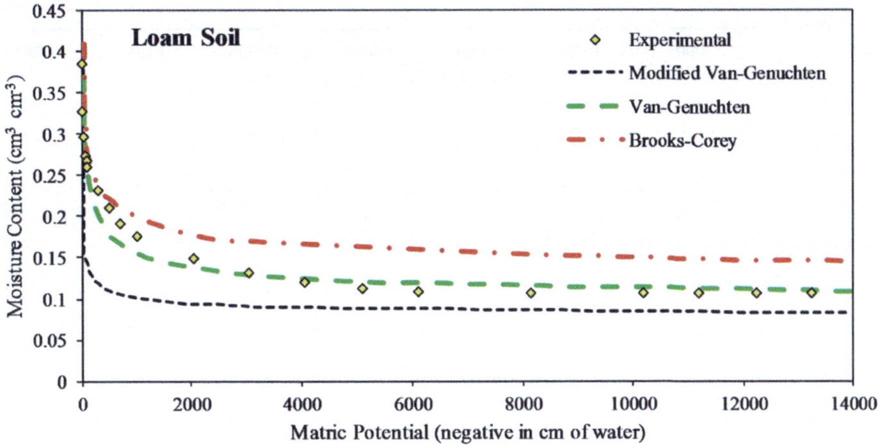

Fig. 4 Comparison of experimental and analytical SMC for loam soil

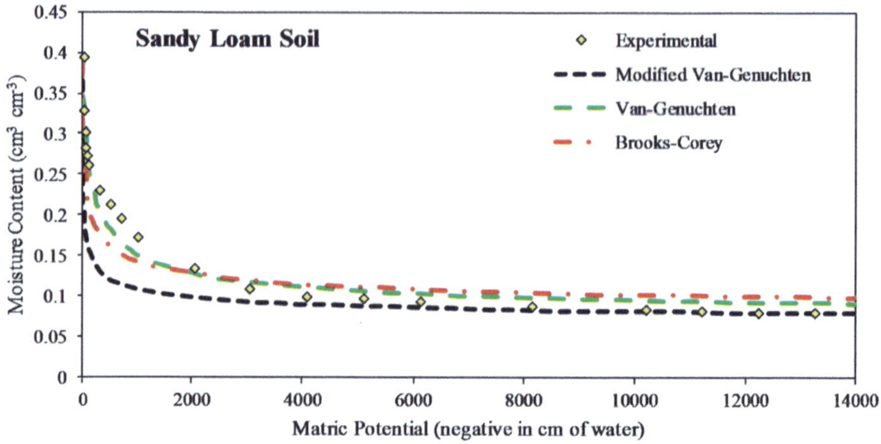

Fig. 5 Comparison of experimental and analytical SMC for sandy loam soil

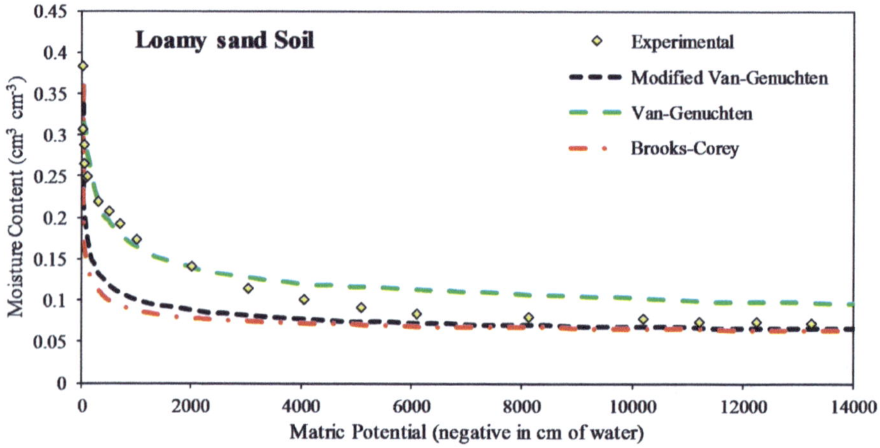

Fig. 6 Comparison of experimental and analytical SMC for loamy sand soil

(COD) and coefficient of variation (COV) as mentioned in Table 3. The numerical formulation of COD and COV can be referred in [21].

4 Conclusion

Following conclusions are drawn from the study:

i. Soil hydraulic parameter estimation using nonlinear regression is a reliable method for obtaining optimal values.
ii. Van Genuchten model simulated SMC was closest to experimental SMC indicating the accuracy of model among other models considered in the study.
iii. In case of sandy loam soil, all the models presented satisfactory results.

Acknowledgements The authors would like to acknowledge the Civil Engineering Department, National Institute of Technology, Hamirpur, for providing necessary facilities related to experimental work for the study. The funding is provided through MoES-NERC funded project "Sustaining Himalayan Water Resources in a Changing Climate".

Appendix

Statistical analysis of Brooks–Corey model for loamy sand soil.

NonlinearRegression Analysis

IterationHistory

I. No	R.S.O. S	Parameter	
		h_b	λ
1.0	.940	1.000	.100
1.1	.098	6.914	.243
2.0	.098	6.914	.243
2.1	.061	11.645	.283
3.0	.061	11.645	.283
3.1	.060	11.782	.277
4.0	.060	11.782	.277
4.1	.060	11.837	.278
5.0	.060	11.837	.278
5.1	.060	11.830	.278
6.0	.060	11.830	.278
6.1	.060	11.831	.278
7.0	.060	11.831	.278
7.1	.060	11.831	.278

The derivatives are numerically computed:
1.Major I. No is shown to left side of decimal and minor I. No to the right side.
2.After 14 model and 7 derivative assessments, the run is terminated due to the fact that there is relative reduction between successive RSOS at most SSCON=1.0E-008.

ParameterEstimates

Parameter	Est.	E_s	95% C. I	
			L.B	U.B
h_b	11.831	1.609	8.450	15.211
λ	.278	.016	.244	.311

Correlations of Parameter Estimates

	h_b	λ
h_b	1.000	.663
λ	.663	1.000

ANOVA

Source	S.O. S	df	M.S
Regression	4.170	2	2.085
Residual	.060	18	.003
T_{unc}	4.230	20	
T_c	1.529	19	

Dependent variable: S[a]
a. R squared = 1 - (Residual Sum of Squares) / (Corrected Sum of Squares) = .961.

References

1. Ojha, C.S., Prasad, K.S., Shankar, V., Madramootoo, C.A.: Evaluation of a nonlinear root-water uptake model. J. Irrig. Drainage Eng. **135**(3), 303–312 (2009)
2. Celia, M.A., Bouloutas, E.T., Zarba, R.L.: A general mass conservative numerical solution for the unsaturated flow equation. Water Resour. Res. **26**, 1483–1496 (1990)
3. Poddar, A., Kumar, N., Shankar, V.: Evaluation of two irrigation scheduling methodologies for potato (*Solanum tuberosum* L.) in north-western mid-hills of India. ISH J. Hydraul. Eng., 1–10 (2018). https://doi.org/10.1080/09715010.2018.1518733
4. Shin, Y., Mohanty, B.P., Ines, A.V.M.: Soil hydraulic properties in one dimensional layered soil profile using layer-specific soil moisture assimilation scheme. Water Resour. Res. **48**, W06529 (2012). https://doi.org/10.1029/2010WR009581
5. Kumar, R., Jat, M.K., Shankar, V.: Evaluation of modeling of water ecohydrologic dynamics in soil-root system. Ecol. Model. **269**, 51–60 (2013)
6. Govindraju, R.S., Or, D., Kavvas, M.L., Rolston, D.E., Biggar, J.: Error analyses of simplified unsaturated flow models under large uncertainty in hydraulic properties. Water Resour. Res. **28**(11), 2913–2924 (1992)
7. Mohanty, B.P., Zhu, J.: Effective averaging schemes for hydraulic parameters in horizontally and vertically heterogeneous soils. J. Hydrometeorol **8**(4), 715–729 (2007). https://doi.org/10.1175/JHM606.1
8. Richards, L.A.: Physical condition of water in soil. In: Black C.A. et al. (eds.) Method of Soil Analysis Part 1, pp. 128–151. American Society Agronomy, Madison, Wisconsin (1965)
9. Stone, L.R., Horton, M.L., Olson, T.C.: Water loss from an irrigated sorghum field: I. Water flux within and below the root zone. Agron. J. **65**, 492–497 (1973)
10. Gupta, S.C., Larson, W.E.: Estimating soil water retention characteristics from particle size distribution, organic matter percent and bulk density. Water Resour. Res. **15**(6), 1633–1635 (1979)
11. Ghosh, R.K.: Estimation of soil-moisture characteristics from mechanical properties of soils. Soil Sci. **130**, 60–63 (1980)
12. Nandagiri, L., Prasad, R.: Relative performance of textural models in estimating soil moisture characteristic. J. Irrig. Drainage **123**, 211–214 (1997)
13. Huang, X., Gao, B.: Review of soil moisture characteristic curve. J. Agric. Technol. Serv. **33** (2016)
14. Brooks, R.H., Corey, A.T.: Hydraulic Properties of Porous Media. In: Hydrology paper no. 3. Civil Engineering Department, Colorado State University, Fort Collins, Colorado (1964)
15. Campbell, G.S.: A simple method for determining unsaturated conductivity from moisture retention data. Soil Sci. **117**, 311–314 (1974)
16. Van Genuchten, M.T.: A closed form equation for predicting the hydraulic conductivity of unsaturated soils. Soil Sci. Soc. Am. J. **44**, 349–386 (1980)
17. Kosugi, K.: Lognormal distribution model for unsaturated soil hydraulic properties. Water Resour. Res. **32**(9), 2697–2703 (1996)
18. Schaap, M.G., Van Genuchten, M.T.: A modified Mualem–van Genuchten formulation for improved description of the hydraulic conductivity near saturation. Vadose Zone J. **5**(1), 27–34 (2006)
19. Trout, T.J., Garcia-Castillas, I.G., Hart, W.E.: Soil Water Engineering: Field and Laboratory Manual. Academic Publishers, Jaipur, India (1982)
20. Simunek, J., Sejna, M., Saito, H., Sakai, M., Van Genuchten, M.T.: Manual for HYDRUS-1D. Department of Environment Sciences, University of California, California (2013)
21. Shankar, V., Hari Prasad, K.S., Ojha, C.S.P., Govindaraju, R.S.: Model for nonlinear root water uptake parameter. J. Irrig. Drainage Eng. **138**(10), 905–917 (2012)

Assessment of Microphysical Parameterization Schemes on the Track and Intensity of Titli Cyclone Using ARW Model

G. Venkata Rao, K. Venkata Reddy and Y. Navatha

Abstract Advanced Weather Research and Forecasting (ARW) model has a wide range of applications for both the operational and research purpose. In this paper, an attempt has been made to investigate the sensitivity of seven Microphysical Parameterization (MP) schemes namely Lin, WSM3, WSM5, WSM6, Ferrier, Morrison, and Thompson schemes in the simulation of Very Severe Cyclonic Storm (VSCS) Titli (2018) occurred in the Bay of Bengal region for the track and rainfall intensity using ARW model. The cyclone track and intensity are simulated in terms of minimum Mean Sea Level Pressure (MSLP), and maximum surface wind. The results are verified with observations provided by the Indian Meteorological Department (IMD). From the results, it was observed that Ferrier scheme has provided best track and intensity forecasts for the selected cyclone.

Keywords Advanced research and weather research forecasting (ARW) · Tropical cyclone · MP schemes

1 Introduction

Over the last few decades, improvements in the computational power allows the usage of Numerical Weather Prediction (NWP) models extensively used at finer resolutions for the simulation of Tropical Cyclones (TCs) [1]. NWP models are composed of many parameterization schemes [3] which introduces errors and uncertainties into forecasts. The representation of physical processes is a key component in the NWP models. During the past two decades, NWP models particularly Advanced Research

G. Venkata Rao · K. Venkata Reddy (✉) · Y. Navatha
Civil Engineering Department, NIT Warangal, Warangal, India
e-mail: kvreddy@nitw.ac.in

G. Venkata Rao
e-mail: venkatg@student.nitw.ac.in

Y. Navatha
e-mail: ynavatha@nitw.ac.in

D. Dutta and B. Mahanty (eds.), *Numerical Optimization in Engineering and Sciences*,
Advances in Intelligent Systems and Computing 979,
https://doi.org/10.1007/978-981-15-3215-3_4

Weather Research Forecasting (ARW) model are simulated using complex physics schemes in both the operational and research applications.

Several studies were conducted to investigate the sensitivity of Microphysical Parameterization (MP) schemes for simulation of Tropical Cyclones (TCs) over the Bay of Bengal (BoB) region [1, 2, 4–9]. Rajeevan et al. [8] compared the performance of four MP schemes for a severe thunderstorm observed over the Gadanki region on 21st May 2008 and found that Thompson scheme is performing reasonably well when compared to other schemes. Rajeevan et al. [6] observed the similar kind of results for the same event. Pattanayak et al. [7] found that Ferrier scheme provided better results in terms of the tract and intensity of a TCs Nargis when compared with WSM3, WSM6, Thompson schemes. Whereas Singh and Bhaskaran [10] observed that Lin scheme was giving the best fit-tract for the cyclones Sidr, Nargis, Thane, Aila, Laila, and Jal. Reshmi et al. [9] has suggested that, the Morrsion scheme simulated the heavy rainfall event over Chennai in close agreement with the observed one than Lin, Thompson, WSM3, WSM6 schemes. From the review, it was observed that TCs are sensitive to MP schemes and are varying from one cyclone to another. So, there is a need to find the best MP scheme for Titli cyclone using ARW model.

2 Data and Methodology

2.1 ARW Model

Advanced Research WRF (ARW) model version 4.0 was used to study the impact of MP schemes on Titli Cyclone (http://www2.mmm.ucar.edu/wrf/users/). ARW model configure with two-way nested domains provided in Fig. 1 with 27 km resolution

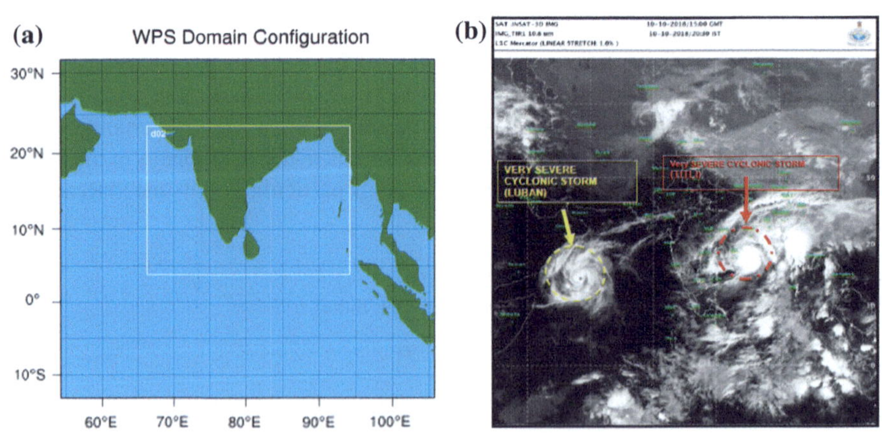

Fig. 1 **a** ARW domain configuration. **b** INSAT-3D imagery at the time of landfall (*source* http://www.rsmcnewdelhi.imd.gov.in/index.php?lang=en)

Table 1 Details of ARW model configuration

Domain center	10° N, 80° E
Number of domains	2 (d01 = 27 km, d02 = 9 km)
Initial and boundary conditions	GFS ANL data 0.5° × 0.5°
Cumulus physics	Parent domain (KF scheme), nested domain (no scheme)
Microphysics	Lin, Ferrier, Morrison, Thompson, WSM3, WSM5, WSM6
Shortwave radiation	Dudhia scheme
Longwave radiation	RRTM
Land surface model	Noah
Surface layer options	Revised MM5 Monin-Obukhov scheme
PBL	YSU scheme

for the outer domain and 9 km resolution for the inner domain [11]. The detailed description about the ARW model configuration is provided in the Table 1.

3 Data Used

The initial and boundary conditions for the ARW model were derived from GFSANL data available at 0.5° × 0.5° resolution (https://nomads.ncdc.noaa.gov/data/gfsanl/). Reports from the Indian Meteorological Department (IMD) were collected for the position of the TCs, MSLP (hPa), and MSW (kts) and are used for validation. Titli cyclone was considered in this paper and the INSAT-3D satellite imagery at the time of landfall is provided in Fig. 1b.

4 Numerical Experiments

Accurate prediction of TCs using ARW models needs suitable parameterizations for planetary boundary layer (PBL), cumulus convection (CC), surface fluxes, and microphysics (MP) because these processes work together to influence the cyclone forecasts. Sensitivity of seven MP schemes namely Lin, Ferrier, Morrison, Thompson, WSM3, WSM5, and WSM6 were evaluated using ARW model.

5 Results and Discussions

The model errors for the parameters MSLP, MSW, and track positions are estimated. Mean error matrix of these parameters are analysed here in order to find out the best microphysical parameterization scheme for track and intensity estimates, the results are analysed.

6 Sensitivity of Microphysics Parameterization Schemes

WRF simulation are conducted with Kain-Fritsch convection for domain1 (D1) and no convection scheme was applied for domain2 (D2), RRTM and Dudhia foe long-wave and shortwave radiation, NOAH and MM5 for surface, YSU for PBL schemes by varying MP schemes as according to Ferrier, Morrison, LIN, Thompson, WSM3, WSM5, and WSM6 schemes. Three hours time series plots for simulated values of MSLP and MSW are presented in Fig. 2. The mean errors of MSLP, MSW, and track positions are presented in Fig. 3.

All the MP schemes overestimated the winds (negative error) and underestimated the MSLP (positive error), thus simulating the over intensification of cyclones from 51-h model integration onwards. Among the seven schemes, the Ferrier scheme produced the minimum error in MSLP and cyclone track position compared with observed IMD records as well as the other schemes. But for MSW, minimum error was observed for WSM6 scheme till 48 h of model integration. After the 48 h Ferrier produced the least error. Of these seven simulations, the Ferrier scheme produced the better intensification of the storm throughout simulation, both in terms of MSLP and MSW.

The mean track error between 12 and 96 h of model integration time varied between 16 and 558 km in the simulations with different MP schemes. Both Ferrier and WSM6 schemes produced least errors in the position of cyclone track till 72 h of model integration, but at the time of landfall the position of cyclone was 18.8 km away for Ferrier scheme with 4.5 h time delay whereas for WSM6 scheme it was 30 km with 7 h of delay. After 72 h of the model integration WSM6 scheme produced

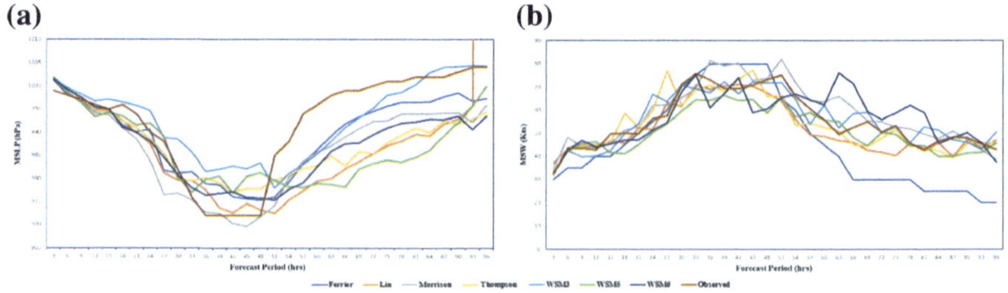

Fig. 2 Temporal variations of simulated values for **a** MSLP (hPa), **b** MSW (kts)

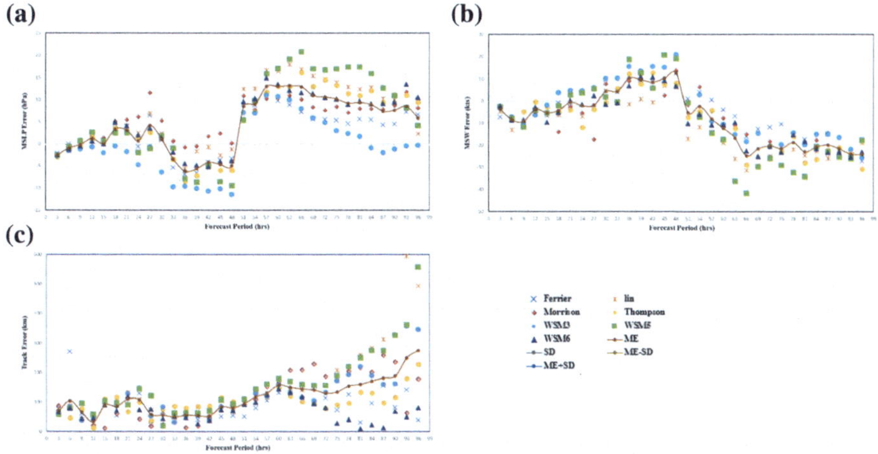

Fig. 3 Temporal variations of mean errors **a** MSLP (hPa), **b** MSW (kts) and **c** cyclone track position

Table 2 Statistical values for SLP

Scheme	Track error (km)	Time delay (h)
Ferrier	18.800	04.5
Lin	130.000	13.0
Morrison	88.000	04.5
Thompson	98.000	13.0
WSM3	90.000	08.0
WSM5	182.000	18.0
WSM6	30.000	07.0

minimum track error. The error in position of the cyclone and time delay at the time of landfall are provided in the Table 2.

7 Track and Intensity Errors

The time series errors (ME and SD) of MSLP, MSW, track position and the simulated tracks are presented in Figs. 3, 4 and 5. The ME for MSLP in the initial stages varies from −3 to +3 hPa up to 30 h of model integration and then increased gradually to 13 hPa at 60 h and reduced to 5 hPa at 96 h of model integration as the storm approaches to land region a gradual increment was observed. Correspondingly ME for MSW varies from −9 to +9 kts up to 48 h of model integration and increases gradually to 24 kts at 96 h, indicating the indicating the higher intensification of storms. The SD value for the MSLP and MSW varying between 0.22 hPa–5 hPa and 1.8 kts–6 kts respectively. ME track error varying marginal between 33 and 110 km

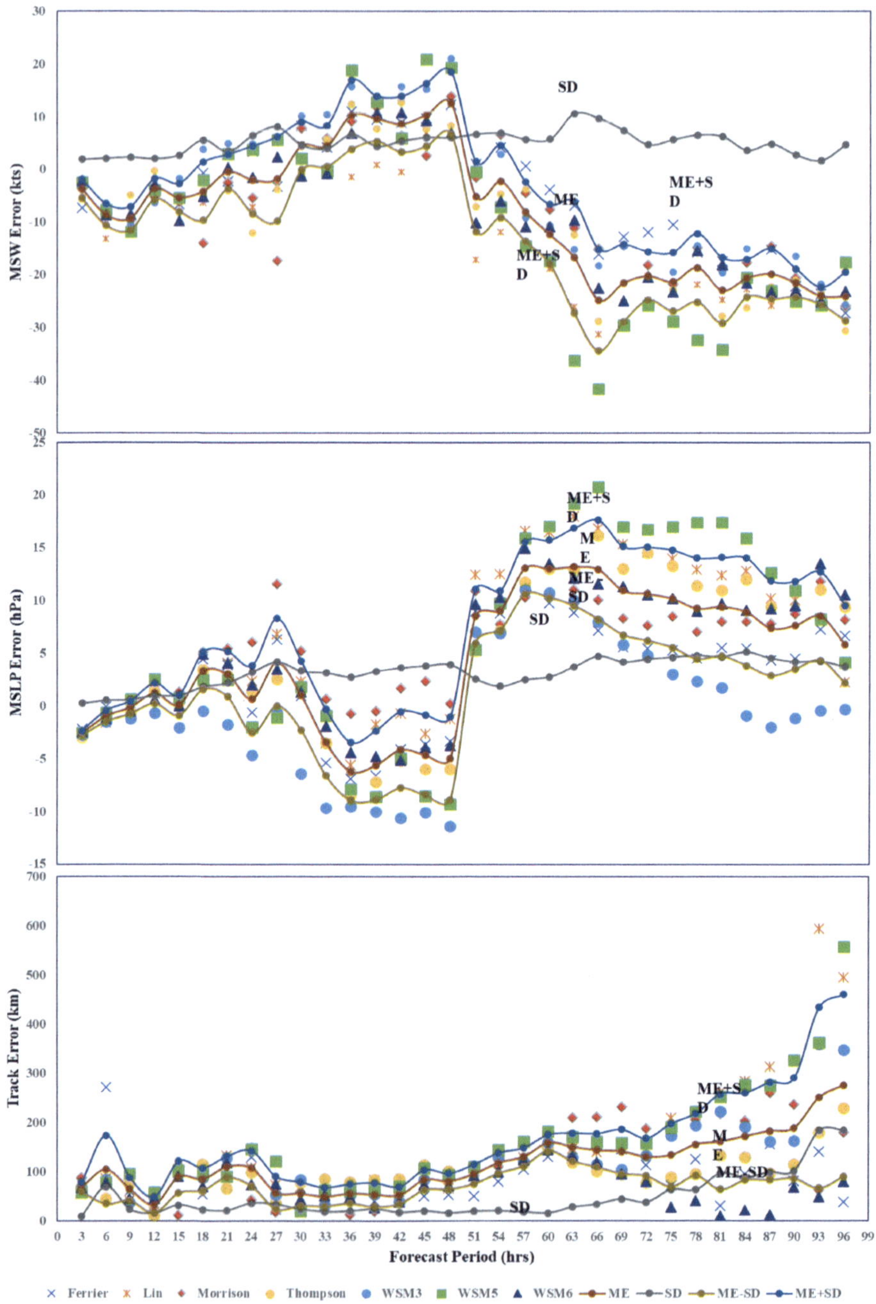

Fig. 4 Mean error (ME), standard deviation (SD), ME + SD, ME − SD of **a** MSW (kts), **b** MSLP (hPa), and **c** track position error (km)

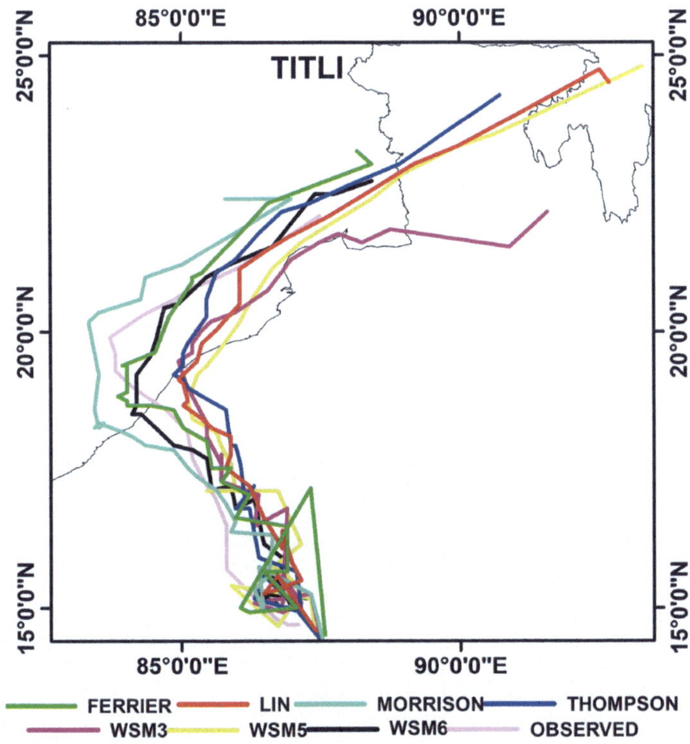

Fig. 5 Observed and simulated tracks of the TITLI cyclone

in the initial stages up to 36 h and then increases gradually from 53 to 275 km at 96 h. The SD values for the track error varying between 16 and 184 km.

Error threshold (ME − SD, ME + SD) values applied for MSLP, MSW, and the track position error time series data to identify the best MP scheme. After applying the threshold value, the least value of ME − SD and the height value of ME + SD are considered as lower and upper limits for selecting the best MP scheme for Titli cyclone. Using this criterion, Ferrier scheme found to be well simulated. It is to be noted that the only one cyclone is considered in the present study, the study is not adequate to derive statistical significance, so the results are used to only to provide qualitative indication.

8 Summary and Conclusions

The present study, the sensitivity to microphysics schemes was investigated to the best track and intensity of VSCS Titli occurred in BoB using ARW mesoscale model Version 4.0. Seven sensitivity experiments were conducted with different MP schemes for the cyclone Titli. The time series mean errors (ME) and SD for the MSLP, MSW, and track position error for all the seven experiments are found to provide the better estimate of track as well as the intensity of the cyclonic storm. Among all the seven schemes, Ferrier scheme provided the best track and storm intensity with least error.

Acknowledgements The authors would like to thank Dr. Satya Prakash Ojha and Dr. Sathiyamoorthy, Space Application Center (SAC—ISRO) Ahmedabad for providing an opportunity to work under SMART programme and the first also thank Indian Meteorological Department (IMD), Global Forecast Systems providing the observed data.

References

1. Choudhury, D., Das, S.: The sensitivity to the microphysical schemes on the skill of forecasting the track and intensity of tropical cyclones using WRF-ARW model. J. Earth Syst. Sci. **126**, 1–10 (2017). https://doi.org/10.1007/s12040-017-0830-2
2. Gunwani, P., Mohan, M.: Sensitivity of WRF model estimates to various PBL parameterizations in different climatic zones over India. Atmos. Res. **194**, 43–65 (2017). https://doi.org/10.1016/j.atmosres.2017.04.026
3. Jandaghian, Z., Touchaei, A.G., Akbari, H.: Sensitivity analysis of physical parameterizations in WRF for urban climate simulations and heat island mitigation in Montreal. Urban Clim. **24**, 577–599 (2018). https://doi.org/10.1016/j.uclim.2017.10.004
4. Karki, R., ul Hasson, S., Gerlitz, L., Talchabhadel, R., Schenk, E., Schickhoff, U., Scholten, T., Böhner, J.: WRF-based simulation of an extreme precipitation event over the Central Himalayas: atmospheric mechanisms and their representation by microphysics parameterization schemes. Atmos. Res. **214**, 21–35 (2018). https://doi.org/10.1016/j.atmosres.2018.07.016
5. Lekhadiya, H.S., Jana, R.K.: Analysis of extreme rainfall event with different microphysics and parameterization schemes in WRF model. Positioning **09**, 1–11 (2018). https://doi.org/10.4236/pos.2018.91001
6. Madala, S., Satyanarayana, A.N.V., Rao, T.N.: Performance evaluation of PBL and cumulus parameterization schemes of WRF ARW model in simulating severe thunderstorm events over Gadanki MST radar facility—case study. Atmos. Res. **139**, 1–17 (2014). https://doi.org/10.1016/j.atmosres.2013.12.017
7. Pattanayak, S., Mohanty, U.C., Osuri, K.K.: Impact of parameterization of physical processes on simulation of track and intensity of tropical cyclone Nargis (2008) with WRF-NMM model. Sci. World J. **2012**, 1–18 (2012). https://doi.org/10.1100/2012/671437
8. Rajeevan, M., Kesarkar, A., Thampi, S.B., Rao, T.N., Radhakrishna, B., Rajasekhar, M.: Sensitivity of WRF cloud microphysics to simulations of a severe thunderstorm event over Southeast India. Ann. Geophys. **28**, 603–619 (2010). https://doi.org/10.5194/angeo-28-603-2010
9. Reshmi Mohan, P., Srinivas, C.V., Yesubabu, V., Baskaran, R., Venkatraman, B.: Simulation of a heavy rainfall event over Chennai in Southeast India using WRF: sensitivity to microphysics parameterization. Atmos. Res. **210**, 83–99 (2018). https://doi.org/10.1016/j.atmosres.2018.04.005
10. Singh, K.S., Bhaskaran, P.K.: Impact of PBL and convection parameterization schemes for prediction of severe land-falling Bay of Bengal cyclones using WRF-ARW model. J. Atmos. Solar Terr. Phys. **165–166**, 10–24 (2017). https://doi.org/10.1016/j.jastp.2017.11.004
11. Srinivas, C.V., Bhaskar Rao, D.V., Yesubabu, V., Baskaran, R., Venkatraman, B.: Tropical cyclone predictions over the Bay of Bengal using the high-resolution advanced research weather research and forecasting (ARW) model. Q. J. R. Meteorol. Soc. **139**, 1810–1825 (2013). https://doi.org/10.1002/qj.2064

Topology Optimization of Concrete Dapped Beams Under Multiple Constraints

V. R. Resmy and C. Rajasekaran

Abstract Topology optimization is now becoming the effective method for solving various problems related to engineering. Optimization is a mathematical method to find the optimum solution by satisfying all the constraints associated with that problem, while topology optimization is a branch of structural optimization as it finds optimum material layout within the given boundary. This study focuses on the topology optimization of concrete dapped beams with various constraints to ensure the applicability of topology optimization during the design phase of structures. Compliance minimization with three different constraints along with volume constraint has been selected to derive the truss-like pattern for beams. To derive a lightweight structure with stress constraint, volume-based topology optimization has adopted. Strut-and-tie modeling (STM) of concrete members has been identified as a powerful method for modeling discontinuity regions within the structural member. Topology optimization can be used as a supporting method for developing more reliable strut-and-tie models.

Keywords Topology optimization · SIMP · Strut-and-tie model · Compliance

1 Introduction

Topology optimization is intended to find the optimum load path associated with a particular load and boundary conditions by satisfying various requirements in the form of objective and constraints. The history of structural optimization has traced from the work of Michell [1] in which he derived lightweight structure for material economy with stress constraint. Vanderplaats [2] identified Schmit's work [3] as the turning point in modern structural optimization as he combines the finite element analysis with nonlinear structural optimization. Topology optimization methods can be divided based on the domain and search method for solution. Starting from the work of Prager [4], Rozvany [5] had adopted optimality criteria methods for different optimization problems. Homogenization method has been widely used in topology

V. R. Resmy (✉) · C. Rajasekaran
National Institute of Technology Karnataka, Mangalore, India
e-mail: resmy.cv16f16@nitk.edu.in

D. Dutta and B. Mahanty (eds.), *Numerical Optimization in Engineering and Sciences*,
Advances in Intelligent Systems and Computing 979,
https://doi.org/10.1007/978-981-15-3215-3_5

optimization [6] in which a microstructure is created in material resulting in a composite structure. Bendsoe [7] introduced a method to derive a non-discrete solution by introducing a density function that allows the discrete variables to vary continuously. This approach is termed as solid isotropic material with penalization (SIMP). Buhl et al. [8] used the SIMP approach along with the method of moving asymptotes (MMA) to minimize various objective functions of geometrically nonlinear structures subject to volume constraints.

Discontinuity regions (D-regions) are the portions of structural member where nonlinear strain distribution occurs as a result of geometry or loading. Strut-and-tie modeling is an accepted method for designing D-regions where Bernoulli's hypothesis is not valid. German civil engineer Ritter [9] designed concrete beams using truss analogy method where concrete beams could carry compressive forces and reinforcing bars could carry tensile forces. The STM method has become popular for designing D-regions after the landmark paper of Schlaich [10]. Conventional methods of STM involve a lot of trial and error procedure which can be overcome by iterative computer programs [11]. Bruggi [12] proposed a methodology that deals with the generation of truss-like designs to derive preliminary strut-and-tie models not only in the established bi-dimensional context but also in a 3D environment. Topology optimization of structures under multiple constraints has been carried out by several researchers to replicate various effects [13, 14]. This paper focuses on the topology optimization of concrete dapped beams with multiple constraints in ABAQUS finite element software.

2 Modeling of Dapped End Beams

Reinforced concrete dapped end beams (RC-DEB) are commonly used in concrete bridge girders and prepared concrete buildings. The use of dapped beams enables the erection of precast members due to its lateral stability of an isolated dapped end beam than that of an isolated beam. In some cases, the nib of a dapped end beam is similar to an inverted corbel. Due to the geometrical nonlinearity, high stress concentration arises at reentrant corners of dapped beams which should be reinforced properly to avoid failure [15]. The aim of the present study is to evolve truss-like pattern of RC-DEBs with two daps under different constraints using topology optimization. The model dimensions and boundary conditions are shown in Fig. 1. Symmetrical loading and boundary conditions are given. Concrete grade of M30 has selected with Poisson's ratio of 0.15. The beam is modeled in ABAQUS with a mesh size of 40 mm. ABAQUS software adopts SIMP material interpolation scheme and the penalty factor entered as 3. The initial design will converge to an optimum topology after finite element analysis, sensitivity analysis, and design variable updating with optimality criteria method. The finite element model with load and boundary condition is given in Fig. 2.

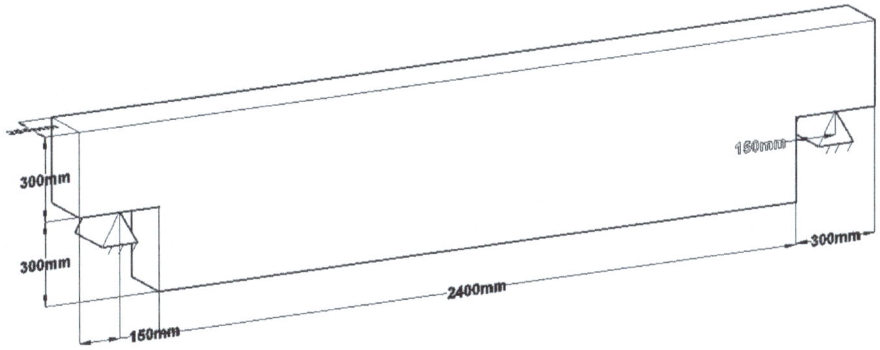

Fig. 1 Dimensions and boundary conditions of dapped beam

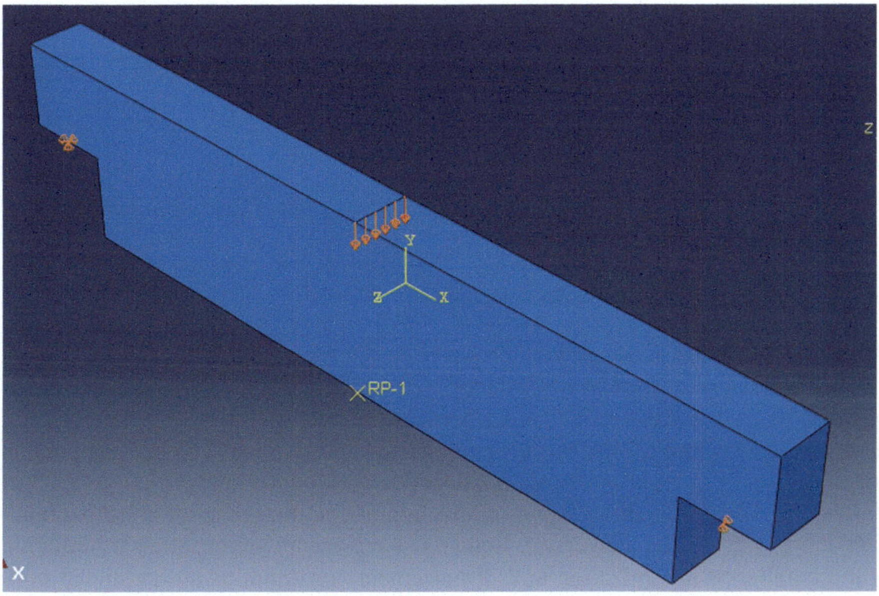

Fig. 2 Finite element model of dapped beam

3 Formulation of Topology Optimization Problems

Four different optimization problems adopted for simulation are stated below. Compliance is an inverse indictor of stiffness. To derive the stiffest layout of a structure for a given set of loadings and boundary conditions, minimization of compliance can be selected as an objective function.

Problem 1:

$$\begin{aligned} &\text{Minimize} \quad C = C(x) = \tfrac{1}{2}u^{\mathrm{T}}Ku \\ &\text{Subject to:} \quad \begin{aligned} &d_j(x) - d^* \le 0 \\ &\frac{V(x)}{V*} = f \end{aligned} \end{aligned} \tag{1}$$

Problem 2:

$$\begin{aligned} &\text{Minimize} \quad C = C(x) = \tfrac{1}{2}u^{\mathrm{T}}Ku \\ &\text{Subject to:} \quad \begin{aligned} &\omega^* - \omega_n(x) \le 0 \\ &\frac{V(x)}{V*} = f \end{aligned} \end{aligned} \tag{2}$$

Problem 3:

$$\begin{aligned} &\text{Minimize} \quad C = C(x) = \tfrac{1}{2}u^{\mathrm{T}}Ku \\ &\text{Subject to:} \quad \begin{aligned} &\sigma_n(x) - \sigma^* \le 0 \\ &\frac{V(x)}{V*} = f \end{aligned} \end{aligned} \tag{3}$$

Problem 4:

$$\begin{aligned} &\text{Minimize} \quad \text{Volume} \\ &\text{Subject to}: \sigma_n(x) - \sigma^* \le 0 \end{aligned} \tag{4}$$

where x is the vector of design variables, u is the displacement vector, K is the global stiffness matrix, C is the mean compliance, d_j is the magnitude of displacement vector of jth node, ω_n is the nth mode natural frequency, σ_n is the stress of nth element, V is the material volume, $V*$ is the design domain volume, f is the prescribed volume fraction with $d*$, $\omega*$, and $\sigma*$ being the imposed constraint values.

The first three problems adopt minimization of compliance as an objective function along with volume constraint. Equation 1 represents optimization formulation with compliance minimization and displacement constraint as an additional constraint while second (Eq. 2) and third problems (Eq. 3) adopt eigen frequency and stress as a constraint. Fourth problem derives a lightweight structure with stress as a constraint.

4 Results and Discussion

In all the compliance minimization problems, volume fraction of 30% is selected as volume constraint. A central load of 60 KN has applied for simply supported concrete dapped beam.

4.1 Problem 1

Displacement constraint ≤1 mm at the center of the bottom surface including all nodes had applied. The optimization history of objective and constraints is shown in Fig. 3. Converged results have obtained with the initial strain energy of 369,000 N mm resulting with final strain energy of 29632.3574 N mm at iteration 40. Volume fraction of 30% with displacement at center of the beam has arrived at a value of 0.77 mm as the final converged value. The final layout of beam is shown in Fig. 4.

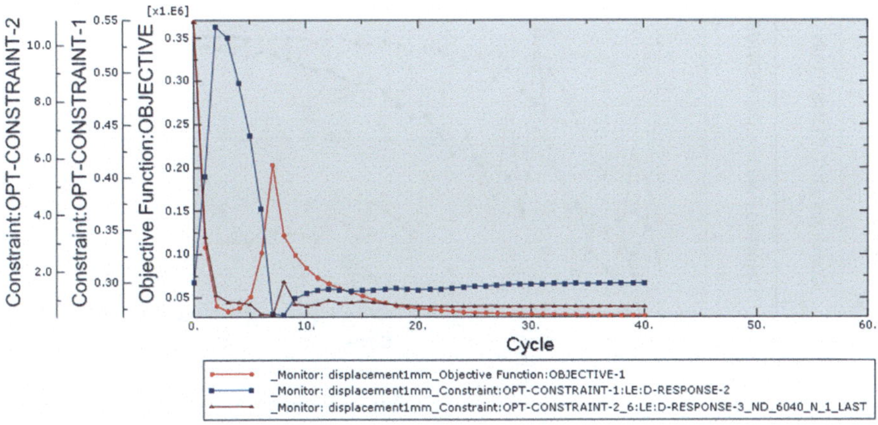

Fig. 3 Optimization history with displacement as constraint-2

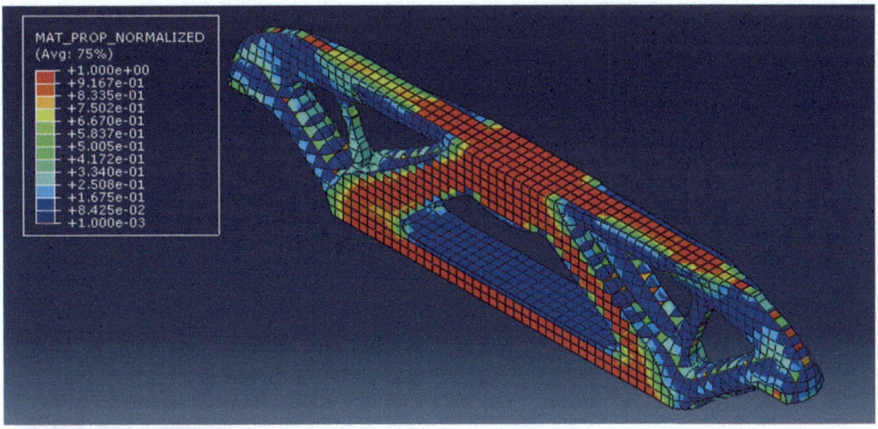

Fig. 4 Optimum material layout when displacement as constraint-2

4.2 Problem 2

Eigen frequency of first mode ≥10 rad/time had applied as a constraint. The initial strain energy of 1.58e5 has converged to 1.48e5 at iteration 34 with a volume fraction of 29.9%. Eigen frequency constraint has arrived to a value of 11.2249 rad/sec in the final stage. Figures 5 and 6 represents the optimization history and optimum material layout.

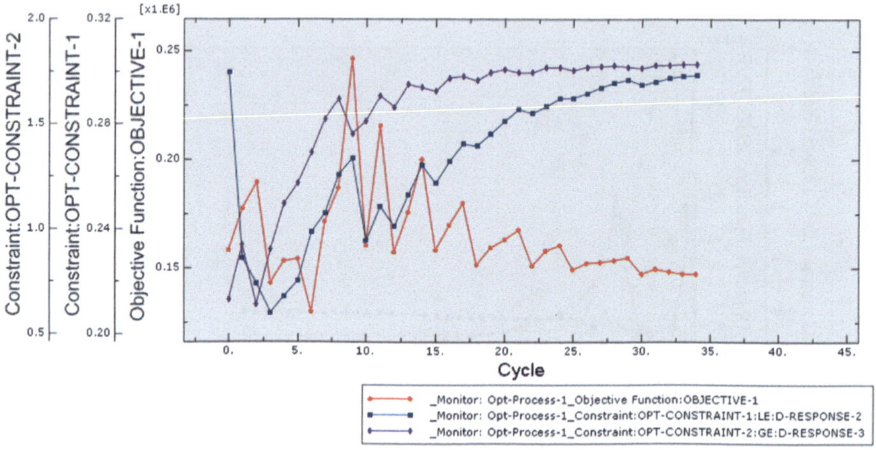

Fig. 5 Optimization history with eigen frequency as constraint-2

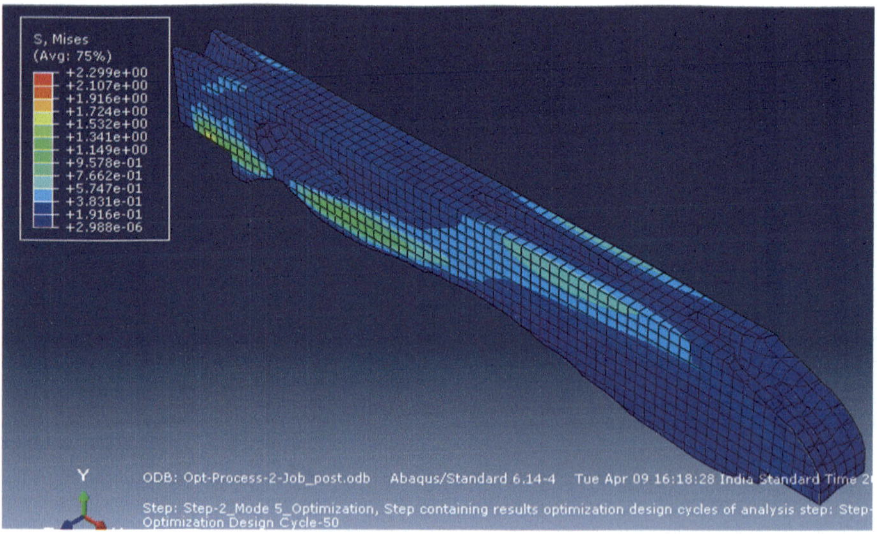

Fig. 6 Optimum material layout when Eigen frequency as constraint-2

4.3 Problem 3

Material strength of concrete is 35 N/mm^2. von Mises stress represents Stress constraint in Problem 3. von Mises stress constraint should not exceed material strength at any point in the beam. At iteration 48, strain energy has changed from 368,500 to 29074.5 N mm with a von Mises stress of 4.256 N/mm^2. Optimization history and material layout are shown in Figs. 7 and 8.

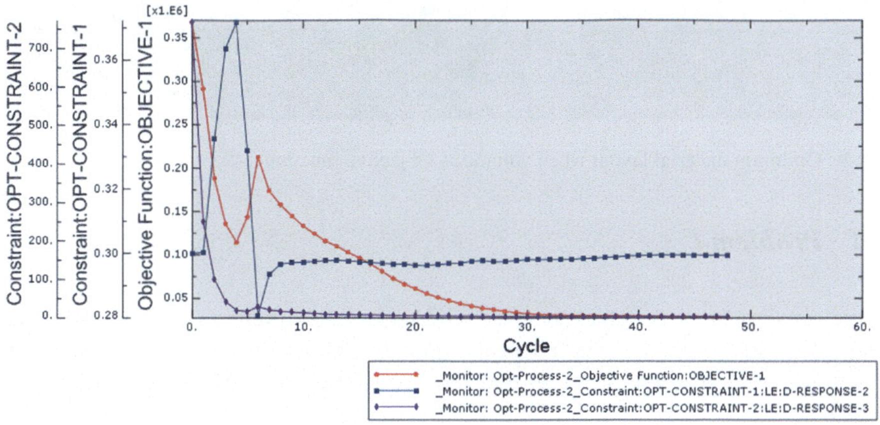

Fig. 7 Optimization history with stress as constraint-2

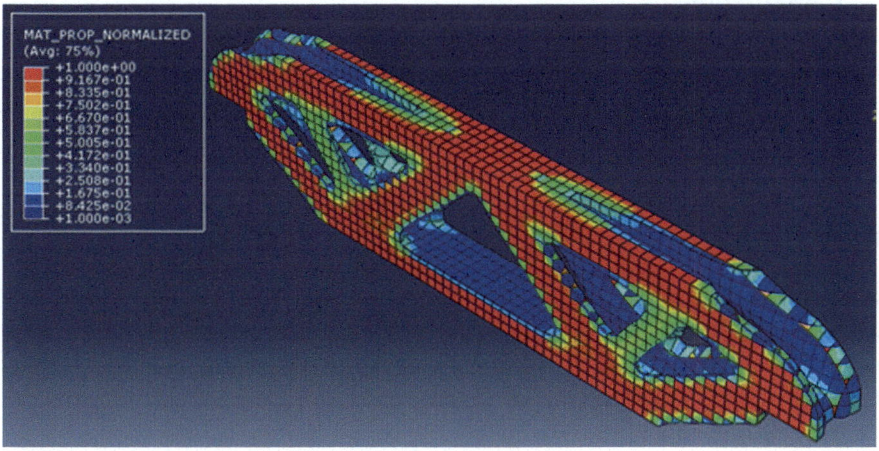

Fig. 8 Optimum material layout when stress as constraint-2

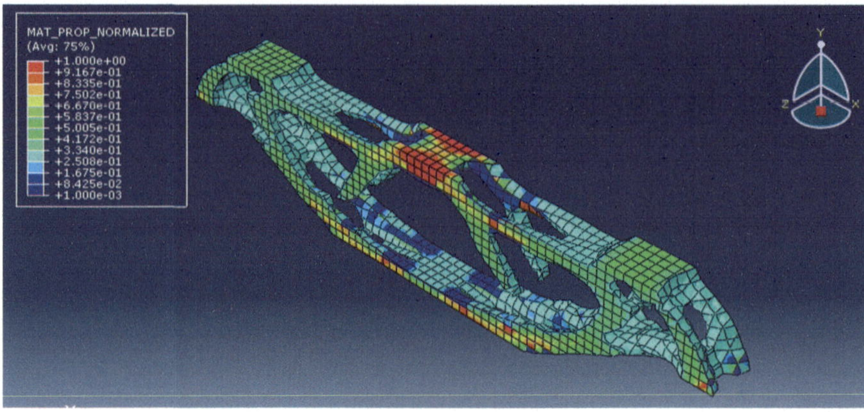

Fig. 9 Optimum material layout when volume as Objective function

4.4 Problem4

Some convergence issues had occurred while selecting volume as an objective function with material strength of 35 Mpa as constraint. However, a reasonable optimum material layout at iteration 56 with a von Mises stress of 30 N/mm^2 is shown in Fig. 9.

5 Conclusions

The present study focuses on the topology optimization of concrete dapped beams with different constraints. As strut-and-tie modeling is an accepted method for modeling discontinuity regions, all inaccuracies related to STM can be avoided with the help of topology optimization which relies on structural mechanics. Different constraints based on the design requirements had adopted in this study to ensure the applicability of topology optimization during the design phase. The constraints selected evaluate the structural performance, which help to save material while satisfying functional constraints in optimization problems. This method results in more accurate output based on load path method which can be utilized for solving various civil engineering problems. As the dimensioning of strut-and-tie model is beyond the scope of this study, it has not presented.

References

1. Michell, A.G.M.: LVIII The limits of economy of material in frame-structures. Lond. Edinburgh Dublin Philos. Mag. J. Sci. **8**(47), 589–597 (1904)
2. Vanderplaats, G.N.: Thirty years of modern structural optimization. Adv. Eng. Softw. **16**(2), 81–88 (1993)
3. Schmit, L.A.: Structural design by systematic synthesis. In Proceedings of the Second National Conference on Electronic Computation, ASCE, Sept 1960
4. Prager, W., Rozvany, G.I.: Optimization of structural geometry. In: Dynamical Systems, pp. 265–293. Academic Press (1977)
5. Rozvany, G.I.: Structural Design Via Optimality Criteria: The Prager Approach to Structural Optimization, vol. 8. Springer Science & Business Media (2012)
6. Bendsøe, M.P., Kikuchi, N.: Generating optimal topologies in structural design using a homogenization method. Comput. Methods Appl. Mech. Eng. **71**(2), 197–224 (1988)
7. Bendsøe, M.P.: Optimal shape design as a material distribution problem. Struct. Optim. **1**(4), 193–202 (1989)
8. Buhl, T., Pedersen, C.B., Sigmund, O.: Stiffness design of geometrically nonlinear structures using topology optimization. Struct. Multi. Optim. **19**(2), 93–104 (2000)
9. Ritter, W.: Die bauweise hennebique. Schweizerische Bauzeitung **33**(7), 59–61 (1899)
10. Schlaich, J., Schäfer, K., Jennewein, M.: Toward a consistent design of structural concrete. PCI J. **32**(3), 74–150 (1987)
11. Marti, P.: Basic tools of reinforced concrete beam design. J. Proc. **82**(1), 46–56 (1985)
12. Bruggi, M.: Generating strut-and-tie patterns for reinforced concrete structures using topology optimization. Comput. Struct. **87**(23–24), 1483–1495 (2009)
13. Da, D., Xia, L., Li, G., Huang, X.: Evolutionary topology optimization of continuum structures with smooth boundary representation. Struct. Multi. Optim. **57**(6), 2143–2159 (2018)
14. Huang, C.W., Chou, K.W.: Volume adjustable topology optimization with multiple displacement constraints. J. Mech. 1–12 (2017)
15. Huang, P.C., Nanni, A.: Dapped-end strengthening of full-scale prestressed double tee beams with FRP composites. Adv. Struct. Eng. **9**(2), 293–308 (2006)

Selecting Optimized Mix Proportion of Bagasse Ash Blended Cement Mortar Using Analytic Hierarchy Process (AHP)

S. Praveenkumar, G. Sankarasubramanian and S. Sindhu

Abstract Durability is an important aspect in determining the performance of concrete structures. The results of the durability properties at 90 days are best suited in determining the performance of the concrete structures and mortars. In this paper, cement mortars blended with sugarcane bagasse ash (BA) are studied for their durability properties at 90 days. One control cement mortar specimen and four specimens at 5, 10, 15 and 20% replacement for cement are prepared, and their properties are compared. However, it is difficult to analyze the performance because of varying factors, and hence, an optimization methodology is adapted. analytic hierarchy process (AHP) is an optimization methodology used in evaluating the performance of the specimens based on durability criteria. In this technique, all the durability criteria for the cement mortar are compared with each other based on their importance in determining the performance. Finally, a single value is obtained by relating all the properties considered, and the optimized mix proportion for replacement of bagasse ash is determined.

Keywords Bagasse ash · Cement mortar · Multi-criteria decision making · Optimization · Analytic hierarchy process

1 Introduction

The mechanical and durability properties of the cement mortar and concrete vary to a large extent when the additives known as supplementary cementing materials (SCMs) are added. However, when the materials are added partially in varying percentages, it becomes difficult to assess the best percentage of replacement. Hence, an optimization technique involving multi-criteria decision making (MCDM) is needed to determine the optimum percentage of replacement of the SCMs. A few examples

S. Praveenkumar (✉) · G. Sankarasubramanian · S. Sindhu
Department of Civil Engineering, PSG College of Technology, Coimbatore 641004, India
e-mail: spk.civil@psgtech.ac.in

G. Sankarasubramanian
e-mail: gsa.civil@psgtech.ac.in

D. Dutta and B. Mahanty (eds.), *Numerical Optimization in Engineering and Sciences*, Advances in Intelligent Systems and Computing 979,
https://doi.org/10.1007/978-981-15-3215-3_6

of the techniques used in MCDM are weighted sum method (WSM), weighted product method (WPM), technique for order preference by similarity to ideal solution (TOPSIS) and analytic hierarchy process (AHP).

The most widely used technique in the decision making is the analytical hierarchy process (AHP) which is developed by Professor Thomas L. Saaty in 1988 [1]. A set of alternative options with a set of evaluation criteria is considered in this process. AHP is a very flexible and powerful tool. It helps us to set priorities and make the best decision when both tangible and non-tangible aspects of decision need to be considered. It not only helps the decision makers to arrive at the best decision but also provides a clear rationale that it is the best. This is because it reduces the decisions of complex nature to a series of one-on-one comparisons, and then, the results are synthesized. Hence, AHP is a tool that is able to translate both qualitative and quantitative evaluations into a multi-criteria ranking and is regarded as the widely used decision-making method. The different studies based on AHP are shown in Table 1.

In this study, the cement in the mortar is replaced by sugarcane bagasse ash (SCBA) bagasse ash, which is a highly pozzolanic SCM. Experiments are conducted to find the various durability (criteria) properties at 90 days on five mix proportions of the cement mortar (alternatives) with 0, 5, 10, 15 and 20% of bagasse ash. These properties are then compared and optimized for all the alternatives using AHP, and the best alternative is selected.

Table 1 Applications based on AHP

Authors	Application
Lamaakchaoui et al. [2]	Helping customers in selecting the best complementary products
Chakladar and Chakraborty [3]	Selection of best non-traditional machining process
Erdebilli and Erkan [4]	Selection of best supplier
Ince et al. [5]	Using combined AHP and TOPSIS for learning object metadata evaluation
Socaciu et al. [6], Lin et al. [7]	Selection of best PCM for vehicles for thermal comfort An adaptive AHP approach that used a soft computing scheme
Mansor et al. [8]	Selection of best fiber-reinforced polymer composites
Hudymacova et al. [9], Praveenkumar et al. [10]	Selection of best supplier Selection of optimum mix proportion of bagasse ash blended high-performance concrete
Karim and Karmaker [11]	Combined AHP and TOPSIS for selecting the best machine
Venkata Rao [12]	Evaluation of best manufacturing systems using TOPSIS and AHP

2 Optimization Methodology

AHP is simple because there is no need of building a complex expert system with the decision maker's knowledge embedded in it.

Three simple steps constitute the AHP:

- Computing the vector of criteria weights.
- Computing the matrix of option scores.
- Ranking the options.

Step 1: Computing the vector of criteria weights

Generating a *pair-wise comparison matrix* **A** is the first step in this process. This is done to find the relative importance of different criteria with respect to the objective. The matrix **A** shown in Eq. (1) is an $m \times m$ real matrix, where m is the number of evaluation criteria considered. Each entry a_{jk} of the matrix **A** represents the importance of the jth criterion relative to the kth criterion as shown in Table 2.

Hence, the pair-wise comparison matrix is generated as follows.

$$A = \begin{bmatrix} a_{11} & \cdots & a_{1n} \\ \vdots & \ddots & \vdots \\ a_{n1} & \cdots & a_{nn} \end{bmatrix} \quad (1)$$

Step 2: Assigning weights for sub-criteria

The normalized weights for the sub-criteria are obtained by Eq. (3) after finding the geometric means of each row in A matrix by Eq. (2) and normalizing them.

$$\mathrm{GM}_i = \{a_{i1} \times a_{i2} \times \cdots \times a_{ij}\}^{1/n} \quad (2)$$

$$W_i = \frac{\mathrm{GM}_i}{\sum_{j=1}^{j=n} \mathrm{GM}_i} \quad (3)$$

Table 2 Nine-point scale of pair-wise comparison by Saaty

Value of a_{jk}	Interpretation
1	j and k are equally important
3	j is slightly more important than k
5	j is more important than k
7	j is strongly more important than k
9	j is absolutely more important than k
2, 4, 6, 8	Intermediate values of relative importance

Step 3: Forming C matrix

It is defined as the product of pair-wise comparison matrix and the column weight matrix given by Eq. (4). It denotes an n-dimensional column vector describing the sum of the weighted values for the importance degrees of the attributes.

$$C = A.W \tag{4}$$

Step 4: Finding the consistency value

The consistency value is given by Eq. (5).

$$\mathrm{CV}_i = \frac{C_i}{w_i} \tag{5}$$

After finding the consistency value, the lambda maximum (λ_{max})is found which the average of the consistency values given by Eq. (6).

$$\lambda_{max} = \frac{\sum_{i=1}^{i=n} CV_i}{n} \tag{6}$$

Step 5: Finding consistency ratio

The consistency ratio (CR) is given by Eq. (7),

$$\mathrm{CR} = \frac{\mathrm{CI}}{\mathrm{RI}} \tag{7}$$

$$CI = \frac{(\lambda_{max} - n)}{n-1} \tag{8}$$

$$\mathrm{RI} = \frac{[1.987 \cdot (n-2)]}{n} \tag{9}$$

where CI is the consistency index as per Eq. (8) and RI is random inconsistency as per Eq. (9). The evaluation of the pair-wise comparison matrix is implied to be perfectly consistent if CI = 0. If the computed CR value is less than 0.1, then the comparison matrix is accepted or else a new comparison matrix is to be constructed.

Step 6: Ranking

The alternatives are then ranked based on overall performance level of each alternative with respect to criteria.

$$P_k = \sum_{i=1}^{i=n} w_i \sum_{j=1}^{j=m} w_{ij}, \quad (10)$$

where w_i are the weights of criteria and w_j are the weights of alternatives, respectively.

3 Methodology for Optimization of Bagasse Ash Blended Cement Mortar

The optimal mix design for the bagasse ash (BA) blended cement mortar is necessary in determining the quality assessment of the mixes. The framework of the methodology is shown in Fig. 1.

The durability criteria considered are saturated water absorption (SWA), porosity (P), sorptivity (S), water permeability (WP), seawater resistance (SWR), acid resistance (ART), drying shrinkage (DS) and air content (AC). Five mortar specimens with cement (C) and sand (S) in the ratio 1:3 and cement being replaced at 0, 5, 10,

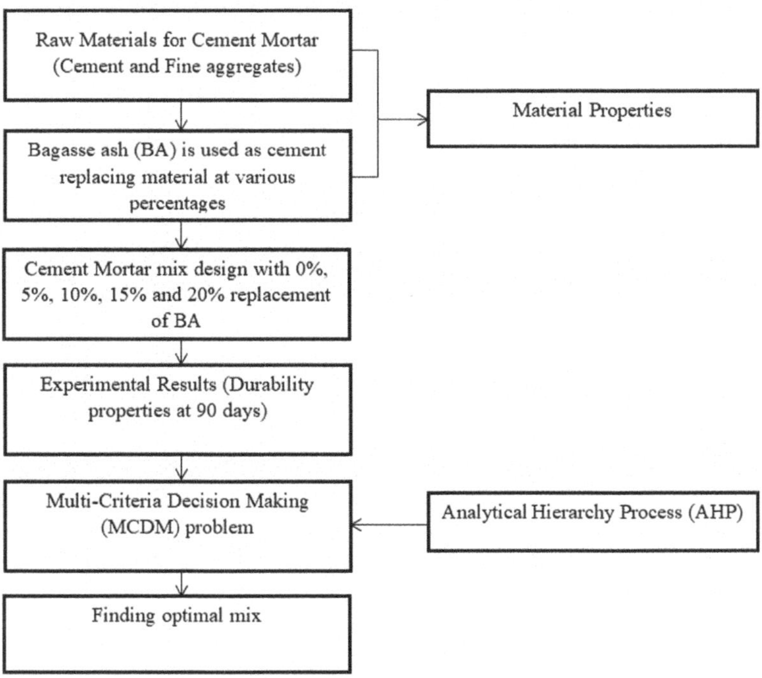

Fig. 1 Methodology framework

15 and 20% of bagasse ash (BA) are prepared. The experimental results are shown in Table 3, and the preference values and importance factor of the results are shown in Table 4.

The hierarchical structure for categorizing the goal, alternatives, criteria and sub-criteria is shown in Fig. 2.

4 Results and Discussions

4.1 Generating Pair-Wise Comparison Matrix

A pair-wise comparison matrix A_1 is generated for durability properties as per Eq. 1. The matrix A_1 is formed based on the preference values in Table 3. For example, it can be seen that SWA is strongly more important than P and S. Hence, a relative importance of 7 is assigned to SWA over P and S (i.e., $a_{12} = a_{13} = 7$), and a relative importance of 1/7 is assigned to P and S over SWA (i.e., $a_{21} = a_{31} = 1/7$).

$$\begin{array}{c} \quad\quad \text{SWA} \; P \; S \; \text{WP} \; \text{SWR} \; \text{ART} \; \text{DS} \; \text{AC} \\ A_1 = \begin{bmatrix} 1 & 7 & 7 & 1 & 3 & 3 & 5 & 9 \\ 1/7 & 1 & 1 & 1/7 & 1/5 & 1/5 & 1/3 & 2 \\ 1/7 & 1 & 1 & 1/7 & 1/5 & 1/5 & 1/3 & 2 \\ 1 & 7 & 7 & 1 & 3 & 3 & 5 & 9 \\ 1/3 & 5 & 5 & 1/3 & 1 & 1 & 3 & 7 \\ 1/3 & 5 & 5 & 1/3 & 1 & 1 & 3 & 7 \\ 1/5 & 3 & 3 & 1/5 & 1/3 & 1/3 & 1 & 6 \\ 1/9 & 1/2 & 1/2 & 1/9 & 1/7 & 1/7 & 1/6 & 1 \end{bmatrix} \begin{matrix} \text{SWA} \\ P \\ S \\ \text{WP} \\ \text{SWR} \\ \text{ART} \\ \text{DS} \\ \text{AC} \end{matrix} \end{array}$$

4.2 Sub-criteria Weights

The weights for the sub-criteria are assigned as per Eq. 3 and are shown below

$$W_1 = \begin{pmatrix} 0.2857 \\ 0.0323 \\ 0.0323 \\ 0.2857 \\ 0.1375 \\ 0.1375 \\ 0.0692 \\ 0.0197 \end{pmatrix} \begin{matrix} \text{SWA} \\ P \\ S \\ \text{WP} \\ \text{SWR} \\ \text{ART} \\ \text{DS} \\ \text{AC} \end{matrix}$$

Table 3 Experimental results at 90 days

Designation	C (g)	S (g)	BA (g)	SWA %	*P* %	*S* (mm/min$^{0.5}$)	WP (m/s × 10^{-12})	SWR %	ART %	DS %	AC %
BA0	200	600	0	4.620	9.71	12.727	1.157	2.067	1.772	0.3392	1.9
BA1	190	600	10	4.842	9.45	9.890	0.955	1.583	1.792	0.3076	1.8
BA2	180	600	20	4.922	9.13	8.132	0.477	1.276	1.193	0.5268	1.6
BA3	170	600	30	5.242	9.06	9.394	0.398	1.707	1.341	0.5408	1.5
BA4	160	600	40	5.523	9.05	12.427	0.308	1.621	1.365	0.5856	1.3

Table 4 Preference values

S. No.	Notation	Properties	Preference	Importance factor
1	AC	Air content (%)	Smaller is better	Slightly important
2	SWA	Saturated water absorption (%)	Smaller is better	Very important
3	*P*	Porosity (%)	Smaller is better	Slightly important
4	*S*	Sorptivity (mm/min$^{0.5}$)	Smaller is better	Slightly important
5	DS	Drying shrinkage (mm)	Smaller is better	Important
6	SWR	Seawater resistance (%)	Larger is better	Important
7	ART	Acid resistance test (%)	Larger is better	Important
8	WP	Water permeability (m/s × 10^{-12})	Smaller is better	Very important

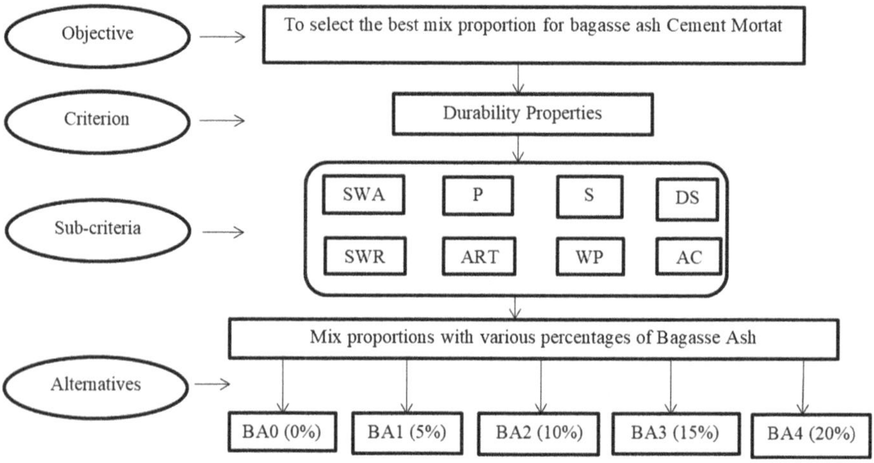

Fig. 2 Hierarchical structure

4.3 *Finding Consistency Ratio*

The consistency ratio (CR) is given by Eq. (7). The λ_{max}(Eq. 6), random inconsistency (Eq. 9) and consistency index (Eq. 8) are found out to be 8.278038, 1.49025 and 0.03972, respectively. A consistency ratio 0.026653 is obtained which is less than 0.1, and hence, the assigned weights are okay.

Table 5 Normalized weights of sub-criteria

Alternatives	SWA	*P*	*S*	WP	SWR	ART	DS	AC
BA0	0.22	0.19	0.16	0.09	0.25	0.24	0.25	0.17
BA1	0.21	0.20	0.21	0.11	0.19	0.24	0.28	0.18
BA2	0.20	0.20	0.25	0.21	0.15	0.16	0.16	0.20
BA3	0.19	0.20	0.22	0.26	0.21	0.18	0.16	0.21
BA4	0.18	0.20	0.16	0.33	0.20	0.18	0.15	0.24

Table 6 Final ranking

Alternatives	Bagasse ash percentage (%)	Final score	Rank
BA4	20	0.226	1
BA3	15	0.210	2
BA2	10	0.193	3
BA0	0	0.187	4
BA1	5	0.185	5

4.4 Normalized Weights and Ranking

After finding the final weights, the normalized weights for each sub-criterion is needed which is shown in Table 5. The final score of the alternatives is then found by Eq. (10), and the alternatives are ranked, as shown in Table 6.

5 Conclusion

The durability properties of the cement mortar specimens with various percentages of bagasse ash are observed at 90 days. Since eight factors are involved in deciding the suitable optimal mix, it is difficult to interpret the results based on the experimental results. Hence, analytical hierarchy process (AHP), a multi-criteria decision-making (MCDM) method, is employed in selecting the optimal mix.

- In this study, eight criteria are considered. They are saturated water absorption, porosity, sorptivity, drying shrinkage, seawater resistance, acid resistance, water permeability and air content are utilized (sub-criteria) for finding the optimal mix proportions.
- The incorporation of bagasse ash up to 10% is found to reduce the water demand due to the presence of high specific surface area, excessive percentage of amorphous silica and calcium oxide, satisfying the predominant requirements of pozzolanic material.

- The results indicate that the increase in percentage of bagasse ash improves the resistance against seawater and acid attack.
- Reduction in depth of penetration was observed, due to the strength effect of pores structures in bagasse ash blended cement mortar.
- Using AHP algorithm, the alternatives are ranked as BA4 > BA3 > BA2 > BA0 > BA1, based on the alternative score obtained.
- In the final step, BA4 has the highest value of 0.226, hence selected as the optimum proportion (20% bagasse ash with 40 g, cement 160 g and sand 600 g with water–cement ratio 0.5).

Hence, AHP is a very effective method in determining a problem like choosing the best optimal mix, considering the influence of various parameters for ranking.

References

1. Saaty, T.L.: What is the Analytic Hierarchy Process? Mathematical Models for Decision Support, pp. 109–121. Springer, Berlin, Heidelberg (1988)
2. Lamaakchaoui, C., Azmani, A., El Jarroudi, M., Laghmari, G.: The selecting of complementary products using the AHP method. ESMB **6**, 1–6 (2016)
3. Chakladar, N.D., Chakraborty, S.: A combined TOPSIS-AHP method based approach for non-traditional machining processes selection. J. Eng. Manuf. **222**(12), 1613–1623 (2008)
4. Erdebilli, B., Erkan, T.E.: Selecting the best supplier using analytic hierarchy process (AHP) method. Afr. J. Bus. Manag **6**(4), 1455–1462 (2012)
5. Ince, M., Yigit, T., Isik, A.H.: AHP-TOPSIS method for learning object metadata evaluation. Int. J. Inf. Educ. Technol. **7**(12), 884–887 (2017)
6. Socaciu, L., Giurgiu, O., Banyai, D., Simion, M.: PCM selection using AHP method to maintain thermal comfort of the vehicle occupants. Energy Procedia **85**, 489–497 (2016)
7. Lin, C.C., Wang, W.C., Yu, W.D.: Improving AHP for construction with an adaptive AHP approach (A3). Auto. Constr. **17**(2), 180–187 (2008)
8. Mansor, M.R., Sapuan, S.M., Zainudin, E.S., Nuraini, A.A., Hambali, A.: Hybrid natural and glass fibers reinforced polymer composites material selection using analytical hierarchy process for automotive brake lever. Mater. Des. **51**, 484–492 (2013)
9. Hudymacova, M., Benkova, M., Poscova, J., Skovranek, T.: Supplier selection based on multi-criterial AHP method. Acta Montan. Slovaca **15**(3), 249–255 (2010)
10. Praveenkumar, S., Sankarasubramanian, G., Sindhu, S.: Selecting optimized mix proportion of bagasse ash blended high performance concrete using analytical hierarchy process (AHP). Comput. Concr. **23**(6), 459–470 (2019)
11. Karim, R., Karmaker, C.L.: Machine selection by AHP and TOPSIS methods. Am. J. Ind. Eng. **4**(1), 7–13 (2016)
12. Venkata Rao, R.: Evaluating flexible manufacturing systems using a combined multiple attribute decision making method. Int. J. Prod. Res. **46**(7), 1975–1989 (2008)

Regional Optimization of Global Climate Models for Maximum and Minimum Temperature

K. Sreelatha and P. AnandRaj

Abstract Thirty-six global climate models (GCMs) monthly maximum and minimum temperature ($T_{\max}$ and $T_{\min}$) datasets from Coupled Model Intercomparison Project 5 (CMIP5) are optimized using observed dataset collected from Indian Meteorological Department (IMD) from 1970 to 2005 (Telangana). Statistical metrics are considered and measured to find the relative dependency (maximum and minimum temperature) of GCMs with IMD. Weights are assigned to each statistical metric using entropy method. GCMs are optimized based on ranking assigned by compromise programming technique for combined statistical metrics. Group decision approach is employed at each grid point for combined ranking. Results conclude that among all statistical metrics, skill score is significant for both ($T_{\max}$ and $T_{\min}$). BCC-CSM1.1(m), MIROC5, CanESM2, CNRM-CM5 and BCC-CSM1.1 for maximum temperature and CanESM2, ACCESS1.0, BCC-CSM1.1, MRI-CGCM3 and CNRM-CM5 for minimum temperature are optimized as suitable models among thirty-six GCMs considered for the study. It has also observed that though the study area is same, the optimized models are varying for the variables considered. This study concludes that evaluation of optimized GCMs varies with variable to variable and region to region. So, individual regional studies of evaluating GCMs give better results for further climate change analysis.

Keywords Global climate models · CMIP5 · Entropy method · Compromise programming · Group decision making

1 Introduction

Global climate models (GCMs) are computer driven models for weather forecasting and projecting climate change. The performance of global climate model with the observed conditions in atmosphere, land and ocean is not clear. Optimizing performance of global climate model is complex as it varies at spatial, temporal and variables of interest. GCMs are the only models available in projecting the present,

K. Sreelatha (✉) · P. AnandRaj
Department of Civil Engineering, National Institute of Technology, Warangal, India
e-mail: koppalasreelatha@student.nitw.ac.in

D. Dutta and B. Mahanty (eds.), *Numerical Optimization in Engineering and Sciences*,
Advances in Intelligent Systems and Computing 979,
https://doi.org/10.1007/978-981-15-3215-3_7

past and future climate. Climate data of GCMs are progressively being used for analyzing impact of climate in meteorological, hydrological and agricultural studies. So, evaluating the performance of GCMs is important at regional scale. Many evaluation studies were dealt in the past decade at different parts of the world at regional and basin scale.

In recent times, to find the performance of GCMs, several measures were improved [1–3]. For evaluating GCMs, simple and significant metrics are recommended [4, 5]. Globally, numerous climate variables like sea surface temperature, precipitation, evapotranspiration are assessed for performance of GCMs. In India, researchers determined the performance of GCMs on basin and regional scale using CMIP3 climate models [6–8]. Many researchers focused on large-scale analysis compared to regional scale around the globe.

Evaluation of GCMs at regional scale helps in assessing reliable projection of future to the regional condition as the impact of climate on variable changes with region to region [9]. Henceforth, this study is focused on performance of GCMs at regional scale, i.e., Telangana region, for maximum and minimum temperature on atmospheric circumstances of CMIP5. These optimized GCMs out of 36 GCMs of this study helps model output users for climate impact assessment in hydrological, meteorological, agricultural and different planning and management studies. Compromise programming technique is used for ranking GCM based on minimum value of distance measure between observed and simulated data of individual GCM at every grid point. Combined ranking is employed using group decision approach to find optimal GCMs considering ranks of all GCMs at all grid points for maximum and minimum temperature.

2 Study Area

Telangana state comes under the drainage basin of both Krishna and Godavari with a catchment of 69 and 79%, respectively. It covers a total geographical area of 112,077 km^2, temperature varying between 22 and 42 °C. The state falls under 77° 16′–81° 43′ E longitude and 15° 46′–19° 47′ N latitude.

3 Observed and Climate Data

Monthly observed and simulated data of T_{max} and T_{min} are obtained from IMD and CMIP5 for the period 1970–2005. Single realizations (r1i1p1) GCMs are considered from CMIP5 datasets. GCM datasets are regridded to common grid points of observed IMD data, i.e., 55 grids (Telangana).

4 Methodology

4.1 Statistical Metrics

Normalized Root Mean Square Error (NRMSE)
NRMSE explains relative error among observed and simulated data. Smaller value near to zero is desirable

$$\text{NRMSD} = \frac{\sqrt{\left[\left(\frac{1}{n}\right)\sum_{i=1}^{n}(o_i - s_i)^2\right]}}{\bar{o}} \tag{1}$$

o_i is the historic value and s_i is the model value, respectively; i represents dataset from $1, 2, \ldots n$.

Correlation Coefficient (CC)
CC represents the relationship between observed and simulated model values. The performance of model is best when value is near to 1.

$$\text{CC} = \frac{\sum_{i=1}^{n}(o_i - \bar{o})(s_i - \bar{s})}{(n-1)\text{sd}_o\text{sd}_s} \tag{2}$$

sd_o, sd_s are standard deviations of observed and simulated model values, respectively; $\bar{o}$, $\bar{s}$ are mean of observed and simulated model values, respectively.

Skill Score (SS)
SS describes similarity between observed and simulated model with probability density function. The performance of the model is good when SS = 1.

SS is expressed as

$$\text{SS} = \sum_{x=1}^{ci} \min(f_o, f_s) \tag{3}$$

where f_o, f_s represent frequencies of observed and simulated data based on intervals and ci is number of intervals selected to find frequencies of probability distribution function within the region. x represents number of metrics.

Nash-Sutcliffe Efficiency (NSE)
NSE correlates the relative magnitude of the residual difference and compares it with measured data [10].

$$\text{NSE} = 1 - \frac{\sum_{i=1}^{n}(o_i - s_i)^2}{\sum_{i=1}^{n}(o_i - \bar{o}_i)^2} \tag{4}$$

where o_i, s_i are the ith observation for observed and simulated data. $\bar{o}$ = mean of observed data.

Compromise Programming
CP is calculated with minimal distance measure from ideal metric value. It is expressed as

$$L_p(a) = \left[\sum_{x=1}^{X} w_x^p \left| f_x^* - f_x(a) \right|^p \right]^{\frac{1}{p}} \quad (5)$$

$L_p(a)$ is the minimal measure for GCM a; w_x is weightage of metric; f_x^* is normalized value of x; $f_x(a)$ = normalized ideal value x; p is distance measure parameter (1, 2, … ∞).

5 Results and Discussion

Statistical metric values (CC, SS, NSE and NRMSE) were calculated for all fifty-five grid (latitude and longitude) points of Telangana region. Weightages were calculated using entropy technique for each statistical metric for all GCMs. Ranks were assigned for all GCMs using minimal value obtained from compromise programming (distance measure $p = 2$) technique. Smaller L_p value represents the best optimized GCM as shown in Table 1. The above process is performed separately for all 55 grid points. For instance, optimized suitable models for $T_{\max}$ and $T_{\min}$ at a grid point, i.e., 16.75° × 80.75° are explained in Table 1, and the same procedure is followed for remaining grid points to obtain suitable optimized models for Telangana region.

Table 1 Suitable GCMs employed at grid point 16.75° × 80.75° for maximum and minimum temperature based on minimum value (L_p)

GCMs occupying	Maximum temperature	Minimal value (L_p)	Minimum temperature	Minimal value (L_p)
Rank 1	BCC-CSM1.1(m)	0.0049	CanESM2	0.0158
Rank 2	MIROC5	0.0057	ACCESS1.0	0.0159
Rank 3	CanESM2	0.0060	BCC-CSM1.1	0.0162
Rank 4	CNRM-CM5	0.0062	MRI-CGCM3	0.0167
Rank 5	BCC-CSM1.1	0.0078	CNRM-CM5	0.0173
Rank 36	IPSL-CSM5B-LR	0.0510	ACCESS1.3	0.0374

5.1 Analysis of Maximum Temperature

For maximum temperature, it is noticed that compared to all statistical metrics, SS (36.07%) is significant than CC (24.38%), NSE (22.32%) and NRMSE (17.23%) percentage-wise obtained from Table 3. To optimize the GCMs for the study, ranks are assigned in such a way that minimum L_p value is ranked as 1. So, the minimum L_p value performed at grid point 16.75° × 80.75° for GCMs is 0.0049 (BCC-CSM1.1(m)) followed by 0.0057 (MIROC5), 0.0060 (CanESM2), 0.0062 (CNRM-CM5) and 0.0078 (BCC-CSM1.1); maximum value is 0.0510 observed for IPSL-CSM5B-LR as mentioned in Table 1.

5.2 Analysis of Minimum Temperature

For minimum temperature, the same procedure is continued and observed that SS (33.78%) is prominent than CC (26.81%), NSE (23.14%) and NRMSE (16.27%), respectively. The minimal L_p values for GCMs observed at grid point 16.75° × 80.75° are 0.0158 (CanESM2), 0.0159 (ACCESS1.0), 0.0162 (BCC-CSM1.1), 0.0167 (MRI-CGCM3) and 0.0173 (CNRM-CM5); maximum value is observed as 0.0374 (ACCESS1.3) presented in Table 1.

The above statistics were calculated to all grid points (fifty-five), and combined ranking is employed, and optimized GCMs are finalized using multi-criteria decision-making approach. For maximum temperature, group decision approach is employed and observed that BCC-CSM1.1(m) is ranked 7 times in the first position subsequently MIROC5(6), CanESM2(6), CNRM-CM5(6) and BCC-CSM1.1(5) are shown in Table 2. In the same way, for minimum temperature, based on the group decision approach, optimized GCMs for the study are resulted as CanESM2(8), BCC-CSM1.1(7), CNRM-CM5(7), ACCESS1.0(6) and MRI-CGCM3(6) as shown in Table 2. It is observed that depending upon a small variation in weightages, model ranks are varying. Weightage distributions of the variables (maximum and minimum temperature) at all grid points are mentioned in

Table 2 Suitable GCMs based on combined ranking obtained for Telangana state from entropy method in parenthesis, value represents number of times that particular GCM occupying rank 1 position at different grids, GCMs occupying rank 1 position not less than 5 times are not mentioned here

Rank 1 position	Maximum temperature	Minimum temperature
GCMs occupying	BCC-CSM1.1(m)(7)	CanESM2(8)
	MIROC5(6)	ACCESS1.0(6)
	CanESM2(6)	BCC-CSM1.1 (7)
	CNRM-CM5(6)	MRI-CGCM3(6)
	BCC-CSM1.1(5)	CNRM-CM5(7)

Table 3. Optimized models for maximum temperature were BCC-CSM1.1(m), MIROC5, CanESM2, CNRM-CM5 and BCC-CSM1.1 whereas for minimum temperature CanESM2, BCC-CSM1.1, CNRM-CM5, ACCESS1.0, MRI-CGCM3, HadGEM2-CC and HadGEM2-ES. Spatial distribution of optimized GCMs for the study is shown in Fig. 1. Grids points occupying first three position for two variables (36 GCMs) are mentioned in Table 4.

Table 3 Distribution of weights for T_{max} and T_{min} over 55 grid points of Telangana region using entropy method

Variables	SspMetrics	Weights in %				
		0–20	20–40	40–60	60–80	80–100
T_{max}	CC	9	34	12	–	–
	SS	7	28	20	–	–
	NSE	33	12	10	–	–
	NRMSE	7	22	26	–	–
T_{min}	CC	–	29	26	–	–
	SS	–	38	17	–	–
	NSE	13	28	14	–	–
	NRMSE	11	23	14	7	–

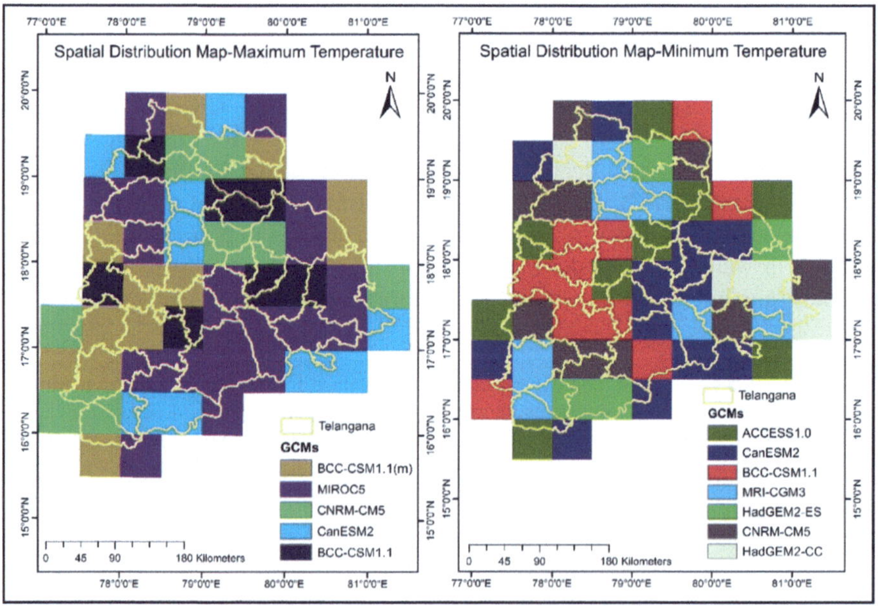

Fig. 1 Spatial distribution map of optimal GCMs for maximum and minimum temperature

Table 4 Number of grid points of maximum and minimum temperature that GCMs occupying the first three positions

GCMs	Maximum temperature			Minimum temperature		
	1	2	3	1	2	3
BCC-CSM1-1	5	6	2	7	5	1
BCC-CSM1.1(m)	7	7	4	1		6
BNU.ESM	2	1	3	2	3	4
CanESM2	6	7	3	8	6	6
CMCC-Cm						
CMCC-CMS	1		3			
CNRM-CM5	6	5	6	7		3
ACCESS1.0	2	3	2	6	5	5
ACCESS1.3						
CSIRO-Mk3.6.0			1			
FIO-ESM	3	1	2		2	2
EC-EARTH		1	1			3
INMCM4.0			3	1		
IPSL-CM5A-LR					3	
IPSL-CM5A-MR	1	2		1	4	3
IPSL-CM5B-LR		2	1			
FGOALS-g2	3		1		1	
MIROC-ESM		2			3	
MIROC-ESM-CHEM			7			
MIROC5	7	6	5		4	
HadGEM2-CC	1		2	4	4	3
HadGEM2-ES	2		3	4	4	4
MPI-ESM-LR						
MPI-ESM-MR						
MRI-CGCM3	1	1		6	6	2
GISS-E2-H						5
GISS-E2-R						
CCSM4	1	2		1		
CESM1(BGC)						
CESM1(CAM5)	2		3	1	1	4
CESM1(WACCM)		2		1		2
NorESM1-M	2	1	3	2		1
HadGEM2-AO	3	4		2	2	3
GFDL-CM3		2			2	
GFDL-ESM2G				1		2
GFDL-ESM2M						

6 Conclusion

Four statistical metrics (CC, SS, NSE and NRMSE) were considered to assess 36 global climate models of simulated data with IMD dataset of maximum and minimum temperature (IMD) at fifty-five grid points (Telangana). Weights calculated using entropy techniques at all grid points effected ranking pattern. The study concludes that evaluation of distance measure value at each grid point helps in finding the optimal GCM in the study region. Statistical metrics, SS (36.07%) was observed as more significant than CC (24.38%), NSE (22.32%) and NRMSE (17.23%) for maximum temperature whereas for minimum temperature, SS (33.78%) is prominent than CC(26.81%), NSE(23.14%) and NRMSE (16.27%), respectively. BCC-CSM1.1(m) is ranked 7 times in position 1, subsequently MIROC5(6), CanESM2(6), for maximum temperature and for minimum temperature, CanESM2(8), BCC-CSM1.1 (7) and CNRM-CM5(7) based on group decision-making approach. The optimal models suggested as BCC-CSM1.1(m), MIROC5, CanESM2, CNRM-CM5 and BCC-CSM1.1 for maximum temperature and CanESM2, BCC-CSM1.1, CNRM-CM5, ACCESS1.0 and MRI-CGCM3 are suitable for minimum temperature. Results conclude that evaluation of optimal GCMs using CP and group decision making helps in various hydrological studies.

References

1. Knutti, R., Abramowitz, G., Collins, M., Eyring, V., Gleckler, P.J., Hewitson, B., Mearns, L.: Good practice guidance paper on assessing and combining multi model climate projections. In: Stocker, T.F., Qin, D., Plattner, G.K., Tignor, M., Midgley, P.M. (eds.) Meeting Report of the Intergovernmental Panel on Climate Change Expert Meeting on Assessing and Combining Multi Model Climate Projections. IPCC Working Group I Technical Support Unit, University of Bern, Bern: Switzerland (2010)
2. Fordham, D.A., Wigley, T.L., Brook, B.W.: Multi-model climate projections for biodiversity risk assessments. Ecol. Appl. **21**, 3316–3330 (2011)
3. Pitman, A.J., Arneth, A., Ganzeveld, L.: Regionalizing global climatic models. Int. J. Clim. **32**, 321–337 (2012)
4. Preethi, B., Kripalani, R.H.: Indian summer monsoon rainfall variability in global coupled ocean-atmospheric models. Clim. Dyn. **35**, 1521–1539 (2010)
5. Lupo, A., Kininmonth, W., Armstrong, J.S., Green, K.: Global climate models and their limitations. http://www.nipccreport.org/reports/ccr2a/pdf/Chapter-1-Models.pdf (2015)
6. Raju, K.S., Nagesh, K.D.: Ranking of global climatic models for India using multicriterion analysis. Clim. Res. **60**, 103–117 (2014)
7. Raju, K.S., Nagesh, K.D.: Ranking of global climate models for India using TOPSIS. J. Water Clim. Change **6**, 288–299 (2014)
8. Tiwari, P.R., Kar, S.C., Mohanty, U.C., Kumari, S., Sinha, P., Nair, A., De, S.: Skill of precipitation prediction with GCMs over north India during winter season. Int. J. Clim. **34**, 3440–3455 (2014)

9. Sreelatha, K., Anand Raj, P.: Ranking of CMIP5-based global climate models using standard performance metrics for Telangana region in the southern part of India. ISH J. Hydraul. Eng. 1–10 (2019)
10. Nash, J.E., Sutcliffe, J.V.: River flow forecasting through conceptual models part I—a discussion of principles. J. Hydrol. **10**(3), 282–290 (1970)

Numerical Optimization of Settlement in Geogrid Reinforced Landfill Clay Cover Barriers

Akshit Mittal and Amit Kumar Shrivastava

Abstract Landfill clay cover barriers are prone to differential settlement, as a result of which the hydraulic conductivity of the barrier increases, thus increasing the amount of water which comes in contact with the waste. This could lead to the development of profuse amount of leachate, which is highly detrimental to the environment. Geogrid reinforcement of landfill cover can be a low cost and highly efficient way of bolstering the shear strength of the cover soil. Such reinforced barriers require computation of settlement response before designing the structure, so as to ensure that the post-construction operations of the landfill can be assessed effectively. For this purpose, triaxial compression tests on geogrid reinforced landfill clay cover barrier were conducted and different parameters such as geogrid properties like tensile strength, tensile modulus and geogrid placement parameters such as number of geogrid layers were tested. Results of the triaxial tests were then used for computation of the elasticity modulus of the soil cover barrier, and a statistical multiple regression model was fitted for different parameters. The regression analysis was performed in R studio. The model with the confidence level of 95% or greater was suggested for computation of settlement in geogrid reinforced landfill covers.

Keywords Geogrid reinforcement · Landfill covers · Elasticity modulus · Regression · Shear behavior

1 Introduction

Efficient design of landfill system is of significant importance to ensure minimum deleterious impacts on the environment. Even though the availability of land is a major concern, with the available land reducing at a very fast rate, due to various

A. Mittal (✉)
Department of Environmental Engineering, Delhi Technological University, New Delhi, Delhi 110042, India
e-mail: akshit97mittal@gmail.com

A. K. Shrivastava
Department of Civil Engineering, Delhi Technological University, New Delhi, Delhi 110042, India
e-mail: aksrivastava@dce.ac.in

D. Dutta and B. Mahanty (eds.), *Numerical Optimization in Engineering and Sciences*, Advances in Intelligent Systems and Computing 979,
https://doi.org/10.1007/978-981-15-3215-3_8

factors such as urbanization, landfilling of waste still remains one of the primary methods for waste disposal [1]. A modern landfill system usually consists of an erosion protection layer, cover barrier, leachate collection and removal layer and a low permeability bottom liner. A cover barrier is usually provided to arrest the flow of water, which might arise from precipitation, so as to prevent its interaction with the stored waste, as higher interaction could lead to the production of profuse amount of leachate. This leachate could infiltrate into the nearby groundwater, thus polluting it and rendering it useless. These cover barriers are usually prone to differential settlement, as a result of which tensile cracks are formed along its structure, thus increasing the hydraulic conductivity of the landfill cover [2]. These cracks are usually associated with low shear strength of landfill clay cover barrier soil, which is usually bentonite blended with a locally available soil or kaolin blended with sand. Many researchers have suggested different methods for reinforcing landfill covers with different materials to enhance its shear strength and maintain its usefulness and integrity at each stage of operation [3–6]. Different reinforcing materials have been suggested for reducing crack development in landfill covers. Reference [7] analyzed the performance of fiber additives in reducing desiccation-induced cracking in Akaboku soil, which is a possible material for landfill covers. The study demonstrated the usefulness of using fiber additives for suppressing cracks developed as a result of desiccation. The volumetric strain reduced with the increase in the fiber content, thus signifying fiber additive as potential material to be used in landfill cover soils. Reference [8] analyzed the potential of utilizing old dump waste for landfill covers. In the past few years, geogrids have emerged as a low cost and a highly effective way of reinforcing landfill covers. Various researchers have conducted extensive studies on analyzing the potential of geogrids in landfill cover. Reference [9] conducted centrifuge study on geogrid reinforced landfill covers. The study demonstrated that the inclusion of geogrid suppressed the cracks, formed as a result of differential settlement, in the landfill cover. Reference [10] conducted stability analysis for geogrid reinforced landfill cover slopes and analyzed different parameters such as slope angles, cover soil cohesion, slope length and interface frictional angles. As aforementioned, low shear strength of the landfill cover is one of the major reasons for occurrence of tensile cracks. Hence, it becomes imperative to analyze this behavior of the landfill cover soil with the inclusion of geogrids. One of the most reliable techniques to analyze the shear behavior of a soil is to conduct triaxial compression tests, as has been employed by different researchers to investigate the performance of different geosynthetics with soil [11–15]. Reference [11] conducted triaxial tests using different forms of reinforcement materials such as planar reinforcements, geocells and discrete fibers on sand. The study proffered that the addition of geosynthetics within sand specimen led to deviation in the failure pattern in sand, with the deformation occurring between the reinforcements, unlike in an unreinforced specimen, wherein the failure occurred along the conventional failure plane. The study also demonstrated that geocells performed the best of the three reinforcement type, whereas planar geosynthetics performed much better than the discrete fibers. Reference [12] conducted large-scale triaxial tests on coarse-grained soil reinforced with geogrid. The study demonstrated that as the number of reinforcements was increased, the deformation behavior in the soil tended toward strain hardening. Also, the study proved that cohesive strength of the soil burgeoned with the inclusion of geogrids. Reference [13] conducted triaxial compression tests on geotextile reinforced sand. The study demonstrated that the peak strength ratio was the highest at lower confining pressure, due to higher sand–geotextile interaction at lower confining pressure. Reference [14] conducted triaxial tests on geotextile reinforced Yamuna sand. Study showed that the woven geotextiles performed much better than the non-woven geotextiles, even though the interface bonding efficiency of the non-woven geotextiles was higher than the woven geotextile. Reference [15] conducted a comparative study of geotextile reinforced siliceous and carbonaceous sand, respectively. The study demonstrated that higher the relative density of the sand, the higher would be the mobilization in the geotextile. In the present analysis, consolidated undrained triaxial tests were performed on geogrid reinforced landfill clay cover soil. Kaolin clay and sand blended in 4:1 proportion, compacted at OMC + 5% was used to simulate landfill clay cover soil. Elasticity modulus was computed at different testing configurations and with different types of geogrid and a regression model included these parameters was established in R package R studio. A novel model with 95% confidence level was suggested for computation of settlement responses in geogrid reinforced landfill covers.

2 Materials Used

2.1 Soil

Soil properties play the most significant role in choosing a suitable material for simulating the characteristics of landfill clay cover barrier. The most significant of these properties is the permeability of the cover soil, which should typically be lower than or equal to 10^{-7} cm/s for MSW type landfills. Other soil properties such as grain size characteristics and plasticity index, which are correlated with the hydraulic conductivity of the soil, need to be considered to ensure a suitable landfill cover barrier. Researchers such as [16] proffered a plasticity index of greater than or equal to 7–15% and percentage fines of greater than or equal to 30–50% for designing an efficient landfill cover. Sand-clay mixtures can also be efficient materials for designing of landfill clay cover barriers as they prevent desiccation-induced cracking in the covers [17]. Desiccation cracking, if occurs, could lead to the expansion and softening of the soil, on the infiltration of water into the soil [18]. Hence, for the simulating the typical characteristics of landfill cover properties reported in the literature, kaolin clay and sand blend in 4:1 proportion was chosen. The use of this material for simulating the characteristics of cover soils has also been reported by [9, 19]. As stated earlier, the low permeability requirement of landfill cover soil is of prime importance for designing an efficient landfill cover. A dispersed

Table 1 Geotechnical properties of landfill clay cover barrier soil

Properties	Soil blend
Specific gravity	2.63
Percentage passing IS sieve [75 μm (%)]	61
Liquid limit (%)	41.2
Plastic limit (%)	20.44
Plasticity index (%)	21.01
Differential free swell (%)	20
Maximum dry unit weight (kN/m^3)	15.41
Optimum moisture content (%)	24.52
Maximum dry unit wight (OMC + 5%) (kN/m^3)	14.02
Coefficient of permeability (cm/s)	8.19×10^{-7}
Coefficient of permeability (cm/s) (OMC + 5%)	1.12×10^{-7}

structure usually can achieve lower permeability as compared to a flocculated soil structure. Hence, the soil was compacted at a moisture content of OMC + 5% [20]. The various geotechnical properties of the soil are reported in Table 1.

2.2 Geogrids

Two geogrids with different mechanical and physical characteristics were chosen for the purpose of the study. The properties of the same were procured from the manufacturer (H.M.B.S. Textiles Pvt. limited). The geogrids were cut in circular shape of suitable diameter, and after its placement within the specimen, the protruding ends were removed to prevent any damage to the latex membrane being used in the triaxial testing procedure. For the purpose of convenience, first and second type of geogrid has been termed using nomenclature as GGR1 and GGR2, respectively. The properties of the same are presented in Table 2.

Table 2 Properties of geogrids

Parameters	GG1		GG2	
	MD	TD	MD	TD
Tensile strength	7.8	8.2	5	5
Tensile modulus	550	350	220	260
Aperture size	30 × 30		40 × 27	

[a]Tensile modulus and tensile strength are in KN/m
[b]MD and TD are machine direction and transverse direction, respectively

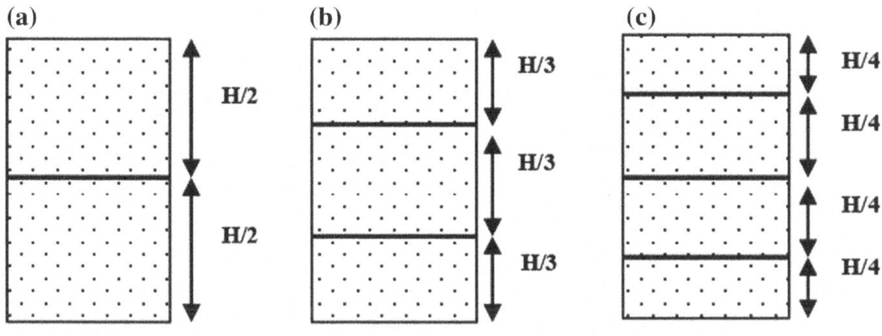

Fig. 1 Different testing configurations at **a** C1, **b** C2 and **c** C3

3 Experimental Testing Procedure

Landfill covers usually experience low overburden pressure, which is usually provided by the erosion protection layer, protection layer and the drainage layer. The overburden pressure is nearly 25 kPa [9]. Hence, for the present analysis, triaxial tests were performed at a confining pressure of 25 kPa. The degree of saturation was assessed using the Skempton's coefficient (B). As B achieved a value greater than or equal to 0.95, desired saturation was believed to be obtained. Before placing the specimen into the triaxial mold, it was made sure that all the perforations and valves were washed properly to prevent plausible blockages. After placing the specimen in the cell, the cell was filled with water at a controlled flow rate. A pneumatic control panel type system was used in which three soil specimens in three different triaxial cells were first placed at suitable confining and back pressures simultaneously. The tests were conducted for each type of geogrid at three different configurations as shown in Fig. 1, consisting of a single layer, double layer and three layer of reinforcement, respectively. For sake of convenience, these have been termed as C1, C2 and C3, respectively.

4 Results and Discussion

4.1 Effect of Number of Layers of Geogrids

The Number of reinforcement layers plays a vital role in designing of a landfill. The tests were thus conducted for $N = 1$, 2 and 3 layers of geogrid reinforcement for both types of geogrids. Figure 2a, b shows the deviatoric stress versus % axial strain at failure for the two geogrids at different configurations. As can be observed, the addition of multiple geogrid layers increases the peak stress of the landfill clay cover barrier. This observation could be to the fact that as the number of layers is increased,

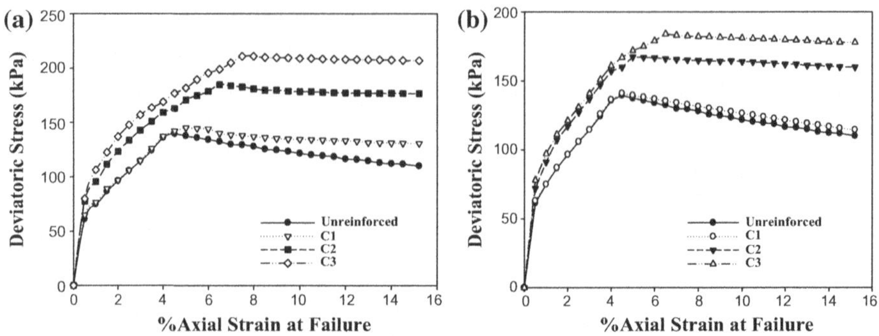

Fig. 2 Deviatoric stress versus % axial strain curve for **a** GGR1 and **b** GGR2

the mass of geogrid increases, which enhances the cushion available to the specimen. The better cushion enables better confinement of soil within the geogrid web, thus increasing the lateral resistance provided to the soil. This increase in lateral resistance prevents the deformation of the soil particles under applied load. Also, the increase in the interface friction as a result of inclusion of multiple layers of reinforcements provides better mobilization of tensile force, which prevents the deformation of the soil specimen under axial load. These observations corroborate the results obtained by [13, 15]. The increase in the stress carried by the specimen is subject to the type of geogrid being used. GGR1 performed better than GGR2, thus signifying that geogrid properties impact the stress-strain characteristics of the specimen under loading.

4.2 *Effect of Type of Geogrid*

For the purpose of this study, two different geogrids have been utilized. The performance of each of the geogrid has been compared by utilizing a parameter called peak strength ratio, which can be defined as the peak deviatoric stress in the reinforced specimen to the peak deviatoric stress in the unreinforced specimen. As can be observed, GGR1 performs much better in comparison to GGR2 to increase both the axial strain at failure and the peak strength ratio. GGR1 has higher tensile strength and tensile modulus in comparison to GGR2, which means that higher load is required to deform GGR1.

Also, the lower aperture size of GGR1 enhances the confinement effect, which is the major reinforcement mechanism in geogrids as was reported by [21]. The selection of a particular geogrid for reinforcement purpose, however, requires the assessment of other parameters such as cost, durability and availability [22] (Fig. 3).

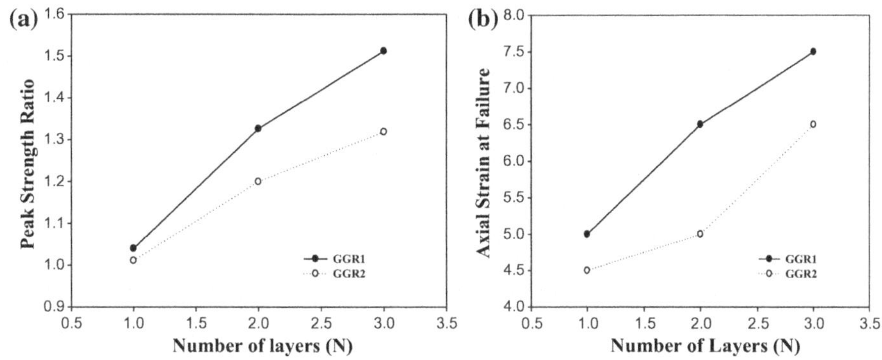

Fig. 3 **a** Peak strength ratio versus number of layers and **b** axial strain at failure versus number of layers for GGR1 and GGR2

5 Regression Analysis

Computation of settlement response of a reinforced soil becomes complex due to additional mechanisms which are provided by the reinforcing material. In landfill cover systems, these computations are essential to ensure that the post-operation mechanical settlement can be estimated accurately. For this purpose, the elasticity modulus of the reinforced soil has been computed using experimental results obtained using triaxial compression tests. A dimensionless parameter E_r/E_{ur} has been defined, where, E_r is the elasticity modulus of the reinforced specimen, whereas, E_{ur} is the elasticity modulus of the unreinforced specimen. As was discussed in the previous section, geogrid tensile strength and tensile modulus are significant parameters for analyzing stress-strain characteristics of landfill clay cover barrier, hence, these parameters have been normalized for the purpose of computation. For the purpose of analysis, the normalization has been carried out at 100 kN/m, which is consistent with normalization factors selected by [23]. It can be numerically represented as

$$J_{\text{Normalized}} = \frac{J\ \text{KN/m}}{100\,\text{kN/m}} \tag{1}$$

where $J_{\text{Normalized}}$ is the normalized tensile modulus and tensile strength, whereas, J is the tensile strength and tensile modulus of the geogrids as per material transfer certificate provided by the manufacturer. The elasticity modulus obtained using experimental results has been shown in Table 3. As can be observed from the table that the value of elasticity modulus decreases with the increase in the number of layers of geogrids. This observation can be attributed to the fact that the landfill clay cover barrier turns ductile as additional a number of reinforcement layers are added. This signifies the importance of number of layers of reinforcements as this characteristic of the landfill clay cover barrier would arrest the development of cracks in the cover soil, thus enhancing its usefulness.

Table 3 Elasticity modulus obtained using experimental study

Parameters	Elasticity modulus (kPa)
Unreinforced	2356
GG1 (N = 1)	2204
GG1 (N = 2)	2163.1
GG1 (N = 3)	2138.1
GG2 (N = 1)	2381.3
GG2 (N = 2)	2244.8
GG2 (N = 3)	2150.9

Regression analysis was carried out using R package R studio, using different parameter configurations as shown in Table 4. As can be observed, two optimized models have been obtained with a confidence level of greater than 95%. For the model consisting of all the three parameters, $|t|$ analysis was carried out and N was obtained as the most significant variable with $|t|$ value of 20.44 and tensile modulus is the least significant variable with $|t|$ value of 0.504 for $N = 1, 2$ and 3 and for stiffness and tensile strength normalized at 100 kN/m. Removal of normalized stiffness from the set of parameter did not impact the confidence level significantly and thus can be ignored while computation of elasticity modulus of geogrid landfill covers. The coefficients and intercept obtained for the optimized model were obtained for the two significant parameters. The obtained equation is as follows

$$\frac{E_{\mathrm{r}}}{E_{\mathrm{ur}}} = -0.0314 \times N - 1.384 \times \text{Normalized Tensile Strength} + 1.091 \quad (2)$$

Knowing the elasticity modulus of the reinforced landfill clay cover barrier, thickness of the landfill clay cover and the overburden pressure can thus help in the computation of settlement in the reinforced landfill cover barrier.

Table 4 Regression analysis using different parameter combinations

Parameters	R^2	R^2 (adjusted)
N, normalized stiffness and normalized tensile strength	0.954	0.951
N and normalized tensile strength	0.953	0.952
N and normalized stiffness	0.945	0.939
N	0.845	0.81
Normalized stiffness and normalized tensile strength	0.82	0.815
Normalized tensile strength	0.44	0.42
Normalized stiffness	0.198	0.147

6 Conclusion

The significant observations obtained from the study are as follows:

- The increase in the number of layers of reinforcement significantly increases stress carrying capacity of the landfill clay cover barrier.
- The properties of the geogrid play an important role in impacting the shear behavior of landfill cover barrier soil.
- Better results are obtained for geogrids with higher tensile strength and lower aperture size.
- Increase in the number of geogrid layers enhances the ductility of the cover soil, which can be significant in reducing tensile cracks in the landfill clay barriers.
- Normalized tensile strength and number of reinforcement layers provide a confidence level of greater than 95% for computation of elasticity modulus of reinforced landfill cover soils and thus this optimized model can be used for evaluating elasticity modulus and further computation of settlement response of reinforced landfill covers.

References

1. Sharholy, M., Ahmad, K., Mahmood, G., Trivedi, R.C.: Municipal solid waste management in Indian cities—a review. Waste Manag. **28**, 459–467 (2008)
2. Cheng, S.C., Larralde, J.L., Martin, J.: Hydraulic conductivity of compacted clayey soils under distortion or elongation conditions. In: Hydraulic Conductivity and Waste Contaminant Transport in Soil. ASTM Int (1994)
3. Viswanadham, B.V.S., Sathiyamoorthy, R., Divya, P.V., Gourc, J.: Influence of randomly distributed geofibers on the integrity of clay-based landfill covers: a centrifuge study. Geosynth. Int. **18**, 255–271 (2011). https://doi.org/10.1680/gein.2011.18.5.255
4. Palmeira, E.M., Viana, H.N.L.: Effectiveness of geogrids as inclusions in cover soils of slopes of waste disposal areas. Geotext. Geomembr. **21**, 317–337 (2003). https://doi.org/10.1016/S0266-1144(03)00030-X
5. Divya, P.V., Viswanadham, B.V.S., Gourc, J.P.: Influence of geomembrane on the deformation behaviour of clay-based landfill covers. Geotext. Geomembr. **34**, 158–171 (2012). https://doi.org/10.1016/j.geotexmem.2012.06.002
6. Bacas, B.M., Cañizal, J., Konietzky, H.: Shear strength behavior of geotextile/geomembrane interfaces. J. Rock Mech. Geotech. Eng. **7**, 638–645 (2015). https://doi.org/10.1016/j.jrmge.2015.08.001
7. Harianto, T., Hayashi, S., Du, Y.-J., Suetsugu, D.: Effects of fiber additives on the desiccation crack behavior of the compacted Akaboku soil as a material for landfill. Water Air Soil Pollut. **194**, 141–149 (2008). https://doi.org/10.1007/s11270-008-9703-2
8. Koda, E.: Anthropogenic waste products utilization for old landfills rehabilitation. Ann. Warsaw Univ. Life Sci. L. Reclam. **44**, 75–88 (2012)
9. Rajesh, S., Viswanadham, B.V.S.: Centrifuge modeling and instrumentation of geogrid-reinforced soil barriers of landfill covers. J. Geotech. Geoenviron. Eng. **138**, 26–37 (2012). https://doi.org/10.1061/(ASCE)GT.1943-5606.0000559
10. Feng, S.-J., Ai, S.-G., Huang, R.-Q.: Stability analysis of landfill cover systems considering reinforcement. Environ. Earth Sci. **75**, 303 (2016). https://doi.org/10.1007/s12665-015-5186-9

11. Latha, G.M., Murthy, V.S.: Effects of reinforcement form on the behavior of geosynthetic reinforced sand. Geotext. Geomembr. **25**, 23–32 (2007). https://doi.org/10.1016/j.geotexmem.2006.09.002
12. Chen, X., Zhang, J., Li, Z.: Shear behaviour of a geogrid-reinforced coarse-grained soil based on large-scale triaxial tests. Geotext. Geomembr. **42**, 312–328 (2014)
13. Nguyen, M.D., Yang, K.H., Lee, S.H., et al.: Behavior of nonwoven-geotextile-reinforced sand and mobilization of reinforcement strain under triaxial compression. Geosynth. Int. **20**, 207–225 (2013). https://doi.org/10.1680/gein.13.00012
14. Mudgal, A., Sarkar, R., Shrivastava, A.K.: Influence of geotextiles in enhancing the shear strength of Yamuna sand. Int. J. Appl. Eng. Res. **13**, 10733–10740 (2018)
15. Goodarzi, S., Shahnazari, H.: Strength enhancement of geotextile-reinforced carbonate sand. Geotext. Geomembr. **47**, 128–139 (2019). https://doi.org/10.1016/j.geotexmem.2018.12.004
16. Koerner, R.M., Daniel, D.E.: Materials. In: Final Covers for Solid Waste Landfills and Abandoned Dumps. pp. 78–79 (1997)
17. Daniel, D.E., Wu, Y.-K.: Compacted clay liners and covers for arid sites. J. Geotech. Eng. **119**, 223–237 (1993)
18. Mitchell, J.K., Soga, K.: Fundamentals of Soil Behavior (1993)
19. Viswanadham, B.V.S., Rajesh, S.: Centrifuge model tests on clay based engineered barriers subjected to differential settlements. Appl. Clay Sci. **42**, 460–472 (2009). https://doi.org/10.1016/j.clay.2008.06.002
20. Benson, C.H., Daniel, D.E., Boutwell, G.P.: Field performance of compacted clay liners. J. Geotech. Geoenviron. Eng. **125**, 390–403 (1999)
21. Chen, Q., Abu-Farsakh, M., Sharma, R., Zhang, X.: Laboratory investigation of behavior of foundations on geosynthetic-reinforced clayey soil. Transp. Res. Rec. J. Transp. Res. Board 28–38 (2007)
22. Shukla, S.K. (ed.): Geosynthetics and Their Applications. Thomas Telford (2002)
23. Abu-farsakh, M., Voyiadjis, G., Chen, Q.: Finite element parametric study on the performance of strip footings on reinforced crushed limestone over embankment soil. Electron. J. Geotech. Eng. **17**, 723–742 (2012)

Optimization of Bias Correction Methods for RCM Precipitation Data and Their Effects on Extremes

P. Z. Seenu and K. V. Jayakumar

Abstract Studies on impact due to climate change on water resources are typically evaluated for regional scale and the evaluation is carried out at site-specific or local scale. General Circulation Models (GCMs) and Regional Climate Models (RCMs) are used to understand future climate changes. RCMs have higher resolution to understand the reliable estimation of local-scale climate variables. RCMs show critical biases in precipitation and therefore it is mandatory that bias correction is to be carried out so that they are usable for research. Six RCMs are considered and analysed in this study to reduce the errors in RCMs during the period from 1970 to 2005 over Amaravati region in Andhra Pradesh, India. Four statistical bias correction techniques, namely linear scaling, cumulative distributive transformation, quantile mapping using parametric transformation and quantile mapping using smooth spline methods are used. These bias-corrected datasets are compared with observed datasets using different relative errors, viz. standard error, mean absolute error, root mean square error and mean square error. Relative errors show the performance of simulated data with observed data. The results show that quantile mapping using parametric transformation technique gave optimum values for the results with minimum error compared to the other three methods. However, there is no generalized optimized technique available, at present, to reduce the bias in the datasets of the RCMs, and there is a need to reduce the errors for reducing the uncertainties in the climate impact studies either at the local or regional scale.

Keywords Regional climate model · Bias correction · Relative error

P. Z. Seenu (✉) · K. V. Jayakumar
Department of Civil Engineering, National Institute of Technology, Warangal, India
e-mail: seenu.pz1503@gmail.com

K. V. Jayakumar
e-mail: kvj@nitw.ac.in

D. Dutta and B. Mahanty (eds.), *Numerical Optimization in Engineering and Sciences*,
Advances in Intelligent Systems and Computing 979,
https://doi.org/10.1007/978-981-15-3215-3_9

1 Introduction

The impact study for hydrological modelling and water resources management applications due to climate change needs higher resolution meteorological data. Even though General Circulation Models (GCMs) are industry-standard tools used in assessing the changes to Earth's climate on a large scale, they are not appropriate choices for the proper representation of climatic conditions of a region due to their relative coarse horizontal distortion. Limited area Regional Climate Models (RCMs) produce finer grid simulations of the climate of a region spanning a particular area being imposed by reanalysis products or lateral boundary conditions of GCM [1]. RCMs are consistent with respect to the parent GCM and which are spatially coherent. RCMs generate the parameters dynamically in which a degree of significant atmospheric parameters is automatically confirmed. Many drought and flood events are triggered by spatial and temporal changes in precipitation patterns, and hence, the simulation studies using RCMs are vital in creating realistic driving data for water resources research. One of the key challenges is to compute the impacts of changes that can occur in the future, that too in extreme rainfall events, while analysing the vulnerability of the impacts of changing climate in hydrological systems.

In spite of the major improvements made by RCMs in replicating the regional climate scenarios, systematic errors are still significantly present in them [2–4]. The precipitation simulations produced by RCMs are known to be biased as the understanding about the process is limited or due to inadequate spatial resolution. It is, therefore, statistically needed to adjust post-processing before it could be used for the assessment of climate variables [5, 6]. Various studies have used daily precipitation data obtained from climate models which are then bias-corrected and matched with the precipitation statistics that are already observed for modelling applications. Appraising the impact of climate change over hydrological systems lies in evaluating the effects of variations in the precipitation frequencies that may occur in the future and this is a major challenge, since the daily precipitation values are not very accurately simulated by climate models always.

Various bias correction methods are used in climate change impact studies. Quantile mapping is one of the widely used methods [7, 8]. Quantile mapping corrects the model variable by creating quantiles of the distribution of model onto those of the observations. Quantile mapping is mostly used for the output of climate model worldwide [8]. Another technique used for bias correction is the cumulative distribution transformation method [9], with the assumption that the modelled and observed CDF mappings affect future data.

This study compares the ability of four selected methods used to decrease the bias in an RCM precipitation output and to determine optimal method among: (i) Linear scaling, (ii) cumulative distributive transformation, (iii) quantile mapping using parametric transformation and (iv) quantile mapping using smooth spline methods.

2 Materials and Methodology

Nine grid points, from in and around Amaravati, the capital city of Andhra Pradesh, are selected for this study. The CORDEX simulated RCM outputs for South Asia such as CNRM-CM5, ACCESS, CCSM4, GFDL-CM3, NORESM1-M and MPI ESM LR and are selected for bias correction. The observed precipitation data obtained from the Indian Meteorological Department (IMD) with a resolution of 0.5° × 0.5° observed gridded data are considered to obtain the optimize method for bias correction. Four statistical bias correction techniques, viz. linear scaling, cumulative distributive transformation, quantile mapping using parametric transformation and quantile mapping using smooth spline methods are used in this study and these are applied to each observation stations separately.

2.1 *Bias Correction Methods*

Linear scaling. In this approach, the RCM daily precipitation data, P, are transferred into $P_{\text{corrected}}$ such that

$$P_{\text{corrected}} = \alpha P_{\text{mod}} \quad (1)$$

where α is a scaling factor, $\alpha = O/P_{\text{mod}}$, O is the observed precipitation values and P_{mod} is modelled values. For every modelled precipitation value, the scaling factor is applied and the corrected time series is generated. Linear scaling methods include delta change methods and factor change methods which belong to the same family [10]. This method is simple and the data requirement is modest. But correction applied in the monthly mean rainfall data can adversely affect the relative change in the distribution of rainfall in different months and may distort other moments of daily rainfall distribution models.

Cumulative distributive transformation function (CDF) The basic principle of this approach is to first arrive at a statistical connection between observations and outputs of models on the basis of historical data. The transfer function, thus derived, is then applied to future model projections and the track of future observations is inferred. The distribution of monthly RCM precipitation variables is mapped onto that of the observed gridded data by the quantile-based mapping method. Quantile-based mapping is a simple and efficient method that has been used successfully in various climate impact studies as well as in hydrology [11, 12]. The method can be mathematically represented for a climate variable y is corrected to $\check{y}_{\text{adjst}}$:

$$\check{y}_{\text{adjst}} = F^{-1}_{\text{obs}-c}\left(F_{\text{obs}-c}\left(y_{m-p}\right)\right) \quad (2)$$

Here, F denotes the CDF of either the modelled values (m) or observations (obs) for the current climate or future projection period (p) or a historic training period (c). The bias correction of future model values is done by first calculating the percentile values for the future projection points in model's CDF for the period. Next, for that CDF, the observed values are traced to arrive at the bias-corrected model values. An important advantage of the method is that the rank correlation between observations and models is maintained and all the moments are adjusted in such a way that the observations used for the training period completely agree with the entire distribution. However, the method is based on an important assumption that there is no significant change in climate distribution over time. In other words, only the mean changes and the skew and variance of the distribution remain constant. If there are variations in the higher moments also, then this will not hold true [13]. It would be better if instead of making the assumption that the historic model distribution can be applied to the future period, information from model projection CDF is incorporated. It is assumed that for a given percentile, the adjustment function is a constant, that is, the difference between the observed and model values also applies to the future.

Quantile mapping parametric transformation (Qmap.P) The quantile-quantile relation can be directly modelled by using parametric transformations. The appropriateness of the parametric transformations given below is analysed.

$$\tilde{P}_{\text{obs}} = a P_{\text{mod}} \tag{3}$$

$$\tilde{P}_{\text{obs}} = b + a P_{\text{mod}} \tag{4}$$

$$\tilde{P}_{\text{obs}} = a P_{\text{mod}}^{c} \tag{5}$$

$$\tilde{P}_{\text{obs}} = a(P_{\text{mod}} - x)^{c} \tag{6}$$

$$\tilde{P}_{\text{obs}} = (b + a P_{\text{mod}}) \left(1 - \mathrm{e}^{-(P_{\text{mod}} - x)/\tau}\right) \tag{7}$$

Here, $\tilde{P}_{\text{obs}}$ denotes the appropriate estimate of P_{obs} and a, b, c, x and τ are free parameters that are subject to calibration. The direct scaling, Eq. (3), is very much related to linear scaling [14, 15] and is often used to correct RCM precipitation [5]. Piani et al. [16] used the transformations, Eqs. (4)–(7), and some of them have been used in some studies that followed [17]. By reducing the residual sum of squares, all parametric mappings are matched to the part of CDF correlated to wet days in observed series ($P_{\text{obs}} > 0$). The modelled values set to zero correspond to dry part of the empirical CDF observed.

Quantile mapping smoothing splines (Qmap.S) Nonparametric regression can also be used in modelling the transformation, Eq. (8). Even though there are other nonparametric methods which are equally efficient, the use of cubic smoothing splines is more popular as it is the only method suited for part of CDF correlated to wet days in observed series. In order to identify the spline's smoothing parameter, generalized cross-matching is used.

$$P_{\mathrm{obs}} = h(P_{\mathrm{mod}}) \tag{8}$$

2.2 *Evaluation Methodology*

By comparing the differences between observed, corrected and RCM datasets, it is possible to analyse the overall performance of each method used for bias correction. But quantitative analysis of the robustness of the methods has not been carried out. Therefore, the assessment of the robustness of each method is done by quantifying the relative errors such as standard error, mean absolute error, root mean square error and mean square error.

3 Results

Bias correction is carried out for nine grid points in and around the study area for monthly average and maximum precipitation time series of RCM dataset. Qmap.P, Qmap.S, CDF and linear scaling transformation are used for the bias correction. Standard error, mean absolute error, root mean square error and mean square error are calculated for each grid point of all six bias-corrected RCMs. As the methodology applied to all RCMs were similar, as an example CNRM-CM5 RCM dataset results are explained further.

Standard error and mean square error (MSE) calculated for the bias-corrected results using Qmap.P, Qmap.S, CDF and linear scaling transformation for average and maximum monthly series are shown in Table 1. Highest standard error for monthly average series using Qmap.P, Qmap.S, CDF and linear scaling methods are 0.44, 0.62, 0.71 and 1.52, respectively. Correspondingly, the maximum MSE values for monthly average series using Qmap.P, Qmap.S, CDF and linear scaling methods are 9.09, 12.87, 13.61 and 31.54, respectively. The lowest values in standard errors calculated for monthly average series using Qmap.P, Qmap.S, CDF and linear scaling methods are 0.32, 0.43, 0.43 and 0.72, respectively, while the lowest MSE values for monthly average series using Qmap.P, Qmap.S transformation, CDF transformation and linear scaling method are 6.60, 8.89, 8.83 and 41.97, respectively. It is evident from the analysis that the linear scaling technique shows higher error values compared to the other three methods and hence it is having less ability for correcting bias. Similarly, for maximum monthly series, the standard error and MSE obtained for Qmap.P, Qmap.S and CDF methods have error values much higher when compared to average monthly series. The highest MSE value for average monthly series is obtained as 31.54 using linear scaling method, whereas the lowest MSE value is

Table 1 Standard error (SE) and mean square error (MSE) for the bias-corrected results using Qmap.P, Qmap.S transformation, CDF and linear transformation for average and maximum monthly series

Error	Grid	For average monthly series				For maximum monthly series		
		Qmap.P	Qmap.S	CDF	Linear scaling	Qmap.P	Qmap.S	CDF
SE	1	0.42	0.43	0.44	0.98	66.91	85.00	90.37
	2	0.44	0.50	0.51	1.08	47.17	58.92	61.04
	3	0.37	0.53	0.54	1.52	67.84	83.58	120.23
	4	0.44	0.46	0.48	0.82	60.08	79.51	83.36
	5	0.36	0.44	0.43	1.10	51.03	69.89	62.97
	6	0.42	0.62	0.71	1.49	61.71	75.80	112.59
	7	0.37	0.44	0.47	0.72	71.98	106.99	135.79
	8	0.32	0.43	0.43	0.89	25.86	42.67	42.85
	9	0.42	0.61	0.59	1.29	42.55	59.97	61.67
MSE	1	8.68	8.96	9.09	20.36	1387.62	1762.68	1873.95
	2	9.06	10.29	10.64	22.40	978.14	1221.76	1265.70
	3	7.61	10.96	11.26	31.54	1406.84	1733.13	2493.14
	4	9.09	9.55	9.88	17.01	1245.91	1648.78	1728.55
	5	7.47	9.08	8.91	22.88	1058.33	1449.20	1305.73
	6	8.81	12.87	13.61	30.99	1279.79	1571.79	2334.71
	7	7.75	9.05	9.66	14.97	1492.66	2218.61	2815.79
	8	6.60	8.89	8.83	18.42	536.292	884.78	888.64
	9	8.76	12.63	12.13	26.72	882.438	1243.55	1278.75

6.60 using Qmap.P transformation. Standard error and MSE are very high for the monthly maximum series, which shows that bias correction techniques are poor for the extreme events.

Mean absolute error (MAE) calculated for the bias-corrected results using the four methods for average monthly series for nine grid points are shown in Fig. 1. Root mean square error (RMSE) calculated for the bias-corrected results using these methods for average and maximum monthly series of nine grid points are shown in Fig. 2. Higher MAE and RMSE are present in the linear scaling transformation method, which shows the inability of linear scaling approach in correcting the bias compared to Qmap.P, Qmap.S transformation and CDF. Out of these four methods, Qmap.P gives least MAE and RMSE, and hence, it is the optimal method for bias correction.

As the error in each grid points vary, the bias correction methods are spatially independent. From the above results, it can be concluded that Qmap.P method is the best among the other bias correction methods. Linear scaling method shows

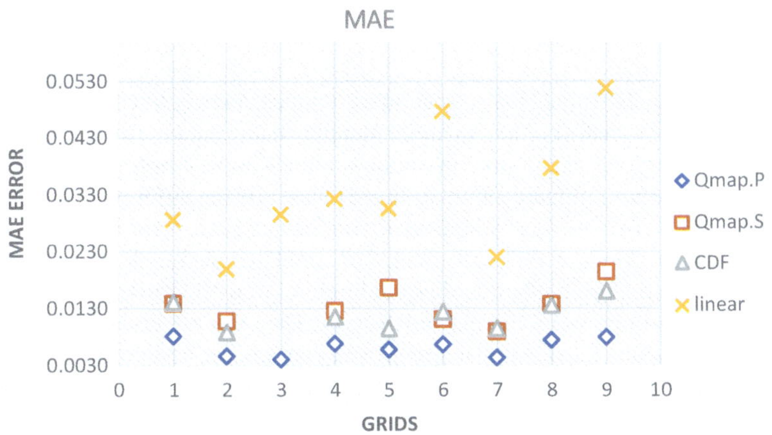

Fig. 1 Mean absolute error (MAE) for the bias-corrected result using Qmap.P, Qmap.S transformation, CDF transformation and linear transformation for average monthly series for nine grid points

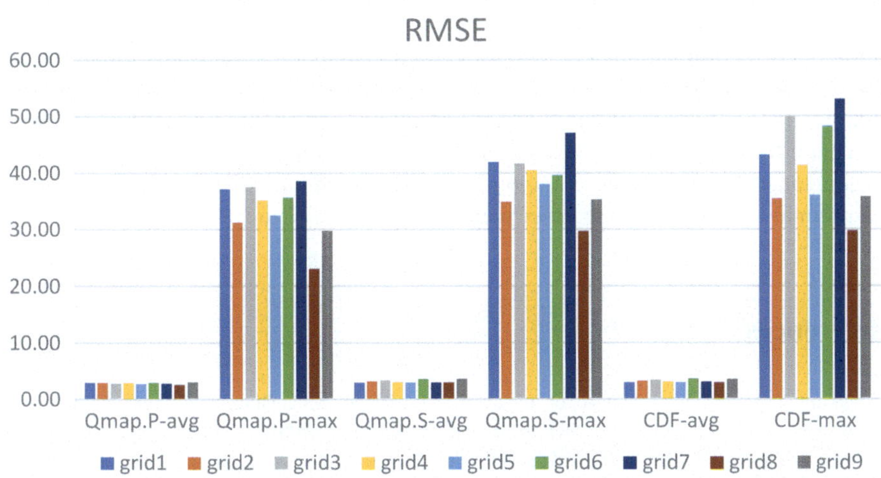

Fig. 2 Root mean square error (RMSE) for the bias-corrected result using Qmap.P, Qmap.S transformation, CDF transformation for average and maximum monthly series of nine grid points

maximum relative error implying its inability for accurate bias correction. Due to higher error values for the monthly maximum bias-corrected series, bias correction methods are inadequate in preserving the extremes.

4 Conclusion

In this study, four statistical bias correction techniques, namely linear scaling, cumulative distributive transformation, quantile mapping using parametric transformation and quantile mapping using smooth spline methods were employed. These bias-corrected datasets are compared with observed datasets using different relative errors, namely standard error, mean absolute error, root mean square error and mean square error. Relative errors show the performance of simulated data with observed data.

As the linear scaling method is designed only to correct the mean, its bias correction results are very poor and not satisfactory. The most effective bias correction method is observed as the quantile mapping using parametric transformation method, which integrates information frequency distribution for the observed and modelled precipitations. It is found that the efficiency of bias correction depends on the spatial variations since the error is varying at each grid point. As the bias correction techniques become more involved, more observed meteorological data are needed. It is, hence, expected that the bias correction procedure will become more accurate as the amount of information available from the observed record increases. Meanwhile, the chances of bias correction procedure being over calibrated to a specific reference data set also increase as the proportion of observed data being utilized for the calculation of correction parameters increases. As the nonparametric transformations do not depend on predetermined functions of any kind, their probability of success is more likely to rely on their flexibility, which allows any quantile-quantile relation to be a good fit. Even though it has less ability in correcting the extreme events, it can be concluded that the quantile mapping approach with parametric transformation is the optimal technique among other methods tried in this study.

References

1. Wang, Y., Leung, L.R., McGregor, J.L., Lee, D.K., Wang, W.C., Ding, Y., Kimura, F.: Regional climate modeling: progress, challenges, and prospects. J. Meteor. Soc. Jpn. **82**, 1599–1628 (2004). https://doi.org/10.2151/jmsj.82.1599
2. Frei, C., Christensen, J.H., Déqué, M., Jacob, D., Jones, R.G., Vidale, P.L.: Daily precipitation statistics in regional climate models: evaluation and intercomparison for the European Alps. J. Geophys. Res. **108**(D3), 4124 (2003). https://doi.org/10.1029/2002jd002287
3. Suklitsch, M., Gobiet, A., Leuprecht, A., Frei, C.: High resolution sensitivity studies with the regional climate model CCLM in the Alpine Region. Meteorol. Z. **17**, 467–476 (2008). https://doi.org/10.1127/0941-2948/2008/0308
4. Suklitsch, M., Gobiet, A., Truhetz, H., Awan, N.K., Göttel, H., Jacob, D.: Error characteristics of high-resolution regional climate models over the Alpine Area. Clim. Dyn. **37**(1–2), 377–390 (2011). https://doi.org/10.1007/s00382-010-0848-5
5. Maraun, D., Wetterhall, F., Ireson, A.M., Chandler, R.E., Kendon, E.J., Widmann, M., Brienen, S., Rust, H.W., Sauter, T., Themeßl, M., Venema, V.K.C., Chun, K.P., Goodess, C.M., Jones, R.G., Onof, C., Vrac, M., Thiele-Eich, I.: Precipitation downscaling under climate change: recent developments to bridge the gap between dynamical models and the end user. Rev. Geophys. **48**, RG3003 (2010). https://doi.org/10.1029/2009rg000314

6. Winkler, J.A., Guentchev, G.S., Liszewska, M., Perdinan, Tan P.N.: Climate scenario development and applications for local/regional climate change impact assessments: an overview for the non-climate scientist—part II: considerations when using climate change scenarios. Geogr. Compass **5**, 301–328 (2011). https://doi.org/10.1111/j.1749-8198.2011.00426.x
7. Panofsky, H.A., Brier, G.W.: Some Application of Statistics to Meteorology, 224 pp. State Univ, University Park, Pa (1968)
8. Thrasher, B., Maurer, E.P., McKellar, C., Duffy, P.B.: Technical note: bias correcting climate model simulated daily temperature extremes with quantile mapping. Hydrol. Earth Syst. Sci. **16**, 3309–3314 (2012). https://doi.org/10.5194/hess-16-3309-2012
9. Michelangeli, P.A., Vrac, M., Loukos, H.: Probabilistic downscaling approaches: application to wind cumulative distribution functions. Geophys. Res. Lett. **36**, L11708 (2009). https://doi.org/10.1029/2009GL038401
10. Hay, L.E., Wilby, R.L., Leavesley, G.H.: Comparison of delta change and downscaled GCM scenarios for three mountainous basins in the United States. J. Am. Water Resour. Assoc. **36**, 387–397 (2000)
11. Cayan, D.R., Maurer, E.P., Dettinger, M.D., Tyree, M., Hayhoe, K.: Climate change scenarios for the California region. Clim. Change **87**(1), 21–42 (2008). https://doi.org/10.1007/s10584-007-9377-6
12. Maurer, E.P., Hidalgo, H.G.: Utility of daily versus monthly large-scale climate data: an intercomparison of two statistical downscaling methods. Hydrol. Earth Syst. Sci. **12**, 551–563 (2008). https://doi.org/10.5194/hess-12-551-2008
13. Meehl, G.A., Thomas, F.S.: Global climate projections, in climate change 2007: the physical science basis. In: Solomon, S. et al. (ed.) Contribution of working Group I to the Fourth Assessment Report of the Intergovernmental Panel on Climate Change, pp. 749–845. Cambridge University Press, New York (2007)
14. Widmann, M., Bretherton, C.S., Salathé, E.P.: Statistical precipitation downscaling over the Northwestern United States using numerically simulated precipitation as a predictor. J. Clim. **16**, 799–816 (2003). https://doi.org/10.1175/1520-0442(2003)016%3c0799:spdotn%3e2.0.co;2
15. Schmidli, J., Frei, C., Vidale, P.L.: Downscaling from GCM precipitation: a benchmark for dynamical and statistical downscaling methods. Int. J. Climatol. **26**, 679–689 (2006). https://doi.org/10.1002/joc.1287
16. Piani, C., Weedon, G., Best, M., Gomes, S., Viterbo, P., Hagemann, S., Haerter, J.: Statistical bias correction of global simulated daily precipitation and temperature for the application of hydrological models. J. Hydrol. **395**, 199–215 (2010). https://doi.org/10.1016/j.jhydrol.2010.10.024
17. Rojas, R., Feyen, L., Dosio, A., Bavera, D.: Improving pan-European hydrological simulation of extreme events through statistical bias correction of RCM-driven climate simulations. Hydrol. Earth Syst. Sci. **15**, 2599–2620 (2011). https://doi.org/10.5194/hess-15-2599-2011

Regional Optimization of Existing Groundwater Network Using Geostatistical Technique

K. SatishKumar and E. Venkata Rathnam

Abstract Groundwater management and optimized monitoring have become more significant to meet the needs and requirements of rapid increase of population. Optimized groundwater networks are rarely designed in most parts of the world. This study presents the optimization of existing network using geostatistical method for 42 observation wells in Warangal District, Telangana State. From 42 observation wells, average groundwater level fluctuation is evaluated with geology, lineament, geomorphology, recharge map, and groundwater level fluctuation map of Warangal District. These parameters are evaluated stochastically with ordinary and universal kriging methods in geographic information system (GIS). Further, semi-variogram has been performed to fit suitable theoretical model compared to the experimental model. Results of experimental models are compared from groundwater level (GWL) data with theoretical models (Gaussian, circular, exponential, and spherical). This study resulted that ordinary kriging method is suitable optimal model, and five observation wells were removed using error variance of monitoring networks. This study also explains the upgradation of existing network using multiple parameters. In this study, the proposed geostatistical method is highly effective in selecting suitable observation wells in complicated geological setup. This study concludes that 37 out of 42 observation wells are adequate for regular monitoring of groundwater.

Keywords Geostatistical method · Semi-variogram · Kriging · Geology · Geomorphology

K. SatishKumar (✉) · E. Venkata Rathnam
Department of Civil Engineering, National Institute of Technology, Warangal, India
e-mail: satish005@student.nitw.ac.in

E. Venkata Rathnam
e-mail: evr@nitw.ac.in

D. Dutta and B. Mahanty (eds.), *Numerical Optimization in Engineering and Sciences*, Advances in Intelligent Systems and Computing 979,
https://doi.org/10.1007/978-981-15-3215-3_10

1 Introduction

Groundwater networks are essential for quantitative and qualitative monitoring of groundwater resources. A properly designed groundwater monitoring system helps in understanding the state of region with respect to space and time. The optimal design of observation well network mainly depends on geographical location, distribution of water levels with respect to space and time. For the addition or removal of groundwater wells in the network, it is essential to know the position of GWLs for potential management. Installation of new observation wells depends on location, its demand, circumstances, and socioeconomic components of specific location [1]. Removal of groundwater wells from the network is based on increase in population, overexploitation, and contamination of groundwater.

Statistical approach is commonly used method for network optimization. This is subdivided into classical statistics, geostatistics, and time series analysis [2]. Geostatistics approach is the extensively used method among three for the network optimization. The main benefit of this approach is to estimate the variance associated with a sampling pattern. This method examines the correlation among the variables using variograms. Many researchers have successfully applied these methods for optimal design of network. Prakash and Singh [3] applied geostatistical technique for the optimal location of groundwater well sites for upper Kongal basin, Andhra Pradesh. Chao [4] used overlay tools for the demonstration of optimal design of groundwater level network. Fahimeh [5] applied NSGA-II algorithm for minimizing the error as well as optimization of observation wells in the existing network. Fisher [6] used kriging-based genetic algorithm method for optimization where forty wells are removed from the monitoring network. Chandan and Yashwant [7] applied geostatistical tool with multiple parameters for the optimization of monitoring network, concluded that 82 number of observation wells are required for the site for better monitoring of observation wells.

The study aims to optimize the current observation well network over Warangal District of Telangana State. As the groundwater levels are decreasing due to overexploitation and varying geological properties, observation wells are to be removed from the network for better utilization and management of groundwater. In this study, to optimize the groundwater network, 42 observation well data are analyzed using geostatistical technique and standard error map.

Network optimization using only observation well data will not be enough. So, several parameters like geology, lineaments, geomorphology, land use, land cover, and precipitation recharge are considered for the effective design of groundwater monitoring network. To optimize the network, few observation wells are removed from the network in such a way that error should be within the limits. To prioritize the suitable site for the removal of observation wells, additional parameters such as geology, lineaments, recharge are used along with groundwater level fluctuation map.

2 Study Area

Warangal District of Telangana State falls in the drainage basins of both Krishna and Godavari rivers with a geographical area of 12,846 km^2. The district is bounded by longitude 78° 49′ E–80° 43′ E and latitude 17° 19′ N–18° 36′ N with a total population of 35,12,576. Piezometric level data from 42 observation wells for the period 2003–2016 obtained from groundwater department are used in the analysis. The groundwater level data is considered for monsoon (June–September) season only.

3 Methodology

In the present study, optimization of existing network is performed using geostatistical approach in consideration with multi-parameter using GIS [8]. In geostatistical approach, ordinary and universal kriging interpolation methods are used. Using these methods, one can estimate the values at unsampled sites along with the error, i.e., unreliability of prediction within the region considered, which acts as root in upgrading the existing network. In geostatistical method, multi-parameters are used in association with the groundwater levels for the optimization of existing observation well network.

3.1 Geostatistical Method

This method helps in understanding the spatial/temporal occurrences and gets most on spatial relationships to model parameters at unsampled locations [9].

Ordinary kriging method

For analyzing ordinary kriging method, an average of regionalized parameter is assumed to be the same all over the region of interest.

$$\widehat{X}(y_o) = \sum_{k=1}^{z} \lambda_k X(y_k) \tag{1}$$

$$\sum_{k=1}^{z} \lambda_k \text{ and } \sum_{k=1}^{z} \lambda_k \gamma(y_k, y_l) - \mu = \gamma(y_k, y) \tag{2}$$

$\widehat{X}(y_o)$ is measured variable at y_o location; λ_k is kriging weight with observed variable $X(y_k)$ over location y_k; z represents samples in dataset; $\gamma(y_k, y_l)$ are the modeled semi-variogram at location y_k and y_l; μ is Lagrange multiplier; $\gamma(y_k, y)$ are modeled semi-variogram at y_k location, and y is predicted location.

Universal kriging method
This method is a complex form of kriging. Spatial distribution of groundwater level (target) is explained as sum of deterministic trend modeled by linear regression on covariates and random component.

$$\widehat{X}(y_0) = n(y_k) + X(y_k) = \sum_{l=0}^{L} a_l f_l(y_k) + X(y_k) \tag{3}$$

$n(y_k)$ is deterministic trend; $X(y_k)$ is random component; a_l is lth drift coefficient; f_l spatial coordinates; L samples in dataset.

3.2 Cross-Validation

Cross-validation test is executed to know unbiasedness prediction in selecting suitable interpolation method (kriging) with semi-variogram model. 42 groundwater level data are considered as an input to cross validate and further used for interpolating using kriging methods. For prediction stage, four semi-variogram models (Gaussian, exponential, circular and spherical) are used to select one suitable method among the kriging methods. The performance of model is checked using statistical metrics like mean square error (MSE), root-mean-square error (RMSE), and average standard error (ASE). These procedures are repeated for ordinary and universal kriging methods.

$$\text{ME} = \frac{1}{k}\sum_{k=1}^{z}\left[x^*(y_k) - x(y_k)\right] \tag{4}$$

$$\text{MSE} = \frac{1}{k}\sum_{k=1}^{z}\left[\frac{x^*(y_k) - x(y_k)}{\sigma_v^2(y_k)}\right] \tag{5}$$

$$\text{RMSE} = \sqrt{\frac{1}{k}\sum_{k=1}^{z}[x^*(y_k) - x(y_k)]^2} \tag{6}$$

$$\text{ASE} = \sqrt{\frac{1}{k}\sum_{k=1}^{z}\sigma_v^2(y_k)} \tag{7}$$

$\sigma_v^2(y_k)$ is variance for the location y_k, $x^*(y_k)$ and $x(y_k)$ are observed and estimated values of variable at that location, respectively.

3.3 Thematic Maps Preparation

ArcGIS 10.2 tool is employed for the processing, analysis, and development of thematic maps. Using observation well data, groundwater level fluctuation (GLF) map is created seasonally. GLF maps are analyzed seasonally and are developed for monsoon season. Using GIS environment observation, well data is processed and interpolated using geostatistical methods for the development of GLF map.

Geology, lineament, and geomorphology are considered as important parameter for groundwater storage, recharge, and discharge. Lineament property is directly related to the geological structures [10]. These datasets are collected from Telangana State Remote Sensing Applications Centre, Telangana. These maps are analyzed in the GIS environment.

Groundwater recharge map is developed for only pre-monsoon season using rainfall data as input. Thomas [11] proposed a relationship between rainfall and recharge as shown below

$$R = 5.732(P - 89.7)^{0.51} \tag{8}$$

R = groundwater recharge and P = average rainfall.

3.4 Estimating Optimum Observation Wells

Using the groundwater level fluctuations from 2003 to 2016, standard error map is prepared for the study region. Standard error (S_{error}) estimates the precision and bias of predicted variable and is calculated by Eqs. 9 and 10.

$$S_{\text{error}} = \frac{\sigma}{\sqrt{n}} \tag{9}$$

$$\sigma = \sqrt{\frac{\sum (x - \bar{x})^2}{n}} \tag{10}$$

where σ is the standard deviation, n is the number of samples, x is the observed GWL, $\bar{x}$ is the mean GWL of the dataset. Standard error map obtained from GLFs and the multi-parameters are compared and separately analyzed with GLF map. From the standard error obtained from GLF map in association with influencing parameters, minimization of observation wells from the network is performed. Initially, few observation wells are removed from the network and standard error is to be computed until the error is bounded within the limits (not exceeding 1). The selection of observation wells that are to be removed will be based on the GLF at that observation well and its influence on multi-parameters.

4 Results and Discussions

4.1 Cross-Validation of GWL Fluctuations

Ordinary and universal kriging interpolation methods were analyzed and computed separately for four semi-variogram models like exponential, Gaussian, circular, and spherical in GIS environment. Among these, the best interpolation technique was finalized using ASE, ME, RMSE, and MSE, and cross-validation results were represented in Table 1. To ensure that the predicted values are consistent, mean error values should be equal to zero.

From Table 1, ordinary and universal kriging methods were compared using four semi-variogram models by ME, RMSE, ASE, and MSE. By using exponential model, ordinary kriging (0.091, 1.321) was performed better than universal kriging (0.109, 1.368) based on ME and RMSE. Among four semi-variograms, exponential model performed well with least mean errors followed by Gaussian model. From Table 1, the cross-validation results suggested that ordinary kriging with exponential model provided accurate estimation of ME (0.091), RMSE (1.321), ASE (0.03), and MSE (1.251) which are least among the kriging methods. Therefore, for further analysis ordinary kriging with exponential semi-variogram model was selected for the optimization process. Using ordinary kriging, GLF map and groundwater recharge map were developed.

4.2 Multi-parameter Impact on GLFs

Many researches have used geostatistical method for the selection of observation well locations either for adding or removal of wells from the existing network [3, 12]. In this method, standard error map was generated for only monsoon season using observation well data and their geographical locations in GIS environment.

Table 1 Ordinary and universal kriging cross-validation

Kriging	Semi-variogram model	ME	RMSE	ASE	MSE
Ordinary	Exponential	0.091	1.321	0.03	1.251
	Gaussian	0.098	1.492	0.031	1.352
	Circular	0.125	1.398	0.036	1.299
	Spherical	0.103	1.499	0.045	1.458
Universal	Exponential	0.109	1.368	0.03	1.239
	Gaussian	0.132	1.476	0.035	1.372
	Circular	0.118	1.654	0.052	1.568
	Spherical	0.112	1.624	0.046	1.503

Using average groundwater level, standard error values were measured. If the error exceeds 1, the site was considered suitable for the addition of observation wells. If the error is <1, removal of existing observation wells from the network can be done. Practically, standard error map obtained from observation well data is insufficient for the site selection. Other than groundwater level fluctuations, there are several other parameters which are considered in geostatistical analysis for this purpose such as groundwater recharge, geology, and lineaments.

From observation well data, groundwater level fluctuation map was generated for monsoon season using ordinary kriging (exponential model) method in the GIS environment as shown in Fig. 1. The GLF map was divided into five classes with the lowest groundwater levels at southwest corner and highest levels in the middle of Warangal. The groundwater levels were varying from 3.82 m below groundwater level (mbgl) to 18.07 mbgl. Other than GLFs, the parameters which influence the selection process directly or indirectly are explained in subsequent sections.

4.3 GLFs with Reference to Geological Features

The study region consists of six classes of geological features as shown in Fig. 2. Most of the area was covered by gneiss–granitoid complex (52.56%) followed by semi-consolidated sediment groups (28.24%). Standard error for the study region exists within the limits ($S_{error} < 1$) as shown in Fig. 5. So, site selection for the removal of observation wells was performed using geological features. Gneiss–granitoid rocks were impervious compared to semi-consolidated group. As the GWFs (13–18 mbgl) were decreasing in the west part of the region (gneiss–granitoid complex), the site was considered for the removal of observation wells. In the central part of the region, groundwater levels (3–6 mbgl) were observed as high and were correlated with rock type (metamorphic rock), so, no observation wells were selected from central part of the region. GLFs (11–14 mbgl) were decreasing in northeast part, where the rock types were semi-consolidated groups. So, observation wells from the northeast part were selected for the removal.

4.4 GLFs with Reference to Lineaments

Lineament specifies potential of permeable zone for groundwater exploitation. It helps in understanding existence and movement of groundwater. The study region comprises of major and minor lineaments varying from meters to kilometers. Although the lineament density was high in west and northeast part of the region, fluctuation of groundwater levels were minimal due to vegetation. The site for the removal of observation wells were decided by including the difference in GLF and lineament density. Thus, the site with low GLF and low lineament density was considered for the removal of observation wells (Fig. 3).

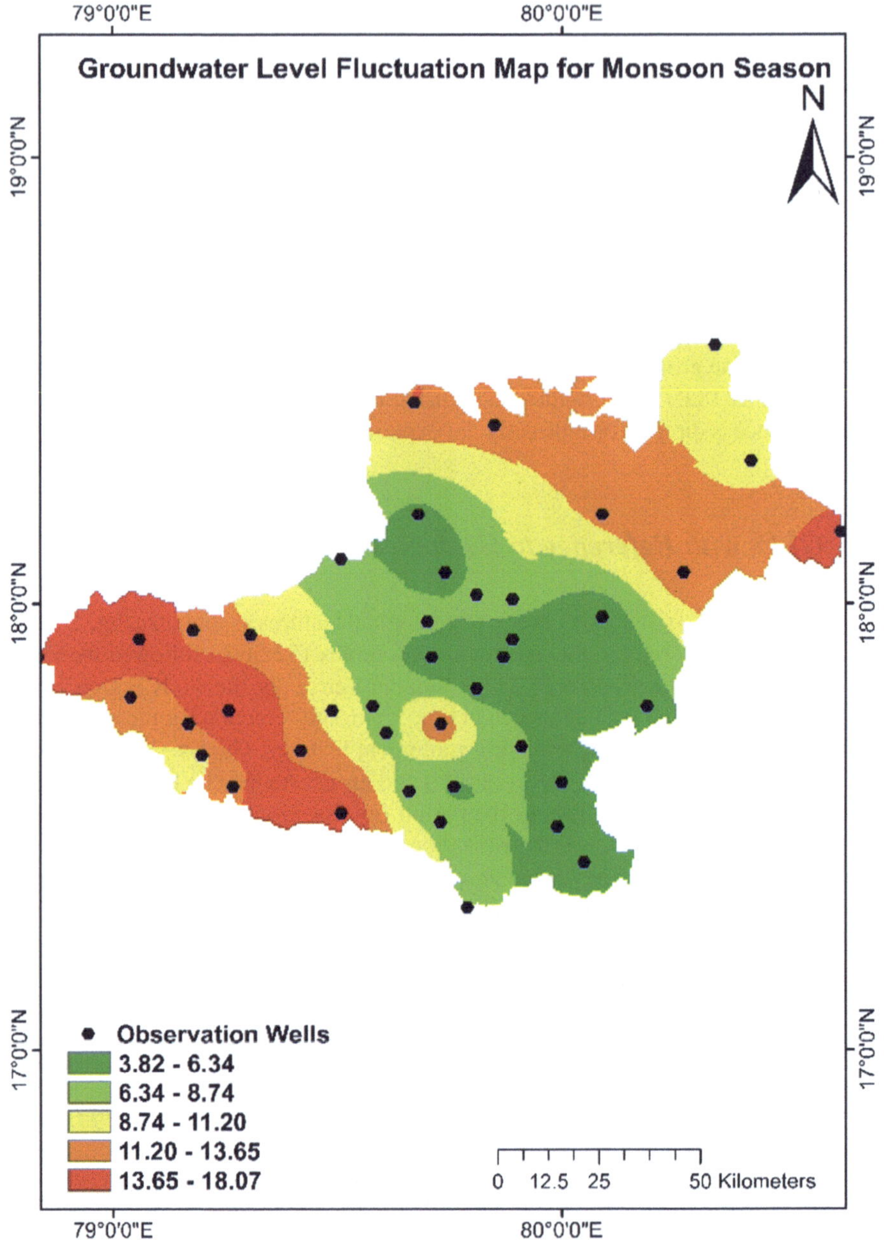

Fig. 1 GLF map for monsoon season

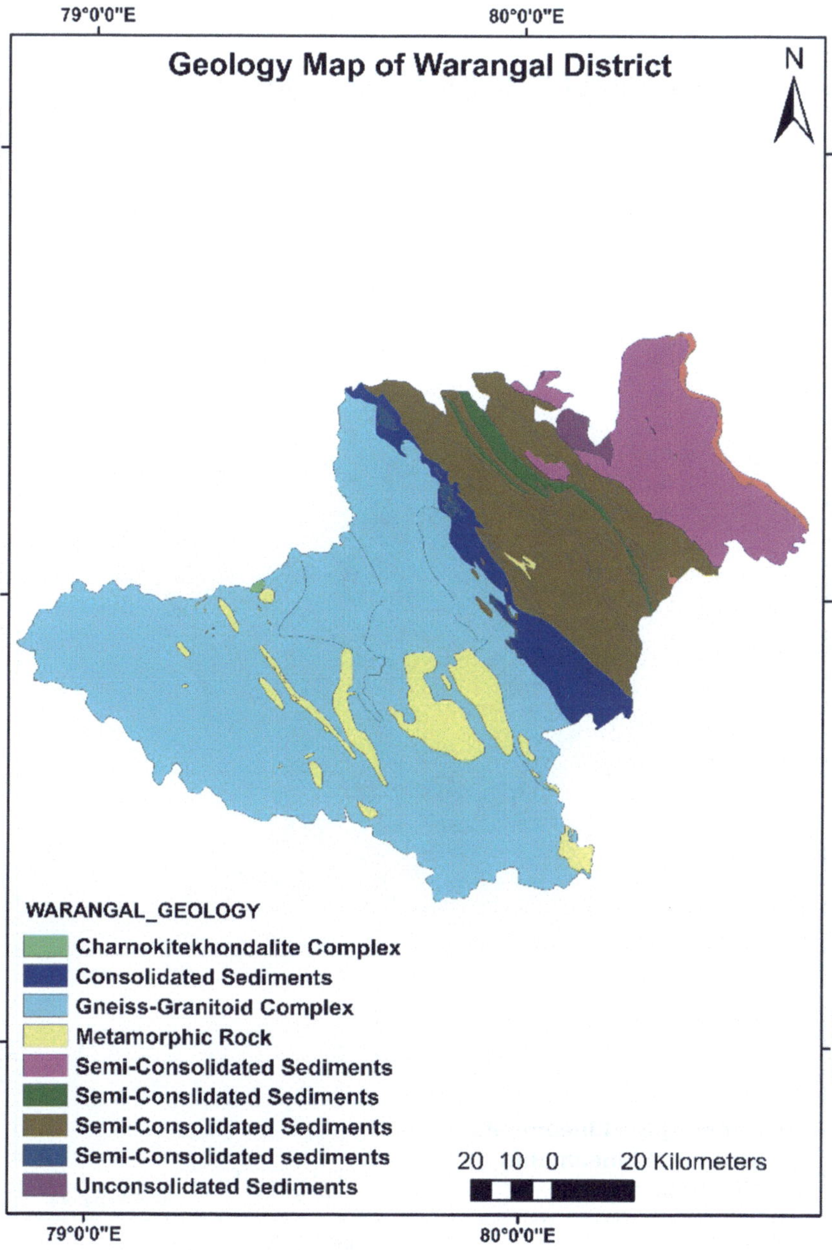

Fig. 2 Geological map of Warangal

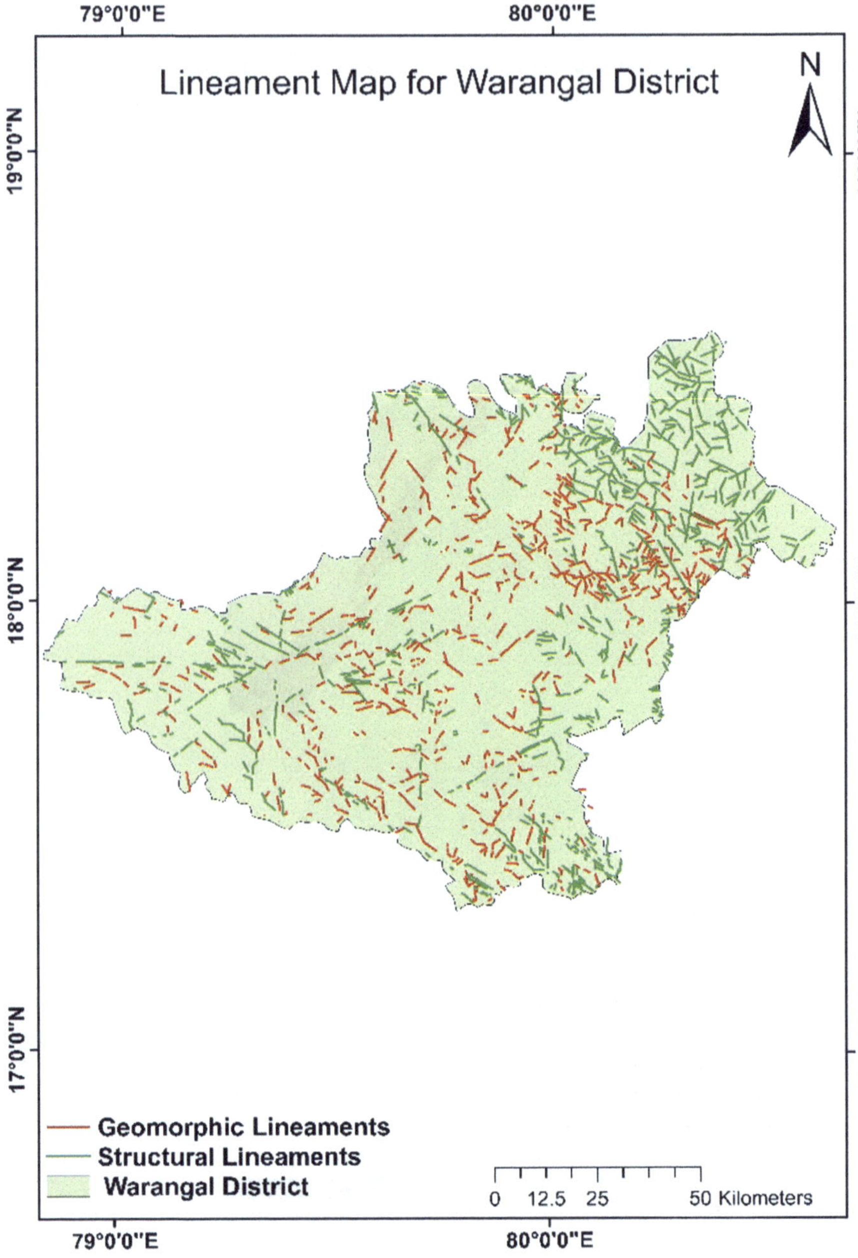

Fig. 3 Lineament map of Warangal

4.5 GLFs with Reference to Groundwater Recharge

Average annual rainfall of the study region is 977 mm. The relationship between annual and monsoon groundwater levels were analyzed for the period 2003–2016. Monsoon GWLs showed high correlation in the previous and current year with rainfall, respectively. Recharge from rainfall was affecting groundwater level, and therefore, it was counted as one of the ways to select observation well location. It was detected that recharge of groundwater was less in southeast region, medium in central region and more in southwest region. From the groundwater recharge map, it was detected that GLF was high at the lowest recharge region. Site with high GLF and high recharge was chosen for the removal of observational wells (Fig. 4).

4.6 Optimization

Standard error map obtained from GLF including lineament, geology, and recharge parameters is used for locating observational wells (add/remove) in the study region. S_{error} was found less than 1 when all 42 observation wells were considered. When observational wells were increased, an increase in error values was observed ($S_{\text{error}} > 1$). So, the removal of observational well sites considering GLF, lineament, geology, and recharge parameters over south east and west parts of the region is performed. To optimize well network, two observational wells were removed from existing network; S_{error} is checked and verified that error is within the limits. In the same way instead of two, four and six observational wells were removed from the network and standard error was evaluated for both the cases. When four observation wells were removed, the error was within the limits. Whereas, in case of six observation wells, the error was exceeding one. Hence, it is analyzed that removal of less than six observational wells was recommended to optimize the well network. So, removal of five observation wells was finalized with minimal error ($S_{\text{error}} > 1$) as shown in Fig. 5. After the removal of five observational wells, error is checked and observed that ASE, MSE, and RMSE of the well values were minimum. The standard error with 42 observation wells was observed as 0.69 which increased to $0.97 < 1$ after the removal of 5 observation wells i.e., with 37 observation wells. Similarly, other errors calculated for 42 to 37 observation wells were observed as—absolute standard error from 1.2812 to 1.6418; mean error— from 0.0128 to 0.0408; mean error—from 0.0264 to 0.072; and root-mean-square error—from 1.1821 to 1.989. Therefore, from the study it is analyzed that 5 wells should be removed from the 42 well networks in the study region.

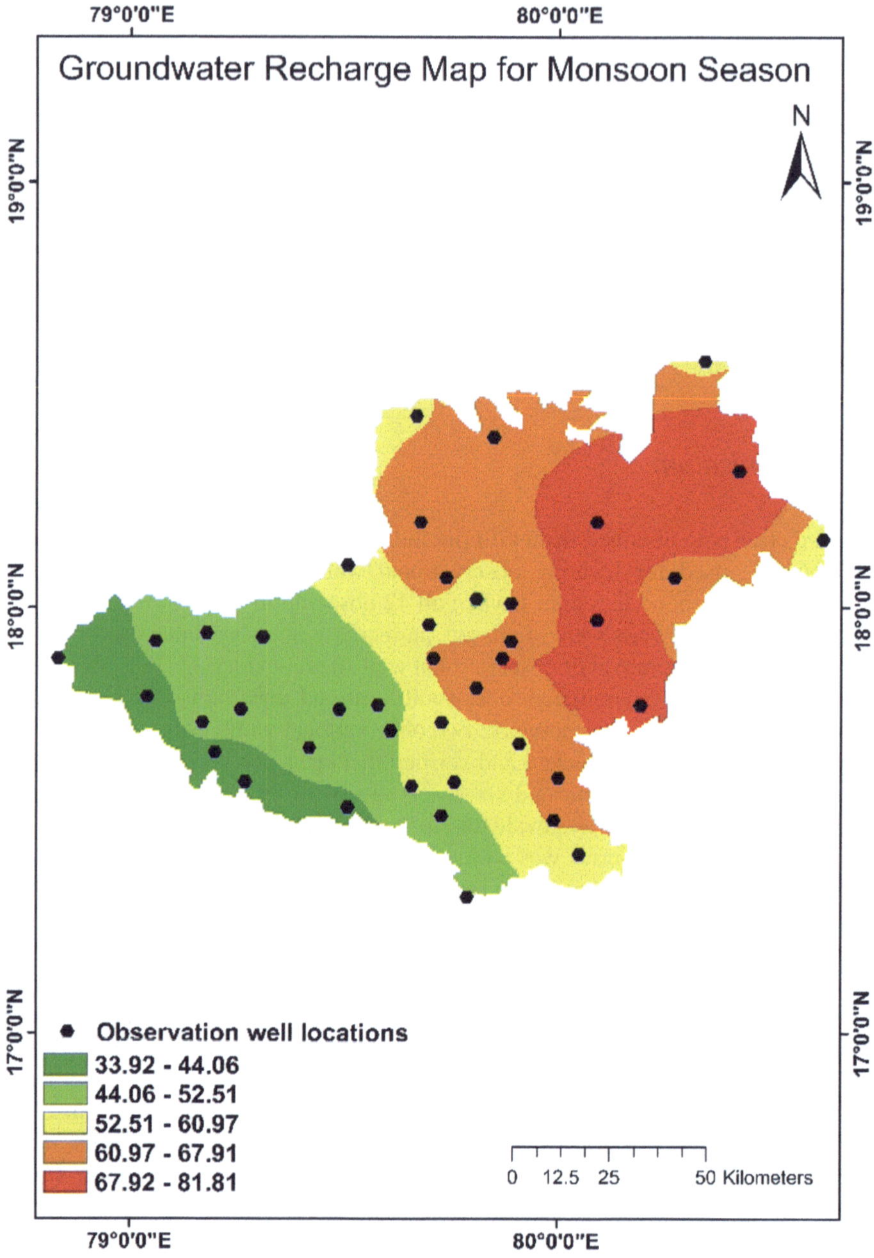

Fig. 4 Groundwater recharge map of Warangal

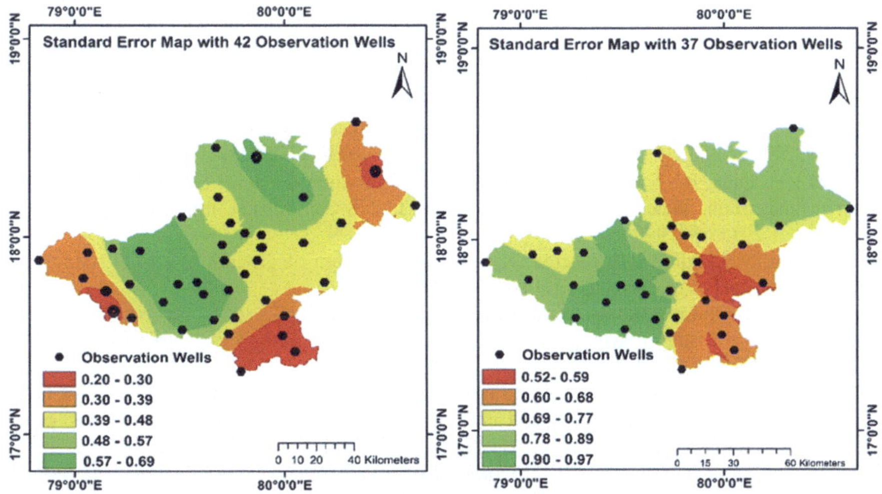

Fig. 5 Standard error map before and after removal of observation wells

5 Conclusion

This study concludes that the use of geostatistical method in combination with different variables will show a profound impact in the optimization of groundwater level network. Different parameters like lineament, recharge, and geology were analyzed for better optimization of the network. This study was carried out to optimize and locate the observation wells among 42 using geostatistical optimization method in the study region. Semi-variogram has been performed to fit a suitable theoretical model compared to experimental model. By using exponential model, ordinary kriging (0.091, 1.321) was performed better than universal kriging (0.109, 1.368) based on ME and RMSE. Standard error map was prepared and checked for the influence of different variables using cross-validation technique. The standard error with 42 observation wells is 0.69 which is increased to 0.97 < 1 with 37 observation wells. Similarly, other errors calculated for 42–37 observational wells were observed as—absolute standard error from 1.2812 to 1.6418; mean error—from 0.0128 to 0.0408; mean error—from 0.0264 to 0.072; and root-mean-square error—from 1.1821 to 1.989. From the study, it is detected that geostatistical method is efficient in locating the observation wells for a better understanding of recharge, availability of groundwater level, and movement of groundwater using other optimization methods.

References

1. IGRAC: Groundwater Monitoring–General Aspects and Design Procedure. Guideline on Groundwater Monitoring for General Reference Purposes. International Groundwater

Resources Assessment Centre. http://www.un-igrac.org/download/file/fid/278 (2006)
2. Uil, H., VanGeer, F.C., Gehrels, J.C., Kloosterman, F.H.: State of the Art on Monitoring and Assessment of Groundwater. Netherlands Institute of Applied Geosciences TNO, The Netherlands, 70pp (1999)
3. Prakash, M.R., Singh, V.S.: Network design for groundwater monitoring: a case study. Environ. Geol. **39**(6), 628–632 (2000). https://doi.org/10.1007/s002540050474
4. Chao, Y., Qian, H., Fang, Y.: Optimum design of groundwater level monitoring network in Yinchuan plain. Water Resour. Environ. Prot. **1**, 278–281 (2011). https://doi.org/10.1109/ISWREP.2011.5892999
5. Fahimeh, M., Omid, B., Loáiciga, H.A.: Optimal design of groundwater-level monitoring Networks. J. Hydroinformatics **19**(6), 920–929 (2017)
6. Fisher, J.C.: Optimization of Water-Level Monitoring Networks in the Eastern Snake River Plain Aquifer Using a Kriging-Based Genetic Algorithm Method. Scientific Investigations Report 2013–5120 (2013)
7. Chandan, K.S., Yashwant, B.K.: Optimization of groundwater level monitoring network using GIS-based geostatistical method and multi-parameter analysis: a case study in Wainganga sub-basin, India. Chin. Geogra. Sci. **27**(2), 201–215 (2017)
8. Zhou, Y.X., Dong, D.W., Liu, J.R.: Upgrading a regional groundwater level monitoring network for Beijing Plain, China. Geosci. Front. **4**(1), 127–138 (2013). https://doi.org/10.1016/j.gsf.2012.03.008
9. Caers, J.: Petroleum Geostatistics. Society of Petroleum Engineers, Richardson (2005)
10. Nag, S.K., Ghosh, P.: Delineation of groundwater potential zone in Chhatna Block, Bankura District, West Bengal, India using remote sensing and GIS techniques. Environ. Earth Sci. **54**(4), 2115–2127 (2012). https://doi.org/10.1007/s12665-012-1713-0
11. Thomas, T., Jaiswal, R.K., Ravi, G.: Development of a rainfall-recharge relationship for a fractured basaltic aquifer in central India. Hydrogeol. J. **23**(15), 3101–3119 (2009). https://doi.org/10.1007/s11269-009-9425-2
12. Ibtissem, T., Moncef, Z., Hamed, B.D.: A geostatistical approach for groundwater head monitoring network optimisation: case of the Sfax superficial aquifer (Tunisia). Water Environ. J. **27**(3), 362–372 (2013). https://doi.org/10.1111/j.1747-6593.2012.00352.x

Water Quality Analysis Using Artificial Intelligence Conjunction with Wavelet Decomposition

Aashima Bangia, Rashmi Bhardwaj and K. V. Jayakumar

Abstract Water is life and is the most precious resource on Earth. Earth consists of 70% of water, 2.5% of freshwater, and 1% of easily accessible freshwater; thus, only 0.007% of Earth's water is accessible. The survival of life on Earth is directly proportional to the presence of water among other important resources. Water remains to be a natural resource with no replacement. In today's era, where science and technology are growing every hour and innovating new technologies and devices to make life easier and comfortable, but no artificial intelligence could either replicate or replace the need for water on Earth. The present study deals with the qualitative exploration of water quality components like potential of hydrogen (pH), chemical oxygen demand (COD); biochemical oxygen demand (BOD); dissolved oxygen (DO) of Yamuna River at different sample sites. Various sample sites designated for highly reported pollutants using artificial intelligence through least squares support vector regression (LSSVR) and hybrid of wavelet and LSSVR. It is observed that hybrid of wavelet and least squares support vector regression (WLSSVR) predicted good quality accurately among the two prototypes simulated on the basis of the simulation errors which are root–mean-square error (RMSE); mean absolute error (MAE); coefficient of determination (R^2); and execution time for both prototypes. RMSE values decrease overall on training and validating via WLSSVR as compared to LSSVR. It is observed that MAE values show a lesser decrease as it is in RMSE; on an average, MAE has lesser variability and R^2 has a greater variability as per simulations. The simulation is carried out to analyze the level of various pollutants in the Yamuna River at different sites for the consideration of the quality of water. The observed pattern from the study may help for the future prediction of the quality of water parameters, so that it prohibits the further decay of water quality which may

A. Bangia
USBAS, GGSIPU, New Delhi 110078, India

R. Bhardwaj (✉)
University School of Basic and Applied Sciences (USBAS), Non-Linear Research Lab, Guru Gobind Singh Indraprastha University (GGSIPU), Dwarka, New Delhi 110078, India
e-mail: rashmib@ipu.ac.in

K. V. Jayakumar
Department of Civil Engineering, National Institute of Technology, Warangal, Telangana, India

D. Dutta and B. Mahanty (eds.), *Numerical Optimization in Engineering and Sciences*, Advances in Intelligent Systems and Computing 979,
https://doi.org/10.1007/978-981-15-3215-3_11

prove to be lethal to the environment. These forecasts may be helpful for the formulation of policies, planning, and execution for the protection of the environment and quality of water.

Keywords Least squares support vector regression · Wavelet decomposition · Water quality · Wavelet LSSVR (WLSSVR) · Coefficient of determination (R^2)

1 Introduction

From Yamunotri to Allahabad, the total length of River Yamuna is 1376 km and total basin area is 366,223 km^2. The river is almost dry from Hathnikund to Delhi, but from the groundwater and small tributaries, etc., some water is added. From Hathnikund, the river reaches Delhi at Palla covering a total distance of 224 km. Earlier, Yamuna was known to have '*crystal blue*' water but the Yamuna has been counted among the most poisoned rivers of the world. Delhi itself dumps about 58% of its waste into waters of the river. Thus, the level of contamination is the highest around Delhi NCR geographically. Toxins are increasing at startling rate in waters. The Yamuna gets most polluted in Delhi, with 18 drainage systems out of 22 that drain directly into the river, and the rest 4 via Agra and Gurgaon channels. River water quality highly depends upon pH, COD, BOD, DO, and physicochemical water parameters. In the last two decades, more than Rs. 6500 crores have been spent on various missions toward sanitizing river. Central Pollution Control Board (CPCB) reported that adulterated expanse of Yamuna had amplified from 500 to 600 km. The permissible level of DO in water for aquatic life to sustain is 4.0 mg/l. Around Wazirabad Barrage that covers the stretch from Delhi to Agra has a range of 0.0–3.7 mg/l. BOD should have a value less than 3 mg/l to avoid contaminating river water, while the most unhygienic section of the Yamuna has BOD concentration ranging 14–28 mg/l. Artificial intelligence-based tools, data mining performance techniques, and wavelet neuro-fuzzy system (WNFS) method for forecasting are discussed [1, 2]. Wavelet and fractal methods for environmental applications are applied for the prediction of pollutants [3]. Bias-free rainfall forecast and temperature trend-based temperature forecast for monsoon season are evaluated [4]. The forecasting of stock is predicted using neuro-fuzzy analysis [5]. Large-scale data is statistically explained using least squares support vector regression [6]. Cross-Wavelet Analysis is studied for Thailand region using rainfall [7]. LSSVM and its applications or acidity prediction in grapes are explained [8]. Bivariate flood frequency analysis of non-stationary flood characteristics is discussed in detail [9]. The river flow growth is evaluated [10]. Forecasting of temperature for India using different nonlinear techniques is developed [11]. Solar radiation from sunshine duration for China is studied [12]. Neuro-fuzzy model is discussed in details [13]. For Tamil Nadu, resistivity is estimated [14]. Water quality is analyzed using statistical analysis [15]. EEG signals are studied [16]. Water component values at different sites of Yamuna River, i.e., Hathnikund, Nizamuddin, Mazawali, Agra, and Juhikha, for 10 years have been considered for detailed study.

At each sample site of River Yamuna, LSSVR and WLSSVR are discussed. In this paper, data set collection is provided in Sect. 2. Discussions on mathematical prototyping of wavelet transformations and artificial intelligence techniques are given in Sect. 3. Different errors during estimation are discussed in Sect. 4 with algorithms for building hybridized models such as LSSVR and WLSSVR. In Sect. 5, results for the hybrid model and the errors are numerically simulated. The comparative performances based on the execution time of the models and the error computations are concluded. As per the literature survey carried out, none of the articles have studied the pollution levels of River Yamuna water through artificial intelligence conjunction with wavelet decomposition (WD) for the sample sites and parameters considered in this paper.

2 Data Collection/Assessment

Yamuna is the largest branch of River Ganga in India. Glacier Yamunotri is located at the height of 6387 m on the southwestern slopes of Bander Pooch Peaks, latitude: 380 59′ N 780 27′ E is believed to be the origin of Yamuna. CPCB observes important water factors for designated sampling sites of River Yamuna. River Yamuna comprises of five fragments: Himalayan fragment (beginning to Tajewala Barrage ranging 172 km); upper fragment (Tajewala Barrage to Wazirabad Barrage ranging 224 km); Delhi fragment (Wazirabad Barrage to Okhla Barrage ranging 22 km); eutrophicated fragment (Okhla Barrage to Chambal confluence 490 km), and diluted fragment (Chambal confluence to Ganga confluence 468 km).

For the present study, five sites have been chosen based on the consumption of river water, i.e., Hathnikund, Nizamuddin, Mazawali, Agra, and Juhikha. The data consists of values for the pollutants pH, BOD, COD, and DO at five monitoring stations Hathnikund (1), Nizamuddin Bridge-MS (2), Mazawali (3), Agra-DS (4), and Juhikha (5) for 10 years as observed by Central Pollution Control Board (CPCB). The pollutants that were highly present in the river water have been studied which are pH, BOD, COD, and DO at all the above-listed stations. In this paper, the pollutant at a particular station is being referred via the pollutant and station number, i.e., pH1, BOD1, COD1, DO1; pH2, BOD2, COD2, DO2; pH3, BOD3, COD3, DO3; pH4, BOD4, COD4, DO4; pH5, BOD5, COD5, DO5 stands for pH, BOD, COD, and DO level at station-1, station-2, station-3, station-4, and station-5, respectively. A total of 120 values for each pollutant at each site have been simulated via two prototypes: LSSVR and WLSSVR in this study. Further, 116 values for each pollutant at each site are being trained and validated in the prototype, and first four target points as responses/inputs for WLSSVR are considered for each pollutant at each site. In this paper, at each site and for each pollutant, 120 data sets have been observed and a total of 2400 data sets have been used for the forecasting of quality of the water. It can be observed that the proposed prototype gives improved efficiency and reduces the error very sharply in comparison with existing classical models. The five monitoring

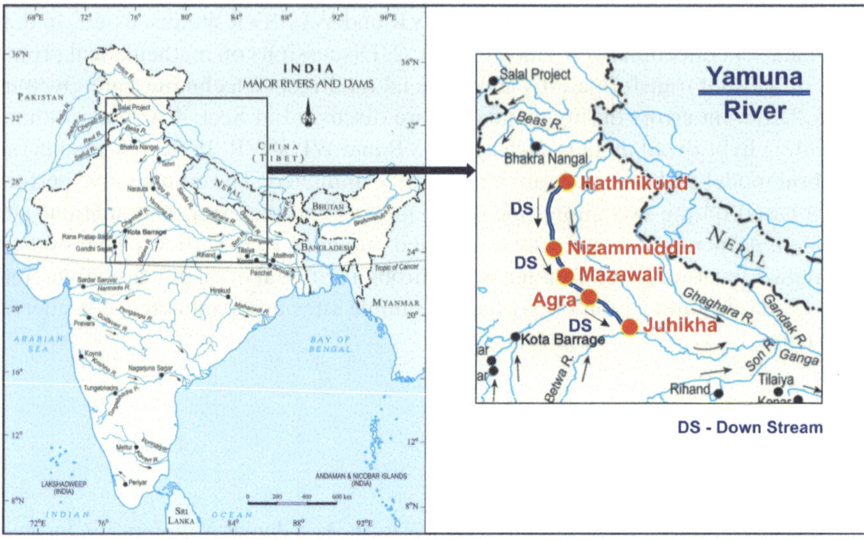

Fig. 1 Yamuna River flow map with different monitoring stations

stations: (1) Hathnikund; (2) Nizamuddin; (3) Mazawali; (4) Agra; and (5) Juhikha have been numbered according to the flow of the river crossing the station and are shown in Fig. 1.

3 Mathematical Prototyping

A new hybrid mathematical model using wavelet analysis and least squares support vector regression (LSSVR) is anticipated which is named as wavelet least squares support vector regression (WLSSVR). Using the proposed prototype, monthly forecast of the water quality parameters at different sites for different parameters is simulated and compared between the two regression models. The different methods used for forecasting are discussed in details.

3.1 Wavelet Analysis

Wavelet is a combination of sine and cosine functions that contain characteristics that fluctuate around zero and restricted in the interval realm. Wavelet function classified in two: the father wavelet—(ϕ)—mother wavelet—(ψ) with the following characteristics: $\int_{-\infty}^{\infty} \phi(t)\,\mathrm{d}t = 1$ and $\int_{-\infty}^{\infty} \psi(t)\,\mathrm{d}t = 0$. Integrating amplified dyadic with integer translations, mother–father wavelets are converted to innovative wavelet

$\phi_{j,k}(t) = 2^{j/2}\phi(2^j t - k)$ and $\psi_{j,k}(t) = 2^{j/2}\psi(2^j t - k)$. Stretching directory, j, is responsible for variation in sustenance and the array of the outcome. It is implied that if the sustenance constricts leading to range widening. The k translation directory stimulates the alteration for wavelet position upon flat league devoid of transforming supporting measurement (width). This sustenance of the function is regarded closure of the set of realm intersections providing function assessment to be nonzero.

3.1.1 Scaling Functions

Foundational structures of wavelets based upon scaling functions are created as:

1. Define a scaling function $\phi(t) = \sum_{k=-\infty}^{\infty} p_k \phi(2t - k)$;
2. Define subspace V of a vector space U, U-collection of elements over the real number R, then $V \subset U$.
3. Given a nested sequence of subspace V_i, V_j is: $V_j = \mathrm{clos}_{L^2}\{\phi_{j,k} | j, k \in Z\}$, where $\phi_{j,k} = \phi(2^j t - k)$, then containment property defined as:

$$\begin{array}{ll} V_0 \subset V_1 \Rightarrow \cdots V_{-1} \subset V_0 \subset V_1 \subset V_2 \subset \cdots \\ \text{Coarser} \quad \leftarrow \qquad \rightarrow \text{Finer} \end{array}$$

4. $\exists$ subspace W_j, orthogonal complement of t V_j in V_{j+1} i.e., $[V_{j+1} = V_j \oplus W_j, j \in Z,]$ and $W_j \perp W_{j'}$, if $j \neq j'$.
5. Since the subspaces V_j are nested, it follows: $V_J = V_j \oplus_{k=0}^{J-j-1} W_{j+k}$ for $j < J$, $L^2(R) = \cdots \oplus W_{-1} \oplus W_0 \oplus W_1 \oplus \cdots$;
6. Given a scaling function ϕ in V_j, $\exists$, ψ in W_0, such that $\{\psi_{j,k} | k \in Z\}$ generates W_j, where $\psi_{j,k} = \psi(2^j t - k)$, $j, k \in Z$.
7. Since $W_0 \subset V_1$ $\exists$ sequence q_k, such that: $\psi(t) = \sum_{k=-\infty}^{\infty} q_k \phi(2t - k)$.

3.1.2 Reconstruction and Decomposition

The following steps are discussed for reconstruction and decomposition of wavelets:

1. Since

$$V_1 = V_0 \oplus W_0, \phi(2t) \text{ and } \phi(2t - 1) \in V_1,$$

$$\phi(2t) = \sum_{k=-\infty}^{\infty} [a_{-2k}\phi(t - k) + b_{-2k}\psi(t - k)];$$

   thus $\phi(2t - 1) = \sum_{k=-\infty}^{\infty} [a_{1-2k}\phi(t - k) + b_{1-2k}\psi(t - k)]$;
2. The decomposition relation is:

$$\phi(2t - l) = \sum\nolimits_{k=-\infty}^{\infty} [a_{l-2k}\phi(t-k) + b_{l-2k}\psi(t-k)],\ l \in Z.$$

3. Reconstruction relation is:

$$\phi(t) = \sum\nolimits_{l=-\infty}^{\infty} p_l\phi(2t-l),\quad \psi(t) = \sum\nolimits_{l=-\infty}^{\infty} q_l\phi(2t-l);$$

4. Given a function f in $L^2(R)$, f can be approximately by $f_n \in V_n$ for some $N \in Z$.

$$f(t) \approx \sum_K c_{N,k}\phi(2^N t - k) = f_N,\ \text{where } c_{N,k} = \langle f(t), \phi_{N,k}\rangle.$$

5. Since $V_j = V_{j-1} \oplus W_{j-1}, f_N$ has a unique decomposition $f_N = f_{N-1} + g_{N-1} = g_{N-1} + g_{N-2} + \cdots + g_{N-M} + f_{N-M}$ where $f_j \in V_j$ and $g_j \in W_j$ for any j;
6. For designating decomposition and reconstruction procedures, f and g_j can be signified as follows:

$$f_j(t) = \sum\nolimits_k c_{j,k}\phi(2^j t - k);\quad g_j(t) = \sum\nolimits_k d_{j,k}\psi(2^j t - k)$$

where

$$c_{j,k} = \langle f_j(t), \phi_{j,k}(t)\rangle,\quad d_{j,k} = \langle f_j(t), \psi_{j,k}(t)\rangle.$$

3.1.3 Wavelet Decomposition Algorithm

It is defined as: $c_{j-1,k} = \sum_l a_{j-1,l-2k}c_{j,l},\quad d_{j-1,k} = \sum_l b_{j-1,l-2k}c_{j,l}$ which are high- and low-pass filters.

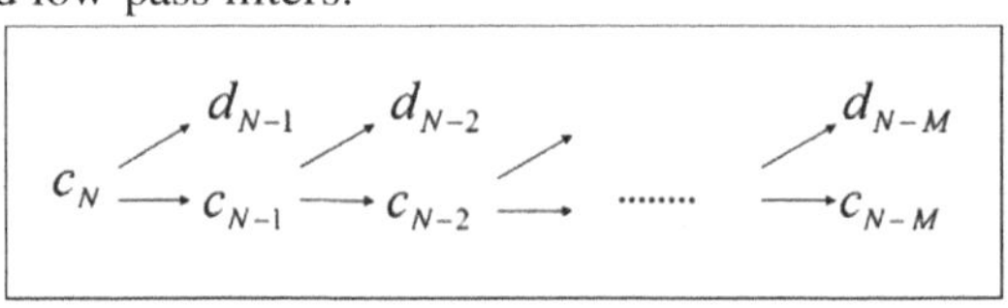

3.1.4 Wavelet Reconstruction Algorithm

It is defined as $c_{j,l} = \sum_k p_{j,l-2k}c_{j-1,k} + \sum_k q_{j,l-2k}d_{j-1,k}$ with filters as:

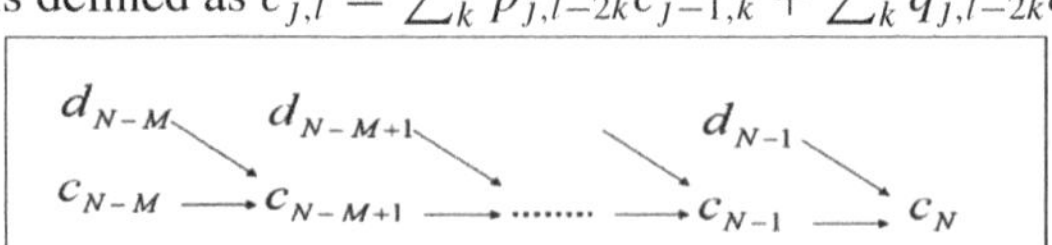

3.1.5 A Wedding of A′ Trous Algorithm and Multiresolution Mallat Algorithm

The algorithm is defined as:

i. A′ Trous and multiresolution decomposition algorithms are two separately motivated implementations.
ii. They are special cases of single filter structure and discrete wavelet transforms (DWT), and their behavior is administrated through the choice of filters.
iii. A′ Trous procedure had been originated to computationally increase the efficient implementation of non-orthonormal multiresolution algorithm for the exact DWTs.
iv. A′ Trous filters and Daubechies filters have a one-to-one mapping in the case of orthonormal wavelets of compact support.
v. In Mallat algorithm, since vectors are orthonormal, basis and dual coincide. Thus, the discrete filter is more fundamental than wavelet themselves. Generally, it is the coefficients that are determined, actual wavelets are rarely computed.

3.1.6 A′ Trous Algorithm

This is a stationary transform. Distance among the trials increases by factor of 2 from measure j to next *j + 1, k* is obtained by: $c_{j+1,k} = \sum_l h(l)c_{j,k+2^j l}$ with: $w_{j+1,k} = \sum_l g(l)c_{j,k+2^j l}$ Generally, wavelet ensuing from difference between 2 consecutive estimates is calculated as: $w_{j+1,k} = c_{j,k} - c_{j+1,k}$. Associated wavelet, $\psi(t)$: $1/2\ \psi(t/2) = \phi(t) - 1/2\ \phi(t/2)$. Reconstruction procedure is immediate: $c_{0,k} = cj, k + \sum_{j=1}^{J} h(l)w_{j,k}$.

3.1.7 Multiresolution Decomposition Algorithm

A sequence of embedded approximation subsets: $0 \leftarrow \ldots V_{j-i} \subset V_j \subset V_{j+1} \ldots \rightarrow R$ with $f(t) \in V_j \cdot f(2t) \in V_{j+1}$

(i) $f(t) \in V_0 \Leftrightarrow f(t-k) \in V_0, k \in Z$; $\varphi(t-k)_{k \in Z}$ form an orthonormal basis and a sequence of orthogonal complements, details' subspaces: W_n such that $V_{n+1} = V_n \oplus W_n$. φ-the scaling function, i.e., a low-pass strainer. The basis $V_{\{j\}}$ is given by: $\varphi_{j,k}(t) = 2^{j/2}\varphi(2^j t - k), k \in Z$. Figure 2 gives the wavelet analysis through decomposition and synthesis through reconstruction of wavelet coefficients.

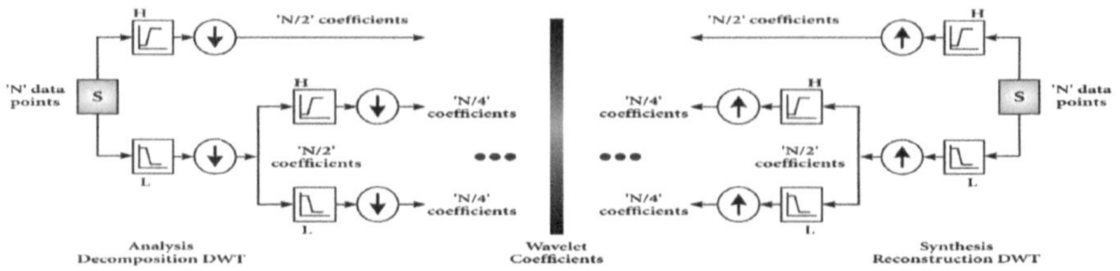

Fig. 2 Wavelet analysis through decomposition and synthesis through reconstruction of wavelet coefficients

3.2 Least Squares Support Vector Regression (LSSVR)

Deliberate a predictor:

$$h(t, u, v, w) = \mu_0 + \mu_1 t + \mu_2 v^2 + \mu_3 vw + \mu_4 t^3 u^2 + \mu_5 uv^4 w^2$$

LSSVRs are developed as an advanced formulation for SVM regression anticipated through Suykens. Objective function remains similar to that of former support vector regression. The difference arises in the epsilon-insensitive loss function being swapped with the classical squared loss job. This means that every b_i coefficient is nonzero. This is referred as assistances of instinctive sparseness lost. Alternatively, model is proficient much more efficiently subsequently creating Lagrange multiplier through resolving linear programming Karush–Kuhn–Tucker (KKT) scheme. Solution of this system is carried out with the help of most standard approaches to solve sets-of-linear equalities example conjugate gradient descent algorithm. The SVM procedure involves fine-tuning of three factors, whereas execution for LSSVM includes the requirement for two factors. Output prediction error is expected to be least through LSSVR. In this paper study, we input in four dimensions taken with different polynomial terms. Then, a normal equation is derived. Similarly, many dimensions of data for estimation using that many coefficients can be considered. The detailed algorithm is defined in (Fig. 3).

3.3 Wavelet LSSVR Prototype

The coefficients of approximations and details of Daubechies wavelet (Db8) as obtained from wavelet procedure will be the inputs for least squares support vector regression. Use LSSVR type 'function estimation' for lapse assessment of response predictors (X–Y), the matrices with training response predictors. Tuning factors are robotically tuned with leave-one-out cross-validation liable on magnitude of data set less or equal than 300 points. Apply Gaussian radial basis kernel. Other options that can be applied are linear kernel and polynomial kernel. Plot outcomes for envisioning regression with best-fit line for pre-arranged data points given in Fig. 4.

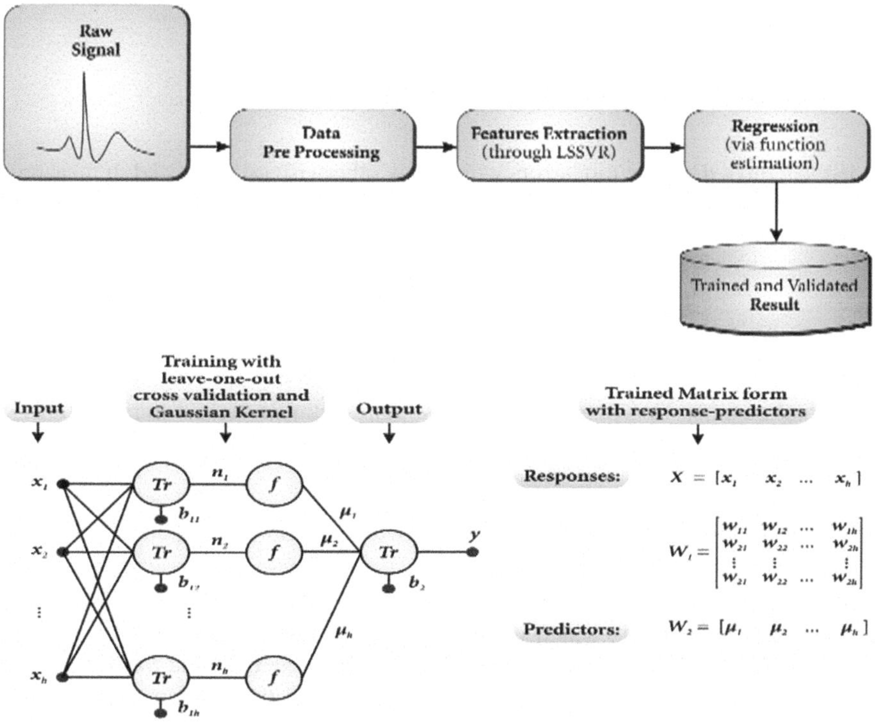

Fig. 3 Training of input responses in LSSVR model

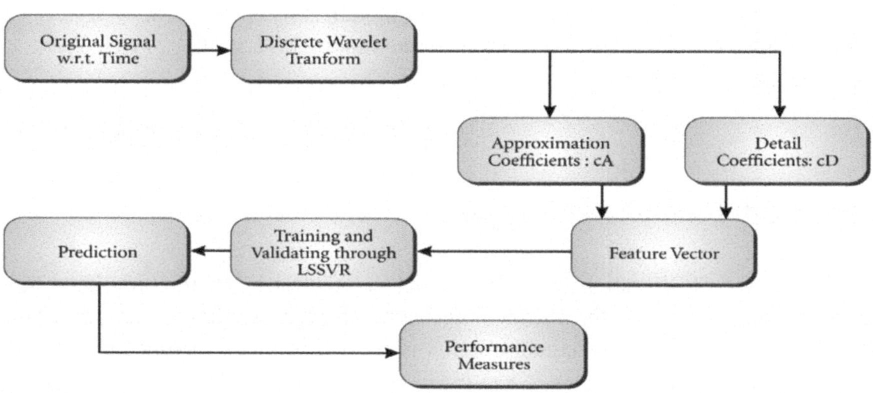

Fig. 4 Flowchart of wavelet LSSVR prototype

4 Simulation Errors

4.1 *Root-Mean-Square Error (RMSE)*

RMSE captures the large rare errors which may otherwise go unnoticed. It measures the magnitude of the inaccuracy that contributes additional weight to the outsized but infrequent errors than the mean. The mean squared error simulated as:

$$\text{RMSE} = \sqrt{\frac{\sum_{i=1}^{n} (y_i - \hat{y}_i)^2}{n}}$$

y_i—actual quantity; $\hat{y}_i$—predicted assessment; n—number of days in prediction.

4.2 *Coefficient of Determination* ($\mathbf{R}^2$)

It calculates the proportion in the dependent variable that is simulated by linear lapse and the predictor variable which is the independent variable. It defines the degree of evaluation ability of a model to envisage or explain an outcome for linear lapse model. To find the value for coefficient of determination, the following algorithm is used:

i. Find the regression line: $y = ax + b$, a and b are real constants.
ii. Find sum of squared errors of regression model, $\text{SSE} = \sum_{i=1}^{n} (y_i - \hat{y}_i)^2$
iii. Find sum of squared errors of our baseline model, $\text{SST} = \sum_{i=1}^{n} (y_i - \overline{y}_i)^2$
iv. Calculate-$R^2 = 1 - \frac{\text{SSE}}{\text{SST}}$.

4.3 *Mean Absolute Error (MAE)*

Absolute value of alteration amid the actual calculations and the simulated values with computation of the mean of the sum of such absolute errors assuming all individual errors acquire equal weights is termed as *mean absolute error*. The mean absolute error demarcated as: $\text{MAE} = \frac{1}{N}\left(\sum_{i=1}^{N} |d_i - y_i|\right)$, d—actual quantity; y—anticipated assessment; N—number of days in prediction.

5 Results and Discussions

A comparative study of techniques—LSSVR and WLSSVR—has been carried out for data monitored by CPCB for the water quality parameters, pH, BOD, COD, and DO at five sample sites. This results in the better performance of the model which proves to be cost-effective. Figure 5 depicts the complexity and simulative analysis through artificial intelligence (AI) techniques.

Table 1 presents the errors MSE, RMSE, and MAE of the two models for pollutants pH, BOD, COD, and DO at all stations. Figs. 6, 7, 8, and 9 show the comparison between the two prototypes: LSSVR and WLSSVR and for the pollutants: pH, COD, DO, and BOD at all of the five stations, respectively.

It can be observed that RMSE values decrease overall on training and validating via WLSSVR as compared to LSSVR. It is observed that MAE values show a lesser decrease as it is in RMSE. So, on an average MAE has lesser variability in error value among LSSVR and WLSSVR. The calculation time is also reducing for WLSSVR model. Thus, the proposed model, WLSSVR, is good for the estimation and the simulation of the data.

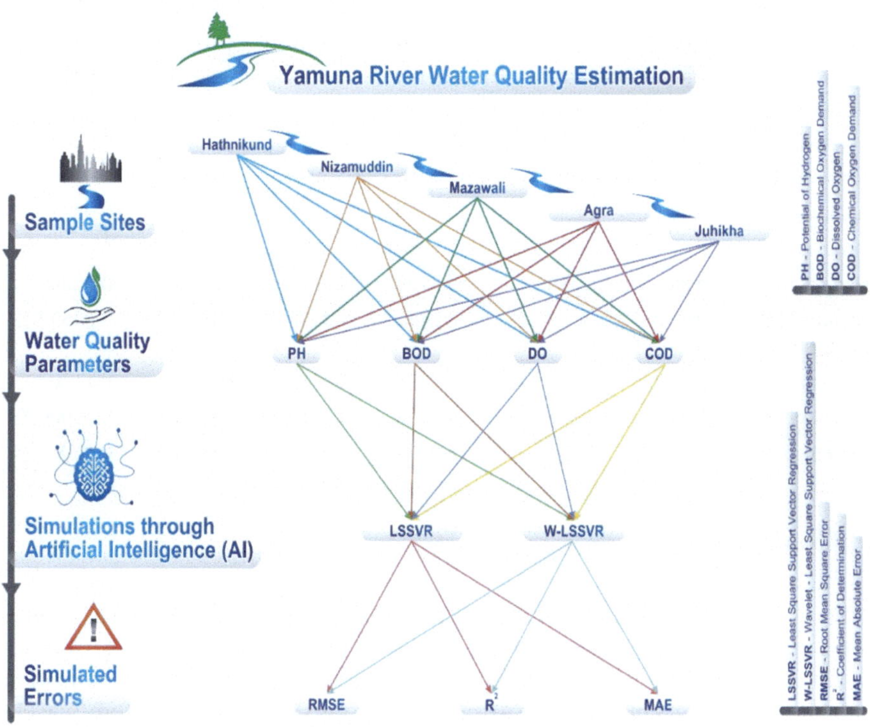

Fig. 5 Schematic diagram of River Yamuna water quality estimation

Table 1 Comparison of RMSE, MAE, R^2 for water quality parameters pH, BOD, COD, and DO

Stations	RMSE of LSSVR	RMSE of WLSSVR	MAE of LSSVR	MAE of WLSSVR	R^2 of LSSVR	R^2 of WLSSVR
Parameter pH						
Hathnikund-1	0.272	0.409	7.7675	7.7659	0.00131	0.00166
Nizamuddin Bridge-MS-2	0.160	0.301	7.4722	7.4721	0.0285	0.0116
Mazawali-3	0.140	0.216	7.7291	7.7273	0.0214	0.0158
Agra-DS-4	0.0998	0.328	7.8223	7.8200	0.000945	0.000153
Juhikha-5	0.3110	0.408	8.0049	8.0005	0.0051	0.00366
Parameter BOD						
Hathnikund-1	0.199	0.323	1.2750	1.2777	0.223	0.0564
Nizamuddin Bridge-MS-2	8.840	16.5	21.5583	22.7264	0.0731	0.0228
Mazawali-3	4.550	9.64	15.3167	15.2639	0.124	0.029
Agra-DS-4	3.690	7.43	16.6000	16.5669	0.026	0.00922
Juhikha-5	0.882	2.31	4.1583	4.1375	0.149	0.0455
Parameter COD						
Hathnikund-1	0.00937	3.19	7.2000	7.2840	0.0718	0.123
Nizamuddin Bridge-MS-2	18.8	35.1	65.0583	65.4363	0.0728	0.0212
Mazawali-3	15.1	29.7	57.6917	57.5528	0.0568	0.0184
Agra-DS-4	19.5	32.0	57.2750	57.3028	0.0223	0.0116
Juhikha-5	0.486	12.5	22.2500	22.1791	0.953	0.0295
Parameter DO						
Hathnikund-1	0.881	1.80	9.3708	9.3622	0.1060	0.0286
Nizamuddin Bridge-MS-2	0.831	1.57	0.6477	1.1375	0.0967	0.0385
Mazawali-3	1.80	3.81	3.8475	4.4188	0.3140	0.1170
Agra-DS-4	1.45	2.62	4.8808	4.9115	0.000199	0.00000701
Juhikha-5	1.27	2.95	9.5504	9.5212	0.0051	0.0006

6 Conclusion

In this work, prediction of pH, BOD, COD, and DO levels of River Yamuna provides increased accuracy on commissioning wavelet decomposition on LSSVR. Comparison clears that the proposed prototype provides accurate outcomes for estimation. Variability of data was incompetent to train through LSSVR alone. Therefore, wavelet decomposition into details plus approximations' coefficient calculation plays a vital role in the simulation of contamination levels for river water. The goodness of fit characteristics, R^2, best explains the prototypes applied in this article. Thus, it can

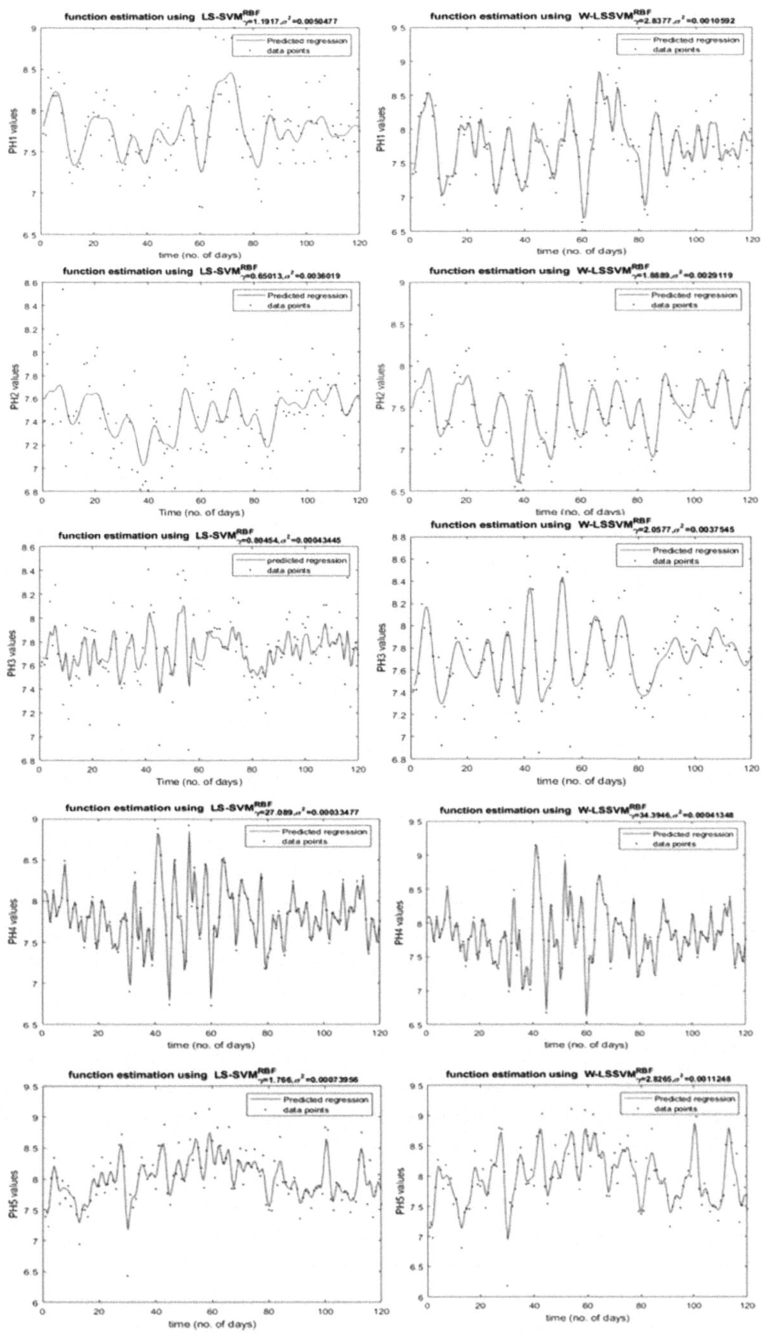

Fig. 6 Comparison of LSSVR and WLSSVR for pH at stations: 1, 2, 3, 4, and 5

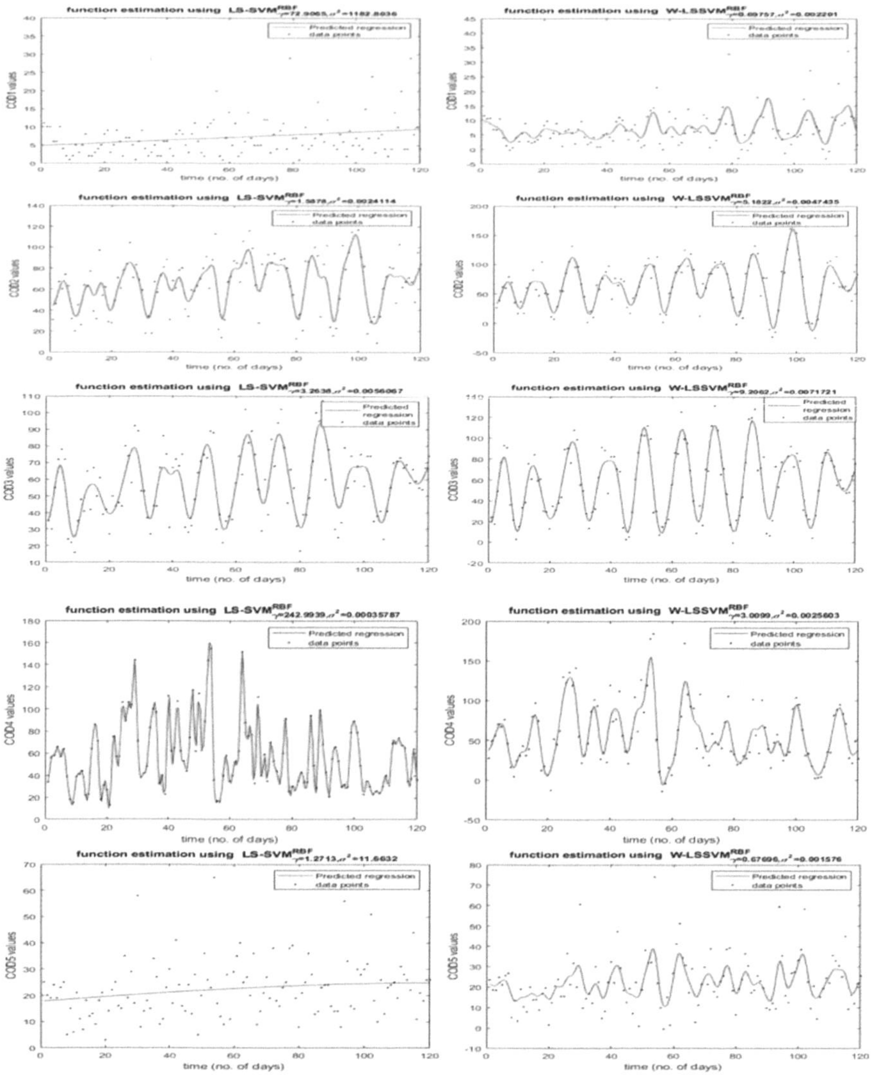

Fig. 7 Comparison of LSSVR and WLSSVR for COD at stations: 1, 2, 3, 4, and 5

be observed that prototype consisting of data trained via wavelet coefficients, i.e., WLSSVR gives better fir for pH, BOD, COD, and DO. It is observed that the predicted prototype proves to be better compared to baseline one as accuracy increases. Also, the execution time decreases on an average considering all sample sites. It can be concluded that this prototype wipes out noises and moderates the computational

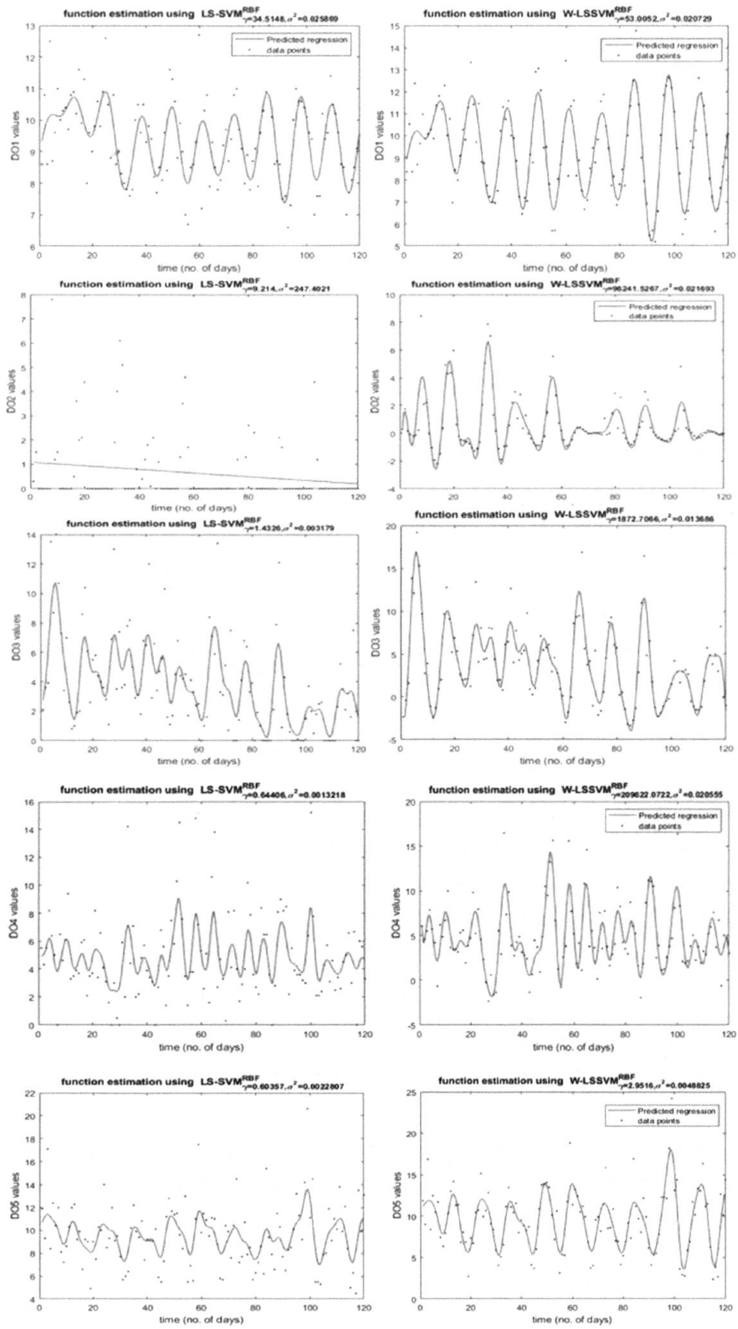

Fig. 8 Comparison of LSSVR and WLSSVR for DO at stations: 1, 2, 3, 4, and 5

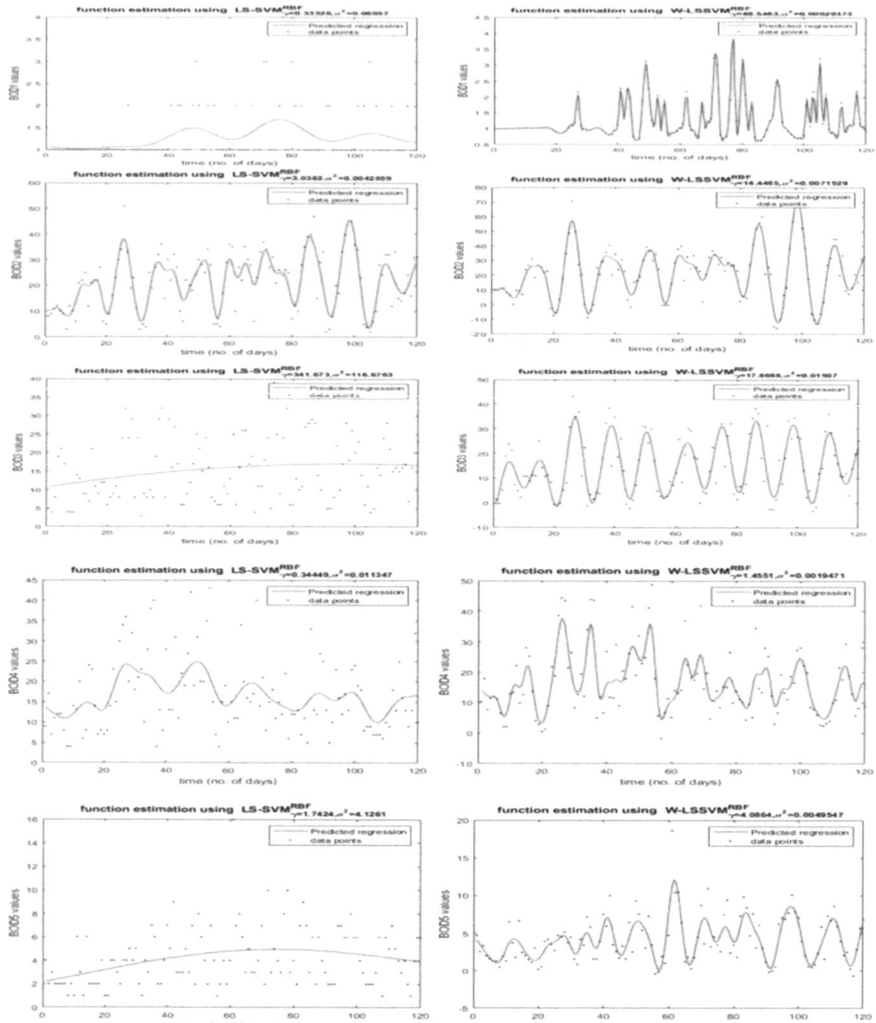

Fig. 9 Comparison of LSSVR and WLSSVR for BOD at stations: 1, 2, 3, 4, and 5

labor. Therefore, for future predictions WLSSVR should be relied upon. It is concluded that as the best forecast is obtained in lesser time, then the proposed model WLSSVR is very helpful in formulating the policy for execution and planning.

References

1. Arasu, B.S., Jeevananthan, M., Thamaraiselvan, N., Janarthanan, B.: Performances of data mining techniques in forecasting stock index evidence from India and US. J. Nat. Sci. Found. Sri Lanka **42**(2), 177–191 (2014)
2. Artha, S.E.M., Yasin, H., Warsito, B., Santoso, R.: Application of Wavelet Neuro-Fuzzy System (WNFS) method for stock forecasting. J. Phys.: Conf. Ser. **1025**, 1–12 (2018)
3. Bhardwaj, R.: Wavelets and fractal methods with environmental applications. In: Siddiqi, A.H., Manchanda, P., Bhardwaj, R. (eds.) Mathematical Models, Methods and Applications, pp. 173–195. Springer, Berlin (2016)
4. Bhardwaj, R., Kumar, A., Maini, P., Kar, S.C., Rathore, L.S.: Bias free rainfall forecast and temperature trend-based temperature forecast based upon T-170 model during monsoon season. Meteorol. Appl. **14**(4), 351–360 (2007)
5. Bhardwaj, R., Bangia, A.: Neuro-fuzzy analysis of demonetization on NSE. In: Bansal, J.C., Das, K.N., Nagar, A., Deep, K., Ojha, A.K. (eds.) Advances in Intelligent Systems and Computing: Soft Computing for Problem Solving, pp. 853–861. Springer, Berlin (2019)
6. Brabanter, K.D.: Least Square Support Vector Regression with Applications to Large-Scale Data: A Statistical Approach. Katholieke Universiteit Leuven, Belgium (2011)
7. Chansaengkrachang, K., Luadsong, A., Aschariyaphotha, N.: A study of the time lags of the Indian Ocean Dipole and Rainfall Over Thailand by using the cross-wavelet analysis. Arab. J. Sci. Eng. **40**(1), 215–225 (2015)
8. Chaucharda, F., Cogdill, R., Roussel, S., Roger, J.M., Bellon-Maurel, V.: Application of LS-SVM to non-linear phenomena in NIR spectroscopy: development of a robust and portable sensor for acidity prediction in grapes. Chemometr. Intell. Lab. Syst. **71**(2), 141–150 (2004)
9. Dong, N.D., Agilan, V., Jayakumar, K.V.: Bivariate Flood frequency analysis of non-stationary flood characteristics. J. Hydrol. Eng. **24**(4), 1–14 (2019)
10. Doyle, M.E., Barros, V.R.: Attribution of the river flow growth in the Plata basin. Int. J. Climatol. **31**, 2234–2248 (2011)
11. Durai, V.R., Bhardwaj, R.: Location specific forecasting of maximum and minimum temperatures over India by using the statistical bias corrected output of global forecasting system. J. Earth Syst. Sci. **123**(5), 1171–1195 (2014)
12. Fan, J., Wu, L., Zhang, F., Cai, H., Zeng, W., Wang, X., Zou, H.: Empirical and machine learning models for predicting daily global solar radiation from sunshine duration: a review and case study in China. Renew. Sustain. Energy Rev. **100**, 186–212 (2019)
13. Jeong, C., Shin, J.Y., Kim, T., Heo, J.H.: Monthly precipitation forecasting with a neuro-fuzzy model. Water Resour. Manag. **26**, 4467–4483 (2012)
14. Mondal, N.C., Devi, A.B., Raj, A.P., Ahmed, S., Jayakumar, K.V.: Estimation of aquifer parameters from surficial resistivity measurement in a granitic area in Tamil Nadu. Curr. Sci. **111**(3), 524–534 (2016)
15. Parmar, K.S., Bhardwaj, R.: Water quality management using statistical analysis and time-series prediction model. Appl. Water Sci. **22**, 397–414 (2014)
16. Subasi, A.: Automatic recognition of alertness level from EEG by using neural network and wavelet coefficients. Expert Syst. Appl. **20**, 1–11 (2004)

Performance Evaluation of Line of Sight (LoS) in Mobile Ad hoc Networks

C. R. Chethan, N. Harshavardhan and H. L. Gururaj

Abstract The era of multimedia communication is a major part and partial of each and every user life each task now depends very much on the Internet. Line-of-sight is the biggest problem many are facing, while using Internet. In particular, the problem of mobile station tracking at various base stations is considered. The data is transferred more quickly in a flat area where there are no obstacles. If there were obstacles between the mobile station and the base station, there will be lot of inefficiency in data transfer. The overview of line-of-sight in mobile ad hoc networks, various problems that arise and impairment concerned with line of sight are deliberated in this paper. The various QoS parameters are considered for the evaluation of LoS in mobile ad hoc networks with and without obstacles.

Keywords LoS · NLOS · Mobile ad hoc networks · Node · Hosts

1 Introduction

Networking facilitates communication between two or more physical device programs. A computer network is a collection of computers that are connected in some way to allow data to be exchanged with others in the network. A digital telecommunications network is a computer network that allows nodes to share resources. Computer networks use connections between nodes to exchange data with each other. The data connections are made via wired or optical cables, wireless media, and other cables. Wired communication refers to the transmission of data through wired technology. Wireless communication is a kind of wireless data communication. The line-of-sight (LOS) is a special type of propagation that transmits and receives data only if the station is visible without a barrier between it and the transmitted or received station. Examples of line of sight communication are FM radio and satellite transmission. Non-line-of-sight (NLOS) is a term commonly used when the radio transmitter and receiver are not on the direct line of vision. The term is used to transmit signals

C. R. Chethan · N. Harshavardhan (✉) · H. L. Gururaj
Department of Computer Science and Engineering, Vidyavardhaka College of Engineering, Mysuru, Karnataka, India
e-mail: vardhandhoni1@gmail.com

D. Dutta and B. Mahanty (eds.), *Numerical Optimization in Engineering and Sciences*, Advances in Intelligent Systems and Computing 979,
https://doi.org/10.1007/978-981-15-3215-3_12

using several pathways. Wired and wireless networks are different in their application areas. The performance of the both of these networks is investigated on the basis of common parameters to know how both of this network behave. The network configuration performance is measured with a computer simulation environment.

2 Literature Survey

The wireless communication on terahertz (THz) indoor channels is reviewed in this paper. The 0.1–10 THz wireless transmission physical mechanisms are extremely molecular absorption and loss of streaming, resulting in very high and frequency-selective path loss for a line of sight (LOS) connection [1, 2]. Depending on shape, material, and surface roughness, wave propagation from THz affects very high degradation of reflection for non-line of sight (NLOS). Taking account of these THz radiation characteristics and using a scattered ray tracing method, a new deterministic channel equivalent model is developed which is responsible for both LOS and NLOS propagation cases [3, 4]. In addition, the proposed model's channel capacity will be examined. Results of simulation show that data rates in the order of terabit a second are obtained for distances of up to 1 m (Tbps) with a transmetric power of 1 W. In addition, NLOS only has a capacity of approximately 100 gigabit-per-second (Gbps). These results motivate the development of future THz wireless systems [5, 6].

In this paper, researchers address the problem of wireless ultra-wideband (UWB), which is necessary to increase the accuracy of applications for radiology and positioning, wireless line-of-sight (LOS) versus non-line-of-sight (NLOS) [7, 8]. A LOS/NLOS probability test approach is used on the basis of exploitation of distinctive channel impulses responses (CIR) statistical characteristics using parameters related to the "skewness" and delay of the root average square (RMS) of the CIR. For the probability densities of the CIR parameters are presented a log-normal fit. Simulation results show that there is measurable difference in their CIR parameters' statistics between different environments (residential, office, outdoor, etc.), which are used to establish the nature of the propagation channels [9]. For most types of environments, correct LOS/NLOS identification channel rates above 90% are demonstrated to be attainable [10]. Further improvements are also achieved by combining statistics on CIR skewness with the delay of RMS [11].

The impact of vehicles as obstructions in the past has been largely neglected in vehicle ad hoc networks (VANETs) [12]. Recent research has shown that additional 10–20 dB losses can be caused by vehicles obstructing the line-of-sight (LOS) path, thus reducing the intercourse [13, 14]. The impact of LOS obstacles in VANET simulations is not being modeled for most traffic mobility models (TMMs) today. This paper analyzes a road scenario for the LOS obstruction caused by other vehicles [15]. First, the model following a car characterizes the movement of cars on a two-lane highway in the same direction [16]. If necessary, vehicles may change lanes.

In accordance with car rules and the lane-changing rules on forward motion, the position of each vehicle is updated. For VANET simulations based on simulated traffic, a simple TMM is provided, which can identify vehicles in the shadow of other vehicles [17, 18]. The mobility model presented, together with the shadow fading track loss model, can take into account the impact of the LOS block in the overall power received multi-lane highway scenarios [19].

Deep ultraviolet (UV) visual blind outdoor non-line communication with different transmitter and receptor geometries up to 100 m range [20]. They recommend and fit the model with a quantitative channel path loss model based on extension measurement. They observe a range-dependent power drop [21, 22]. They compare the single dispersion model and show that the single dispersion hypothesis results in a model not precise for small apex angles. Their model will then be used to study basic interactions between the transmitted optical power, range, connection geometry, data rate, and bit rate. Bound detection performance is taken into account for both weak and strong solar background radiation scenarios. These findings provide guidance for the design of the system.

Simulation was an important method for the evaluation of vehicle ad hoc network (VANET) applications and protocols [23]. Such simulations often work at the connection level with simulation time far from wall time, making it difficult to deploy real hardware analysis in hardware-in-the-loop (HIL) environments or prototypes. Probabilistic models of communication can help, but often suffer from much less accuracy than simulations at connection level [24]. It has been shown that the quality of probabilistic communication models greatly increases the determination that a communication link is of a line of sight (LOS) or non-line of sight (NLOS).

This paper presents a LOS model of VANETs in public and private environments. They find that typical urban environments like rural, urban, and industrial areas have similar characteristics in various cities in Germany and in Europe. They use this to derive a probabilistic LOS model that reliably predicts and performs associated work [25]. Therefore, our model makes it possible to develop accurate package arrival rates models in real-time environment to study VANET technology.

Networks of line-of-sight are a network model lately implemented by Frieze et al. It sees wireless networks where there are a big number of barriers in the underlying setting and communication can only take place between objects that are near to each other in space and are in line of sight [26]. In order to capture the primary characteristics of this model, Frieze et al. suggested a fresh model of random networks in which nodes are randomly positioned on a grid and a node can interact with all nodes at a maximum set range r in the same row or column. In the classical ad hoc radio communication model adapted to random line-of-sight networks, we present effective algorithms for two basic communication issues of broadcasting and gossiping.

3 Methodology

The line of sight in mobile ad hoc networks with various scenarios are elaborated in this section. The methodology can be depicted in the following modules.

3.1 Two Host Communicative Wirelessly

In first step, create a network with two hosts and a host is able to wirelessly send an UDP data stream as shown in Fig. 1. Our objective is to make the lower-layer model protocol in the physical layer as simple as possible. A pattern contains a 600 × 750 m sized playground with a space of 500 meters for two hosts. In addition to the host, the modules present in the network are responsible for tasks such as display, IP layer setup and physical radio modeling. Host typically in INET is the NED-type standard host, a generic TCP/IP host template. It contains UDP, TCP, IP, application model plugging slots, and different network interfaces.

An Ipv4NetworkConfigurator module that is displayed as the configurator subsidiary module in the network assigns IP addresses to hosts. The hosts must be aware of the MAC addresses of each other to communicate, which are managed by using GlobalArp modules per host instead of real ARP within this model. Host A generates UDP packets in the model that the host B receives. To this end, host A is set to contain an UdpBasicApp module that generates 1000-byte UDP messages with exponential distribution, averaging 12 ms at random intervals. The app will therefore generate 100 kbps UDP traffic and not count overhead protocols. Host B has an application UdpSink which only discards packets received.

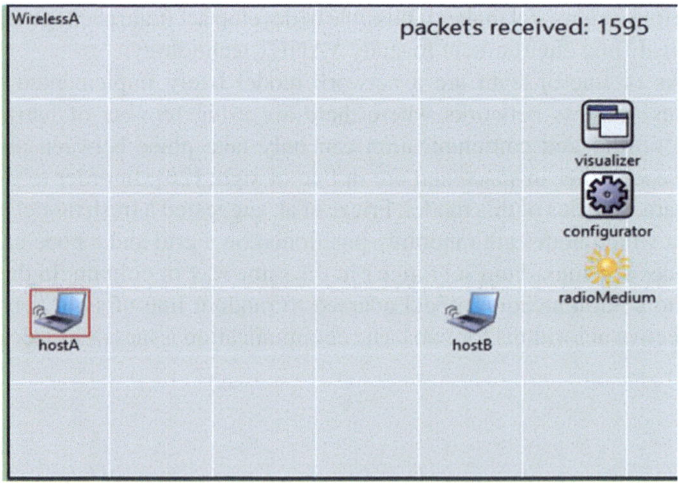

Fig. 1 Two hosts communicating in wireless network

Then focus on the radioMedium module. A radio medium module is needed for every wireless simulation in INET. This module is the common physical medium for communication. It takes into account signal propagation, attenuation, interference, and other physical phenomena. The simplest model UnitDiskRadioMedium is used. It applies a variation of the disk radio unit, so that physical phenomena such as signal attenuation are ignored and the range of communication is simply defined in meters. In-range transmissions are always received correctly, unless there are collisions.

3.2 Adding More Nodes and Decreasing the Communication Range

In this step, model is converted into an ad hoc network and routing experiment. Two original hosts A and host B can communicate with each other directly, and add another three more wireless nodes and reduce the range of communication. By reducing the all hosts to 250 m in communication and their distance is 400 m, so direct communication between two hosts A and B is not possible. The additional hosts are located in the right positions for data transmission from two hosts, A and B, but routing is still not set up. It is therefore impossible to communicate with hosts A and B.

3.3 Establishment of Static Routing

To set routing for packets to flow from host A to host B in this step. The intermediate nodes must act as routers at this stage. IPv4 transmission must be allowed for the recently added hosts to act as routers. The Ipv4NetworkConfigurator module provides static IPv4 configuration, with address assignment and routes adding. The configurator assigns IP addresses within the range of 10.0.0.x and creates routes based on estimated link error rates. NetworkRouteVisualizer module for the visualizer submodule that is capable of rendering packet paths in this network. This module shows paths where a packet was recently sent between the two end hosts network layers. The path is shown to the visited host as a colored arrow. The path continues to disappear and after some time unless it is strengthened by a different packet. Then set the disk radio interference range to 500 m, twice the communication range. In the UnitDiskRadio receiver section, set the ignoreInterference parameter to false to enable interference modeling. Interference range is the UnitDiskRadio transmitter part interference range parameter set to 500 m.

Then, activate the recognition by triggering CsmaCaMac's useAcks parameter. The change on the recipient side is quite simple: if the MAC receives a data frame addressed correctly, it answers after a fixed longitudinal gap (SIFS) with an ACK frame. If ACK is not received properly by the originator of the dataset within the due time, a transmission is initiated.

3.4 Power Consumption

Hosts contain an energy storage component which models a source of energy like a battery or a power supply. In the energyStorageType parameter of the host, INET contains several models to store energy. The energy consumption model of the radio is also preconfigured to draw energy from the energy storage system of the host. (Are also possible hosts with more than one energy storage component).

In this model, IdealEpEnergyStorage in hosts is used. IdealEpEnergyStorage offers an endless amount of energy that cannot be fully loaded or exhausted. By using IdealEpEnergyStorage, it is not only focused on storage, but also on energy consumption.

.host.wlan[0].radio.energyConsumer.typename = "StateBasedEpEnergyConsumer"
.host.wlan[0].radio.energyConsumer.receiverReceivingPowerConsumption = 10mW
.host.wlan[0].radio.energyConsumer.transmitterIdlePowerConsumption = 2mW
.host.wlan[0].radio.energyConsumer.transmitterTransmittingPowerConsumption = 100mW
.host.energyStorage.typename = "IdealEpEnergyStorage".

3.5 Configuring Node Movements

The mobility submodule of hosts under the INET Framework manages node mobility. There are various types of mobility module that can be connected to a host. Here, Linear Mobility is installed in the intermediate nodes. LinearMobility implements movement along a line with parameters of heading and speed. Set the nodes to move north at a speed of 12 m/s.

.hostR.mobility.typename = "LinearMobility"
.hostR.mobility.speed = 12 mps
.hostR.mobility.initialMovementHeading = 270 deg.

3.6 Configuring ad hoc Routing (AODV)

The hosts are converted into AodvRouter instances. AodvRouter is like Wi-Fi, but with an additional submodule AodvRouting. Every node becomes an AODV router. The Ad hoc Distance Vector stands for AODV. Routes are laid down as needed in AODV. When a route has been established, it is kept as long as necessary.

The network silences in AODV until a connection is necessary. The network node requiring a connection at that point transmits a connection request. This message is forwarded to other AODV nodes, which record the node they heard from and create a temporary track explosion back to the needed node. If a node receives a message and has a route to the desired node, a message on a temporary route will be sent back to the requested node. The need node then begins with the route via other nodes with the least hops. Unused entries are recycled after a while in the routing tables.

AODV defines Route Request (RREQ), Route Response (RERP), and Route error (RERRs). Messaging types AODV defines.

*.configurator.addStaticRoutes = false
.host.typename = "AodvRouter"
*.hostB.wlan[0].radio.displayCommunicationRange = true
.visualizer.dataLinkVisualizer.packetFilter = "AODV".

3.7 Adding Obstacles to the Environment

Objects such as walls, trees, buildings, and hills are in reality an obstacle to the spread of radio signals. They absorb and reflect radio waves, lower the quality of the signal and reduce the risk of success. In this step, a concrete wall is fitted in between the hosts A and R1. Use of the perfect radio and wireless medium models that do not allow physical phenomena, modeling obstacles will be very simple: every barrier absorbs radio signals completely, making reception impossible behind them.

Obstacles are described in an XML file. An obstacle is defined by its shape, location, orientation, and material. The XML format allows us to use a pre-defined form like a cuboid, prism, a polyhedron or a globe, and also define new shapes that are used for any number of obstacles. This may also have a name and can define how it should be rendered (color, line width, opacity…) similar to materials: pre-defined materials are available such as concrete, brick, wood, glass, and new materials can also be defined. The physical characteristics of a material such as resistivity, relative permittivity, and relative permeability are defined. These characteristics are used in dielectric loss tangent estimations, refractive index, and signal propagating speeds, and ultimately in signal loss estimations.

3.8 Changing to a More Realistic Radio Model

In this step, UnitDiskRadio will be replaced by ApskScalarRadio. ApskScalarRadio designs a radio that uses the modulation scheme amplitude and phase-shift keying (APSK). It uses BPSK by default, but it can also configure QPSK, QAM-16, QAM-64, QAM-256, and many other modulations. The 'disk unit' type of abstraction, defines the carrier frequency, signal bandwidth, and the transmission capacity of the radios. (Modulation is the transmitter component parameter.) They will allow the radio station and receiver models, along with other basic parameters, to calculate the loss in track, SNIR, bit error rates, and other values and eventually determine the recipient's success. ApskScalarRadio also provides realism by simulating a preamble and a physical layer header preceding the data. Its lengths are parameters too (and if not required, they can be set to zero).

3.9 Configuring a More Accurate Path Loss Model

Figure 2 shows the medium uses a model for free-space loss of paths that assumes a line of sight path without causing any obstacles in close proximity.

As wireless hosts go on the ground, the two-ray floor reflection model calculating with a reflection from the ground would be a more accurate path loss model. FlatGround is used as ground model to enter into the physical environment module. The height of the floor is FlatGround's height parameter. The parameter is set to 0 m.

*.physicalEnvironment.ground.typename = "FlatGround"

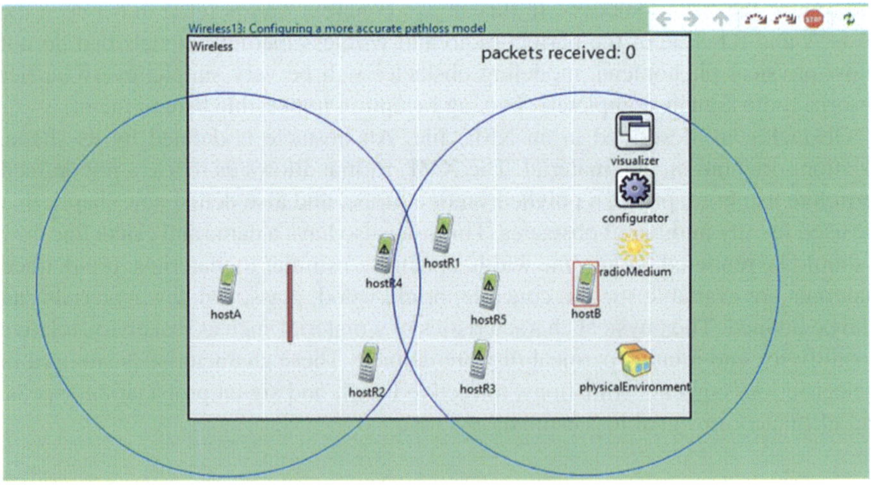

Fig. 2 Configuring a more accurate path loss model

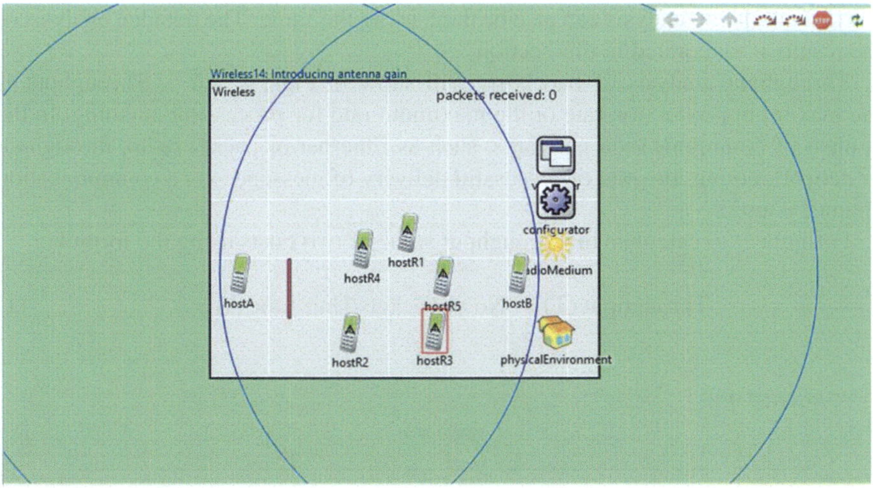

Fig. 3 Introducing antenna gain

*.physicalEnvironment.ground.elevation = 0
*.radioMedium.pathLoss.typename = "TwoRayGroundReflection".

3.10 Introducing Antenna Gain

In the previous phases, a radio antenna with an increase of 1 (0 dB) is used. To improve the simulation, ConstantGainAntenna is used to increasing the antenna gain.

Here, configure the hosts ConstantGainAntenna is an abstraction is shown in Fig. 3. An antenna that has a constant gain in the directions relevant for simulation, no matter what the real-life application of such antenna may take. For example, if all nodes on the same plane on a simulation wireless network, ConstantGainAntenna could be equivalent to an omnidirectional dipole antenna.

4 Result Analysis

Here, chosen the version 5.4 of OMNeT++ (Objective Modular Network Testbed). The INET Framework for OMNeT++ Object-oriented modular event simulation network framework with the 4.1.0-ae90ecd release. Using OMNeT++, Communication networks and other distributed systems simulation is performed INET consists of a number of application models for simulations. It is an open-source tool which follows the fundamental language of object-oriented language called C++. The trace file and graphical analysis can also be done using the same. The detailed analysis of the results is elaborated in this section.

Throughput analysis of two hosts will show in Figs. 4 and 5, Throughput is the maximum production rate or the maximum rate for processing anything. In the context of communication networks, such as Ethernet or packet radio, throughput or network output, the rate of successful delivery of messages via a communication channel is used.

Formula 1, to calculate the throughput value of two hosts using the formula,

$$\text{Throughput (T)} = \text{No of packets/Unit time (kbps)}. \tag{1}$$

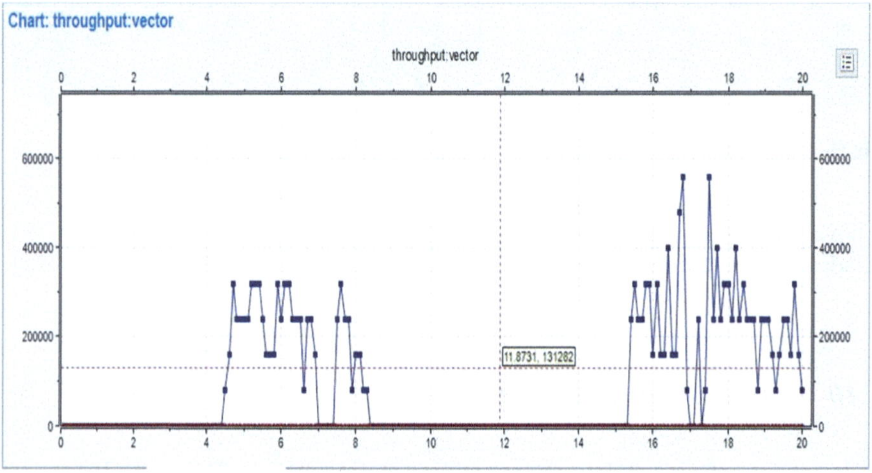

Fig. 4 Throughput vector graph of two hosts when obstacles between them

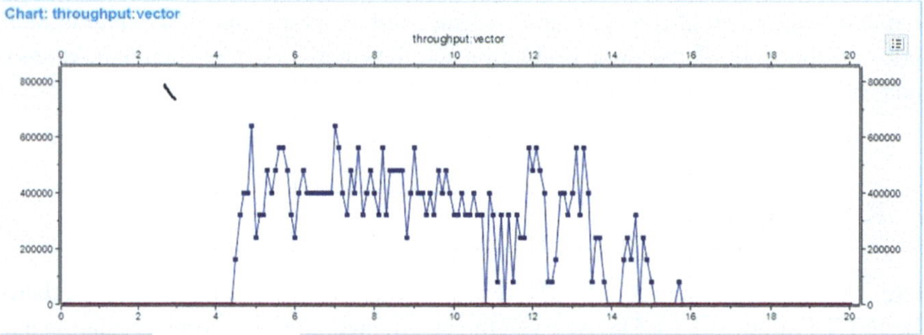

Fig. 5 Throughput vector for antenna gain for line of sight communication

Figures 6 and 7 depict the end-to-end delay, it refers to the time to it takes to transmit a packet through a network. It is a common term in IP network monitoring and differs in that only the path from source to destination in one direction is measured from round-trip time.

Figures 8 and 9 show the power consumption of two hosts in terms of both consumption and capacity growth, the rate of growth of the internet means that it is not likely to achieve a realistic objective to actually cut its overall energy consumption. Energy efficiency refers to the amount of data that can be transmitted by the energy consumed by the network from one end to another.

Figures 10 and 11 depict the transmission state of the hosts in terms when the transmission of data is the transmission of digital or analog data to one or more electronical devices via a communication medium. It enables devices in the point-by-point, point-by-multipoint, and multipoint environment to be transferred and communicated. The transmission of data is also known as digital or digital transmission.

For each packet transmitted and each packet received, a node loses a particular amount of energy. The initial energy value is therefore reduced in a node. After receiving or transmitting packets the current energy value in a node is the residual energy.

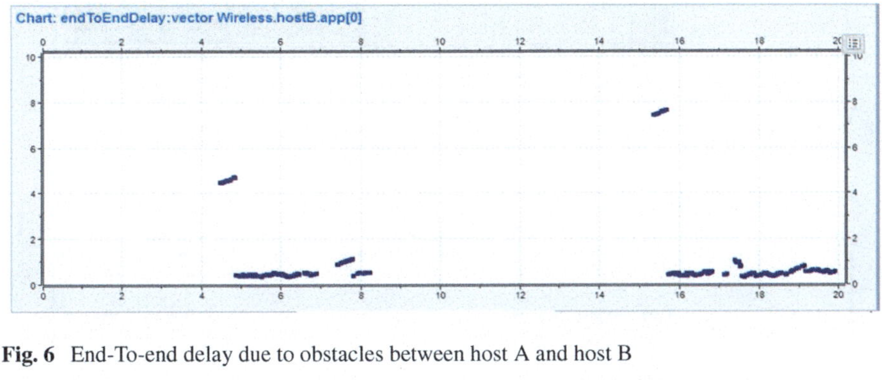

Fig. 6 End-To-end delay due to obstacles between host A and host B

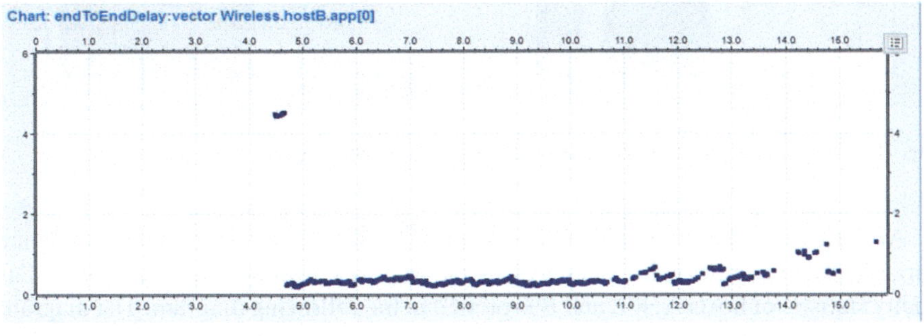

Fig. 7 End-To-end delay due to obstacles between host A and host B after antenna gain

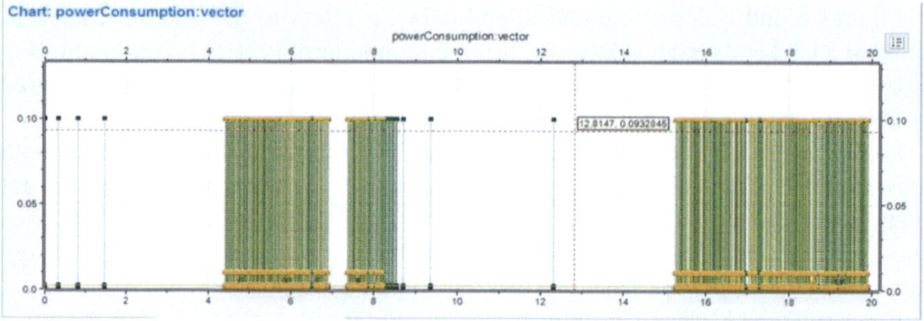

Fig. 8 Total power consumption of host A and host B due to obstacles between them

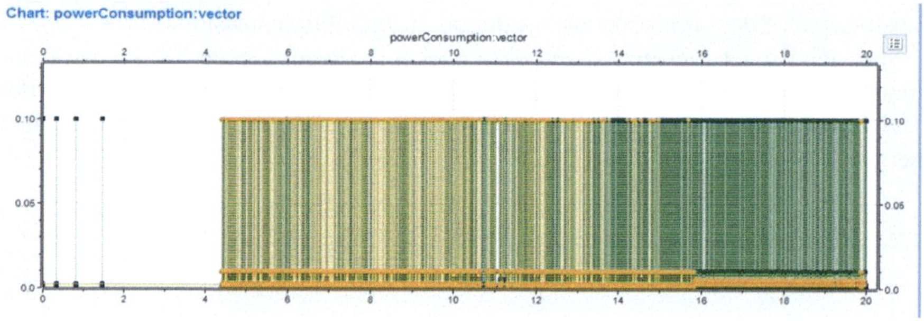

Fig. 9 Total power consumption of host A and host B after the antenna gain

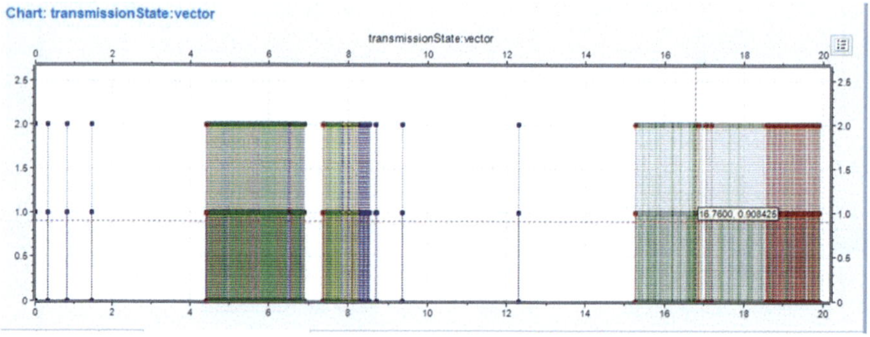

Fig. 10 Transmission state of hosts, when obstacles between them

At the end of the simulation, Figs. 12 and 13 show the host an energyBalance variable. The negative value of energy means energy consumption. The residualCapacity statistic of hosts A, R1, and B is plotted in the following diagram. The diagram shows that host A has consumed the most power because it transmitted more than the other nodes.

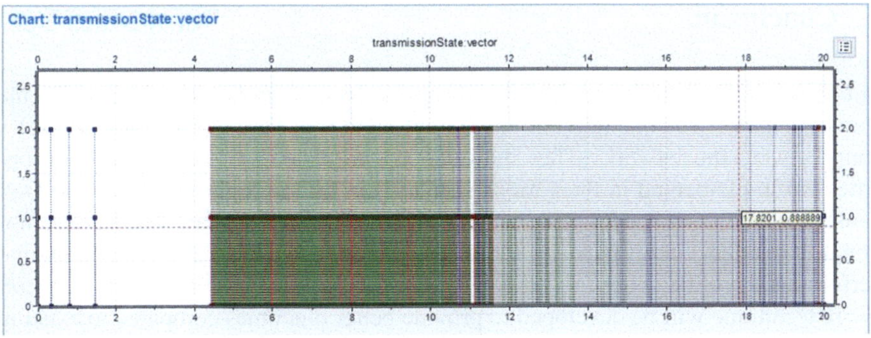

Fig. 11 Transmission state of hosts, when obstacles between them after increase antenna gain

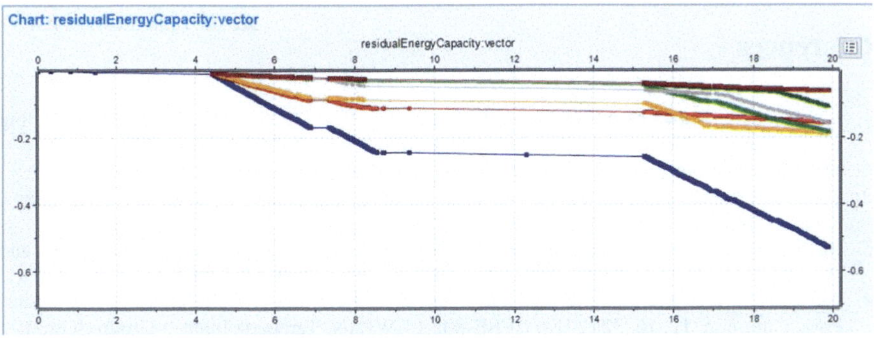

Fig. 12 Residual energy capacity of hosts when pass loss of hosts due to obstacles

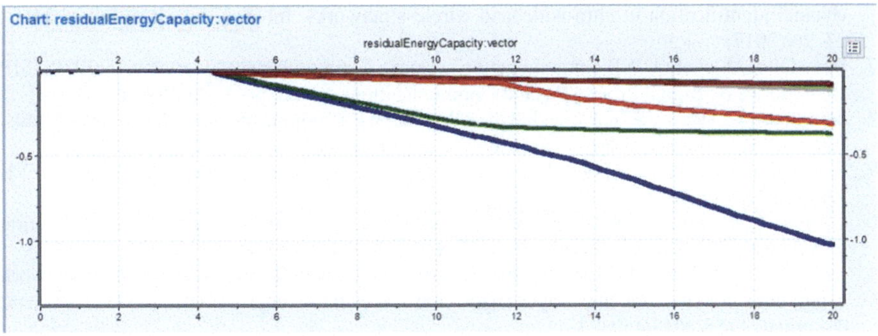

Fig. 13 Residual energy capacity of hosts when pass loss of hosts due to obstacles after increase the antenna gain

5 Conclusion

In this paper, they analyzed the line of sight in mobile ad hoc network. From the above scenario, we conclude that the line of sight will be lower if more obstacles come between the networks. They showed every problem and how the network properties are implemented in the various scenarios in line-of-sight problem. The QoS parameters such as throughput, transmission rate, end-to-end delay, power consumption, and residual energy capacity for the evaluation of line of sight. It is evident that performance will be better in a case here obstacles will be less. In the future, a LoS aware technique will be developed to provide better outcomes compared to existing methods for live multimedia transmission.

References

1. Moldovan, A., Ruder, M.A., Akyildiz, I.F., Gerstacker, W.H.: LOS and NLOS channel modeling for terahertz wireless communication with scattered rays. In: Globecom 2014 Workshop—Mobile Communications in Higher Frequency Bands, pp. 386–392
2. Song, H., Nagatsuma, T.: Present and future of terahertz communications. IEEE Trans. Terahertz Sci. Technol. **1**(1), 256–263
3. Rappaport, T., Murdock, J., Gutierrez, F.: State of the art in 60-GHz integrated circuits and systems for wireless communications. Proc. IEEE **99**(8), 1390–1436 (2011)
4. Akyildiz, I., Jornet, J., Han, C.: Terahertz band: next frontier for wireless communications. Phys. Commun. **12**, 16–32 (2014) (IEEE-802.15-WPAN, Terahertz Interest Group (IGthz))
5. Jornet, J., Akyildiz, I.: Channel modelling and capacity analysis for electromagnetic wireless Nano networks in the terahertz band. IEEE Trans.
6. Landolsi, M.A., Almutairi, A.F.: Reliable line-of-sight and non-line-of-sight propagation ratio channel identification in ultra-wideband wireless networks. Int. Sch. Sci. Res. Innov. **11**(1), 23–26 (2017)
7. Benedetto, M. et al.: UWB communication systems: a comprehensive overview. In: EURASIP Book Series on Signal Processing and Communications, vol. 5, pp. 5–25 (2006)
8. Sahinoglu, Z., Gezici, S., Guvenc, I.: Ultra-Wideband Positioning Systems: Theoretical Limits, Ranging Algorithms and Protocols. Cambridge University Press, Cambridge (2008)
9. Soganci, H., Gezici, S., Poor, H.: Accurate positioning in ultra-wideband systems. IEEE Wireless Comm. **18**(2), 19–27 (2011)
10. Guvenc, I., Chong, C.: A survey on TOA-based wireless localization and NLOS mitigation techniques. IEEE Comm. Surveys Tutorials **11**(3), 107–124 (2009)
11. Abbas, T., Tufvesson, F.: Line-of-sight obstruction analysis for vehicle-to-vehicle network simulations in a two-lane highway scenario. Int. J. Antennas Propag. (2013) [459323]. https://doi.org/10.1155/2013/459323
12. Gozalvez, J., Sepulcre, M., Bauza, R.: Impact of the radio channel modeling on the performance of VANET communication protocols. Telecommun. Syst. 1–19 (2010)
13. Abbas, T., Karedal, J., Tufvesson, F.: Measurement-based analysis: the effect of complementary antennas and diversity on vehicle-to-vehicle communication. IEEE Antennas Wirel. Propag. Lett. **12**(1), 309–312 (2013)
14. Boban, M., Vinhoza, T., Ferreira, M., Barros, J., Tonguz, O.: Impact of vehicles as obstacles in vehicular ad hoc networks. IEEE J. Sel. Areas Commun. **29**(1), 15–28 (2011)
15. Meireles, R., Boban, M., Steenkiste, P., Tonguz, O., Barros, J.: Experimental study on the impact of vehicular obstructions in VANETs. In: 2010 IEEE Vehicular Networking Conference (VNC), pp. 338–345 (Dec. 2010)

16. Varga, A., Hornig, R.: An overview of the OMNeT++ simulation environment. In: Proceedings of the 1st International Conference on Simulation Tools and Techniques for Communications, Networks and Systems & Workshops, ser. Simutools '08. ICST, Brussels, Belgium, Belgium: ICST (Institute for Computer Sciences, Social Informatics and Telecommunications Engineering), pp. 60:1–60:10 (2008) [Online]
17. https://dl.acm.org/doi/10.5555/1416222.1416290
18. Chen, G., Xu, Z., Ding, H., Sadler, B.M.: Path loss modelling and performance trade-off study for short-range non-line-of-sight ultraviolet communications. EURASI J. Wireless Commun. Netw. (Dec. 2010)
19. Henderson, T.R., Roy, S., Floyd, S., Riley, G.F.: ns-3 project goals. In: Proceeding from the 2006 Workshop on ns-2: The IP Network Simulator, ser. WNS2 '06. ACM, New York, NY (2006)
20. Stadler, C., Gruber, T., Germanz, R., Eckhoffz, D.: A line-of-sight probability model for VANETs (2017)
21. Fernandes, P., Nunes, U.: Platooning with IVC-enabled autonomous vehicles: strategies to mitigate communication delays, improve safety and traffic flow. IEEE Trans. Intell. Transp. Syst. **13**(1), 91–106 (2012)
22. Eckhoff, D., Sommer, C.: Simulative performance evaluation of vehicular networks. In: Chen, W. (ed.) Vehicular Communications and Networks: Architectures, Protocols, Operation and Deployment, pp. 255–274. Elsevier, Amsterdam (2015)
23. On-Board System Requirements for V2V Safety Communications, Society of Automotive Engineers Std. SAE J2945/1 (2016)
24. Davis, D., Enameller, P.: System and method for providing simulated hardware-in-the-loop testing of wireless communications networks. November 2006, US Patent 7,136,587
25. Stepanov, I., Rothermel, K.: On the impact of a more realistic physical layer on MANET simulations results. Ad Hoc Netw. **6**(1), 61–78 (2008). (Elsevier)
26. Czumaj, A., Wang, X.: International Symposium on Stochastic Algorithms SAGA 2007: Stochastic Algorithms: Foundations and Applications, pp. 70–81

Activeness Based Propagation Probability Initializer for Finding Information Diffusion in Social Network

Ameya Mithagari and Radha Shankarmani

Abstract Information diffusion is the process of spreading information from one node to another over the network. To calculate the information diffusion coverage, it is important to assign propagation probability to every edge in the social network graph. Most popular models of information diffusion use Uniform Activation (UA) and Degree Weighted Activation (DWA) to calculate propagation probabilities. However, the results obtained by these methods are non-realistic. Therefore, we propose a new Activeness based Propagation Probability Initializer (APPI) model to obtain realistic information diffusion. This is achieved by assigning propagation probabilities based on activeness value inferred using topological node behavior. The experimental results show that APPI provides balanced and meaningful information diffusion coverage when compared with UA and DWA.

Keywords Diffusion · Cascade · Topological · Social network

1 Introduction

Online social networks are rapidly evolving these days, thereby increasing the market base for any new product. The process of spreading of information, ideas, views, product promotions, advertisements, etc. over a network is called Diffusion. For example, information about a product getting viral over the network can be termed as information diffusion in the network [1].

Initially, only a few people accept the product or an idea. As people observe that their neighbors have accepted the new product or an idea, they tend to accept it. As a result, a cascade of information is formed, thereby generating an indirect recommendation system. Hence information diffusion in the social network is studied by many researchers [2].

A. Mithagari (✉) · R. Shankarmani
Sardar Patel Institute of Technology, Andheri, India
e-mail: ameya.mithagari@spit.ac.in

D. Dutta and B. Mahanty (eds.), *Numerical Optimization in Engineering and Sciences*, Advances in Intelligent Systems and Computing 979,
https://doi.org/10.1007/978-981-15-3215-3_13

In this work, a novel model is proposed, named 'Activeness based Propagation Probability Initializer' for a temporal network. A temporal graph is a snapshot of the network taken at regular intervals of time and is represented as a graph. The time period can be monthly, quarterly, semi-annually or annually.

The paper is organized as follows. In the background section, we discuss various methods of information diffusion and the research gap in the algorithms to assign propagation probabilities. The next section consists of a proposed model for assigning propagation probabilities based on activeness. Experimental results and comparison with various methods are discussed in Sect. 4. Finally, the insights obtained by the research are summarized in the Conclusion.

2 Background

The process of acceptance of a product or an idea is modeled by the following diffusion models:

Liner Threshold model (LT): At every iteration, the node accepts the product or an idea if 'sum of the weights of connected edges to its neighbors who have already accepted the product or an idea' crosses a certain 'threshold (t)' [3].

Independent cascade model (ICM): At every iteration, The user (u) accepts the product or an idea with 'Propagation probability ($p_{u,v}$), every time one of it's neighbor (v) accepts the product or an idea. When no node accepts the product the iteration stops [4].

Propagation probability of an edge is the probability by which one node can affect the other node along the edge. Therefore, success of acceptance of the product or an idea depends on propagation probability. The propagation probability is assigned by the following two universally accepted models [5–7]:

Uniform Activation (UA): Same propagation probability is assigned to all the edges.

Degree Weighted Activation (DWA): Propagation probability assigned to edge (u, v) is the reciprocal of the number of neighbors of node v.

2.1 Research Problem

Proper justification to assign propagation probability is not provided by both UA and DWA. Different results can be obtained for different propagation probabilities when UA is considered. In DWA, more the number of neighbors, lesser is the chance of the node accepting the product or an idea. For example, if there is only one neighbor of a node v, then the propagation probability of the edge (u, v) is calculated as 1. This means the node will surely accept the product or an idea. However, this conclusion is incorrect. To obtain realistic information diffusion, we propose a model that uses node topological behavior to initialize propagation probability.

3 Activeness Based Propagation Probability Initializer (APPI)

The input to Activeness based Propagation Probability Initializer (APPI) are snapshot of same network taken at different time instances represented as Graph G with vertices V and edges E, $G(V, E)$, $G_1(V_1, E_1)$, $G_2(V_2, E_2)$ to $G_n(V_n, E_n)$.

Following are the assumptions to deploy the model:

- Unweighted graphs are considered.
- Deleted accounts are represented as isolated nodes.

3.1 Activeness Value Finder

The two sequential snapshots, previous graph G_{i-1} and current graph G_i where i ranges from 1 to n are considered. Then the 'lists of neighbors' for each node in both the graphs are generated simultaneously. Comparing these lists per each node, below conditions are evaluated for inferring Activeness value of the node based on the node's topological behavior:

- *Increase in the number of neighbors*: Set Activeness value of the node to the total number of newly formed connections of the node.
- *No change in the number of neighbors*:
  - If elements in the list of neighbors changes, then Activeness value is set to the total number of changed connection.
  - If there is no change, Activeness value is set to zero.
- *Decrease in number of neighbors*: If the total number of lost connections is greater than 'average degree of the graph', then Activeness value is set to the total number of lost connections else Activeness value is set to zero because over the specified time range the node may be inactive then the loss of connections are due to other users.
- *New users in the network*: The Activeness value is exaggerated due to newly formed connections. Therefore, to compensate for that, the Activeness value is reduced by 75% and rounded off to the next whole number as they can be active nodes but the trust value won't be developed at the initial stages of the node in the network.

The change of neighbors may be negative or positive but the activeness value is always positive. The activeness value is always a whole number.

3.2 *Propagation Probability Initializer*

The propagation probability $p_{u,v}$ is assigned to every edge (u, v) in the current graph G_i. Activeness value of both the nodes u and v connected to an edge e is considered in assigning $p_{u,v}$. Below cases are considered to initialize the propagation probability $p_{u,v}$.

- *If the activeness of node u or node v* ≤ 1 :
  The propagation probability $p_{u,v}$ for edge (u, v) is set to 0.01.
- *If the activeness of node u and node v* > 1:
  The propagation probability $p_{u,v}$ for edge (u, v) is calculated as below:

$$p_{u,v} = \left(1 - \frac{1}{\text{Activeness}(u)}\right) * \left(1 - \frac{1}{\text{Activeness}(v)}\right) - \epsilon \tag{1}$$

The propagation probability $p_{u,v}$ is directly proportional to activeness of both the nodes u and v. A node may be active or inactive and it may or may not spread the information further hence the propagation probability is 0.5. We define ϵ as the limiting factor ranging from 0.20 to 0.24 because even if we are assuming lower probability range by subtracting ϵ, higher probabilities are obtained only if both the nodes connected to the edge are highly active. Thereby generating an upper bound to the propagation probabilities. Therefore $p_{u,v}$ ranges from 0.01 and tends to 0.8 depending upon the ϵ value.

4 Experimental Results and Discussion

4.1 *Implementation*

Different snapshot graphs of the same network at different time interval are input to the model. At a time only two graphs are considered: the previous snapshot graph and the current snapshot graph. A page rank algorithm is applied to the current snapshot graph to obtain rank value. Top k nodes based on rank value are selected as the seed set. These are the influential nodes considered for starting the diffusion process using ICM. Before that, the propagation probabilities of edges in the current snapshot graph are assigned. We compare our model (APPI) with the following propagation probability assignment algorithms: Uniform Activation (UA) and Degree Weighted Activation (DWA).

Propagation Probability assignment: A node may be active or inactive and it may or may not spread the information further hence the propagation probability assigned to every edge using UA is 0.5. According to DWA, reciprocal of the degree of the node is calculated and assigned as propagation probability to incoming edges of that node. The conditions discussed in Sects. 3.1 and 3.2 are incorporated to form an

APPI algorithm which assigns propagation probability to every edge in the current snapshot graph. The limiting factor (ϵ) is set to 0.2.

The information diffusion coverage is considered for evaluation which is obtained by Independent Cascade Model (ICM) after assigning propagation probabilities to the edges in the current snapshot graph by each algorithm separately.

4.2 Results for Synthetic Network

Dataset used are two randomly generated synthetic graphs. One of them is considered as previous snapshot graph with 10 nodes and 24 edges and the other as current snapshot graph with 10 nodes and 29 edges. The experiment is conducted according to the procedure in Sect. 4.1 and the seed set size is considered as 2. The diffusion always starts from the influential nodes found by page rank algorithm which are node '4' and node '9' in current snapshot graph. Green nodes depicts the information diffusion. Results depicted in Fig. 1 shows over-estimation of diffusion when UA is used with ICM. Showcasing higher uncertainty of results. Signifying UA as more of an optimistic approach.

Results depicted in Fig. 2 shows under-estimation of information diffusion when DWA is used with ICM, signifying DWA as more of a conservative approach.

Following are the Activeness values of nodes that are calculated using Activeness value finder discussed in Sect. 3.1: {'0': 2, '1': 3, '2': 0, '3': 0, '4': 5, '5': 3, '6': 3,

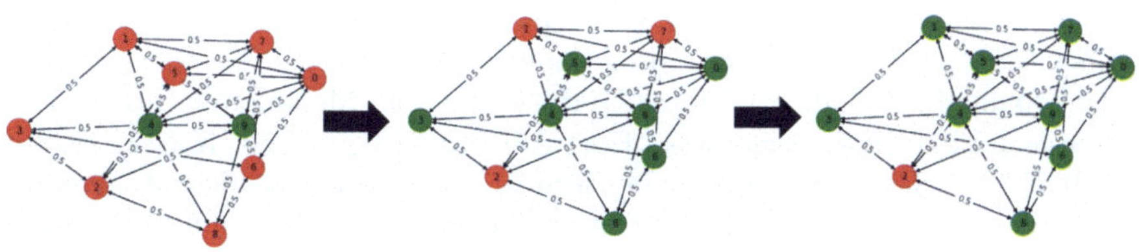

Fig. 1 Information diffusion in different iterations of ICM when UA is used to assign Propagation probabilities

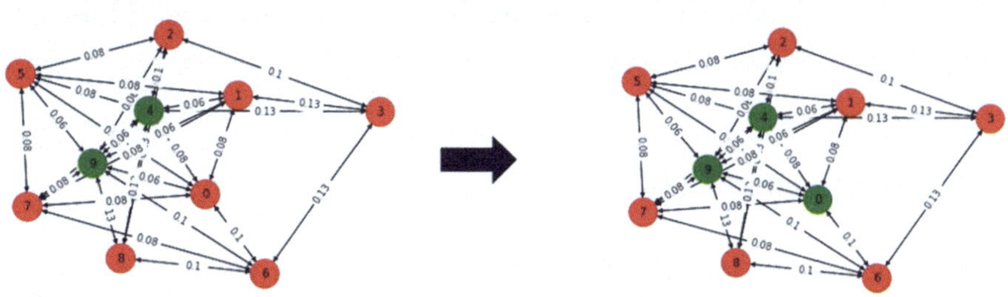

Fig. 2 Information diffusion in different iterations of ICM when DWA is used to assign Propagation probabilities

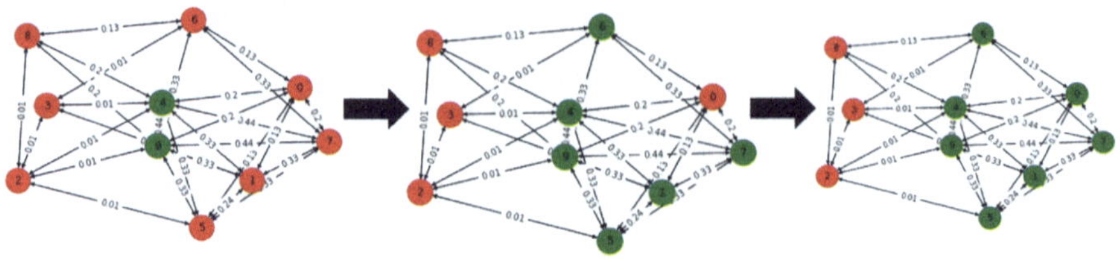

Fig. 3 Information diffusion in different iterations of ICM when our model (APPI) is used to assign propagation probabilities

'7': 5, '8':2, '9': 5}. The Propagation probabilities of edges are calculated according to Sect. 3.2, following is an example, the activeness value of node '5' and node '9' is 3 and 5 respectively therefore, propagation probability $p_{5,9}$ of edge ('5', '9') is calculated using Eq. (1) $p_{5,9} = (1 - \frac{1}{3}) * (1 - \frac{1}{5}) - 0.2 = 0.33$. Taking activeness into consideration, we are obtaining de-escalated propagation probabilities for active nodes by subtracting ϵ i.e 0.2. Hence the confidence of result is increased. Results depicted in Fig. 3 shows balanced and meaningful information diffusion when APPI is used with ICM.

4.3 Real-World Network

Real world Facebook data-set archived in The Max Planck institute for software systems [8] is considered. Initially the graphs are developed in GraphML format in an interval of 3 months using time-stamp of the activities. Number of nodes in 1st Month, 4th Month, 7th Month, 10th Month, 13th Month, 16th Month, 19th Month and 22nd Month snapshots of network are 9101, 12,360, 15,632, 20,090, 25,301, 30,994 and 37,769 respectively. Similar experiment is conducted according to the procedure in Sect. 4.1 with the facebook dataset and 25 most influential nodes found using page rank are considered as seed set. The results are depicted in Fig. 4. The results using UA indicates over estimation of Information diffusion whereas the results using DWA indicates under estimation of information diffusion. Results using APPI (proposed model) shows a balanced and meaningful information diffusion in real-world network.

5 Conclusion and Future Work

Independent Cascade Model (ICM) is generally used for testing information diffusion in Influence Maximization problem. However, the standard algorithms such as UA and DWA used with ICM does not provide justified and realistic results. This problem is efficiently solved using the proposed Activeness based Propagation Probability

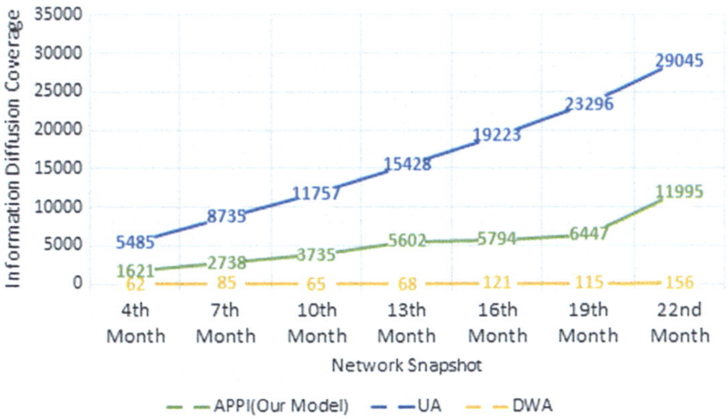

Fig. 4 Information diffusion comparison results

Initializer (APPI) model. Since this model calculates the propagation probability of every edge in the graph based on the activeness of its connected nodes, the generated results have good confidence. Both the experiment results convey that APPI model gives balanced and meaningful information diffusion because it is based on topological node behavior which is not the case with earlier existing models. The future work is to calculate the confidence value and support of the propagation probabilities assigned by APPI.

References

1. Leskovec, J., Adamic, L.A., Huberman, B.A.: The dynamics of viral marketing. In: ACM Transactions on the Web (TWEB). ACM, New York, NY (2007)
2. Rogers, E.M.: Diffusion of Innovations, 4th edn. Free Press, New York (1995)
3. Granovetter, M.: Threshold models of collective behavior. Am. J. Sociol. **83**(6), 1420–1443 (1978)
4. Kempe, D., Kleinberg, J.M., Tardos, E.: Maximizing the spread of influence through a social network. In: Proceedings of the 9th ACM SIGKDD International Conference on Knowledge Discovery and Data Mining, pp. 137–146. ACM, New York, NY (2003)
5. Zhuang, H., Sun, Y., Tang, J., Zhang, J., Sun, X.: Influence maximization in dynamic social networks. In: 2013 IEEE 13th International Conference on Data Mining, pp. 1313–1318. IEEE, Dallas, TX (2013)
6. Song, G., Li, Y., Chen, X., He, X., Tang, J.: Influential node tracking on dynamic social network: an interchange greedy approach. IEEE Trans. Knowl. Data Eng. **29**(2), 359–372 (2017)
7. Li, W., Bai, Q., Zhang, M.: SIMiner: a stigmergy-based model for mining influential nodes in dynamic social networks. IEEE Trans. Big Data (2018)
8. Social Computing Research Homepage. http://socialnetworks.mpi-sws.org. Last accessed 6 April 2019

Solving Multi-attribute Decision-Making Problems Using Probabilistic Interval-Valued Intuitionistic Hesitant Fuzzy Set and Particle Swarm Optimization

Kajal Kumbhar and Sujit Das

Abstract This paper proposes a multi-attribute decision-making (MADM) approach using probabilistic interval-valued intuitionistic hesitant fuzzy set (P-IVIHFS) and particle swarm optimization (PSO) approach. Here, the assessing values of the alternatives regarding the attributes are given by probabilistic interval-valued intuitionistic hesitant fuzzy values (P-IVIHFVs) and weights of the attributes are given by interval-valued intuitionistic fuzzy values (IVIFVs). Firstly, the proposed method uses a P-IVIHFS-based score and an accuracy function to convert the assessment matrix into transformed assessment matrix. Then, PSO is used to determine the optimal weights of the attributes using the transformed assessment matrix. We use interval-valued intuitionistic hesitant fuzzy weighted geometric (IVIHFWG) operator to aggregate the values of each alternative. Next, the ranking of the alternatives is determined using the transformed values which is computed from the aggregated values of each alternative. Finally, the proposed study is validated using a numerical example.

Keywords Multi-attribute decision-making · P-IVIHFS · PSO · Score function

1 Introduction

In order to solve the real-life complicated decision-making problems, researchers have combined fuzzy sets and its various extensions with the optimization techniques in the last few years, which is observed as a successful investigation in the context of uncertain decision-making paradigm. Extending the idea of fuzzy set [1], intuitionistic fuzzy set (IFS) [2] was introduced, where each element present in the set has a degree of membership and non-membership value in [0,1] which are basically

K. Kumbhar · S. Das (✉)
Department of Computer Science and Engineering, National Institute of Technology Warangal, Warangal 506004, India
e-mail: sujit.das@nitw.ac.in

K. Kumbhar
e-mail: kajalmanda392@gmail.com

D. Dutta and B. Mahanty (eds.), *Numerical Optimization in Engineering and Sciences*, Advances in Intelligent Systems and Computing 979,
https://doi.org/10.1007/978-981-15-3215-3_14

crisp or exact values. To represent these values in interval, interval-valued intuitionistic fuzzy sets (IVIFSs) were proposed by Attanassov and Gargov [3], where the degree of membership and non-membership of each element in the set are given in the form of interval between 0 and 1. Among the many other optimization techniques, the application of particle swarm optimization (PSO) [4] is well known in decision-making problems [5]. Koohi and Groza [6] optimized the PSO algorithm by adjusting PSO parameters to get better results. In [7], a fuzzy optimization model is presented to solve multi-criteria decision-making (MCDM) problems using fuzzy analytic hierarchy process (AHP). Yang and Simon [8] studied a new particle swarm optimization (NPSO) method. Chen and Huang [9] explored a MADM method based on accuracy function of IVIFVs to get accuracy value, IIFWGA operator and PSO to find the optimal weights for the attributes. Chen et al. [10] proposed a new MADM method using IVIFS, evidential reasoning methodology to define objective functions and PSO to compute optimal weights of attributes. Torra and Narukawa [11] introduced hesitant fuzzy set (HFS) for embedding multiple number of membership grades for each element in a set where a number of membership grades may vary for each element, and applied it in decision-making. Rodriguez et al. [12] investigated HFS and studied its future research directions. Wu et al. [13] mentioned probabilistic interval-valued hesitant fuzzy set (P-IVHFS) and defined probabilistic interval-valued hesitant fuzzy operations and probabilistic interval-valued hesitant fuzzy operators. Zhai et al. [14] defined probabilistic interval-valued intuitionistic hesitant fuzzy set (P-IVIHFS) and used it to analyze medical data. Yue et al. [15] introduced new aggregation operators for solving decision-making problems using probability and HFS. As found in the study mentioned above, researchers have used probability with fuzzy sets for the purpose of including an extra parameter to handle more uncertainty in decision-making problems [16–22] and PSO is used to optimize some relevant parameters. However, no study is found combining both P-IVIHFS and PSO, where the benefit of both can be considered.

Motivated by this finding, in this study we propose a decision-making approach combining both P-IVIHFS and PSO, where P-IVIHFS is used to express the values of the alternatives/objects corresponding to the attributes and PSO is used to compute the optimal weight from a given interval of weights. To present and validate the proposed approach, the rest of the paper is structured as follows. A brief review of the relevant ideas associated with the proposed approach is mentioned in Sect. 2. The proposed approach is described in stepwise manner in Sect. 3. Section 4 presents an illustrative numerical. Finally, we conclude in Sect. 5.

2 Preliminaries

This section reviews IVIFS, HFS, interval-valued intuitionistic hesitant fuzzy set (IVIHFS), P-IVIHFS, PSO and some relevant ideas associated with this study. In fuzzy set, the belongingness of an element in a set is simply expressed by membership grades, whereas in IFS, the same is expressed by membership and non-membership grade and the sum of these two grades is equal to one. IVIFS presents these two grades in intervals to consider more impreciseness. In HFS, multiple numbers of membership grades are attached with each element in the set and the number of membership grades may vary for different elements. These multiple numbers of membership grades are presented in intervals when we consider IVHFS. However, in IVIHFS, which is an extension of IVHFS, multiple numbers of membership and non-membership grades are attached with each element which are expressed in intervals.

Definition 1 [3] An IVIFS P in the universal set of X is stated as $P = \{\langle x_i, \mu_p(x_i), \nu_p(x_i)\rangle | x_i \in X\}$, where $\mu_p(x_i) = \left[\mu^l_p(x_i), \mu^u_p(x_i)\right]$ and $\nu_p(x_i) = \left[\nu^l_p(x_i), \nu^u_p(x_i)\right]$, $0 \leq \mu^l_p(x_i) \leq \mu^u_p(x_i) \leq 1$, $0 \leq \nu^l_p(x_i) \leq \nu^u_p(x_i) \leq 1$, $0 \leq \mu^u_p(x_i) + \nu^u_p(x_i) \leq 1$ and $1 \leq i \leq m$. Here, $\mu_p(x_i)$ and $\nu_p(x_i)$ indicate the interval for membership and a non-membership function, respectively. For a given $x \in X$, an object $\left[\mu^l_p(x_i), \mu^u_p(x_i)\right]$, $\left[\nu^l_p(x_i), \nu^u_p(x_i)\right]$ is called interval-valued intuitionistic fuzzy number (IVIFN). The largest range of IVIFN $\widetilde{\alpha} = ([a_\alpha, b_\alpha], [c_\alpha, d_\alpha])$ is defined as

$$\widehat{\alpha} = [a_\alpha, 1 - c_\alpha]. \tag{1}$$

Definition 2 [11] A HFS Z on X is given as $Z = \{(x, h_Z(x)) | x \in X\}$, where $h_Z(x)$ is the set of membership values of the element $x \in X$ to the set Z and values in the set are given in the form of interval between 0 and 1.

Definition 3 [14] The IVIHFS H on X is defined as $H = \{(x, \mathrm{IVI}_{\mu_s,\nu_s}(x)) | x \in X\}$, where $\mathrm{IVI}_{\mu_s,\nu_s}$ is IVIFN. The elements of IVIHFS are IVIFNs which are denoted as $(([\mu^l_1, \mu^u_1], [\nu^l_1, \nu^u_1]), \ldots, ([\mu^l_{L(\mathrm{IVIHFE})}, \mu^u_{L(\mathrm{IVIHFE})}], [\nu^l_{L(\mathrm{IVIHFE})}, \nu^u_{L(\mathrm{IVIHFE})}]))$, where $L(\mathrm{IVIHFE})$ denotes the number of elements in interval-valued intuitionistic hesitant fuzzy element (IVIHFE). Score function of the IVIHFE is obtained as

$$S(\mathrm{IVIHFE}) = \frac{1}{\#h} \sum_{k=1}^{L(\mathrm{IVIHFE})} \left(\frac{\mu^{(l)k} - \nu^{(l)k} + \mu^{(u)k} - \nu^{(u)k}}{2} \right).$$

Let $X = (x_1, x_2, x_3, x_4)$ be a given set, then the corresponding IVIHFS is given below.

$$\left\{ \begin{array}{l} (x_1, [[0.3, 0.5], [0.2, 0.4]], [[0.2, 0.4], [0.1, 0.5],], [[0.4, 0.6], [0.2, 0.3]]), \\ (x_2, [[0.6, 0.8], [0.1, 0.2]], [[0.3, 0.7], [0.1, 0.2]]), \\ (x_3, [[0.5, 0.6], [0.1, 0.2], 0.3]), \\ (x_4, [[0.2, 0.4], [0.3, 0.5]], \\ \quad [[0.1, 0.5], [0.2, 0.4]], [[0.1, 0.5], [0.2, 0.4]], [[0.0, 0.3], [0.1, 0.5]]) \end{array} \right\}$$

Definition 4 [14] The P-IVIHFS H_p on X is defined as $H_p = \{(x, \left[\left(([\mu^l_p, \mu^u_p], [\nu^l_p, \nu^u_p]), p\right)\right](x) | x \in X\}$, where P-IVIHFS is a set of probabilistic interval-valued intuitionistic hesitant fuzzy elements (P-IVIHFEs) and each P-IVIHFE contains an IVIFN and a probability measure. Here, the probability denotes the possible occurrence degree of IVIHFN. When P-IVIHFE is finite, then we can denote it as $\left[\left(([\mu^l_p, \mu^u_p], [\nu^l_p, \nu^u_p]), p\right)\right](x)$, $0 \leq p_i \leq 1$ and $\sum_{i=1}^{L(\text{PIVIHFE})} p_i \leq 1$, where L(P-IVIHFE) denotes the number of elements in P-IVIHFE. The accuracy function of P-IVIHFE is computed as

$$\Phi(a) = \sum_{k=1}^{L}\left(\left(\left[\frac{(\mu_p^{(l)k} - \nu_p^{(l)k}) + (\mu_p^{(u)k} - \nu_p^{(u)k})}{2}\right] - S(h)\right)^2 \times p^k\right), \tag{2}$$

where L is the set length of P-IVIHFE, $0 \leq \mu_p^{(l)k} \leq \mu_p^{(u)k} \leq 1$, $0 \leq \nu_p^{(l)k} \leq \nu_p^{(u)k} \leq 1$, $0 \leq \mu_p^{(u)k} + \nu_p^{(u)k} \leq 1$ and $0 \leq p^k \leq 1$. $S(h)$ is the score function and calculated as

$$S(h) = \frac{1}{\#h}\sum_{k=1}^{L}\left(\left[\frac{(\mu_p^{(l)k} - \nu_p^{(l)k}) + (\mu_p^{(u)k} - \nu_p^{(u)k})}{2}\right] \times p^k\right), \tag{3}$$

where $\#h$ is the number of P-IVIHF values in h and L is the length of P-IVIHFE.

Definition 5 [23] Let $h_i (i = 1, 2, \ldots, n)$ be a set of IVIHFEs whose aggregated value is calculated using the IVIHFWG operator which is shown below.

$$\begin{aligned}&\text{IVIHFWG}(h_1, h_2, \ldots, h_n) = \\ &\left\{\left(\left[\prod_{i=1}^{n}(\mu^l_{\alpha_i})^{w_i}, \prod_{i=1}^{n}(\mu^u_{\alpha_i})^{w_i}\right], \left[1 - \prod_{i=1}^{n}(1 - \nu^l_{\alpha_i})^{w_i}, 1 - \prod_{i=1}^{n}(1 - \nu^u_{\alpha_i})^{w_i}\right]\right)\right\} \\ &|\alpha_1 \in h_1, \alpha_2 \in h_2, \ldots, \alpha_n \in h_n\end{aligned}$$

where w_i is the weight associated with the given attributes.

PSO is presented in [4], where position and velocity of particle pd of dimension n are, respectively, denoted by position vector $X_{pd} = [x_{pd,1}, x_{pd,2}, \ldots, x_{pd,n}]$ and velocity vector $V_{pd} = [v_{pd,1}, v_{pd,2}, \ldots, v_{pd,n}]$, where

$$v_{pd,j} = v_{pd,j} + 2 \times \text{rand}() \times (p_{pd,j} - x_{pd,j}) + 2 \times \text{Rand}() \times (x_{gBest,j} - x_{pd,j}) \tag{4}$$

$$x_{pd,j} = x_{pd,j} + v_{pd,j}, \tag{5}$$

where $p_{pd,j}$ is the jth element of the personal best position vector $P_{pd} = [p_{pd,1}, p_{pd,2}, \ldots, p_{pd,n}]$ of particle pd, $x_{gBest,j}$ is the jth element in the global best position vector $X_{gBest} = [x_{gBest,1}, x_{gBest,2}, \ldots, x_{gBest,n}]$, rand() and Rand() are considered as random numbers which are uniformly distributed in [0,1], and $1 \leq j \leq n$.

3 Proposed Algorithm

This section presents the proposed decision-making algorithm using the concept of P-IVIHFS and PSO, where P-IVIHFS captures more dimensions of uncertainty and PSO determines the optimized weight of the attributes. Consider the set of m alternatives $\{E_1, E_2, \ldots, E_m\}$ and n attributes $\{A_1, A_2, \ldots, A_n\}$. The assessment matrix supplied by decision-maker is denoted as $D = (d_{ij})_{m\times n} = ([\mu^l_{ij}, \mu^u_{ij}], [\nu^l_{ij}, \nu^u_{ij}] | p_{ij})_{m\times n}$, where $(d_{ij})_{m\times n}$ denotes P-IVIHFV of alternative E_i corresponding to attribute A_j, $0 \le \mu^l_{ij} \le \mu^u_{ij} \le 1, 0 \le \nu^l_{ij} \le \nu^u_{ij} \le 1, 0 \le \mu^u_{ij} + \nu^u_{ij} \le 1, 0 \le p_{ij} \le 1, 1 \le i \le m$, $1 \le j \le n$. Let $\widetilde{w}_j = ([\mu^l_j, \mu^u_j], [\nu^l_j, \nu^u_j]), 1 \le j \le n$ represent the weight of the attribute A_j, and it is represented using IVIFN. Below, we present the proposed approach in stepwise manner.

Step 1. For each attribute A_j, the biggest range $\overline{w_j}$ of the weight $\widetilde{w}_j$ is computed in the form of IVIFV using (1).

Step 2. Using the score function of P-IVIHFV mentioned in (3), accuracy function of P-IVIHFV mentioned in (2) and the assessment matrix $D = (d_{ij})_{m\times n}$, the transformed assessment matrix $\widetilde{D} = (\widetilde{d}_{ij})_{m\times n}$ is computed, where $\widetilde{d}_{ij} \in [-1, 1]$ is the accuracy value of d_{ij}.

Step 3. Objective function F is considered as linear programming model. F is obtained by using the optimal weight w^* of the attribute A_j and transformed values of the assessment matrix $(\widetilde{d}_{ij})_{m\times n}$ as given below.

$$\max F = \sum_{i=1}^{m}\sum_{j=1}^{n}(w^*_j \times \Phi(d_{ij})), \quad \text{such that} \begin{cases} \mu^l_j \le w^*_j \le 1 - \nu^l_j \\ \sum\limits_{j=1}^{n} w^*_j = 1 \\ 0 \le w^*_j \le 1 \end{cases} \tag{6}$$

Step 4. The respective optimal weights $w^*_1, w^*_2, \ldots, w^*_n$ for the attributes $A_1, A_2, \ldots, A_n$ are obtained using PSO techniques [9, 24] to optimize (maximize) the objective function F. To apply PSO, we randomly generate s particles, where each particle is of dimension n. Consider kth particle has position vector $X_k = [w_{k,1}, w_{k,2}, \ldots, w_{k,n}]$ of dimension n, where $\mu^l_j \le w_{k,j} \le 1 - \nu^l_j$ and velocity vector $V_k = [v_{k,1}, v_{k,2}, \ldots, v_{k,n}]$ of dimension n, where $\frac{-0.2\times[(1-\nu^l_j)-\mu^l_j]}{2} \le v_{k,j} \le \frac{0.2\times[(1-\nu^l_j)-\mu^l_j]}{2}$, $1 \le j \le n$, $1 \le k \le s$. To get the best position of the kth particle, personal best position vector $P_{Best,k} = [p_{k,1}, p_{k,2}, \ldots, p_{k,n}]$ is defined. Initially, let the $P_{Best,k}$ of the kth particle be similar to its initial position vector X_k. The most important point to be noted is that $\sum_{j=1}^{n} w_{k,j} = 1$; therefore to normalize weights, we have used

$$w_{k,j} = \frac{w_{k,j}}{w_{k,1} + w_{k,2} + \cdots w_{k,n}}, \mu^{l}_{j} \leq w_{k,j} \leq 1 - \nu^{l}_{j}.$$

Step 5. Aggregation is performed based on IVIHFWG operator, assessment matrix $(d_{ij})_{m\times n}$ and the updated optimal weights of the attributes. The optimal weights are upgraded using probability and represented as $v_j = (\beta \times p_{ij}) + ((1-\beta) \times w^*_j)$, where $\beta \in [0, 1]$ and v_j is the weight that combines probabilities and optimal weights of P-IVIHFVs.

Step 6. The score value $S(t_i)$ is calculated for the obtained IVIHFE $t_i = ([\rho^l_i, \rho^u_i], [\tau^l_i, \tau^u_i])$ of each alternative E_i as

$$S(t_i) = \frac{1}{\#h} \sum_{k=1}^{L} \left(\frac{(\rho_i^{(l)k} - \tau_i^{(l)k}) + (\rho_i^{(u)k} - \tau_i^{(u)k})}{2} \right),$$

$$\rho^l_i = \prod_{j=1}^{n} (\mu^l_{ij})^{v_j}, \rho^u_i = \prod_{j=1}^{n} (\mu^u_{ij})^{v_j}, \tau^l_i = (1 - \prod_{j=1}^{n} (1 - \nu^l_{ij})^{v_j})$$

$$\tau^u_i = (1 - \prod_{j=1}^{n} (1 - \nu^u_{ij})^{v_j}),$$

where $0 \leq \rho^l_i \leq \rho^u_i \leq 1, 0 \leq \tau^l_i \leq \tau^u_i \leq 1, 0 \leq \rho^u_i + \tau^u_i \leq 1$.

Step 7. Alternatives with higher $S(t_i)$ values are ranked first.

4 Numerical Example

We have used Jupyter Notebook using Python on a Core i5 × 86-64 personal computer to implement the proposed method. This section presents a numerical example to demonstrate the proposed approach.

Let E_1, E_2, E_3, E_4 and E_5 be the alternatives and A_1, A_2, A_3 and A_4 be the attributes. Attributes' weights $\widetilde{w}_1$, $\widetilde{w}_2$, $\widetilde{w}_3$ and $\widetilde{w}_4$ are expressed in the form of IVIFV as ([0.1, 0.2], [0.5, 0.7]), ([0.2, 0.3], [0.4, 0.5]), ([0.6, 0.7], [0.3, 0.3]) and ([0.2, 0.4], [0.1, 0.3]), respectively. $D = (d_{ij})_{m\times n}$ represented by P-IVIHFS is shown as $D = (d_{ij})_{m\times n} = ([\mu^l_{ij}, \mu^u_{ij}], [\nu^l_{ij}, \nu^u_{ij}] | p_{ij})_{m\times n}$ given below.

$$D=(d_{ij})_{5\times 4}=\left(\begin{array}{lllll} & A_1 & A_2 & A_3 & A_4 \\ E_1 & \begin{pmatrix}[[0.3,\ 0.5],[0.2,\ 0.4],\ 0.5],\\ [[0.2,\ 0.4],[0.1,\ 0.5],\ 0.2],\\ [[0.4,\ 0.6],[0.2,\ 0.3],\ 0.2]\end{pmatrix} & \begin{pmatrix}[[0.6,\ 0.8],[0.1,\ 0.2],\ 0.4],\\ [[0.3,\ 0.7],[0.1,\ 0.2],\ 0.5]\end{pmatrix} & ([[0.5,\ 0.6],[0.1,\ 0.2],\ 0.3]) & \begin{pmatrix}[[0.2,\ 0.4],[0.3,\ 0.5],\ 0.2],\\ [[0.1,\ 0.5],[0.2,\ 0.4],\ 0.2],\\ [[0.1,\ 0.5],[0.2,\ 0.4],\ 0.4],\\ [[0.0,\ 0.3],[0.1,\ 0.5],\ 0.1]\end{pmatrix} \\ E_2 & \begin{pmatrix}[[0.5,\ 0.8],[0.1,\ 0.2],\ 0.6],\\ [[0.1,\ 0.4],[0.0,\ 0.5],\ 0.4]\end{pmatrix} & \begin{pmatrix}[[0.2,\ 0.4],[0.1,\ 0.3],\ 0.3],\\ [[0.1,\ 0.6],[0.2,\ 0.3],\ 0.6]\end{pmatrix} & \begin{pmatrix}[[0.3,\ 0.5],[0.1,\ 0.4],\ 0.1],\\ [[0.2,\ 0.7],[0.1,\ 0.2],\ 0.3],\\ [[0.6,\ 0.9],[0.0,\ 0.1],\ 0.3]\end{pmatrix} & ([[0.1,\ 0.4],[0.2,\ 0.5],\ 0.1]) \\ E_3 & \begin{pmatrix}[[0.8,\ 0.9],[0.0,\ 0.1],\ 0.4],\\ [[0.4,\ 0.6],[0.2,\ 0.3],\ 0.1],\\ [[0.4,\ 0.7],[0.1,\ 0.2],\ 0.4]\end{pmatrix} & ([[0.3,\ 0.7],[0.1,\ 0.2],\ 0.4]) & \begin{pmatrix}[[0.4,\ 0.5],[0.1,\ 0.3],\ 0.5],\\ [[0.3,\ 0.6],[0.2,\ 0.3],\ 0.4]\end{pmatrix} & \begin{pmatrix}[[0.1,\ 0.7],[0.0,\ 0.1],\ 0.3],\\ [[0.3,\ 0.6],[0.2,\ 0.3],\ 0.4],\\ [[0.4,\ 0.8],[0.1,\ 0.2],\ 0.2]\end{pmatrix} \\ E_4 & \begin{pmatrix}[[0.4,\ 0.5],[0.1,\ 0.4],\ 0.5],\\ [[0.2,\ 0.6],[0.2,\ 0.3],\ 0.1]\end{pmatrix} & \begin{pmatrix}[[0.4,\ 0.5],[0.1,\ 0.3],\ 0.5],\\ [[0.3,\ 0.5],[0.2,\ 0.3],\ 0.4],\\ [[0.1,\ 0.4],[0.2,\ 0.5],\ 0.1]\end{pmatrix} & \begin{pmatrix}[[0.3,\ 0.4],[0.2,\ 0.5],\ 0.2],\\ [[0.5,\ 0.6],[0.1,\ 0.2],\ 0.8]\end{pmatrix} & \begin{pmatrix}[[0.2,\ 0.4],[0.1,\ 0.5],\ 0.3],\\ [[0.1,\ 0.2],[0.3,\ 0.7],\ 0.5]\end{pmatrix} \\ E_5 & \begin{pmatrix}[[0.7,\ 0.8],[0.0,\ 0.1],\ 0.1],\\ [[0.6,\ 0.7],[0.1,\ 0.3],\ 0.3],\\ [[0.3,\ 0.4],[0.1,\ 0.3],\ 0.5]\end{pmatrix} & \begin{pmatrix}[[0.5,\ 0.6],[0.2,\ 0.3],\ 0.7],\\ [[0.4,\ 0.5],[0.2,\ 0.4],\ 0.2]\end{pmatrix} & \begin{pmatrix}[[0.2,\ 0.6],[0.1,\ 0.3],\ 0.1],\\ [[0.4,\ 0.5],[0.3,\ 0.4],\ 0.2],\\ [[0.5,\ 0.6],[0.1,\ 0.3],\ 0.3],\\ [[0.1,\ 0.3],[0.5,\ 0.6],\ 0.4]\end{pmatrix} & ([[0.2,\ 0.3],[0.4,\ 0.5],\ 0.5]) \end{array}\right)$$

Step 1 We find the biggest range $\overline{w}_j$ of weights $\widetilde{w}_j$ using (1), where $\overline{w_1} = [0.1, 0.5]$, $\overline{w_2} = [0.2, 0.6]$, $\overline{w_3} = [0.6, 0.7]$ and $\overline{w_4} = [0.2, 0.9]$.

Step 2 Using score and accuracy function of P-IVIHFS, we obtain transformed assessment matrix $\widetilde{D} = (\widetilde{d}_{ij})_{5\times 4}$ as

$$\widetilde{D} = (\widetilde{d}_{ij})_{5\times 4} = \begin{pmatrix} 0.0118 & 0.0613 & 0.0235 & 0.0037 \\ 0.0825 & 0.0027 & 0.1178 & 0.0008 \\ 0.1619 & 0.0176 & 0.0147 & 0.0504 \\ 0.0111 & 0.0243 & 0.0512 & 0.0368 \\ 0.0766 & 0.0229 & 0.0917 & 0.0050 \end{pmatrix}.$$

Step 3 The linear programming model is obtained as

$$\max F = \sum_{i=1}^{5}\sum_{j=1}^{4} w_j^* \times \Phi(d_{ij})_{5\times 4},$$

Step 4 Using PSO, we get the optimal weights of the attributes $w_1^* = 0.2176$, $w_2^* = 0.1489$, $w_3^* = 0.4532$ and $w_4^* = 0.1803$. In this example, we consider s to be 20 particles and number of iterations to be 150, $c_1 = 2.05$ and $c_2 = 2.05$, and let maximum and minimum inertia weight be $\omega_{\max} = 0.9$ and $\omega_{\min} = 0.4$. The convergence of the values of the objective function F to get optimal weights is shown in Fig. 1.

Step 5 Using the original assessment matrix and the obtained optimal weights, we get the aggregated value as follows:

$$t_1^1 = ([\rho_1^{(l)1}, \rho_1^{(u)1}], [\tau_1^{(l)1}, \tau_1^{(u)1}]) = ([0.3333, 0.522], [0.1894, 0.3587]),$$
$$t_1^2 = ([\rho_1^{(l)2}, \rho_1^{(u)2}], [\tau_1^{(l)2}, \tau_1^{(u)2}]) = ([0.2713, 0.5582], [0.1181, 0.3116])$$
$$t_1^3 = ([\rho_1^{(l)3}, \rho_1^{(u)3}], [\tau_1^{(l)3}, \tau_1^{(u)3}]) = ([0.2123, 0.5448], [0.1690, 0.3161]),$$
$$t_1^4 = ([\rho_1^{(l)4}, \rho_1^{(u)4}], [\tau_1^{(l)4}, \tau_1^{(u)4}]) = ([0.0, 0.5767], [0.1231, 0.2659])$$
$$t_2^1 = ([\rho_2^{(l)1}, \rho_2^{(u)1}], [\tau_2^{(l)1}, \tau_2^{(u)1}]) = ([0.2879, 0.5389], [0.1094, 0.3357]),$$

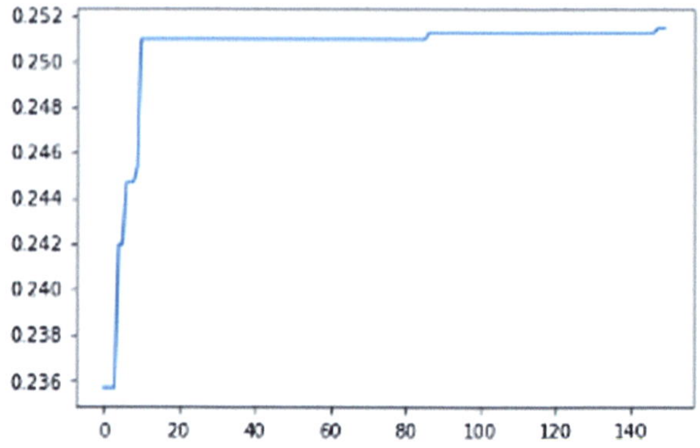

Fig. 1 Convergence process

$$t_2^2 = ([\rho_2^{(l)2}, \rho_2^{(u)2}], [\tau_2^{(l)2}, \tau_2^{(u)2}]) = ([0.0496, 0.4205], [0.1737, 0.4620])$$
$$t_2^3 = ([\rho_2^{(l)3}, \rho_2^{(u)3}], [\tau_2^{(l)3}, \tau_2^{(u)3}]) = ([0.0919, 0.5098], [0.1438, 0.3960]),$$
$$t_3^1 = ([\rho_3^{(l)1}, \rho_3^{(u)1}], [\tau_3^{(l)1}, \tau_3^{(u)1}]) = ([0.2672, 0.5912], [0.0999, 0.2413])$$
$$t_3^2 = ([\rho_3^{(l)2}, \rho_3^{(u)2}], [\tau_3^{(l)2}, \tau_3^{(u)2}]) = ([0.3166, 0.6180], [0.1833, 0.2826]),$$
$$t_3^3 = ([\rho_3^{(l)3}, \rho_3^{(u)3}], [\tau_3^{(l)3}, \tau_3^{(u)3}]) = ([0.2722, 0.6281], [0.1621, 0.2760])$$
$$t_4^1 = ([\rho_4^{(l)1}, \rho_4^{(u)1}], [\tau_4^{(l)1}, \tau_4^{(u)1}]) = ([0.2504, 0.3814], [0.1429, 0.4797]),$$
$$t_4^2 = ([\rho_4^{(l)2}, \rho_4^{(u)2}], [\tau_4^{(l)2}, \tau_4^{(u)2}]) = ([0.2262, 0.3960], [0.2063, 0.4212])$$
$$t_4^3 = ([\rho_4^{(l)3}, \rho_4^{(u)3}], [\tau_4^{(l)3}, \tau_4^{(u)3}]) = ([0.1579, 0.3287], [0.2307, 0.5192]),$$
$$t_5^1 = ([\rho_5^{(l)1}, \rho_5^{(u)1}], [\tau_5^{(l)1}, \tau_5^{(u)1}]) = ([0.2404, 0.4481], [0.2477, 0.3945])$$
$$t_5^2 = ([\rho_5^{(l)2}, \rho_5^{(u)2}], [\tau_5^{(l)2}, \tau_5^{(u)2}]) = ([0.3577, 0.4653], [0.2730, 0.4146]),$$
$$t_5^3 = ([\rho_5^{(l)3}, \rho_5^{(u)3}], [\tau_5^{(l)3}, \tau_5^{(u)3}]) = ([0.2530, 0.3570], [0.2701, 0.4479])$$
$$t_5^4 = ([\rho_5^{(l)4}, \rho_5^{(u)4}], [\tau_5^{(l)4}, \tau_5^{(u)4}]) = ([0.1089, 0.2353], [0.429, 0.5831])$$

Step 6 Based on score function of IVIHFS, we obtain transformed value $\widetilde{t_i}$ for each alternative as follows:

$$\widetilde{t_1} = 0.2917, \widetilde{t_2} = 0.0927, \widetilde{t_3} = 0.4827, \widetilde{t_4} = -0.0865 \text{ and } \widetilde{t_5} = -0.1486.$$

Step 7 Based on the transformed value $\widetilde{t_i}$, $i = 1, 2, \ldots, 5$, the preference order of alternatives is $E_3 > E_1 > E_2 > E_4 > E_5$.

5 Conclusion

This study proposes a MADM approach based on P-IVIHFS, IVIHFWG operator and PSO techniques. P-IVIHFS is used to characterize the assessing values of the alternative regarding the attributes. PSO calculates the best weights of the attributes from the mentioned interval. The time complexity of the proposed approach can be obtained as $O(pnlog(n))$, where n is the initial population and p is the no. of iterations. We have considered the largest range of weights to reach the global optimum quite smoothly. Considering the weights of attributes in interval is much helpful when the necessary information about the weights is not available. The descriptive numerical example shows the workability of the proposed approach. In the future, researchers may apply probability with many other types of fuzzy sets such as picture fuzzy set, neutrosophic fuzzy set and Pythagorean fuzzy set and then investigate the decision-making process with the suitable optimization algorithms.

References

1. Zadeh, L.A.: Fuzzy sets. Inf. Control **8**, 338–353 (1965)
2. Attanassov, K.T.: Intuitionistic fuzzy sets. Fuzzy Sets Syst. **20**(1), 87–96 (1986)
3. Attanassov, K.T., Gargov, G.: Interval-valued intuitionistic fuzzy sets. Fuzzy Sets Syst. **31**(3), 343–349 (1989)
4. Kennedy, J., Eberhart, R.: Particle swarm optimization. In: Proceedings of the IEEE International Conference on Neural Networks, vol. 4, pp. 1942–1948 (1995)
5. Esfandiary, M.J., Sheikholarefin, S., Bondarabadi, H.A.R.: A combination of particle swarm optimization and multi-criterion decision-making for optimum design of reinforced concrete frames. Int. J. Optim. Civil Eng. **6**(2), 245–268 (2016)
6. Koohi, I., Groza, V.Z.: Optimizing particle swarm optimization algorithm. In: 2014 IEEE 27th Canadian Conference on Electrical and Computer Engineering (CCECE). https://doi.org/10.1109/ccece.2014.6901057
7. Javanbarg, M.B., Scawthorn, C., Kiyono, J., Shahbodaghkhan, B.: Fuzzy AHP-based multi-criteria decision-making systems using particle swarm optimization. Expert Syst. Appl. **39**, 960–966 (2012)
8. Yang, C., Simon, D.: A new particle swarm optimization technique. In: Proceedings of the 18th International Conference on Systems Engineering (ISCEng'05), pp. 164–169 (2005)
9. Chen, S.M., Huang, Z.C.: Multi attribute decision-making based on interval-valued intuitionistic fuzzy values and particle swarm optimization techniques. Inf. Sci. **397–398**, 206–218 (2017)
10. Chen, S.M., Chiou, C.H.: Multiattribute decision-making based on interval-valued intuitionistic fuzzy sets, PSO techniques, and evidential reasoning methodology. IEEE Trans. Fuzzy Syst. **23**(6), 1905–1916 (2015)
11. Torra, V., Narukawa, Y.: On hesitant fuzzy sets and decision. In: FUZZ-IEEE, pp. 20–24 (2009)
12. Rodríguez, R.M., Martínez, L., Torra, V., Xu, Z.S., Herrera, F.: Hesitant fuzzy sets: state of the art and future directions. Int. J. Intell. Syst. **29**(6), 495–524 (2014)
13. Wu, W., Li, Y., Ni, Z., Jin, F., Zhu, X.: Probabilistic interval-valued hesitant fuzzy information aggregation operators and their application to multi-attribute decision-making. Algorithms **11** (0120) (2018)
14. Zhai, Y., Xu, Z., Liao, H.: Measures of probabilistic interval-valued intuitionistic hesitant fuzzy sets and the application in reducing excessive medical examinations. IEEE Trans. Fuzzy Syst. **26**(3), 1651–1670 (2018)

15. Yue, L., Sun, M., Shao, Z.: The probabilistic hesitant fuzzy weighted average operators and their application in strategic decision-making. J. Inf. Comput. Sci. **10**(12), 3841–3848 (2013)
16. Roy, J., Das, S., Kar, S., Pamučar, D.: An extension of the CODAS approach using interval-valued intuitionistic fuzzy set for sustainable material selection in construction projects with incomplete weight information. Symmetry **11** (3) (2019). https://doi.org/10.3390/sym11030393
17. Das, S., Malakar, D., Kar, S., Pal, T.: A brief review and future outline on decision-making using fuzzy soft set. Int. J. Fuzzy Syst. applications **7**(2), 1–43 (2018)
18. Das, S., Kar, M.B., Kar, S., Pal, T.: An approach for decision-making using intuitionistic trapezoidal fuzzy soft set. Ann. Fuzzy Math. Inform. **16**(1), 85–102 (2018)
19. Das, S., Kumar, S., Kar, S., Pal, T.: Group decision making using neutrosophic soft matrix: an algorithmic approach. J. King Saud Univ. Comput. Inf. Sci. (2017). https://doi.org/10.1016/j.jksuci.2017.05.001
20. Krishankumar, R., Ravichandran, K.S., Premaladha, J., Kar, S., Zavadskas, E.K., Antuchevi-ciene, J.: A decision framework under a linguistic hesitant fuzzy set for solving multi criteria group decision making problems. Sustainability **10**(8), 2608–2628 (2018)
21. Krishankumar, R., Ravichandran, K.S., Kar, S., Gupta, P., Mehlawat, M.K.: Interval-valued probabilistic hesitant fuzzy set for multi-criteria group decision-making. Soft. Comput. (2018). https://doi.org/10.1007/s00500-018-3638-3
22. Raghunathan, K., Saranya, R., Nethra, R.P., Ravichandran, K.S., Kar, S.: A decision-making framework under probabilistic linguistic term set for multi-criteria group decision-making problem. J. Intell. Fuzzy Syst. (2018). https://doi.org/10.3233/JIFS-181633
23. Zhang, Z.: Interval-valued intuitionistic hesitant fuzzy aggregation operators and their application in group decision-making. J. Appl. Math. Article ID 670285 (2013). http://dx.doi.org/10.1155/2013/670285
24. Umapathy, P., Venkataseshaiah, C., Arumugam, M. S.: Particle swarm optimization with various inertia weight variants for optimal power flow solution. Discrete Dyn. Nat. Soc. Article ID 462145 (2010)

Assessment of Stock Prices Variation Using Intelligent Machine Learning Techniques for the Prediction of BSE

Rashmi Bhardwaj and Aashima Bangia

Abstract Significance of this research is to accomplish tentative study on highs and lows of particular S&P BSE stock prices using proposed intelligent models, multivariate adaptive regression spline (MARS) and M5 prime regression tree-(M5'). Anticipated models work to predict as there exists vitality for price instability. Daily highs and lows of the stock price data have been considered as the data set. This article discusses about computational ability of the MARS and M5' regressions during the time period and also how better accuracy can be attained. M5' constructs in two phases: growing and pruning which smoothen regression tree at nodes. MARS builds complex configuration of correlation among multiple responses. This can be helpful for investors to predict significant statistics for trading stocks listed on BSE.

Keywords Bombay Stock Exchange (BSE) · Multivariate adaptive regression spline (MARS) · M5 prime regression tree (M5') · Stock prices

1 Introduction

Financial market is a place which caters to the shares of pubic listed companies that can be traded. The main characteristics include having transparency in prices, basic regulations on trading plus related costs plus forces that would govern the amounts for securities. In most cases nowadays, firms raise investments through financial institutions. Primary market allows companies to float shares towards general public through initial public offering (IPO) to advance funds. It issues new securities on every exchange. Firms and governments among various entities get finances with the aid of debt-/equity-based securities. When newer securities are sold in the primary market, they have been transferred towards the secondary market. This is when an investor can consume shares from another investor at prevalent market prices.

R. Bhardwaj (✉)
University School of Basic and Applied Sciences (USBAS), Non-Linear Research Lab, Guru Gobind Singh Indraprastha University (GGSIPU), Dwarka, New Delhi 110078, India
e-mail: rashmib22@gmail.com

A. Bangia
USBAS, GGSIPU, Dwarka, New Delhi 110078, India

D. Dutta and B. Mahanty (eds.), *Numerical Optimization in Engineering and Sciences*, Advances in Intelligent Systems and Computing 979,
https://doi.org/10.1007/978-981-15-3215-3_15

Secondary markets are open for other securities also, that can be funds, investment banks, etc. In secondary market, during trading cash finances go towards the investor rather than to the concerned firm directly. Security and Exchange Board of India (SEBI) governs both of the above described markets in India.

Bombay Stock exchange (BSE) assists stockbrokers in trading stocks plus other securities. Stocks that are listed on this exchange are only invested upon. Thus, it serves as a meeting place for buyers and sellers. Least-squares support vector machines are studied in detail [1]. M5' regression model tree toolbox is developed [2]. Support vector to classify power system disturbances has been implemented [3]. Scour depth spillways have been forecasted through classification and regression trees [4]. Continuous classes through learning have been studied [5]. Stock market analysis to study the impact of demonetization has been carried out [6]. Suspended sediment load using regression trees and model trees approaches have been estimated [7]. Statistical learning theory is explained in detail [8]. Wavelet with support vector machine has been devised [9]. Model trees for predicting continuous classes have been introduced [10]. Machine learning tools and techniques with applications have been studied [11].

None of the authors have described the highs and lows of the S&P stock prices under BSE involving intelligent models MARS and M5' regression for assessments of future investments.

2 Methodology

2.1 Data Collection

Bombay Stock Exchange data on a daily basis of high and low prices is the data set taken into consideration from 01 April 2016 to 31 July 2017 in this study. A total of 330 days' data is processed as mentioned in Table 1. Out of this sample, the data are used to train have been applied as the input values.

Table 1 Data sets involved in simulation of the six hybrid models

Cross-validation	Data sets	Time period	
		From	To
Data type 1	High	1 April 2016	31 July 2017
Data type 2	Low	1 April 2016	31 July 2017

2.2 *M5 Prime Regression Tree (M5')*

M5 algorithm was extended to provide M5' which is regarded as a substitute to conventional regression structured by hierarchies which graph the *if-then* rules. The prototype is designed for segregating response domain into many sub-domains; in addition, a linear lapse model is fitted for every sub-domain. M5' constructs regression hierarchy through recursive splitting founded on discussing standard deviation for class values which influence node and measure error at the node. Attributes which maximize predictable inaccuracy that decrease get opted for splitting at the node. Standard deviation reduction (SDR) is deliberated as follows:

$$\mathrm{SDR} = \mathrm{sd}(T) - \sum_{i} \frac{|T_i|}{|T|} \times \mathrm{sd}(T_i)$$

sd—standard deviation; T—set of instances which touch node; T_i—sets consequential from splitting node in accordance with particular characteristic with split value. Splitting process dismisses as and when outcomes for every instance which touch the node vary only slightly else if some instance remains.

2.3 *Multivariate Adaptive Regression Splines (MARS)*

MARS algorithm utilizes nonparametric models free from considerations concerning forms of the relationship among the predictors and responses. It is able to describe interactions of explanatory–response variables which prove missing in other methods' output. MARS modelling divides the parametre hyperspace of responses into disjoint hyper regions. The influence of predictors upon responses is described through linearity within boundaries of the hyper regions. *Knots* are intersections wherever incline drifts among hyper regions. Assortment of knots recognized by this model generates basis functions splines which are single-variable alterations and multivariable interactions as per design.

3 Results and Discussions

The machine learning models, MARS and M5', are compared based on S&P BSE stock data of 330 days. This has been conferred through error calculations. Figure 1 shows MARS regression fit for high prices and low prices, respectively. Figure 2 shows M5' regression fit line for high prices and low prices, respectively. Figure 3 shows M5' regression tree for high prices and low prices, respectively. Table 2 shows calculation of various errors in data sets for intelligent models with execution time (in seconds) of each model for various data sets.

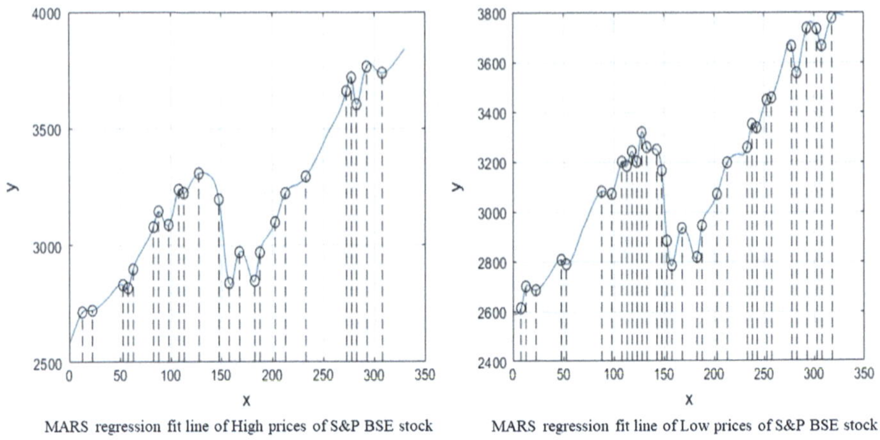

Fig. 1 Comparison of MARS regression fit for **a** high prices **b** low prices

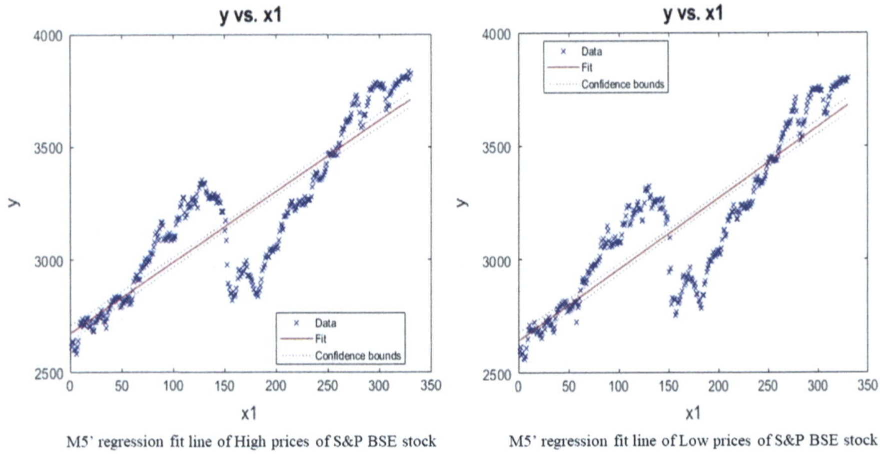

Fig. 2 Comparison of M5' regression fit line for **a** high prices **b** low prices

4 Classification and Regression Tree (CART)

For CART graph of BSE high, the data are divided into 15 nodes with a set of rules that divides the data at each node with respect to time. At node 2, if time in [1, 256.5[then BSE-High = 3050 in 78.1% of cases. At node 3, if time in [256.5, 330[then BSE-High = 3702 in 21.9% of cases. At node 4, if time in [1, 174.5[then BSE-High = 2985 in 53.1% of cases. At node 5, if time in [174.5, 256.5[then BSE-High = 3186 in 25% of cases. At node 6, if time in [256.5, 293.5[then BSE-High = 3627 in 10.9% of cases. At node 7, if time in [293.5, 330[then BSE-High = 3777 in 10.9% of cases. At node 8, if time in [293.5, 317.5[then BSE-High = 3758 in

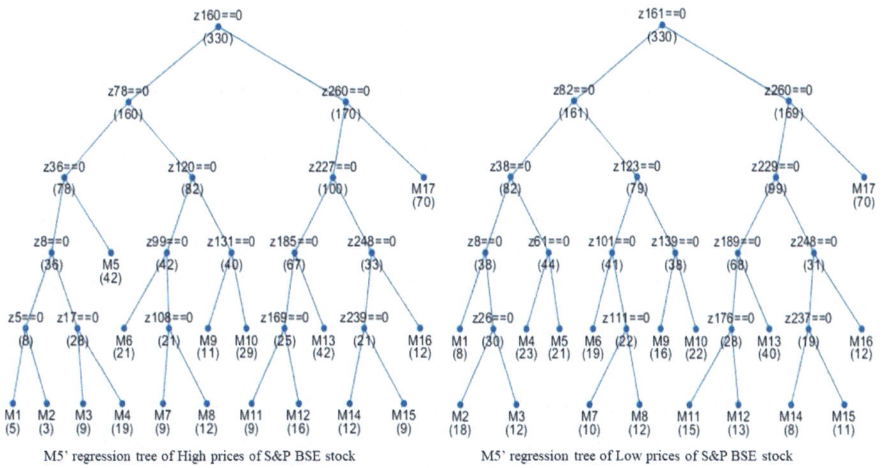

Fig. 3 Comparison of M5' regression tree for **a** high prices **b** low prices

Table 2 Calculation of various errors and execution time in datasets for intelligent models

Errors	Cross-validation data set	Intelligent models	
		MARS	M5'
RMSE	Data type 1	25.5578	346.8343
	Data type 2	29.8138	368.9377
MAE	Data type 1	20.4566	289.4927
	Data type 2	22.4019	310.0380
MSE	Data type 1	660.8043	1.2088e+05
	Data type 2	914.8941	1.3855e+05
R^2	Data type 1	0.9939	−0.0897
	Data type 2	0.9914	−0.2366
Execution time (in seconds)	Data type 1	0.99	0.08
	Data type 2	1.02	0.09

6.9% of cases. At node 9, if time in [317.5, 330[then BSE-High = 3809 in 4.1% of cases. At node 10, if time in [293.5, 314.5[then BSE-High = 3754 in 5.9% of cases. At node 11, if time in [314.5, 317.5[then BSE-High = 3780 in 0.9% of cases. At node 12, if time in [293.5, 311.5[then BSE-High = 3753 in 5% of cases. At node 13, if time in [311.5, 314.5[then BSE-High = 3760 in 0.9% of cases. At node 14, if time in [317.5, 319.5[then BSE-High = 3794 in 0.6% of cases. At node 15, if time in [319.5, 330[then BSE-High = 3812 in 3.4% of cases. The nonlinear regression graph shows the data points and the model regression line. It can be clearly seen that the nonlinear model is unable to fit the data points through the model. Further, it can be clearly observed through the errors RMSE, MSE and R^2 values.

For CART of BSE low, the data are divided into 15 nodes with a set of rules that divides the data at each node with respect to time. At node 2, if time in [1, 255.5[then BSE-Low = 3012 in 77.2% of cases. At node 3, if time in [255.5, 330[then BSE-Low = 3665 in 22.8% of cases. At node 4, if time in [1, 174.5[then BSE-Low = 2953 in 53.1% of cases. At node 5, if time in [174.5, 255.5[then BSE-Low = 3142 in 24.1% of cases. At node 6, if time in [255.5, 294.5[then BSE-Low = 3593 in 12.2% of cases. At node 7, if time in [294.5, 330[then BSE-Low = 3747 in 10.6% of cases. At node 8, if time in [294.5, 312.5[then BSE-Low = 3715 in 5.3% of cases. At node 9, if time in [312.5, 330[then BSE-Low = 3778 in 5.3% of cases. At node 10, if time in [294.5, 311.5[then BSE-Low = 3715 in 5% of cases. At node 11, if time in [311.5, 312.5[then BSE-Low = 3719 in 0.3% of cases. At node 12, if time in [312.5, 317.5[then BSE-Low = 3757 in 1.6% of cases. At node 13, if time in [317.5, 330[then BSE-Low = 3787 in 3.8% of cases. At node 14, if time in [317.5, 318.5[then BSE-Low = 3776 in 0.3% of cases. At node 15, if time in [318.5, 330[then BSE-Low = 3788 in 3.4% of cases. The nonlinear regression graph shows the data points and the model regression line. It can be clearly seen that the nonlinear model is unable to fit the data points through the model. Further, it can be clearly observed through the errors RMSE, MSE and R^2 values.

The nonlinear regression plots for BSE high and BSE low are shown in Fig. 4, and the errors are shown in Table 3. CART for BSE high and BSE low are shown in Figs. 5 and 6, respectively.

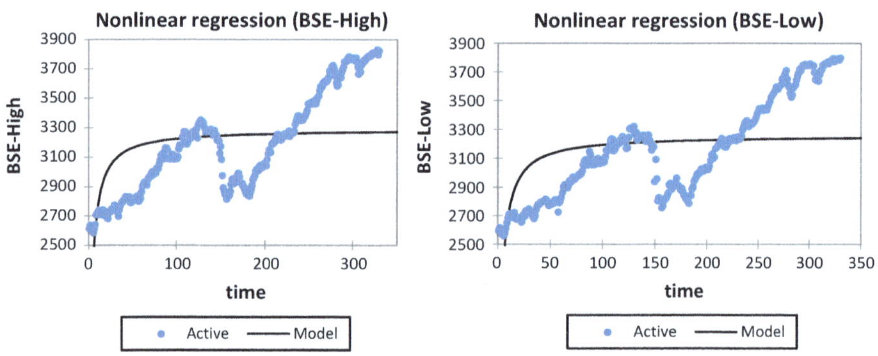

Fig. 4 Comparison of nonlinear regression for **a** high prices **b** low prices

Table 3 Calculation of various errors in datasets for nonlinear intelligent models

Model	Cross-validation data set	Errors		
		RMSE	MSE	R^2
Nonlinear regression model	Data type 1	307.014	94,257.457	0.212
	Data type 2	308.072	94,908.362	0.210

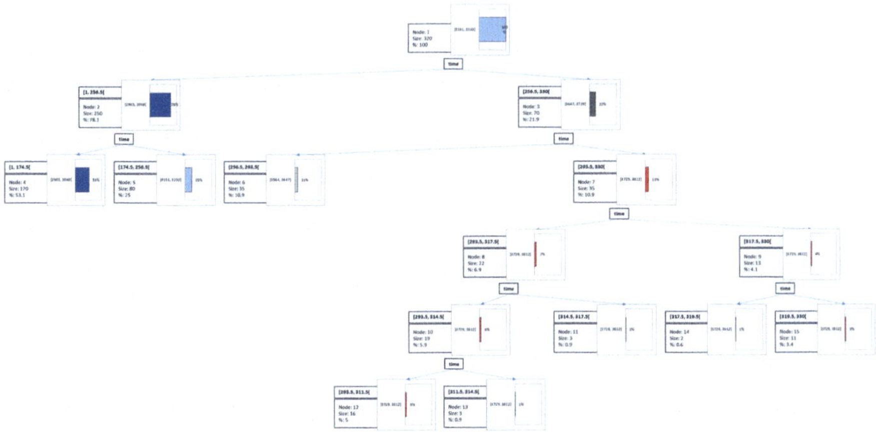

Fig. 5 CART representation for BSE high

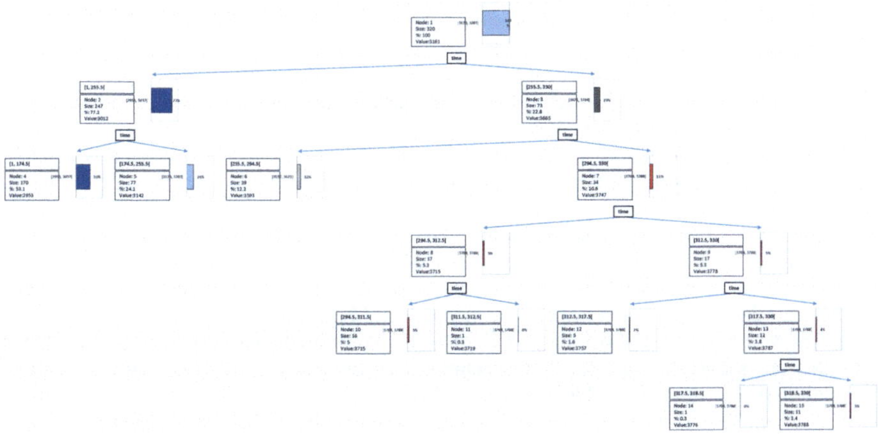

Fig. 6 CART representation for BSE low

From figures and tables, it is observed that BSE high and BSE low show more inclination towards linear regression as compared to nonlinear regression. The trend in linear regression is more predictive as compared to nonlinear regression.

5 Conclusion

The accuracy of MARS and M5’ tree regression model has been investigated for forecasting daily high and low BSE stock prices in this current training. Using MARS and M5’ tree regression prototype, the nonlinear nature of high and low stock prices is appropriately observed. Execution time for MARS and M5' of data type 1 is 0.99 s and 0.08 s, respectively. For data type 2, MARS and M5' have execution time 1.02 s and 0.09 s, respectively. Clearly, M5' has lesser execution time compared to MARS prototype. Comparison of the models has been discussed with the help of the error computations for linear regression and nonlinear regression. Results attained from this study approve that these regression models based on machine learning are sufficient for predicting the highs and lows of the ever-changing stocks, and linear regression is more predictive as compared to nonlinear regression.

Acknowledgements Authors are thankful to Guru Gobind Singh, Indraprastha University, for providing financial support and research facilities.

References

1. Suykens, J.A.K., Van Gestel, T., De Brabanter, J., De Moor, B., Vandewalle, J.: Least Squares Support Vector Machines. World Scientific Publishing, Singapore (2002)
2. Jekabsons, G.: M5 prime lab: M5' regression tree, model tree, and tree ensemble toolbox for Matlab/Octave (2016). Available at http://www.cs.rtu.lv/jekabsons/
3. Ekici, S.: Classification of power system disturbances using support vector machines. Expert Syst. Appl. **36**(6), 9859–9868 (2009)
4. Samadi, M, Jabbari, E, Azamathulla, H. Md.: Assessment of M5 model tree and classification and regression trees for prediction of scour depth below free overfall spillways. Neural Comput. Appl. **24**, 357–366 (2014)
5. Quinlan J.R.: Learning with continuous classes. In: Proceedings of 5th Australian Joint Conference on Artificial Intelligence, pp. 343–348. World Scientific, Singapore (1992)
6. Patel, R., Hiral, M., Parikh, S.: Experimental study on stock market to analyse the impact of the latest demonetization in India. Artif. Intell. Syst. Mach. Learn. **9**(6), 106–110 (2017)
7. Talebi, A., Mahjoobi, J., Dastorani, M.T., Moosavi, V.: Estimation of suspended sediment load using regression trees and model trees approaches (Case study: Hyderabad drainage basin in Iran). ISH J. Hydraul. Eng. **23**(2), 212–219 (2017)
8. Vapnik, V.: The Nature of Statistical Learning Theory. Springer, New York (1995)
9. Zhang, L., Zhou, W., Jiao, L.: Wavelet support vector machine. IEEE Trans. Syst. Man Cybernet. Part B Cybernet. **34** (2004)
10. Wang, Y., Witten, I.H.: Induction of model trees for predicting continuous classes. In: Proceedings of the 9th European Conference on Machine Learning Poster Papers, pp. 128–137. Prague (1997)
11. Witten, I.H., Frank, E.: Data mining: practical machine learning tools and techniques, 2nd edn. Morgan Kaufmann, San Francisco (2005)

Short-Term Electricity Load Forecast Using Hybrid Model Based on Neural Network and Evolutionary Algorithm

Priyanka Singh and Pragya Dwivedi

Abstract Electricity load forecast needs to ensure minimum load wastage and requires intelligent decision-making systems to accurately predict future load demand. Learning capability, robustness and ability to solve nonlinear problems make ANN widely acceptable. For accurate short-term load forecast (STLF), an ensemble soft computing approach, namely ANN-PSOHm, composed of an artificial neural network (ANN) and particle swarm optimization (PSO) with homeostasis based mutation is presented in this article. To enhance the learning strength of ANN, PSO undergoes homeostasis mutation to bring greater diversity among solutions by exploring in wider search space. To demonstrate the effectiveness of ANN, three case studies on load data of NEPOOL region (courtesy ISO New England) are performed. The experimental results show that ANN-PSOHm improves accuracy by 11.57% MAPE over ANN-PSO.

Keywords Electricity load forecast · Artificial neural network · Particle swarm optimization · Homeostasis mutation

1 Introduction

The electricity load forecasting is one of the major research areas in the field of forecasting. Accurate prediction helps in proper planning and operation of electric utilities and load scheduling [1]. Models like support vector machine [10], Bayesian [7], neural network [4, 14, 17] and regression [3] are already available for a STLF. Among these, ANN has attracted the most and is capable of finding greater co-relation among forecasting variables. The over-fitting problem, a chance of being trapped in local minima and no proper rule for selecting network architecture, makes ANN

P. Singh (✉) · P. Dwivedi
Department of Computer Science and Engineering, MNNIT, Allahabad 211004, India
e-mail: 2015rcs53@mnnit.ac.in

P. Dwivedi
e-mail: pragyadwi86@mnnit.ac.in

D. Dutta and B. Mahanty (eds.), *Numerical Optimization in Engineering and Sciences*, Advances in Intelligent Systems and Computing 979,
https://doi.org/10.1007/978-981-15-3215-3_16

undesirable. To overcome these difficulties, ANN has been integrated with other techniques to generate an accurate prediction [8, 16]. Population-based methods have generated interest among researchers due to their evolution and combined them with the intelligence-based method to overcome the problem of STLF.

Unlike other population-based methods, particle swarm optimization (PSO) solves the stagnation problem due to its huge search space and diversity maintaining ability. It helps in finding an optimal solution in multi-dimensional search space by learning through self-experience and global experience [12]. PSO performs well with nonlinear problems and converges faster in problems where analytical methods fail. But, results generated by PSO are not good, and PSO still lacks its original problem of being trapped in a local minimum with regard to a multi-modal problem that has many sharp peaks [6]. To overcome this limitation of PSO, certain improvement can be made in PSO. It is more effective to improve existing heuristic with mutations, rather than iteratively executing an improved one starting from solutions generated randomly [5]. In [2], a new mutation feature has been combined with ant colony optimization (ACO) to increase the convergence rate and decrease computation time. In [21], PSO with Gaussian and adaptive mutation combined with SVM has been used for power load forecasting. Similarly, in [11], neural network (NN) based model has been implemented to find prediction intervals with mutated PSO to bring higher diversity. Findings of this literature show that mutation brings diversity within the population-based algorithms and results in a higher convergence rate. Based on this, we have used newly proposed homeostasis mutation in our study. Homeostasis mutation helps particles to adjust the condition that is optimal for their existence [19]. Homeostasis mutation has been integrated with differential evolution for software cost estimation. Results obtained show that homeostasis mutation helps in generating more promising results.

From the literature, it can be observed that an appropriate learning algorithm and optimized NN structure may enhance the learning capability of NN. To overcome the shortcomings of neural network, a new approach to ANN is presented by combining PSO with homeostasis mutation for solving STLF problem.

The rest of the article is structured as follows. The basics of artificial neural network, particle swarm optimization and homeostasis mutation are described in Sect. 2. Section 3 shows the implementation of ANN-PSOHm on STLF, and Sect. 4 illustrates its effectiveness over traditional approaches. Lastly, Sect. 5 concludes the article with some future remarks.

2 Background Details

ANN is a computing system composed of several small processing elements called neurons. These neurons process information received from external inputs by its dynamic state response. ANN has great learning capability, which is inspired by the working of a human brain system [9, 15]. ANN map relationship between input and output vector and output F_j at node j of a network is the weighted sum of its n input values ($X = x_1, x_2, x_3, \dots x_k$) given by Eq. 1.

$$F_j = \frac{1}{1 + e^{-ay_j}} \quad j = 1, 2, 3, \ldots, M \tag{1}$$

where a is a slope of the sigmoid function, Y_j is the output of a single node j and M is the number of output nodes in output layer.

PSO is population-based search algorithm, every particle moves in a multi-dimensional search space with velocity and this velocity dynamically get adjusted according to its own moving experience and its companions moving experience which continually falls towards the optimal solution [13, 20]. While implementing PSO, the position of each particle is randomly initialized and their velocity and position are updated by Eqs. 2 and 3.

$$V_{\text{new}} = w * V_{\text{old}} + c_1 * r_1() * (P_{\text{best}} - P_{\text{old}}) + c_2 * r_2() * (P_{\text{gbest}} - P_{\text{old}}) \tag{2}$$

$$P_{\text{new}} = \delta * V_{\text{new}} + P_{\text{old}} \tag{3}$$

where V_{old} is velocity of the particle in previous iteration, P_{old} is position of the particle in previous iteration, V_{new} is velocity of the particle in current iteration, P_{new} is position of the particle in current iteration, P_{best} is particle's own best position, G_{best} is global best position, w is inertial weight, r_1 and r_2 are random values between 0 and 1, c_1 and c_2 are constants, δ is retardation factor and P_{old} is particle's previous position.

Homeostasis is a self-regulating process by which stability is maintained inside biological bodies by adjusting to the conditions that are optimal for their own survival by multiplying different parameter values depending on the nature of the problem and available counterbalancing resources [18, 19]. If homeostasis process is successful, then the life of the biological body continues else death occurs. With this concept, a new mutant vector is generated to maintain diversity within the population. Homeostasis mutation vector is defined by Eq. 4.

$$\gamma_{i,G} = \alpha_{\text{best},G} + \delta_1 \cdot (\alpha_{r_i^1,G} * H_v - \alpha_{r_i^2,G} * H_v) \tag{4}$$

where α_{best} is best vector of current population, $\alpha_{r_i^1,G}$ and $\alpha_{r_i^2,G}$ are random individuals generated from the entire search space, H_v is homeostasis value that lies between 0.01 and 0.1 and δ_1 is a random value between 0 and 1.

3 Short-Term Electricity Load Forecast

Load forecasting plays a very significant role in decisions of purchasing and generating electric load. Hence, it is important to have simple, feasible, fast and precise load forecasters. ANN is one of these forecasters, but sometimes it fails due to the problem of over-fitting and over-training. To overcome this problem, evolutionary-based algorithms are combined to optimize network parameters such as PSO. Although

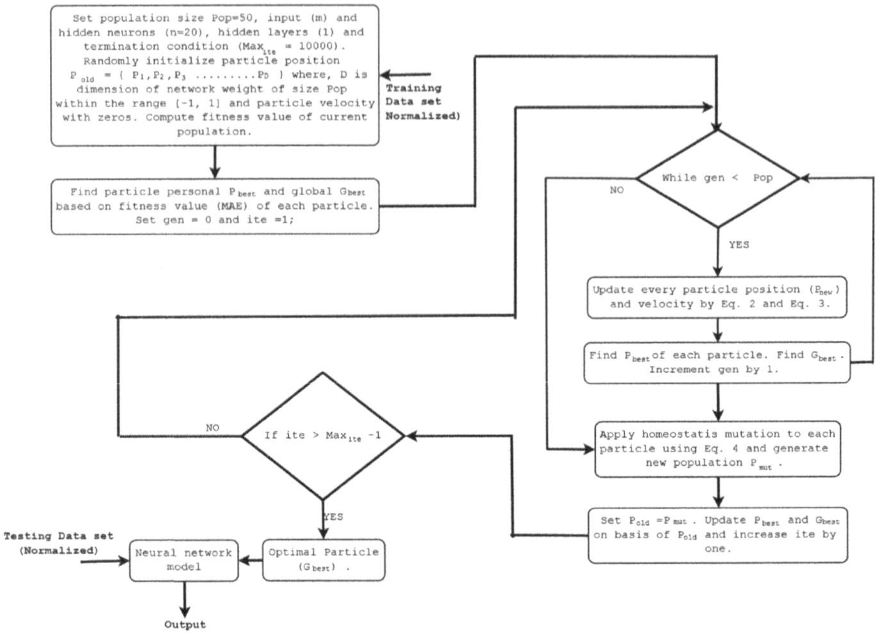

Fig. 1 Sequence of building ANN-PSOHm load forecast model

PSO is a simple search algorithm, it suffers from partial optimism. One of the solutions to this problem is to bring genetic diversity within the population by mutation. Figure 1 presents a flow chart of our load forecasting approach. The steps involved during load forecast are as follows:

- **Data splitting**: Split load data set into training and testing set and normalize them. The training set is used to train the neural network, and testing set generates the final performance of the proposed approach.
- **Initialization**: Choose a multi-layer feed-forward neural network with one hidden layer, 20 hidden neurons. Set maximum numbers of generation as 10,000. Input neurons are the same as input variables chosen for network learning. Initialize random weights of neural network which represents particle position. Initialize particle velocity with zeros.
- **Velocity and position update**: Updated velocity and position of the particles by exchanging their findings with each other (Shi et. al. [13]). These are updated by Eqs. 5 and 6, respectively.
- **Mutation operator**: Homeostasis mutation is performed to generate diversity within the population which is integrated with PSO.
- **Update P_{best} and G_{best}**: P_{best} is the personal best value of each particle, and G_{best} is the best value within the swarm. Update P_{best} and G_{best} on the basis of fitness value of generated child particles.
- **Training termination**: Training stops when maximum generation is reached.

– **Test and evaluation**: An optimal solution (G_{best}) is chosen for model testing. The forecasted load is again scaled back with the same factor with which training data is normalized to get its actual load value.

4 Experiments and Result Analysis

This section presents forecasting evaluation metrics and results of our proposed approach to verify the effectiveness of our proposed approach. Following input parameters are considered in our case study for load prediction: dry bulb temperature, dew point temperature, an hour of the day, day of the week, holiday/weekend indicator, previous 24 h average load, 24 h lagged load and 168 h (previous week) lagged load. Further, load data set is grouped in four different categories depending on England seasons, namely autumn (September to November), winter (December to February), Spring (March to May) and Summer (June to August). Load data from the year 2004–2007 is chosen as training samples and data for the year 2008 and 2009 as testing samples.

All statistical metrics used in this research article are shown in Table 1. MAE generates total absolute forecasting error. RMSE is more stable and less sensitive to noise. NMSE estimates the overall deviations between predicted and measured values. MAPE describes forecasting error which indicates accuracy as a percentage. Daily peak MAPE is a mean value of percentage error change in every 24 h.

Result Analysis To investigate the performance of ANN-PSOHm, we have compared it with linear regression, back-propagation neural network (BPNN), generalized regression neural network (GRNN), ANN-PSO and ANN-PSOm (ANN-PSO with Gaussian mutation).

Table 1 Evaluation metrics

S. No.	Evaluation metric	Metrics equation
1	Mean absolute error	$\text{MAE} = \frac{1}{N}\sum_{j=1}^{N} \lvert y_j - y'_j \rvert$
2	Root of mean squared error	$\text{RMSE} = \sqrt{\frac{1}{N}\sum_{j=1}^{n} (y_j - y'_j)^2}$
3	Normalized mean squared error	$\text{NMSE} = \frac{1}{\Delta^2 N}\sum_{j=1}^{N} (y_j - y'_j)^2$, $\Delta^2 = \frac{1}{N-1}\sum_{j=1}^{N} (y_j - \bar{y})^2$
4	Mean absolute percentage error	$\text{MAPE} = \frac{1}{N}\sum_{j=1}^{N} \frac{\lvert y_j - y'_j \rvert}{y_j} * 100$
5	Daily peak mean absolute per cent error	Daily peak MAPE = $\frac{1}{N} * \frac{\lvert \max(TL_m) - \max(FL_m) \rvert}{\max(TL_m)} * 100$

Note y_j = actual value of day j, y'_j = predicted value of day j, $\bar{y}$ = mean of actual value, $\bar{y}'$ = mean of predicted value, FL_m = forecast load value for every 24 h, TL_m = target load for every 24 h, N = number of elements in training data

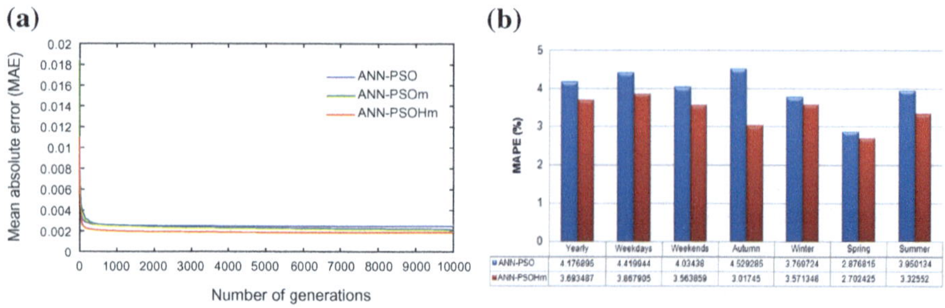

Fig. 2 **a** Convergence rate shown by ANN-PSO, ANN-PSOm and ANN-PSOHm with increasing number of generations and **b** Graph showing comparison between ANN-PSO and ANN-PSOHm using MAPE as an error metric

After successful implementation of forecasting approaches, rigorous analysis has been done to show the effectiveness of our proposed approach. Figure 2a shows the learning ability of neural network with PSO, PSOm and PSOHm. This figure demonstrates that PSO with homeostasis mutation-based ANN approach network has a faster convergence rate along with minimized error compared to the one without a mutation and with traditional mutation (Gaussian mutation). Homeostasis mutation operator creates greater diversity among the population than that of Gaussian mutation, which helps in increased exploration and exploitation capability of PSO.

Table 2 shows the comparative analysis of error metrics produced by the proposed approach (ANN-PSOHm) along with other traditional approaches for yearly, weekly and seasonal calendars. The analysis shows that ANN-PSOHm performs better with the least value of 3.69% MAPE and 548.46 (MWh) MAE, while BPNN generates higher error percentage. ANN-PSOHm lowers the MAE of ANN-PSOm by 0.1979% and ANN-PSO by 0.4838% because of higher diversity maintaining the ability for the yearly forecast.

Table 2 also shows the result of the load forecast predicted over weekdays and weekends. Result reveals that ANN-PSOHm produced the best forecasting results with 0.2165% MAPE variation over ANN-PSOm and 0.552% variation over ANN-PSO during weekdays. It shows that ANN-PSOHm achieved the most accurate prediction value with a variation of 0.1144% MAPE over ANN-PSOm and 0.1634% over ANN-PSO on weekends.

Results obtained in all four seasons show that ANN-PSOHm achieved the most predictive values. In the autumn season, the load forecast of ANN-PSOHm overcomes all approaches by generating the optimal results, with 3.0174% MAPE and 424.1457 (MWh) MAPE. In winter, ANN-PSOHm offered the least error than any other state-of-the-art algorithm, with a MAPE value of 3.5713% and MAE value of 537.4397 (MWh). Apart from the optimal solution produced by ANN-PSOHm, the results also present that the error has been reduced by the MAPE value of 0.0985% and 0.1984% for ANN-PSOm and ANN-PSO, respectively. For the spring season, ANN-PSOHm produced the ideal forecasting results than any other mentioned approach,

Table 2 Comparison of error metrics for year 2008–2009

Calendar		Algorithm	RMSE (MWh)	NMSE	MAE (MWh)	MAPE (%)	Daily peak MAPE (%)
Yearly	2008–2009	Linear regression	876.1726	0.096165	645.5463	4.347394	3.979897
		GRNN	1186.348	0.176303	840.4085	5.614846	4.497605
		BPNN	1129.586	0.159836	852.0643	5.917566	4.978817
		ANN-PSO	850.1101	0.090529	623.4493	4.176895	3.850884
		ANN-PSOm	793.091	0.078792	580.3469	3.891394	3.683529
		ANN-PSOHm	***746.1335***	***0.069738***	***548.4582***	***3.693487***	***3.616976***
Weekly	Weekdays	Linear regression	900.5164	0.097886	664.8451	4.377995	3.753069
		GRNN	1186.511	0.169934	819.543	5.362099	4.392156
		BPNN	1216.229	0.178553	903.3511	5.880937	5.397314
		ANN-PSO	912.4786	0.100504	665.1454	4.419944	3.675826
		ANN-PSOm	861.6829	0.089626	618.0008	4.080443	3.743331
		ANN-PSOHm	***785.8354***	***0.074542***	***576.6551***	***3.867905***	***3.35671***
	Weekends	Linear regression	743.238	0.095049	561.3386	4.020267	4.877239
		GRNN	1031.779	0.183175	757.3971	5.337287	4.598976
		BPNN	915.8744	0.144333	702.525	5.21507	4.449637
		ANN-PSO	765.5896	0.100852	566.2481	4.03438	4.161539
		ANN-PSOm	753.5384	0.097702	555.2023	3.957178	4.112578
		ANN-PSOHm	***687.7966***	***0.081398***	***502.7183***	***3.563859***	***3.998135***
Season	Autumn	Linear regression	772.3644	0.099265	575.005	4.106392	3.958938
		GRNN	1085.279	0.19599	760.4547	5.399025	4.028724
		BPNN	1009.285	0.169504	758.4348	5.378196	5.167414
		ANN-PSO	820.0898	0.111911	625.9252	4.529285	4.356279
		ANN-PSOm	753.6649	0.094517	568.1676	4.057046	4.061379
		ANN-PSOHm	***587.1167***	***0.057359***	***424.1457***	***3.017453***	***3.514169***

(continued)

Table 2 (continued)

Calendar		Algorithm	RMSE (MWh)	NMSE	MAE (MWh)	MAPE (%)	Daily peak MAPE (%)
	Winter	Linear regression	749.0536	0.100129	574.0396	3.825044	3.207751
		GRNN	1184.201	0.250256	894.4198	5.801034	4.174389
		BPNN	865.4497	0.133665	682.3763	4.463057	5.139326
		ANN-PSO	735.6062	0.096566	566.1826	3.769724	2.753557
		ANN-PSOm	715.8424	0.091447	551.4715	3.669858	***2.617679***
		ANN-PSOHm	***694.8657***	***0.086166***	***537.4397***	***3.571348***	***2.65845***
	Spring	Linear regression	587.7354	0.069575	438.0351	3.188387	2.684402
		GRNN	777.7608	0.121838	565.8317	4.101719	3.716093
		BPNN	751.2703	0.11368	574.2589	4.173754	4.042743
		ANN-PSO	525.9699	0.05572	393.9254	2.876815	2.48476
		ANN-PSOm	495.574	0.049466	373.4645	2.705296	2.681057
		ANN-PSOHm	***482.9654***	***0.046981***	***367.3192***	***2.702425***	***2.355562***
	Summer	Linear regression	891.4899	0.067415	700.9039	4.668146	3.78347
		GRNN	1571.863	0.20958	1149.011	7.183112	6.020333
		BPNN	1458.383	0.180411	1075.106	6.53912	6.223644
		ANN-PSO	758.27	0.048772	595.4153	3.950134	3.003364
		ANN-PSOmm	640.0093	0.034745	501.8638	3.319555	2.930752
		ANN-PSOHm	***633.3212***	***0.034023***	***496.0001***	***3.32552***	***2.664822***

with MAPE value of 2.7024% and MAE value of 367.3192 (MWh). The optimal solution produced by ANN-PSOHm shows that the error of ANN-PSOm and ANN-PSO has been reduced by MAPE value of 0.0028% and 0.1744%. For the summer season, ANN-PSOHm provides greater improvement in forecasting accuracy than any other mentioned approach, with a MAPE value of 3.3255% and MAE value of 496.0001 (MWh).

Remark Summary of all three case studies has been presented in Fig. 2b. This figure shows the comparison of MAPE generated by ANN-PSO and ANN-PSOHm for all calendar effects. Graph reveals that ANN-PSOHm performed better in all the three case studies and is independent of calendar effects. There is 11.57% improvement of MAPE in ANN-PSOHm over ANN-PSO on yearly basis, 12.48 and 11.66% improvement of MAPE in ANN-PSOHm over ANN-PSO on weekdays and weekends, respectively. The improvement of 33.37, 5.26, 6.06 and 15.81% MAPE in autumn, winter, spring and summer for ANN-PSOHm over ANN-PSO has been seen in the study.

5 Conclusion

Forecasting is a prediction or estimation of future events. In this paper, an ensemble soft computing approach composed of ANN and PSO with homeostasis based mutation is presented for forecasting the electricity load. Results are drawn from numerous experiments based on weekly, seasonal and yearly. It can be stated that homeostasis mutation provides great diversity among the population and accurate results than that of the algorithms without mutation and with Gaussian mutation when applied to short-term load forecast application. Homeostasis mutation supplies a great combination of exploration and exploitation within a population, resulting in a higher convergence rate and wider search space. Further, this approach can be used in other applications such as wind forecasting and stock market prediction.

References

1. AlFuhaid, A., El-Sayed, M., Mahmoud, M.: Cascaded artificial neural networks for short-term load forecasting. IEEE Trans. Power Syst. **12**(4), 1524–1529 (1997)
2. Ayari, K., Bouktif, S., Antoniol, G.: Automatic mutation test input data generation via ant colony. In: Proceedings of the 9th Annual Conference on Genetic and Evolutionary Computation, pp. 1074–1081. ACM (2007)
3. Berk, K., Hoffmann, A., Müller, A.: Probabilistic forecasting of industrial electricity load with regime switching behavior. Int. J. Forecast. **34**(2), 147–162 (2018)
4. Dimoulkas, I., Mazidi, P., Herre, L.: Neural networks for gefcom2017 probabilistic load forecasting. Int. J. Forecast. (2018)
5. Dorigo, M., Gambardella, L.M.: Ant colony system: a cooperative learning approach to the traveling salesman problem. IEEE Trans. Evol. Comput. **1**(1), 53–66 (1997)
6. Higashi, N., Iba, H.: Particle swarm optimization with gaussian mutation. In: Proceedings of the 2003 IEEE Swarm Intelligence Symposium, 2003. SIS'03, pp. 72–79. IEEE (2003)
7. Kiartzis, S., Kehagias, A., Bakirtzis, A., Petridis, V.: Short term load forecasting using a bayesian combination method. Int. J. Electr. Power Energy Syst. **19**(3), 171–177 (1997)
8. Liao, G.-C.: Hybrid chaos search genetic algorithm and meta-heuristics method for short-term load forecasting. Electr. Eng. **88**(3), 165–176 (2006)
9. Lungu, M., Lungu, R.: Neural network based adaptive control of airplanes lateral-directional motion during final approach phase of landing. Eng. Appl. Artif. Intell. **74**, 322–335 (2018)

10. Pai, P.-F., Hong, W.-C.: Support vector machines with simulated annealing algorithms in electricity load forecasting. Energy Convers. Manag. **46**(17), 2669–2688 (2005)
11. Quan, H., Srinivasan, D., Khosravi, A.: Short-term load and wind power forecasting using neural network-based prediction intervals. IEEE Trans. Neural Netw. Learn. Syst. **25**(2), 303–315 (2014)
12. Shi, Y., Eberhart, R.C.: Empirical study of particle swarm optimization. In: Proceedings of the 1999 Congress on Evolutionary Computation 1999. CEC 99. vol. 3, pp. 1945–1950, IEEE (1999)
13. Shi, Y., et al.: Particle swarm optimization: developments, applications and resources. In: Proceedings of the 2001 Congress on Evolutionary Computation, 2001, vol. 1, pp. 81–86, IEEE (2001)
14. Singh, P., Dwivedi, P.: Integration of new evolutionary approach with artificial neural network for solving short term load forecast problem. Appl. Energy **217**, 537–549 (2018)
15. Singh, P., Dwivedi, P.: A novel hybrid model based on neural network and multi-objective optimization for effective load forecast. Energy **182**, 606–622 (2019)
16. Singh, P., Dwivedi, P., Kant, V.: A hybrid method based on neural network and improved environmental adaptation method using controlled gaussian mutation with real parameter for short-term load forecasting. Energy **174**, 460–477 (2019)
17. Singh, P., Mishra, K., Dwivedi, P.: Enhanced hybrid model for electricity load forecast through artificial neural network and jaya algorithm. In: 2017 International Conference on Intelligent Computing and Control Systems (ICICCS), pp. 115–120. IEEE (2017)
18. Singh, S.P., Kumar, A.: Software cost estimation using homeostasis mutation based differential evolution. In: 2017 11th International Conference on Intelligent Systems and Control (ISCO). pp. 173–181, IEEE (2017)
19. Singh, S.P., Kumar, A.: Multiobjective differential evolution using homeostasis based mutation for application in software cost estimation. Appl. Intell. **48**(3), 628–650 (2018)
20. Trelea, I.C.: The particle swarm optimization algorithm: convergence analysis and parameter selection. Nformation Process. Lett. **85**(6), 317–325 (2003)
21. Wu, Q.: Power load forecasts based on hybrid pso with gaussian and adaptive mutation and wv-svm. Expert. Syst. Appl. **37**(1), 194–201 (2010)

Diagnostics Relevant Modeling of Squirrel-Cage Induction Motor: Electrical Faults

SSSR Sarathbabu Duvvuri

Abstract In this paper, simplified SCIM models are formulated based on stationary, rotor, and synchronous reference frames. All these models are compared and analyzed in terms of their diagnostic relevance to major electrical faults (stator inter-turn short-circuit and broken rotor bars). Ability to develop distinct residual signatures is a key for any model-based fault diagnosis method. The performance of various models in terms of their ability to generate a distinct residual and best-suited model is recommended based on discriminatory ability index proposed in this manuscript. Extended Kalman filter is the most commonly used estimator for nonlinear systems. The SCIM, being a nonlinear system, extended Kalman filter is considered for state estimation. As an extension, parameter sensitivity analysis is carried out for the best-suited model. Efforts are made to convey which parameters have a significant effect in case there is plant-model mismatch. Analytical computations are carried out for a 3 kW SCIM motor using MATLAB software. The results show that the most effective squirrel-cage induction motor model for model-based fault diagnostics.

Keywords Extended Kalman filter (EKF) · Reference frame theory · Squirrel-cage induction motor (SCIM)

1 Introduction

While earlier, ease of computation carried significant emphasis, nowadays relatively computationally demanding methodologies, such as observer-based method, are gaining popularity [1–4] among academic researchers as well as industry. Primary reason for this change is increased speeds as well as multi-core architecture of processor. The sensitivity of the innovations may vary with the chosen model. For effective diagnosis, the model should be simple yet accurate as well as computationally efficient. Model-based diagnosis is becoming more and more popular in the area of electrical machine fault diagnosis as well [1–4]. A large number of models, ranging from very detailed finite element-based models to simple models

S. Sarathbabu Duvvuri (✉)
Indian Institute of Technology Hyderabad, Kandi, TS 502285, India
e-mail: ee11p1007@iith.ac.in

D. Dutta and B. Mahanty (eds.), *Numerical Optimization in Engineering and Sciences*, Advances in Intelligent Systems and Computing 979,
https://doi.org/10.1007/978-981-15-3215-3_17

such as models based on the reference frame theory, are available and proposed. However, as stated earlier, the task of FDD requires computationally non-expensive yet accurate model. As models based on finite element, multiple-coupled circuit, and magnetic equivalent circuit is complex and simulating these models in real time is computationally intensive; these models will not be effective for the purpose of FDD [1–4]. Therefore simple models based on reference frame theory proposed in, are better suited for machine fault diagnosis. SCIM models can be formulated in stationary, rotor, synchronous, and arbitrary reference frames. Similarly, there are different physical variables, which may be chosen as a state variable for an SCIM model. As will be shown later, there are multiple ways in which a physical variable form of an extended state-space model for SCIM can be modeled. This poses some unique questions pertinent to SCIM fault diagnosis.

- Are all the SCIM models are similar?
- Which reference one should prefer for fault diagnostics?
- Which SCIM is diagnostically most relevant?
- Which parameter is highly sensitive in the chosen model in case of plant-model mismatch?

The aim of this paper is to answer all these questions and find the model that is most relevant for SCIM electrical fault diagnosis. Electrical faults, i.e., stator inter-turn short-circuit and broken rotor bars, account for nearly 45–50% of total faults in SCIM [1] and thus constitute majority of the faults in SCIM. Since extended Kalman filter is extensively used nonlinear observer, the same is considered for state estimation of SCIM [1–11]. A detailed performance evaluation was carried out to analyze the performance of various SCIM models and select the most appropriate model that can generate distinctive residuals and thus are better suited to discriminate. Further, once a model is chosen, the next task is to identify key parameters for the same. Toward this end, a parameter sensitivity analysis is carried out and robustness of an SCIM model to plant-model mismatch is analyzed. It is expected that such detailed analysis will provide a benchmark inappropriate model selection as well as parameter identification for an SCIM model.

The paper is presented as follows: state-space SCIM models are presented in detail in Sect. 2 followed by standard extended Kalman filter design, and a performance index is defined to compare performance in Sect. 3. Validated simulation results and corresponding discussions are placed in Sect. 4, and the manuscript is finalized by concluding remark in Sect. 5.

2 Extended State-Space Model of SCIM

In this section, an extended SCIM model in physical variable form is formulated. For the development of SCIM models, assumptions presented in [1–11] are considered for further analysis.

$$\begin{bmatrix}\dot{\mathbf{i}}^s_{abc}\\ \dot{\mathbf{i}}^r_{abc}\end{bmatrix}=\begin{bmatrix}\mathbf{L}^s_{abc} & \mathbf{L}^{sr}_{abc}\\ \mathbf{L}^{rs}_{abc} & \mathbf{L}^r_{abc}\end{bmatrix}^{-1}\begin{bmatrix}\mathbf{v}^s_{abc}\\ 0\end{bmatrix}-\begin{bmatrix}\mathbf{L}^s_{abc} & \mathbf{L}^{sr}_{abc}\\ \mathbf{L}^{rs}_{abc} & \mathbf{L}^r_{abc}\end{bmatrix}^{-1}\begin{bmatrix}\mathbf{R}^s_{abc} & \dot{\mathbf{L}}^{sr}_{abc}\\ \dot{\mathbf{L}}^{rs}_{abc} & \mathbf{R}^r_{abc}\end{bmatrix}\begin{bmatrix}\mathbf{i}^s_{abc}\\ \mathbf{i}^r_{abc}\end{bmatrix}$$

$$\dot{\omega}_r=\frac{1}{J}\left(n_p\left(\mathbf{i}^s_{abc}\right)^T\frac{\partial}{\partial\theta_r}\left[\mathbf{L}^{sr}_{abc}\right]\left(\mathbf{i}^r_{abc}\right)-T_l-n_pB_l\omega_r\right)$$

$$\dot{T}_l=0 \tag{1}$$

Changes in the inductance are neglected when rotor bar breaks [1–11]. The rotor bars are electrically insulated. The effect of friction is neglected as the friction coefficient is imperceptible. The induction motor dynamic equations in *abc* machine variables are presented in [1–11] as follows.

The eighth-order SCIM model presented in (1) is computationally intensive. Therefore, the model presented in (1) is transformed using reference frame theory [5–7] as follows:

$$\begin{aligned}\dot{\mathbf{i}}^s_{qd0}&=\mathbf{F}^s_{qd0}\left(\mathbf{v}^s_{qd0}-\mathbf{R}^s_{qd0}\mathbf{i}^s_{qd0}-\omega_s\mathbf{K}_\omega\left(\mathbf{L}^s_{qd0}\mathbf{i}^s_{qd0}+\mathbf{L}^{sr}_{qd0}\mathbf{i}^r_{qd0}\right)\right)\\&\quad+\mathbf{F}^{sr}_{qd0}\left(-\mathbf{R}^r_{qd0}\mathbf{i}^r_{qd0}-\left(\omega_s-n_p\omega_r\right)\mathbf{K}_\omega\left(\mathbf{L}^{rs}_{qd0}\mathbf{i}^s_{qd0}+\mathbf{L}^r_{qd0}\mathbf{i}^r_{qd0}\right)\right)\\ \dot{\mathbf{i}}^r_{qd0}&=\mathbf{F}^{rs}_{qd0}\left(\mathbf{v}^s_{qd0}-\mathbf{R}^s_{qd0}\mathbf{i}^s_{qd0}-\omega_s\mathbf{K}_\omega\left(\mathbf{L}^s_{qd0}\mathbf{i}^s_{qd0}+\mathbf{L}^{sr}_{qd0}\mathbf{i}^r_{qd0}\right)\right)\\&\quad+\mathbf{F}^r_{qd0}\left(-\mathbf{R}^r_{qd0}\mathbf{i}^r_{qd0}-\left(\omega_s-n_p\omega_r\right)\mathbf{K}_\omega\left(\mathbf{L}^{rs}_{qd0}\mathbf{i}^s_{qd0}+\mathbf{L}^r_{qd0}\mathbf{i}^r_{qd0}\right)\right)\\ \dot{\omega}_r&=\frac{1}{J}\left(\frac{3}{2}n_pL_m\left(i^s_qi^r_d-i^s_di^r_q\right)-T_l-n_pB_l\omega_r\right)\\ \dot{T}_l&=0\end{aligned} \tag{2}$$

The eighth-order SCIM model based on stator currents and rotor currents given in (2) (***Model d***) can be represented in various physical variable forms, i.e., there are many physical variables, which can be chosen as states. One can select either currents or fluxes on either rotor side or stator side of the machine. Thus, there are four sets of variables available to model the SCIM: (i) three stator currents $\mathbf{i}^s_{qd0}$; (ii) three rotor currents $\mathbf{i}^r_{qd0}$; (iii) three stator fluxes $\boldsymbol{\lambda}^s_{qd0}$; and (iv) three rotor fluxes $\boldsymbol{\lambda}^r_{qd0}$. One can select any two from the above set of variables, which gives rise to $^4C_2$, i.e., six possible combinations.

2.1 SCIM Models

As stated earlier, the SCIM models presented in can also be written in one of the three reference frames, i.e., stationary, synchronous, or rotor reference frame as mentioned below.

- ***Model a*: Stator fluxes and rotor fluxes**

$$\dot{\boldsymbol{\lambda}}^s_{qd0}=\mathbf{v}^s_{qd0}-\mathbf{R}^s_{qd0}\left(\mathbf{F}^s_{qd0}\boldsymbol{\lambda}^s_{qd0}+\mathbf{F}^{sr}_{qd0}\boldsymbol{\lambda}^r_{qd0}\right)-\omega\mathbf{K}_\omega\boldsymbol{\lambda}^s_{qd0}$$

$$\begin{aligned}
\dot{\boldsymbol{\lambda}}^{r}_{qd0} &= -\mathbf{R}^{r}_{qd0}\left(\mathbf{F}^{rs}_{qd0}\boldsymbol{\lambda}^{s}_{qd0} + \mathbf{F}^{r}_{qd0}\boldsymbol{\lambda}^{r}_{qd0}\right) - (\omega - n_p\omega_r)\mathbf{K}_{\omega}\boldsymbol{\lambda}^{r}_{qd0} \\
\dot{\omega}_r &= \frac{1}{J}\left(\frac{3}{2}n_p\frac{L_m}{(L_sL_r - L_m^2)}\left(\lambda^s_q\lambda^r_d - \lambda^s_d\lambda^r_q\right) - T_l\right) \\
\dot{T}_l &= 0
\end{aligned} \tag{3}$$

- ***Model b*: Stator fluxes and rotor currents**

$$\begin{aligned}
\dot{\boldsymbol{\lambda}}^{s}_{qd0} &= \mathbf{v}^{s}_{qd0} - \mathbf{R}^{s}_{qd0}\left(\left(\mathbf{L}^{s}_{qd0}\right)^{-1}\boldsymbol{\lambda}^{s}_{qd0} - \left(\mathbf{L}^{s}_{qd0}\right)^{-1}\mathbf{L}^{sr}_{qd0}\mathbf{i}^{r}_{qd0}\right) - \omega\mathbf{K}_{\omega}\boldsymbol{\lambda}^{s}_{qd0} \\
\dot{\mathbf{i}}^{r}_{qd0} &= \mathbf{F}^{rs}_{qd0}\left(\mathbf{v}^{s}_{qd0} - \mathbf{R}^{s}_{qd0}\left(\left(\mathbf{L}^{s}_{qd0}\right)^{-1}\boldsymbol{\lambda}^{s}_{qd0} - \left(\mathbf{L}^{s}_{qd0}\right)^{-1}\mathbf{L}^{sr}_{qd0}\mathbf{i}^{r}_{qd0}\right) - \omega\mathbf{K}_{\omega}\boldsymbol{\lambda}^{s}_{qd0}\right) \\
&\quad - \mathbf{F}^{r}_{qd0}\mathbf{R}^{r}_{qd0}\mathbf{i}^{r}_{qd0} - \left(\mathbf{F}^{r}_{qd0}(\omega - n_p\omega_r)\mathbf{K}_{\omega}\left(\mathbf{L}^{rs}_{qd0}\left(\mathbf{L}^{s}_{qd0}\right)^{-1}\boldsymbol{\lambda}^{s}_{qd0}\right.\right. \\
&\quad \left.\left. + \left(\mathbf{L}^{r}_{qd0} - \mathbf{L}^{rs}_{qd0}\left(\mathbf{L}^{s}_{qd0}\right)^{-1}\mathbf{L}^{sr}_{qd0}\right)\mathbf{i}^{r}_{qd0}\right)\right) \\
\dot{\omega}_r &= \frac{1}{J}\left(\frac{3}{2}n_p\frac{L_m}{L_s}\left(\lambda^s_q i^r_d - \lambda^s_d i^r_q\right) - T_l\right) \\
\dot{T}_l &= 0
\end{aligned} \tag{4}$$

- ***Model c*: Stator currents and stator fluxes**

$$\begin{aligned}
\dot{i}^s_q &= -\left(\frac{r_sL_r + r_rL_s}{L_sL_r - L_m^2}\right)i^s_q - (\omega - n_p\omega_r)i^s_d + \frac{r_r}{L_sL_r - L_m^2}\lambda^s_q \\
&\quad - \frac{L_r}{L_sL_r - L_m^2}n_p\omega_r\lambda^s_d + \frac{L_r}{L_sL_r - L_m^2}v^s_q \\
\dot{i}^s_d &= -\left(\frac{r_sL_r + r_rL_s}{L_sL_r - L_m^2}\right)i^s_d + (\omega - n_p\omega_r)i^s_q + \frac{r_r}{L_sL_r - L_m^2}\lambda^s_d \\
&\quad + \frac{L_r}{L_sL_r - L_m^2}n_p\omega_r\lambda^s_q + \frac{L_r}{L_sL_r - L_m^2}v^s_d \\
\dot{i}^s_0 &= \left(\frac{v^s_0 - r_si^s_0}{L_{ls}}\right) \\
\dot{\boldsymbol{\lambda}}^{s}_{qd0} &= \mathbf{v}^{s}_{qd0} - \mathbf{R}^{s}_{qd0}\mathbf{i}^{s}_{qd0} - \omega\mathbf{K}_{\omega}\boldsymbol{\lambda}^{s}_{qd0} \\
\dot{\omega}_r &= \frac{1}{J}\left(\frac{3}{2}n_p\left(\lambda^s_d i^s_q - \lambda^s_q i^s_d\right) - T_l\right) \\
\dot{T}_l &= 0
\end{aligned} \tag{5}$$

- ***Model e*: Stator currents and rotor fluxes**

$$\begin{aligned}
\dot{\mathbf{i}}^{s}_{qd0} &= \mathbf{F}^{s}_{qd0}\left(\mathbf{v}^{s}_{qd0} - \mathbf{R}^{s}_{qd0}\mathbf{i}^{s}_{qd0} - \omega\mathbf{K}_{\omega}\left(\left(\mathbf{L}^{s}_{qd0} - \mathbf{L}^{sr}_{qd0}\left(\mathbf{L}^{r}_{qd0}\right)^{-1}\mathbf{L}^{rs}_{qd0}\right)\mathbf{i}^{s}_{qd0}\right.\right. \\
&\quad \left.\left. + \mathbf{L}^{sr}_{qd0}\left(\mathbf{L}^{r}_{qd0}\right)^{-1}\boldsymbol{\lambda}^{r}_{qd0}\right)\right) + \mathbf{F}^{sr}_{qd0}\left(-\mathbf{R}^{r}_{qd0}\left(\left(\mathbf{L}^{r}_{qd0}\right)^{-1}\boldsymbol{\lambda}^{r}_{qd0}\right.\right.
\end{aligned}$$

$$
\begin{aligned}
&- \left(\mathbf{L}^{r}_{qd0}\right)^{-1}\mathbf{L}^{rs}_{qd0}\mathbf{i}^{s}_{qd0}\Big) - (\omega - n_p\omega_r)\mathbf{K}_{\omega}\boldsymbol{\lambda}^{r}_{qd0}\Big) \\
\dot{\boldsymbol{\lambda}}^{r}_{qd0} &= -\mathbf{R}^{r}_{qd0}\Big(\left(\mathbf{L}^{r}_{qd0}\right)^{-1}\boldsymbol{\lambda}^{r}_{qd0} - \left(\mathbf{L}^{r}_{qd0}\right)^{-1}\mathbf{L}^{rs}_{qd0}\mathbf{i}^{s}_{qd0}\Big) - (\omega - n_p\omega_r)\mathbf{K}_{\omega}\boldsymbol{\lambda}^{r}_{qd0} \\
\dot{\omega}_r &= \frac{1}{J}\left(\frac{3}{2}n_p\frac{L_m}{L_r}(i^s_q\lambda^r_d - i^s_d\lambda^r_q) - T_l\right) \\
\dot{T}_l &= 0
\end{aligned} \tag{6}
$$

- ***Model f*: Rotor fluxes and rotor currents**

$$
\begin{aligned}
\dot{\boldsymbol{\lambda}}^{r}_{qd0} &= -\mathbf{R}^{r}_{qd0}\mathbf{i}^{r}_{qd0} - (\omega - n_p\omega_r)\mathbf{K}_{\omega}\boldsymbol{\lambda}^{r}_{qd0} \\
\dot{i}^r_q &= -\left(\frac{r_sL_r + r_rL_s}{L_sL_r - L_m^2}\right)i^r_q - \omega i^r_d + \frac{r_s}{L_sL_r - L_m^2}\lambda^r_q \\
&\quad + \frac{L_s}{L_sL_r - L_m^2}n_p\omega_r\lambda^r_d - \frac{L_m}{L_sL_r - L_m^2}v^s_q \\
\dot{i}^r_d &= -\left(\frac{r_sL_r + r_rL_s}{L_sL_r - L_m^2}\right)i^r_d + \omega i^r_q + \frac{r_r}{L_sL_r - L_m^2}\lambda^r_d \\
&\quad - \frac{L_s}{L_sL_r - L_m^2}n_p\omega_r\lambda^r_q - \frac{L_m}{L_sL_r - L_m^2}v^s_d \\
\dot{i}^r_0 &= -\frac{r_r}{L_{lr}}i^r_0 \\
\dot{\omega}_r &= \frac{1}{J}\left(\frac{3}{2}n_p(\lambda^r_q i^r_d - \lambda^r_d i^r_q) - T_l\right) \\
\dot{T}_l &= 0
\end{aligned} \tag{7}
$$

For the model to be diagnostic relevant, the residues must be sensitive to faults and robust to disturbances and load changes.

2.2 *Key Parameters for SCIM Models*

The SCIM models derived in the previous subsection require seven parameters: (i) stator phase resistance (r_s); (ii) rotor phase resistance (r_r); (iii) stator leakage inductance (L_{ls}); (iv) rotor leakage inductance (L_{lr}); (v) magnetizing inductance (L_{ms}); (vi) magnetizing inductance (L_{mr}); and (vii) moment of inertia (J). As the models are nonlinear, effect of parameter variation or plant-model mismatch on diagnostic performance may be different for different parameters. Therefore, the sensitivity of the SCIM model parameters to inaccuracy or changes in parameters is investigated. The influence of different parameters on the residues generated from the model is evaluated in this manuscript.

3 Squirrel-Cage Induction Motor State Estimation Using Extended Kalman Filter and Discriminatory Ability Index for Model-Based Fault Diagnosis

The state estimates for SCIM is obtained from extended Kalman filter (EKF) [1–11]. The residual or innovations are generated as follows:

$$\boldsymbol{\gamma}(k) = \mathbf{y}(k) - \mathbf{C}(k)\hat{\mathbf{x}}(k|k-1) \tag{8}$$

Under normal operating conditions, innovations or residuals presented in follow Gaussian distribution. However, when a rotor inter-turn fault occurs, these innovations become non-white. Here, out of the fifteen various models for SRIM, which model(s) is(are) diagnostically relevant has to be determined. For finding a suitable model, a new discriminatory ability index (DAI) based on residual (innovations) sum of squares (RSS, denoted by Γ) is proposed in

$$\Gamma = \sum_{k=t}^{t+N} \boldsymbol{\gamma}(k)^{\mathrm{T}}\boldsymbol{\gamma}(k), \overline{\Gamma} = \sum_{i} \frac{\Gamma(i)}{m}, \mathrm{DA} = \begin{cases} \frac{\Gamma_f - \overline{\Gamma}}{\overline{\Gamma}} \times 100 & \Gamma_f > \Gamma_{\max} \\ 0 & \Gamma_f < \Gamma_{\max} \end{cases} \tag{9}$$

where $\Gamma_{\max}$ denotes empirically obtained maximum value of Γ during normal operation. A proposed discriminatory ability index can be used to compare relative performance of various models.

4 Main Simulation Results and Observations

Simulations were carried out for fifteen different state-space models of SCIM. Diagnostic ability index is calculated for the proposed different induction motor models and is presented in this section.

4.1 Stator Inter-Turn Fault and Rotor Inter-Turn Fault

In this simulation study, covariance matrices are considered in diagonal form for state estimation and the sampling time is $T = 0.1$ ms. The performance of various models was computed with one percent stator inter-turn short-circuit and one broken rotor bar fault severity [6–9]. In simulations, an electrical fault presented in is introduced at $t = 0.2$ s (Table 1).

As can be seen from Table 2, not all models show the same amount of discriminatory ability toward stator fault. The models best suited for stator fault diagnosis are:

Table 1 SCIM parameters used in simulations

Parameter	Variable	Value
Rated power	P_{rated}	3 kW
Rated stator voltage	V_l	380 V
Rated speed	N_r	1430 rpm
Rated stator frequency	f_s	50 Hz
Number of stator turns	N_s	312
Number of rotor bars	N_b	28
Stator phase resistance	r_s	2.283 Ω
Rotor phase resistance	r_r	2.133 Ω
Magnetizing inductance	L_{ms}	146.7 mH
Stator leakage inductance	L_{ls}	11.1 mH
Moment of inertia	J	0.06 kg m^2

Table 2 Performance index for different SCIM models

Model	State variables	Reference frame	DA	
			Stator Inter-turn	Broken rotor bars
a	Stator fluxes and rotor fluxes	Stationary	0	60.08
		Synchronous	11.0628	43.82
		Rotor	0	53.84
b	Stator fluxes and rotor currents	Stationary	0	33.54
		Synchronous	0	0
		Rotor	0	0
c	Stator currents and stator fluxes	Stationary	101.185	0
		Synchronous	0	0
		Rotor	0	0
d	**Stator currents and rotor currents**	**Stationary**	**105.61**	**111.71**
		Synchronous	27.0767	0
		Rotor	0	0
e	Stator currents and rotor fluxes	Stationary	82.2875	0
		Synchronous	0	0
		Rotor	0	0

- SCIM model with state variables as $(\mathbf{i}^s_{qd0}, \mathbf{i}^r_{qd0}, \omega_r, T_l)$, i.e., ***Model d*** in the stationary reference frame.

The results are also validated by observing the autocorrelation coefficients as shown in Figs. 1 and 2. As can be seen from the figure, when ***Model d*** in the stationary reference frame is used for EKF, residues become non-Gaussian. The same is also

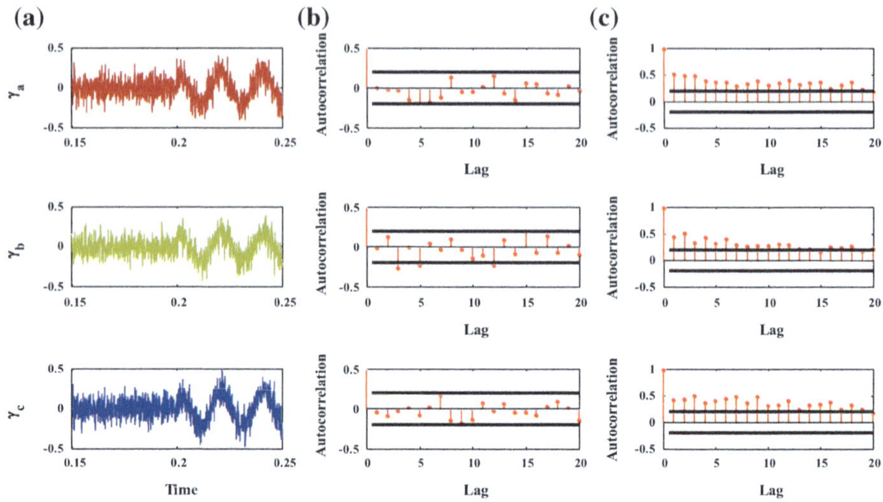

Fig. 1 **a** Innovations for ***Model d*** in stationary reference frame; **b** autocorrelation coefficients at $t = 0.15$–0.20 s (normal operation); **c** autocorrelation coefficient at $t = 0.20$ s (stator inter-turn fault operation)

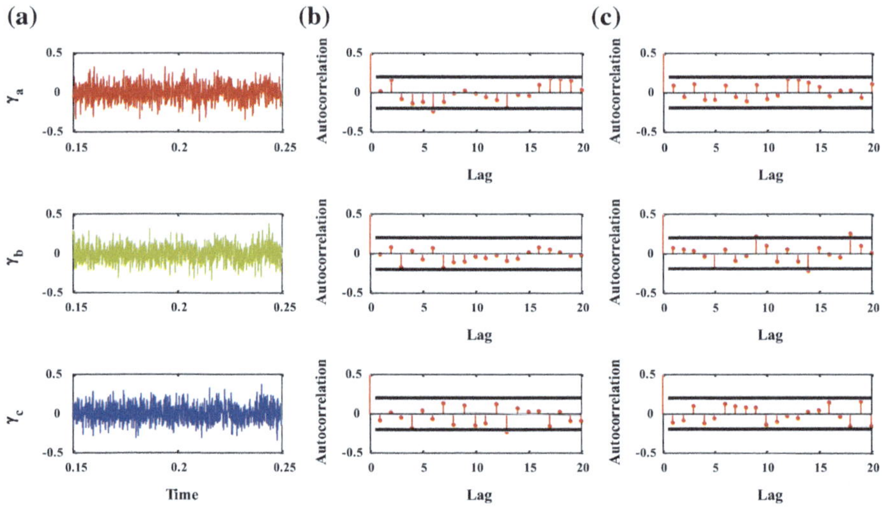

Fig. 2 **a** Innovations for ***Model e*** in synchronous reference frame; **b** autocorrelation coefficients at $t = 0.15$–0.20 s (normal operation); **c** autocorrelation coefficient at $t = 0.20$ s (stator inter-turn fault operation)

validated by the autocorrelation coefficients, which suggest that the residues have become non-Gaussian after the introduction of a stator fault and rotor fault [6–9].

4.2 Robustness to Parameter Variations

As ***Model d*** in the stationary reference frame is the most suitable model for fault diagnosis, only that model is considered for further analysis. Any system may undergo degradation of performance over time. This degradation would mean parameter of the system may vary. Therefore, it is important to analyze sensitivity of residues to parameter variation or plant-model mismatch. Each of the seven key parameters was varied, and simulations were carried out. Similar to the previous subsection, the best-suited SCIM model is fed from a balanced supply, and state estimation is carried out using an extended Kalman filter.

In simulations, percentage change in one of the key parameters is introduced at $t = 0.2$ s. The change in parameter will lead to plant-model mismatch, which should deviate residues from zero-mean Gaussian white noise sequence. The purpose of the analysis carried out in this subsection is to identify the percentage range of the key parameters such that plant-model mismatch is not detected as a fault. The limits for inaccuracies of the parameters in terms of percentages are presented in Table 3. If the mismatch between the actual motor and the model parameters is larger than what is indicated in Table 3, the innovations will become non-Gaussian, causing a residual-based fault diagnosis method to generate false alarms (Figs. 3 and 4).

As seen from Table 3, not all the parameters show the same amount of parameter sensitivity. The parameters may be classified into three categories as shown below:

- Highly sensitive parameters: Magnetizing inductances L_{ms} and L_{mr}.
- Moderately sensitive parameters: Stator leakage inductance L_{ls} and stator phase resistance r_s.
- Highly insensitive parameters: Rotor leakage inductance L_{lr}, rotor phase resistance r_r and moment of inertia J (Fig. 5).

Table 3 Parameter sensitivity index for SCIM model

Serial No.	Key Parameters	Variable	Range of model inaccuracies in terms of %
1	**Magnetizing inductance**	L_{ms}	**±3**
2	**Magnetizing inductance**	L_{mr}	**±3**
3	*Stator leakage inductance*	L_{ls}	***±10***
4	*Stator phase resistance*	r_s	***±30***
5	Rotor phase resistance	r_r	±60
6	Rotor leakage inductance	L_{lr}	±80
7	Moment of inertia	J	±90

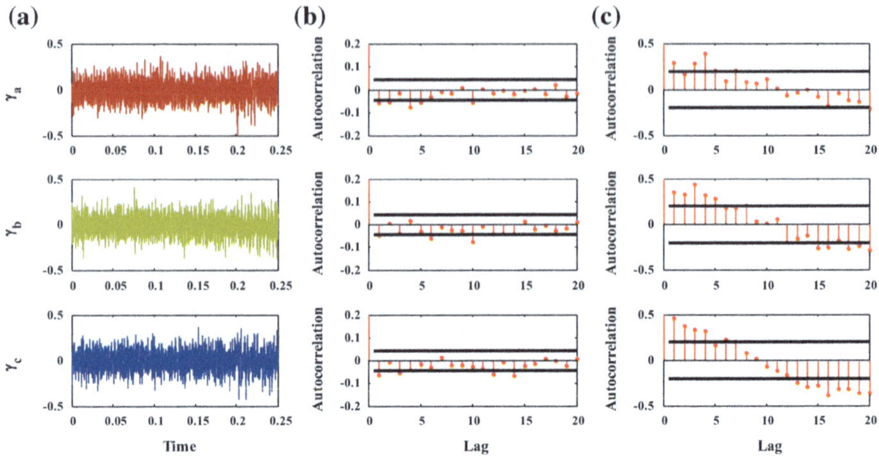

Fig. 3 **a** Innovations for ***Model d*** in stationary reference frame; **b** autocorrelation coefficients at $t = 0$–0.20 s (normal operation); **c** autocorrelation coefficient at $t = 0.20$–0.25 s (3% increase in (L_{ms}))

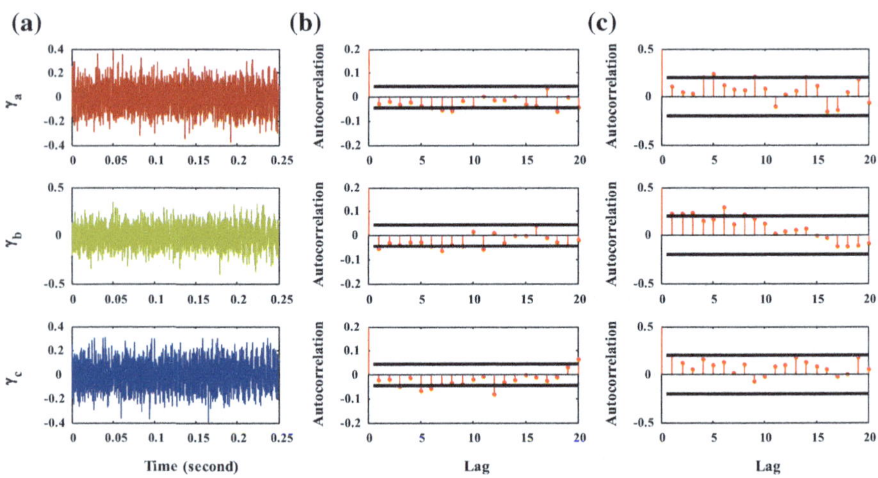

Fig. 4 **a** Innovations for ***Model d*** in stationary reference frame; **b** autocorrelation coefficients at $t = 0$–0.20 s (normal operation); **c** autocorrelation coefficient at $t = 0.20$–0.25 s (10% increase in (L_{ls}))

5 Conclusion

It was found that not all SCIM models could generate diagnostically relevant residuals and thus have varying discriminatory capabilities. In order to compare diagnostic relevance of various models, a discriminatory ability index is proposed in this paper.

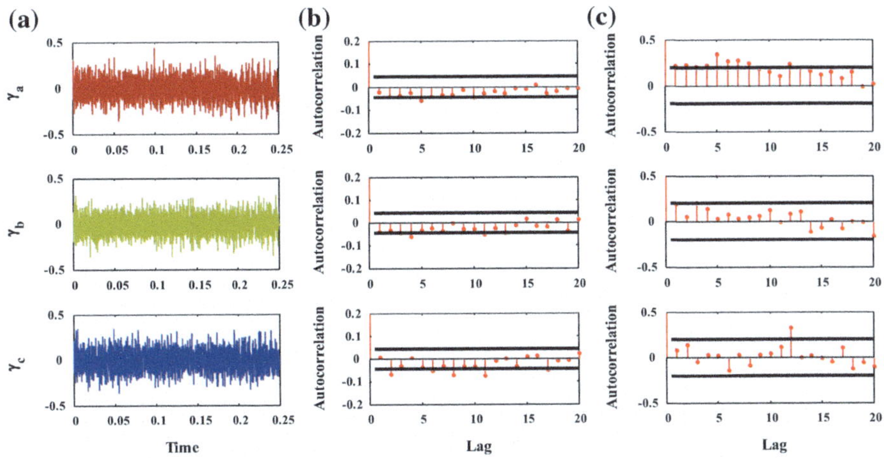

Fig. 5 **a** Innovations for ***Model d*** in stationary reference frame; **b** autocorrelation coefficients at $t = 0$–0.20 s (normal operation); **c** autocorrelation coefficient at $t = 0.20$–0.25 s (10% increase in (r_s))

It can be concluded that ***Model d*** in the stationary reference frame is best suited or diagnostically most relevant for model-based FDD. This has also been validated through exhaustive simulation studies.

The parameter sensitivity analysis for the best-suited model is done, and acceptable percentage inaccuracies of the model parameters for robust fault diagnosis are provided.

Acknowledgements The author expresses gratitude to Dr. Ketan Detroja, Associate Professor, Indian Institute of Technology Hyderabad for his valuable suggestions and discussions for the successful completion of this research work.

Appendix

All the parameters of the SCIM model are given here. Also, $\theta_l = \left(\theta_e - n_p\theta_r\right)$. Here, θ_e and θ_r are transformation angular position and rotor angle, respectively.

$$\mathbf{K}_s = \frac{2}{3}\begin{bmatrix} \cos\theta_e & \cos(\theta_e - 2\pi/3) & \cos(\theta_e + 2\pi/3) \\ \sin\theta_e & \sin(\theta_e - 2\pi/3) & \sin(\theta_e + 2\pi/3) \\ 1/2 & 1/2 & 1/2 \end{bmatrix},$$

$$\mathbf{K}_r = \frac{2}{3}\begin{bmatrix} \cos\theta_l & \cos(\theta_l - 2\pi/3) & \cos(\theta_l + 2\pi/3) \\ \sin\theta_l & \sin(\theta_l - 2\pi/3) & \sin(\theta_l + 2\pi/3) \\ 1/2 & 1/2 & 1/2 \end{bmatrix}$$

$$\mathbf{L}_{abc}^{s} = \begin{bmatrix} (L_{ls} + L_{ms}) & -\frac{1}{2}L_{ms} & -\frac{1}{2}L_{ms} \\ -\frac{1}{2}L_{ms} & (L_{ls} + L_{ms}) & -\frac{1}{2}L_{ms} \\ -\frac{1}{2}L_{ms} & -\frac{1}{2}L_{ms} & (L_{ls} + L_{ms}) \end{bmatrix},$$

$$\mathbf{L}_{abc}^{r} = \begin{bmatrix} (L_{lr} + L_{mr}) & -\frac{1}{2}L_{mr} & -\frac{1}{2}L_{mr} \\ -\frac{1}{2}L_{mr} & (L_{lr} + L_{mr}) & -\frac{1}{2}L_{mr} \\ -\frac{1}{2}L_{mr} & -\frac{1}{2}L_{mr} & (L_{lr} + L_{mr}) \end{bmatrix}$$

$$\mathbf{K}_{\omega} = \begin{bmatrix} 010 \\ -100 \\ 000 \end{bmatrix}, \mathbf{v}_{qd0}^{s} = \mathbf{K}_{s}\mathbf{v}_{abc}^{s}, \mathbf{R}_{qd0}^{s} = r_{s}\mathbf{I}_{3\times 3}, \mathbf{R}_{qd0}^{r} = r_{r}\mathbf{I}_{3\times 3}$$

$$\mathbf{L}_{abc}^{sr} = L_{sr}\begin{bmatrix} \cos(n_p\theta_r)\cos(n_p\theta_r + 2\pi/3)\cos(n_p\theta_r - 2\pi/3) \\ \cos(n_p\theta_r - 2\pi/3)\cos(n_p\theta_r)\cos(n_p\theta_r + 2\pi/3) \\ \cos(n_p\theta_r + 2\pi/3)\cos(n_p\theta_r - 2\pi/3)\cos(n_p\theta_r) \end{bmatrix},$$

$$\mathbf{L}_{abc}^{rs} = \left(\mathbf{L}_{abc}^{sr}\right)^{\mathrm{T}}$$

$$\dot{\mathbf{L}}_{abc}^{sr} = -\omega_r L_{sr}\begin{bmatrix} \sin(n_p\theta_r)\sin(n_p\theta_r + 2\pi/3)\sin(n_p\theta_r - 2\pi/3) \\ \sin(n_p\theta_r - 2\pi/3)\sin(n_p\theta_r)\sin(n_p\theta_r + 2\pi/3) \\ \sin(n_p\theta_r + 2\pi/3)\sin(n_p\theta_r - 2\pi/3)\sin(n_p\theta_r) \end{bmatrix},$$

$$\dot{\mathbf{L}}_{abc}^{rs} = \left(\dot{\mathbf{L}}_{abc}^{sr}\right)^{\mathrm{T}}$$

$$T_{em} = -n_p L_{ms}\left\{\begin{array}{l} \left(i_{as}\left(i_{ar} - \frac{1}{2}i_{br} - \frac{1}{2}i_{cr}\right) + i_{bs}\left(i_{br} - \frac{1}{2}i_{ar} - \frac{1}{2}i_{cr}\right)\right. \\ \left. + i_{cs}\left(i_{cr} - \frac{1}{2}i_{br} - \frac{1}{2}i_{ar}\right)\right)\sin(n_p\theta_r) \\ + \frac{\sqrt{3}}{2}[i_{as}(i_{br} - i_{cr}) + i_{bs}(i_{cr} - i_{ar}) + i_{cs}(i_{ar} - i_{br})]\cos(n_p\theta_r) \end{array}\right\} \text{(or)}$$

$$T_{em} = n_p\left(\mathbf{i}_{abc}^{s}\right)^{T}\frac{\partial}{\partial\theta_r}\left[\mathbf{L}_{abc}^{sr}\right]\left(\mathbf{i}_{abc}^{r}\right).$$

References

1. Zhang, P., Du, Y., Habetler, T.G., Lu, B.: A survey on condition monitoring and protection methods for medium-voltage induction motors. IEEE Trans. Ind. Appl. **47**(1), 34–46 (2011)
2. Nandi, S., Toliyat, H.A., Li, X.: Condition monitoring and fault diagnosis of electrical motors-a review. IEEE Trans. Energy Convers. **20**(4), 719–729 (2005)
3. Bellini, A., Filippetti, F., Tassoni, C., Capolino, G.A.: Advances in diagnostic techniques for induction machines. IEEE Trans. Ind. Electron. **55**(12), 4109–4126 (2008)
4. Filippetti, F., Bellini, A., Capolino, G.A.: Condition monitoring and diagnosis of rotor faults in induction machines: state of art and future perspectives. In: Proceedings of the 2013 International Workshop on Electrical Machines Design Control, pp. 196–209. IEEE (2013)

5. Krause, P.C., Wasynczuk, O., Sudhoff, S.D.: Analysis of electric machinery and drive systems, 3rd edn. IEEE Press, New York (2002)
6. Arkan, M., Perovic, D.K., Unsworth, P.J.: Modelling and simulation of induction motors with inter-turn faults for diagnostics. Elect. Power Syst. Res. **75**(1), 57–66 (2005)
7. Chen, S., Zivanovic, R.: Modelling and simulation of stator and rotor fault conditions in induction machines for testing fault diagnostic techniques. Euro. Trans. Elect. Power **20**(1), 611–629 (2009)
8. Duvvuri, S.S.S.R.S., Detroja, K.: Model-based stator interturn fault detection and diagnosis in induction motors. In: 2015 7th IEEE Conference on Information Technology and Electrical Engineering, pp. 167–172 (2015)
9. Duvvuri, S., Detroja, K.: Model-based broken rotor bars fault detection and diagnosis in squirrel-cage induction motors. In: 2016 3rd IEEE Conference on Control and Fault-Tolerant Systems, pp. 537–539 (2016)
10. Duvvuri, S.S., Detroja, K.: Stator inter-turn fault diagnostics relevant modeling of squirrel-cage induction motor. In: 2016 55th IEEE Conference on Decision and Control, pp. 1279–1284 (2016)
11. Duvvuri, S.S.S.R.S., Dangeti, L.K., Garapati, D.P., Meesala, H.: State estimation for wound rotor induction motor using discrete-time extended Kalman filter. In: 2018 2nd IEEE Conference on Power, Instrumentation, Control and Computing, pp. 1–6 (2018)

Comparative Study of Perturb & Observe (P&O) and Incremental Conductance (IC) MPPT Technique of PV System

Kanchan Jha and Ratna Dahiya

Abstract The International Solar Alliance aims to efficiently exploit solar energy in sunshine countries (Suryaputra) between the tropics. Similar pledges, like Sustainable Development Goal 7 (affordable and clean energy), to increase the contribution from other non-conventional sources are in place to meet the objectives set out in UNFCCC. However, availability of renewable energy is an intermittent process and thus requires both efficiency and effectiveness in the processes used to harness them. The present paper aims to bring out a comparative study of the two MPPT methods globally used to maximize the solar energy trapped: The two techniques are P&O and IC. PV module consisting of DC–DC boost converter, MPPT controller, inverter, etc., has been used to analyze the methods. Two-level voltage source inverter has been incorporated to enhance the simulation and eliminate the harmonics. All simulations have been performed in MATLAB–Simulink platform.

Keywords Solar PV module · DC–DC boost converter · PWM technique · MPPT techniques · Three-phase VSI inverter

1 Introduction

With the world becoming more sensitive to impacts of climate change, the share of renewable energy in total energy demand has been increasing. However, much needs to be done to efficiently harness the solar energy, as it is intermittent and varies with the movement of the Earth and other climatic factors [1, 2]. This paper is presenting an equivalent circuit of solar PV cell which is applied to develop 200 W PV module which is shown in Sect. 2. Several MPPT techniques have been proposed in the literature to find MPP of solar PV module [3]. MPPT, maximum power point tracking, is a method to effectively track the point at which the maximum energy can be trapped. Two methods—the P&O and the IC methods—are most commonly used for this purpose [4, 5]. MPPT uses DC–DC boost converter so that it can match solar PV source and the load by controlling the duty cycle [6]. This paper further

K. Jha · R. Dahiya (✉)
National Institute of Technology Kurukshetra, Haryana, India
e-mail: ratna_dahiya@yahoo.co.in

D. Dutta and B. Mahanty (eds.), *Numerical Optimization in Engineering and Sciences*, Advances in Intelligent Systems and Computing 979,
https://doi.org/10.1007/978-981-15-3215-3_18

presents three-level inverter to supply AC load. LC filter is used in this paper to reduce harmonics and to get desired output voltage waveform [7, 8].

2 Solar Power Generation

In this paper, we have designed solar panels to convert solar energy into electrical energy. We have designed it as a combination of series and parallel connections of photovoltaic modules of solar. Solar PV module is designed by solar cell which is manufactured by semiconductor devices [2]. A typical solar cell is a representation of a current source connected in parallel path with a diode and two resistors [1] (Fig. 1 and Table 1).

The modeling equation of the solar cell is described

$$I_{pv} = N_p I_{\text{ph}} - N_p I_{\text{sh}} \left[\exp\left\{ q \frac{(V_{pv} + I_{pv} R_{\text{sh}})}{N_s A K T_c} \right\} - 1 \right] - \frac{(V_{pv} + I_{pv} R_s)}{R_{\text{sh}}} \quad (1)$$

where I_{sh} is the diode saturation current.

I_{sh} depends on temperature, and expression for temperature dependency is

$$I_{\text{sh}} = I_{rs} \left[\frac{T_c}{T_{\text{ref}}} \right]^3 \left[\exp\left\{ \frac{q E_g \left(\frac{1}{T_{\text{ref}}} - \frac{1}{T_c} \right)}{A K} \right\} \right] \quad (2)$$

where I_{rs} is the reverse diode current given by

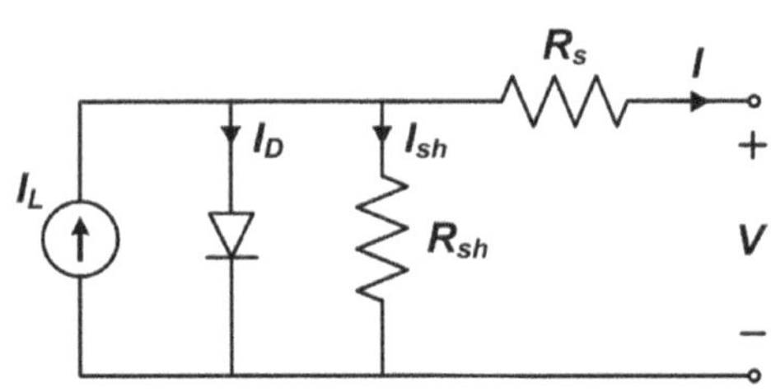

Fig. 1 Equivalent circuit of an ideal solar cell

Table 1 Specification of solar module

Parameter	Specification
Maximum power (P_{mpp})	200.1430 W
Maximum voltage (V_{mpp})	26.3 W
Maximum current (I_{mpp})	7.61 A
Short-circuit current (I_{SC})	8.21 A
Open-circuit voltage (V_{OC})	32.9 V

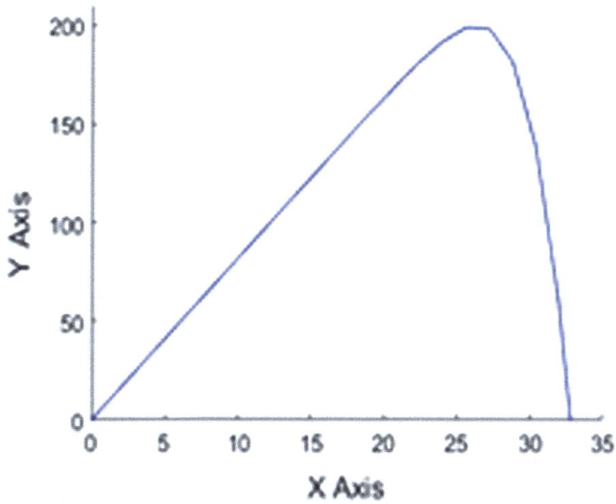

Fig. 2 P(Y)-V(X)-axis curve of solar cell

$$I_{rs} = \frac{I_{sc}}{\left[\exp\left\{\frac{qV_{oc}}{N_s AKT_c}\right\} - 1\right]} \quad (3)$$

$$I_{pv} = [I_{sc} + K_i(T_c - T_{ref})]\left[\frac{G}{1000}\right] \quad (4)$$

where

I_{pv} is the output power of PV module.

N_p are the PV cells arranged in parallel, and N_s are the PV cells arranged in series.

I_{ph} is the photocurrent, I_{sh} is the diode saturation current, q is the charge of electron, R_s is the series resistance (Ω), R_{sh} is the shunt resistance (Ω), T_c is the actual cell temperature (°C), K is the Boltzmann constant ($1.381e^{-23}$ J/Watt), G is the irradiance of sun $\left(\frac{\omega}{m^2}\right)$ and A = 1.3 is the ideality factor (depends on PV cell) (Figs. 2 and 3).

3 Maximum Power Point Tracking (MPPT) Algorithm

This paper presents two MPPT techniques. These techniques are used to extract maximum power from PV module under mentioned conditions.

A. ***Perturb & Observe Technique (P&O)***

Two conditions are required in P&O-type MPPT technique. First condition is it can operate at high sampling rate. The output values of voltage waveform and current waveform should be able to make the pattern of the output power waveform if any changes occur in the reference signal. This signal is used for the power converter of the maximum power point tracking. Second condition is the response time of

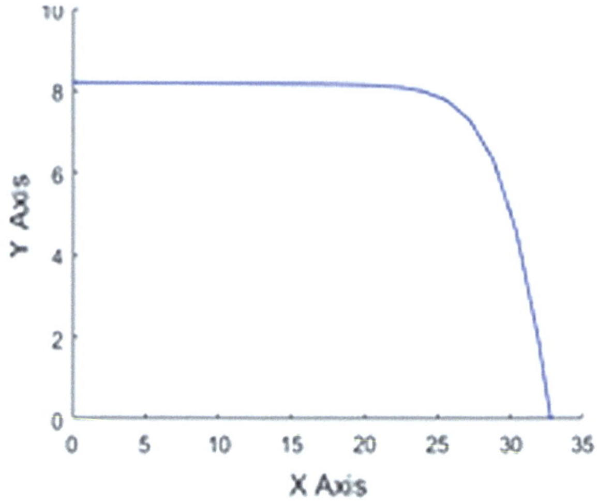

Fig. 3 I(Y)-V(X)-axis curve of solar cell

the converter must be rapid while maintaining the switching losses (frequency) as minimum as possible. The boost converter (DC–DC) switch is turned on when the actual current matches the reference current [6]. Hence, the reference current can be perturbed (increased/decreased) in every switching cycle [3].

B. ***Incremental Conductance Method***

The P&O method is tracking peak power, but it is not accurate and the response is slow with more oscillations. To overcome this problem, we have further used incremental conductance MPPT technique. This method can determine that the maximum power point tracking is reaching the MPP and subsequently stops perturbing the point of operation [5]. This condition is mandatory; if it is not met, then the perturbation direction is determined by the relation between dl/dV and –I/V. Mathematically, we can derive this equation with the fact that dP/dV is negative when the maximum power point technique is to the right of the MPP and it will be positive when present in the left of the maximum power point [4]. DC–DC boost converter is used to step up the voltage according to the demand, and VSI is used to convert DC to AC to supply AC load cascaded with LC filter.
where

$$L=\frac{\sqrt{2}Rl}{w0}, C = \frac{1}{\sqrt{2}w0Rl}, L \text{ and } C \text{ are the values of LC filter}$$

4 Simulation Result: Comparison and Discussion

The P&O and IC MPPT algorithms are simulated, and we have analyzed the results by keeping the conditions identical. We have seen that when atmospheric conditions are stable, the perturb & observe MPPT oscillates close to maximum power point, but incremental conductance finds the maximum power point accurately and its response is more fast. We have calculated these values corresponding to irradiance of 1000 w/m^2 and an ambient temperature of 25 °C (Figs. 4, 5, 6, 7, 8, 9, 10, 11, 12 and 13).

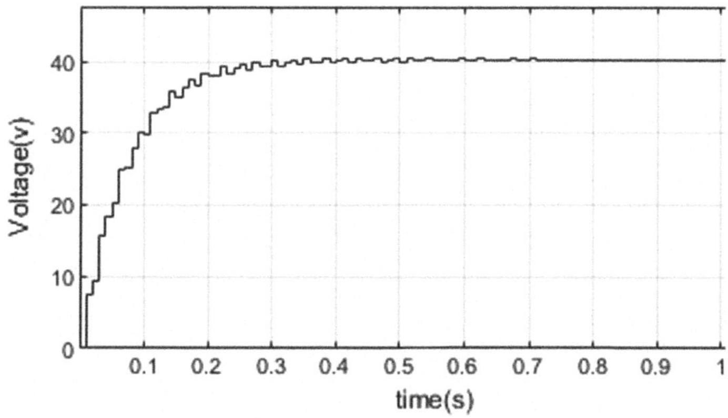

Fig. 4 Output voltage of P&O MPPT technique

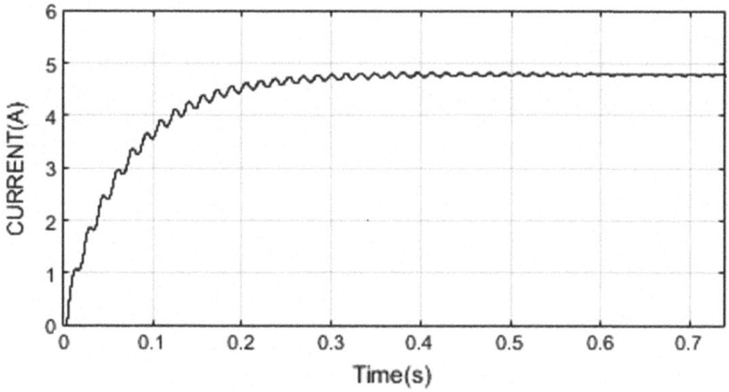

Fig. 5 Output current of P&O MPPT technique

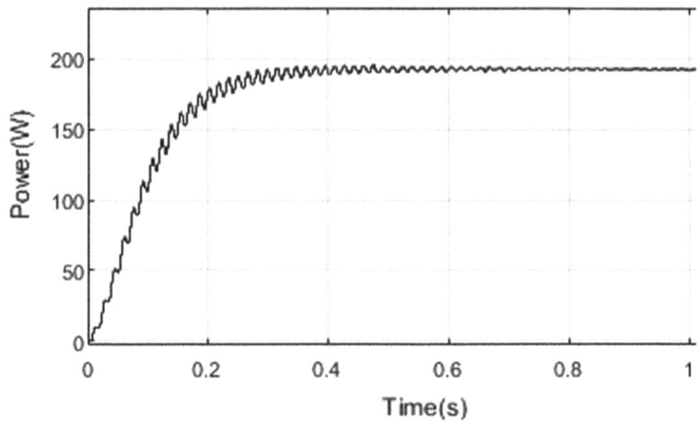

Fig. 6 Output power of P&O MPPT technique

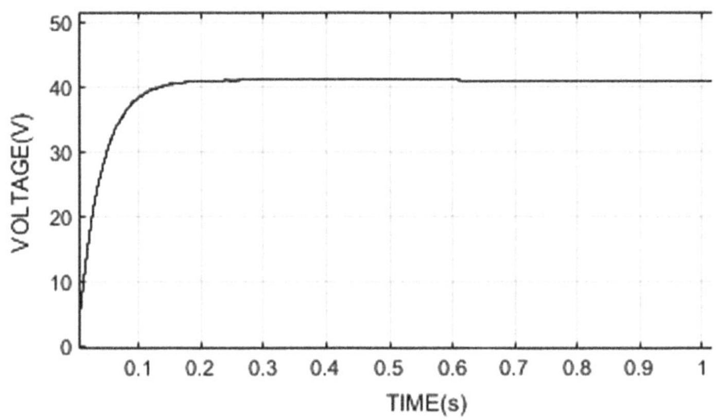

Fig. 7 Output voltage of IC MPPT technique

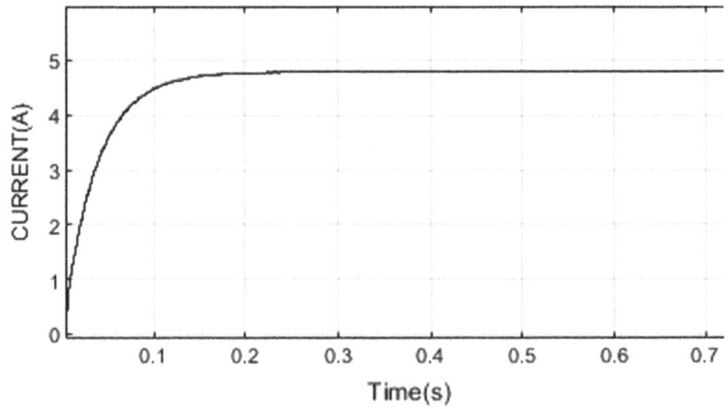

Fig. 8 Output current of IC MPPT technique

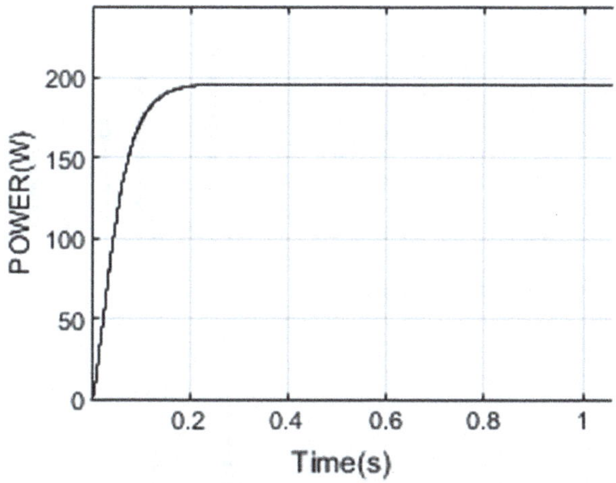

Fig. 9 Output power of IC MPPT technique

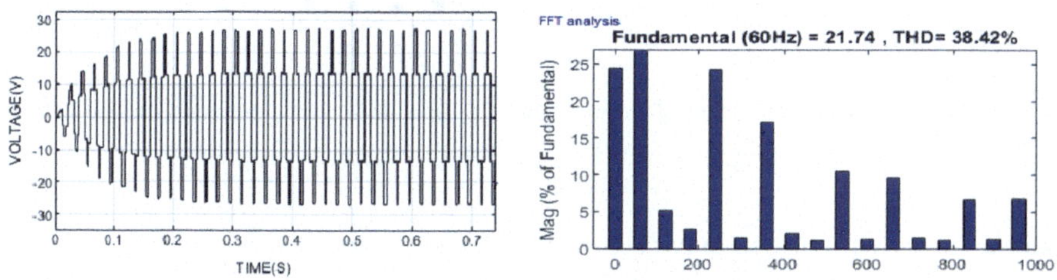

Fig. 10 Output voltage and THD of inverter in P&O MPPT technique

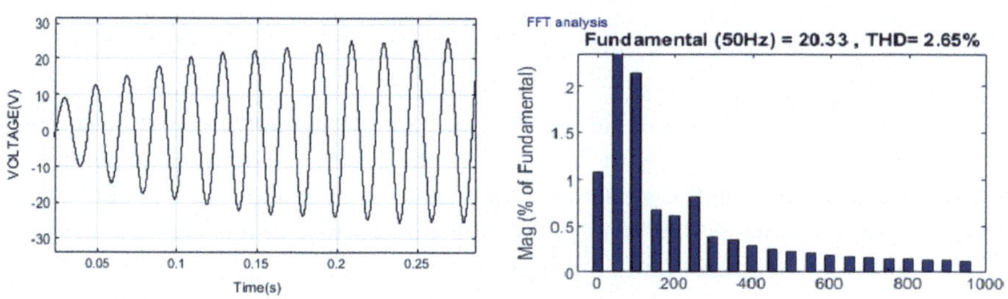

Fig. 11 Output voltage and THD of inverter with LC filter in P&O MPPT technique

5 Conclusion

The IC method is more advantageous over perturb & observe because it is able to determine the MPP with less harmonics around this value. It can also perform MPPT technique under very fast variation in irradiation with more accuracy than the

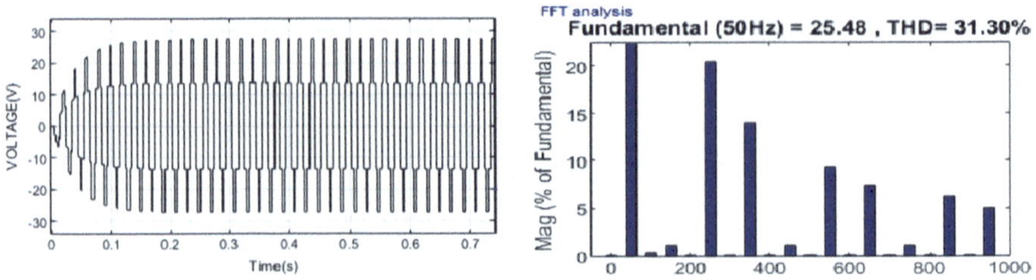

Fig. 12 Output voltage and THD of inverter in incremental conductance MPPT technique

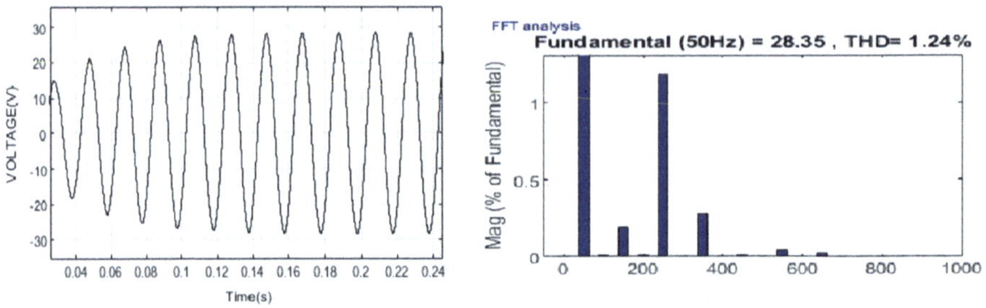

Fig. 13 Output voltage and THD of inverter with LC filter in incremental conductance MPPT technique

Table 2 Comparative analysis of P&O versus IC MPPT technique

S. no.	MPPT technique	THD without filter (%)	THD with filter (%)	Settling time (s)
1	P&O	38.42	2.65	0.3
2	IC	31.30	1.24	0.15

P&O method which is shown in Table 2. We have eliminated lower-order harmonics using VSI inverter cascaded with LC filter. In the future, some modifications can be explored in incremental conductance method so as to make it more robust. The complexity can be checked by using fuzzy logic and artificial intelligence methods so as to make the technique more intelligible.

References

1. Santhosh, N., Prasad, B.: Efficiency improvement of a solar PV-panel through spectral sharing by combination of different panels. In: 2016 IEEE Students' Conference on Electrical, Electronics and Computer Science (SCEECS), pp. 1–4. Bhopal, (2016)
2. Harini, K., Syama, S.: Simulation and analysis of incremental conductance and Perturb and Observe MPPT with DC-DC converter topology for PV array. In: 2015 IEEE International

Conference on Electrical, Computer and Communication Technologies (ICECCT), pp. 1–5. Coimbatore (2015)
3. Raiker, G.A.: Dynamic response of maximum power point tracking using perturb and observe algorithm with momentum term. In: 2017 IEEE 44th photovoltaic specialist conference (PVSC), pp. 3073–3076. Washington, DC (2017)
4. Sundareswaran, K., Kumar, V.V., Simon, S.P., Nayak, P.S.R.: Cascaded simulated annealing/perturb and observe method for MPPT in PV systems. In: 2016 IEEE International Conference on Power Electronics, Drives and Energy Systems (PEDES), pp. 1–5. Trivandrum (2016)
5. Hossain, M.J., Tiwari, B., Bhattacharya, I.: An adaptive step size incremental conductance method for faster maximum power point tracking. In: 2016 IEEE 43rd Photovoltaic Specialists Conference (PVSC), pp. 3230–3233. Portland, OR (2016)
6. Nakajima, A., Masukawa, S.: Study of boost type DC-DC converter for single solar cell. In: IECON 2018—44th Annual Conference of the IEEE Industrial Electronics Society, pp. 1946–1951. Washington, DC (2018)
7. Nauman, M., Hasan, A.: Efficient implicit model-predictive control of a three-phase inverter with an output LC filter. IEEE Trans. Power Electron. **31**(9), 6075–6078 (2016)
8. Liu, Y., Peng, J., Wang, G., Wang, H., See, K.Y.: THD and EMI performance study of foil-wound inductor of LCL filter for high power density converter. In: 2016 IEEE 8th International Power Electronics and Motion Control Conference (IPEMC-ECCE Asia), pp. 3467–3471. Hefei (2016)
9. Islam, M.A., Merabet, A., Beguenane, R., Ibrahim, H.: Modeling solar photovoltaic cell and simulated performance analysis of a 250 W PV module. In: 2013 IEEE Electrical Power & Energy Conference, pp. 1–6, Halifax, NS (2013)

Conceptualization of Finite Capacity Single-Server Queuing Model with Triangular, Trapezoidal and Hexagonal Fuzzy Numbers Using α-Cuts

K. Usha Prameela and Pavan Kumar

Abstract We present a limited extent single-server lining model with triangular, trapezoidal and hexagonal fuzzy numbers separately. The entry rate and overhauled rate are fuzzy in description, and the arithmetic of interval numbers is enforced. The execution proportions are fuzzified and after that assessed by utilizing α-cut and DSW (Dong, Shah and Wong) calculation. Relating to each sort of fuzzy number, the numerical precedent is encapsulated to check the achievability of this miniature. A comparative investigation relative to individual fuzzy numbers is accomplished for various estimations of α.

Keywords Performance measures · Triangular · Trapezoidal and hexagonal fuzzy numbers · α-cuts · DSW algorithm · Interval analysis

1 Introduction

Fuzzy queuing model was initiated by Lie and Lee [1] in 1989 and created by Buckley [2] in 1990, Negi and Lee [3] in 1992. Chen [4] has estimated fuzzy queues utilizing Zadeh's [5] development standard. Kao et al. accomplished the membership functions of the system attributes for fuzzy lines utilizing parametric linear programming. Lately, Chen [4] advanced (FM/FM/1): (∞/FCFS) and (FM/FM/k): (∞/FCFS) where FM signifies fuzzified exponential time found on queuing hypothesis. Here, fuzzy service rate is described by semantic terms very substantial, substantial, depressed and conservative. Gupta et al. [6] in 2007 discussed various queuing models in his book "Operations Research." Rose [7] in 2010 recommended a fuzzy logic with engineering applications.

Voskoglou et al. [8] proposed utilization of triangular fuzzy model to the appraisal of analogical thinking abilities. Subbotin et al. [9] studied the trapezoidal fuzzy logic

K. U. Prameela (✉) · P. Kumar
Department of Mathematics, Koneru Lakshmaiah Education Foundation (KLEF), Vaddeswaram, Guntur, AP 522502, India
e-mail: kushaprameela@gmail.com

P. Kumar
e-mail: pavankmaths@gmail.com

D. Dutta and B. Mahanty (eds.), *Numerical Optimization in Engineering and Sciences*, Advances in Intelligent Systems and Computing 979,
https://doi.org/10.1007/978-981-15-3215-3_19

model for learning assessment. Zimmermann [10] presented the fuzzy and linear programming with several objective functions. Yovgav [11] proposed a portrayal of the expansion guideline.

In this paper, we narrate the FM/FM/1 queuing model with limited capacity and FCFS regulation using triangular, trapezoidal and hexagonal fuzzy numbers under α-cut regulation through DSW algorithm to handle with uncertain parameters. To explore membership function of execution proportions in single-server fuzzy queuing model, DSW algorithm is utilized.

In Sect. 2, some essential ideas and definitions are presented. In Sect. 3, the documentations and suspicions are portrayed. In Sect. 4, the proposed queuing model is given. In Sect. 5, the solution approach to the present model is reported. In Sect. 6, three numerical examples are solved. In Sect. 7, comparison of three queuing models is interpreted. In Sect. 8, results and discussions are conferred, and in Sect. 9, some limitations are discussed. In Sect. 10, the model is concluded.

2 Essential Ideas and Definitions

2.1 Fuzzy Number [5]

A fuzzy set $\bar{D}$ portrayed on the arrangement of real numbers R is articulated to be fuzzy number if $\bar{D}$ has the accompanying qualities, for example, normal, convex set, and the support of it is closed as well as limited.

2.2 α-Cut [5]

An α-cut of a fuzzy set is a crisp set $\bar{D}\,\alpha$ that contains every one of the components of the universal set X that have a participation grade in $\bar{D}$ greater than or equivalent to the determined estimation of α. Thus,

$$\bar{D}_\alpha = \left\{x \in X : \mu_{\bar{D}}(x) \geq \alpha, 0 \leq \alpha \leq 1\right\}$$

2.3 Triangular Fuzzy Number [8]

A triangular fuzzy number $\bar{D}$ is portrayed by (d_1, d_2, d_3) where $d_i \in R$ and $d_1 \leq d_2 \leq d_3$. Its membership function is

$$\mu_{\bar{D}}(x) = \begin{cases} \frac{x-d_1}{d_2-d_1} & \text{for } d_1 \le x \le d_2 \\ \frac{d_3-x}{d_3-d_2} & \text{for } d_2 \le x \le d_3 \\ 0 & \text{otherwise} \end{cases}$$

2.4 Trapezoidal Fuzzy Number [9]

A trapezoidal fuzzy number $\bar{D}$ is portrayed by (d_1, d_2, d_3, d_4) where $d_i \in \text{R}$ and $d_1 \le d_2 \le d_3 \le d_4$. Its membership function is

$$\mu_{\bar{D}}(x) = \begin{cases} \frac{x-d_1}{d_2-d_1} & \text{for } d_1 \le x \le d_2 \\ 1 & \text{for } d_2 \le x \le d_3 \\ \frac{d_4-x}{d_4-d_3} & \text{for } d_3 \le x \le d_4 \\ 0 & \text{otherwise} \end{cases}$$

2.5 Hexagonal Fuzzy Number [10]

A fuzzy number $\bar{D}$ is a hexagonal fuzzy number portrayed by $(d_1, d_2, d_3, d_4, d_5, d_6)$ where $d_1, d_2, d_3, d_4, d_5, d_6$ are real numbers. Its membership function is given below.

$$\mu_{\bar{D}}(x) = \begin{cases} 0 & \text{for } \; x < d_1 \\ \frac{1}{2}\left(\frac{x-d_1}{d_2-d_1}\right) & \text{for } d_1 \le x \le d_2 \\ \frac{1}{2} + \frac{1}{2}\left(\frac{x-d_2}{d_3-d_2}\right) & \text{for } d_2 \le x \le d_3 \\ 1 - \frac{1}{2}\left(\frac{x-d_4}{d_5-d_4}\right) & \text{for } d_4 \le x \le d_5 \\ \frac{1}{2}\left(\frac{d_6-x}{d_6-d_5}\right) & \text{for } d_5 \le x \le d_6 \\ 0 & \text{otherwise} \end{cases}$$

2.6 Arithmetic for Interval Analysis [12]

The two interval numbers assigned by ordered pairs of real numbers with lower and furthest cutoff points be

$$G = [a_1, a_2]_2, a_1, \le a_2; \text{ and } H = [b_1, b_2], b_1 \le b_2$$

The math property is signified generally with the image *, where * = [+ , −, × , ÷] and the movement is portrayed by $G * H = [a_1, a_2] * [b_1, b_2]$ where

$$[a_1 a_2] + b_1, b_2] = [a_1 + b_1, a_2 + b_2]$$
$$[a_1, a_2] - b_1, b_2] = [a_1 - b_2, a_2 - b_1]$$

$$[a_1, a_2] * [b_1, b_2] = [\min(a_1 b_1, a_1 b_2, a_2 b_1, a_2 b_2), \max(a_1 b_1, a_1 b_2, a_2 b_1, a_2 b_2)]$$

$[a_1, a_2] \div [b_1, b_2] = [a_1, a_2] * [1/b_2, 1/b_1]$ gave that 0 does not have a place with $[b_1, b_2]$

$\alpha[a_1, a_2] = [\alpha\, a_1, \alpha\, a_2]$ for $\alpha > 0$ and $[\alpha\, a_2, \alpha\, a_1]$ for $\alpha < 0$.

3 The Documentations and Suspicions

3.1 Suspicions

(i) Queuing model with one server and limited capacity.
(ii) Inter-arrival time taken Poisson dispersion and overhauled time taken exponential circulation.
(iii) Arrival (entry) rate, service (overhauled) rate are fuzzy numbers.

3.2 Documentations

λ Average number of clients arriving per unit of time.
μ Average number of clients being overhauled per unit of time.
L_S The normal number of clients in the system.
L_q The average number of clients holding up in the line.
W_S The normal holding up time of clients in the system.
W_q Average holding up time of clients in the line.
X Set of the entomb entry time.
Y Set of the administration time.
A Inter-entry time.
S Overhauled times.

4 Formulation of Proposed Lining Miniature

We propose a solitary server limited limit, with first started things out served (FCFS) regulation queuing model, indicated as (FM/FM/1): (N/FCFS), where the inter-entry time and the overhauled time ensue Poisson and exponential dispersions with fuzzy parameters λ and μ.

The execution proportions of the above model are given beneath:

(a) Expected number of clients in the framework

$$L_s = \rho \frac{\left[1 - (N+1)\rho^N + N\left(\rho^{N+1}\right)\right]}{(1-\rho)\left(1-\rho^{N+1}\right)}$$

(b) Expected number of clients holding up in the line

$$L_q = \rho^2 \frac{\left[1 - (N)\rho^{N-1} + (N-1)\left(\rho^N\right)\right]}{(1-\rho)\left(1-\rho^{N+1}\right)}$$

(c) Average time a client spends in the framework

$$W_s = \rho \frac{\left[1 - (N+1)\rho^N + N\left(\rho^{N+1}\right)\right]}{\lambda(1-\rho)\left(1-\rho^{N+1}\right)}$$

(d) Average holding up time of a customer in the line

$$W_q = \rho^2 \frac{\left[1 - (N)\rho^{N-1} + (N-1)\left(\rho^N\right)\right]}{\lambda(1-\rho)\left(1-\rho^{N+1}\right)}$$

5 Solution Approach

DSW (Dong, Shah and Wong) is an estimated strategy which utilized the intervals at various α-cut estimations in establishing membership functions. The entry time A and overhauled time S are joined by accompanying fuzzy sets:

$$\bar{A} = \left\{\left(a, \mu_{\tilde{A}}(a)\right)/a \in X\right)\right\}; \bar{S} = \{(s, \mu_{\check{s}}(s))/s \in Y)\}$$

Here, X is the arrangement of the inter-entry time and Y is the arrangement of the overhauled time. $\mu_{\tilde{A}}(a)$ is membership function of the inter-entry time. $\mu_{\check{s}}(s)$ is enrollment capacity of the overhauled time, and their α-cuts are communicated as:

$$\bar{A}(\alpha) = \{a \in X/\mu_A(\alpha) \geq \alpha\}: \bar{S}(\alpha) = \{s \in Y/\mu\check{s}(s) \geq \alpha\}$$

The DSW calculation contains the accompanying advances:

Stage 1. Accept α-cut an incentive from $0 \leq \alpha \leq 1$.
Stage 2. The intervals in the information enrollment capacities comparing to the above α-cut esteems are to be processed.
Stage 3. Decide the interim for the yield participation work for the picked α-cut estimations.
Stage 4. Iterate stages 1–3 for various estimations of α to finish the α-cut portrayal of the arrangement.

6 Numerical Illustrations

Example 1 Acknowledge both the entry rate and overhauled rate as triangular fuzzy numbers symbolized by $\lambda = [3, 4, 5]$ and $\mu = [13, 14, 15]$. The greatest limit of the system, i.e., $N = 2$. The interval of certainty at probability level α as $[3 + \alpha, 5 - \alpha]$ and $[13 + \alpha, 15 - \alpha]$. Where $x = [3 + \alpha, 5 - \alpha]$ and $y = [13 + \alpha, 15 - \alpha]$ (Table 1; Fig. 1).

Example 2 Take the pair entry and overhauled rate as trapezoidal fuzzy numbers communicated by $\lambda = [1, 2, 3, 4]$ and $\mu = [11, 12, 13, 14]$. The interval of confidence at probability level α as $[1 + \alpha, 4 - \alpha]$ and $[11 + \alpha, 14 - \alpha]$, the most extreme limit of the system, i.e., $N = 2$. Where $x = [1 + \alpha, 4 - \alpha]$ and $y = [11 + \alpha, 14 - \alpha]$ (Table 2; Fig. 2).

Example 3 Recognize both the entry and overhauled rate as hexagonal fuzzy numbers constituted by $\lambda = [4, 5, 6, 7, 8, 9]$ and $\mu = [20, 21, 22, 23, 24, 25]$. The maximum limit of the system, i.e., $N = 2$. The interval of certainty at possibility level α as $[4 + \alpha, 9 - \alpha]$ and $[20 + \alpha, 25 - \alpha]$. Where $x = [4 + \alpha, 9 - \alpha]$ and $y = [20 + \alpha, 25 - \alpha]$ (Table 3; Fig. 3).

Table 1 The α-cuts of L_q, L_s, W_s and W_q at α values (Example 1)

α	L_q	L_s	W_s	W_q
0	[0.01364, 0.1906]	[0.1442, 0.6585]	[0.0095, 0.0485]	[0.0009, 0.0146]
0.1	[0.0162, 0.1707]	[0.1586, 0.6146]	[0.0105, 0.0451]	[0.0010, 0.0130]
0.2	[0.01925, 0.1527]	[0.1737, 0.5738]	[0.0116, 0.0420]	[0.0013, 0.0111]
0.3	[0.0225, 0.1365]	[0.1897, 0.5357]	[0.0127, 0.0392]	[0.0015, 0.0102]
0.4	[0.0263, 0.1219]	[0.2065, 0.5002]	[0.0140, 0.0364]	[0.0018, 0.0091]
0.5	[0.0305, 0.1087]	[0.2242, 0.4669]	[0.0152, 0.0339]	[0.0021, 0.0080]
0.6	[0.0351, 0.0968]	[0.2428, 0.4357]	[0.0166, 0.0315]	[0.0024, 0.0071]
0.7	[0.0403, 0.0860]	[0.2625, 0.4064]	[0.0181, 0.0272]	[0.0028, 0.0062]
0.8	[0.0461, 0.0763]	[0.2832, 0.3788]	[0.0196, 0.0272]	[0.0032, 0.0055]
0.9	[0.0525, 0.0676]	[0.3051, 0.3528]	[0.0213, 0.0253]	[0.0037, 0.0048]
1	[0.0597, 0.0597]	[0.3283, 0.3283]	[0.0230, 0.0234]	[0.0042, 0.0042]

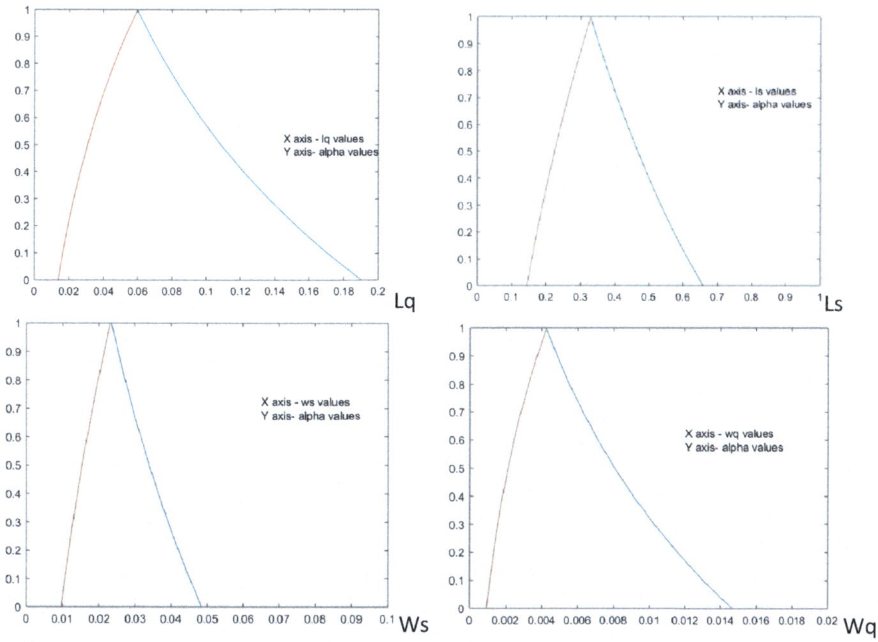

Fig. 1 Graph of L_q, L_s, W_s and W_q at α values (Example 1)

Table 2 The α-cuts of L_s, L_q, W_s and W_q at α values (Example 2)

α	L_s	L_q	W_s	W_q
0	[0.0464, 0.6187]	[0.0015, 0.2159]	[0.0033, 0.0562]	[0.0001, 0.0196]
0.1	[0.0541, 0.5797]	[0.0020, 0.1920]	[0.0038, 0.0522]	[0.0001, 0.0172]
0.2	[0.0624, 0.5433]	[0.0027, 0.1706]	[0.0045, 0.0485]	[0.0001, 0.0152]
0.3	[0.0713, 0.5093]	[0.0035, 0.1515]	[0.0052, 0.0450]	[0.0002, 0.0134]
0.4	[0.0806, 0.4775]	[0.0044, 0.1345]	[0.0059, 0.0418]	[0.0003, 0.0117]
0.5	[0.0906, 0.4476]	[0.0056, 0.1192]	[0.0067, 0.0389]	[0.0004, 0.0103]
0.6	[0.1011, 0.4195]	[0.0069, 0.1056]	[0.0075, 0.0361]	[0.0005, 0.0091]
0.7	[0.1122, 0.3930]	[0.0084, 0.0933]	[0.0084, 0.0335]	[0.0006, 0.0079]
0.8	[0.1239, 0.3680]	[0.0102, 0.0824]	[0.0093, 0.0311]	[0.0007, 0.0069]
0.9	[0.1362, 0.3443]	[0.0123, 0.0726]	[0.0103, 0.0289]	[0.0009, 0.0061]
1	[0.1491, 0.3219]	[0.0147, 0.0638]	[0.0114, 0.0268]	[0.0011, 0.0053]

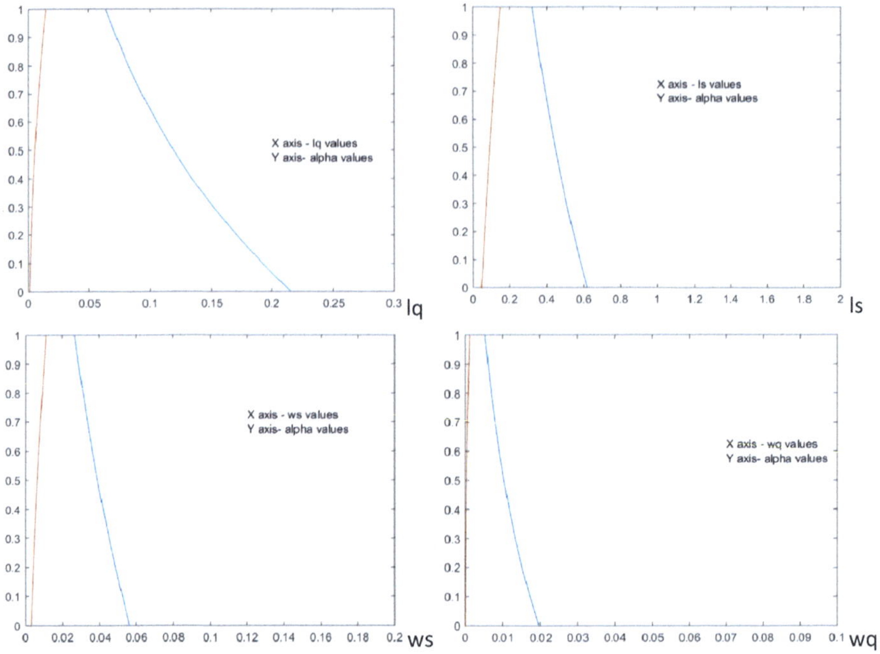

Fig. 2 Graph of L_q, L_s, W_s and W_q at α values (Example 2)

Table 3 The α-cuts of L_s, L_q, W_s and W_q at α values (Example 3)

α	L_s	L_q	W_s	W_q
0	[0.0766, 0.9951]	[0.0038, 0.3575]	[0.0030, 0.0457]	[0.0001, 0.0178]
0.1	[0.0833, 0.9504]	[0.0046, 0.3339]	[0.0033, 0.0437]	[0.00018, 0.0166]
0.2	[0.0902, 0.9082]	[0.0054, 0.312]	[0.0036, 0.0417]	[0.0002, 0.0154]
0.3	[0.0973, 0.8682]	[0.0063, 0.2915]	[0.0039, 0.0399]	[0.0002, 0.0143]
0.4	[0.1047, 0.8302]	[0.0074, 0.2723]	[0.0042, 0.0381]	[0.0003, 0.0133]
0.5	[0.1124, 0.7942]	[0.0085, 0.2544]	[0.0045, 0.0364]	[0.0003, 0.0124]
0.6	[0.1203, 0.7600]	[0.0097, 0.2377]	[0.0049, 0.0348]	[0.0003, 0.0115]
0.7	[0.1285, 0.7274]	[0.0110, 0.2220]	[0.0052, 0.0333]	[0.0004, 0.0107]
0.8	[0.1369, 0.6964]	[0.0124, 0.2073]	[0.0056, 0.0319]	[0.0005, 0.0099]
0.9	[0.1457, 0.6668]	[0.0139, 0.1936]	[0.0060, 0.0305]	[0.0005, 0.0092]
1	[0.1547, 0.6386]	[0.0155, 0.1807]	[0.0064, 0.0292]	[0.0006, 0.0086]

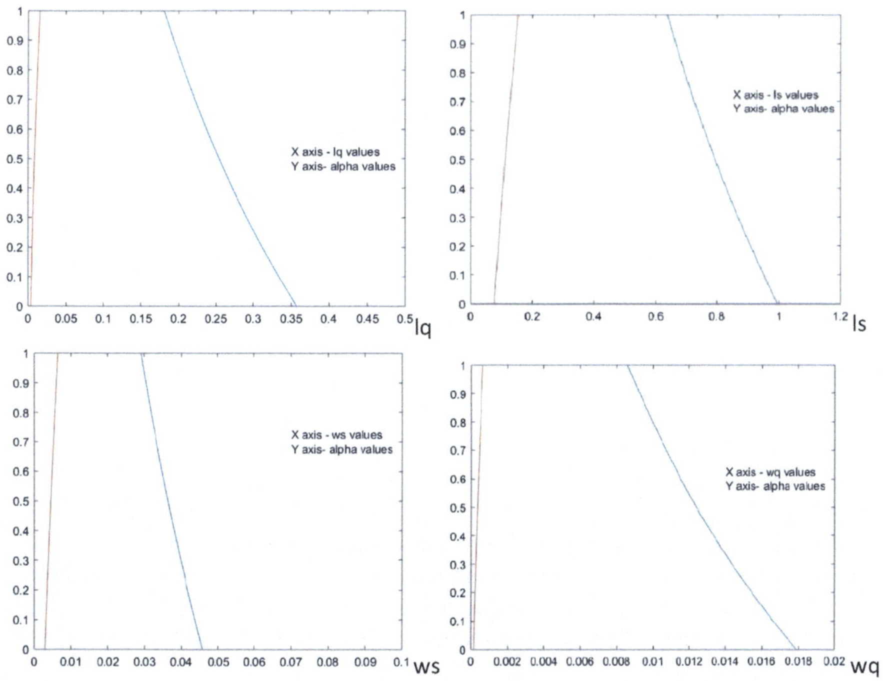

Fig. 3 Graph of L_q, L_s, W_s and W_q at α values (Example 3)

7 Comparison of Triangular, Trapezoidal and Hexagonal Fuzzy Numbers at Various α Values

See Tables 4 and 5.

Table 4 The α-cuts of L_s and L_q

α	L_s			L_q		
	Tri	Trap	Hexa	Tri	Trap	Hexa
0	[0.1442, 0.6585]	[0.0464, 0.6187]	[0.0766, 0.9951]	[0.0136, 0.1906]	[0.0015, 0.2159]	[0.0038, 0.5375]
0.5	[0.2242, 0.4669]	[0.0906, 0.4476]	[0.1124, 0.7942]	[0.0305, 0.1087]	[0.0056, 0.1192]	[0.0085, 0.2544]
1	[0.3283, 0.3283]	0.1491, 0.3219]	[0.1547, 0.6386]	[0.0351, 0.1087]	[0.0147, 0.0638]	[0.0155, 0.1807]

Table 5 The α-cuts of W_s and W_q

α	W_s			W_q		
	Tri	Trap	Hexa	Tri	Trap	Hexa
0	[0.0095, 0.0485]	[0.0033, 0.0562]	[0.0030, 0.0457]	[0.0009, 0.0146]	[0.0001, 0.0196]	[0.0001, 0.0178]
0.5	[0.0152, 0.0339]	[0.0067, 0.0389]	[0.0045, 0.0348]	[0.0021, 0.0080]	[0.0004, 0.0103]	[0.0003, 0.0124]
1	[0.0230, 0.0234]	[0.0114, 0.0268]	[0.0064, 0.0292]	[0.0042, 0.0042]	[0.0011, 0.0053]	[0.0006, 0.0086]

8 Results and Discussions

Utilizing MATLAB software package, we achieve α-cuts of entry rate, overhauled rate and fuzzy anticipated number of livelihoods in line at eleven variation levels: 0, 0.1, 0.2, 0.3, 0.4, 0.5, 0.6, 0.7, 0.8, 0.9, 1. Crisp intervals for fuzzy anticipated number of businesses in line at various probability α levels are represented in Tables 1, 2 and 3. The execution estimates, for example, anticipated number of clients in the framework (L_s), anticipated length of line (L_q), the normal holding up time in the framework (W_s) and the normal holding up time of clients in the line (W_q), are likewise inferred in Tables 1, 2 and 3.

From Table 1

(1) Expected number of clients in the queue lies in [0.0351, 0.1087] and infeasible falls outside [0.0136, 0.1906].
(2) Expected number of clients in the framework is 0.3283 and infeasible falls outside [0.1442, 0.6585].
(3) Average holding up time of a client in the queue is 0.0042 and infeasible falls outside [0.0009, 0.0146].
(4) Average holding up time of a client in the framework lies in [0.0230, 0.0234] and infeasible falls outside [0.0095, 0.0485].

From Table 2

(5) Anticipated number of clients in the line is 0.0638 and infeasible falls outside [0.0015, 0.2159].
(6) Expected number of clients in the system is 0.3219 and infeasible falls outside [0.0464, 0.6187].
(7) Average holding up time of a client in the queue is 0.0053 and infeasible falls outside [0.0001, 0.0196].
(8) Average holding up time of a client in the system is 0.0268 and infeasible falls outside [0.0033, 0.0562].

From Table 3

(9) Expected number of clients in the queue lies in [0.0155, 0.1807] and infeasible falls outside [0.0038, 0.3575].

(10) Expected number of clients in the system lies in [0.1547, 0.6386] and infeasible falls outside [0.0766, 0.9951].
(11) Average holding up time of a client in the queue lies in [0.0001, 0.0178] and infeasible falls outside [0.0006, 0.0086].
(12) Average holding up time of a client in the system lies in [0.0064, 0.0292] and infeasible falls outside [0.0030, 0.0457].

9 Limitations of the Proposed Model

There are several limitations of the proposed model. One of them is that limitation is the possibility that the waiting space may in fact be restricted. The second possibility is that arrival rate may be considered as state dependent. Many situations in industry and service are multi-channel queuing problems. When a customer has been attended to and the service provided, it may still have to get some other service from another service and may have to fall in queue once again. In such situations, the problem becomes still more difficult to analyse.

10 Conclusion

We deduce that fuzzy personality has been activated to limited lining miniature by utilizing diverse fuzzy numbers. The inter-entry time and overhauled time are fuzzy in description. The execution estimates, for example, framework length, line length, framework time, line time, etc., are so forth and are likewise fuzzified. Numerical model demonstrates the flexibility of DSW calculation. Here, it is noticed that the achievement of queuing model can be upgraded by expanding the number of variables. The suggested model can invigorate the ventures, wholesalers and retailers in unequivocally deciding the ideal execution proportions of the queuing model.

References

1. Li, R.J., Lee, E.S: Analysis of fuzzy queues. Comput. Math. Appl. **17**(7), 1143–1147 (1989)
2. Buckley, J.J.: Elementary queuing theory based on possibility theory. Fuzzy Sets Syst. **37**, 43–52 (1990)
3. Negi, D.S., Lee, E.S.: Analysis and simulation of fuzzy queue. Fuzzy Sets Syst. **46**, 321–330 (1992)
4. Chen, S.P.: Parametric nonlinear programming approach to fuzzy queues with bulk, service. Eur. J. Oper. Res. **163**, 434–444 (2005)
5. Jadeh, L.A.: Fuzzy sets. Inf. Control **8**, 338–353 (1965)
6. Gupta, P., Hira, D.S.: Operations Research, 884–885 (2007)
7. Rose, T.: Fuzzy Logic and Its Applications to Engineering. Wiley Eastern Publishers (2005)

8. Voskoglou, M.G., Subbotin, I.Y.: An application of the triangular fuzzy model to assessment of analogical reasoning skills. Am. J. Appl. Math. Stat., 3–1, 1–6 (2015)
9. Subbotin, I.Y.: Trapezoidal Fuzzy Logic Model for Learning Assessment. arXiv:1407.0823[math.gm] (2014)
10. Zimmermann, H.J: Fuzzy programming and linear programming with several objective functions. Fuzzy Sets Syst. **I**, 45–55 (1978)
11. Yovgav, R.R.: A characterization of the extension principle. Fuzzy Sets Syst. **18**, 71–78 (1986)
12. Moore, R., Lodwick, W.: Interval analysis and fuzzy set theory. Fuzzy Sets Syst. **135**(1), 5–9 (2003)
13. Prameela, K.U., Kumar, P.: Execution proportions of multi server queuing model with pentagonal fuzzy number: DSW algorithm approach. Int J Innov. Technol. Explor. Eng. **8**(7), 1047–1051 (2019)

A Deteriorating Inventory Model with Uniformly Distributed Random Demand and Completely Backlogged Shortages

Pavan Kumar and D. Dutta

Abstract This paper proposes a deteriorating inventory model with random demand with shortages. Shortages are permitted and treated as to be fully backlogged. Deterioration rate is constant. Demand of the items is a nonstop irregular variable, which pursues the uniform likelihood dispersion. Based on these assumptions, a mathematical model for the optimization of average total cost (ATC) function of the problem is formulated. Two numerical examples are presented. The convexity of the average total cost function is verified graphically. To measure the flexibility of the present model; the post-optimal analysis is applied by changing parameters.

Keywords Inventory · Deterioration · Random variable · Uniform distribution · Complete backlog

1 Introduction

Inventory is the stock of goods to run the business smoothly and efficiently. Fuzzy set theory has been applied by some researchers in inventory control problems. Kao et al. [1] presented a single-period inventory model with fuzzy demand. The impact of deterioration is a key factor in many inventory control problems. It is seen that a portion of the physical things or items disintegrates over a timeframe. For the inventory models consisting of such as vegetables, foodstuffs, natural products and so on, experience the ill effects of consumption by direct deterioration while kept in distribution center. The same situation replicates for highly volatile liquids which undergo physical depletion over course of time by the process of evaporation.

P. Kumar (✉)
Department of Mathematics, Koneru Lakshmaiah Education Foundation (KLEF), Vaddeswaram, Guntur, Andhra Pradesh 522502, India
e-mail: pavankmaths@gmail.com

D. Dutta
Department of Mathematics, National Institute of Technology, Warangal, Telangana 506004, India
e-mail: ddutta@nitw.ac.in

D. Dutta and B. Mahanty (eds.), *Numerical Optimization in Engineering and Sciences*, Advances in Intelligent Systems and Computing 979,
https://doi.org/10.1007/978-981-15-3215-3_20

In real-life cases, during the formulation of an inventory model, the deterioration rate, demand rate, unit cost, etc., probably all of them, will have some little fluctuations for each cycle. So, all things considered, cases, if these amounts are treated as fuzzy factors; at that point, it will be considerably more sensible. Samanta et al. [2] proposed a creation stock model with falling apart things and deficiencies. Holding cost comprises of all expenses brought about in conveying the merchandise to distribution center, additionally capacity and dealing with expenses. The holding cost somewhere in the range of few models has been considered as a variable one expense since utilization of crude material depends on the creation limit as the higher the generation, higher the variable expense. Alfares [3] examined a stock model with stock-level ward request rate with variable holding cost.

Pan and Yang [4] formulated some integrated inventory models in a supply chain considering the interest and creation rate as fluffy numbers. Zhou et al. [5] contemplated a store network coordination considering the interest rate as relying upon the dimension of stock. Tripathy and Mishra [6] displayed a stock model for Weilbull weakening things with value subordinate interest and time-fluctuating holding cost. Patra [7] proposed a request level stock model for crumbling items considering the fractional multiplying and incomplete lost deals. Singh and Singh [8] displayed an incorporated inventory network show for the short-lived items under fuzzy vulnerability as fuzzy generation rate and fuzzy interest rate.

Jaggi et al. [9] examined a fuzzy stock model for crumbling item following the time-differing request and deficiencies. Maihami and Nakhai Kamalabadi [10] proposed a joint evaluating and stock control for non-prompt breaking down items considering the time and value subordinate interest and incomplete buildup. Giri and Maiti [11] exhibited a store network show for a falling apart item with time-subordinate interest rate and generation rate.

Dutta and Kumar [12] contemplated an inventory problem for deteriorating items under fully backlogged shortages by introducing the fuzzy type uncertainty. Later, Dutta and Kumar [13, 14] examined a stock model with fractional multiplying for breaking down things, time-subordinate interest rate and holding cost by applying the established advancement procedure and an interim number methodology. Sustenance et al. [15] exhibited a monetary request amount (EOQ) demonstrate with slope type request rate and unit creation cost considering consistent crumbling rate. Zhao et al. [16] proposed an ideal approach for a generation stock model with an incorporated multi-organize inventory network and time-shifting interest.

Shabani et al. [17] proposed a model for two-distribution center stock issue with fuzzy weakening rate and fuzzy interest rate following the restrictive admissible deferral in installment. Palanivel et al. [18] displayed a coordinated model for generation stock issue with special exertion, variable creation cost and considering the probabilistic decay rate. Saha [19] examined a financial generation amount (EPQ) show for disintegrating items with probabilistic interest considering the variable creation rate.

In this paper, we propose a crumbling stock model with consistently circulated irregular interest and totally accumulated deficiencies. This paper is composed as pursues: In Sect. 2, the documentations and presumptions embraced to define the proposed model are given. In Sect. 3, the scientific model is defined. In Sect. 4, we present the algorithm for the solution procedure. In Sect. 5, two numerical examples are solved. In Sect. 6, the post-optimal analysis is performed for various changes in the parameters. In Sect. 7, some main observations are reported. In the last, we conclude in Sect. 8.

2 Documentations and Assumptions

The accompanying documentations and suspicions are considered for the proposed stock model as:

2.1 Notations

C_{o}	Ordering cost per order.
C_{h}	Holding cost per unit per unit time.
C_{s}	Shortage cost per unit time.
C_{p}	Purchasing cost per unit per unit time
U	Random interest rate of the clients.
$f(x)$	Probability density function (pdf), and $\int_0^\infty f(x)\mathrm{d}x = 1$.
$E(X)$	Expectation of random variable X.
θ	Deterioration rate in (0, 1).
T	Length of each ordering cycle (choice variable).
t_1	Time when shortages start (choice variable).
$I(t)$	On-hand inventory level at any time t, $0 \le t \le T$.
Q	Order quantity per unit.
ATC (t_1, T)	Average total cost of the stock expense per unit time.

2.2 Assumptions

(i) The inventory model consists of a single item.

(ii) Demand rate U is a continuous irregular likelihood thickness capacity of uniform conveyance

$$f(x) = \begin{cases} \frac{1}{b-a} & \text{when } a \le x \le b \\ 0 & \text{otherwise} \end{cases}$$

(iii) Lead time is zero and replenishment rate is infinite.

(iv) Allowed shortages are completely backlogged.

(v) θ is assumed to be constant.
(vi) There is no fix or substitution of the crumbled things amid the cycle.

3 Mathematical Model

Consider that the order quantity (Q) is purchased at the beginning of each period with fulfilling the backorders. Due to various reasons such as the demand in market and the weakening of things, the stock dimension of the thing continuously diminishes during $[0, t_1]$ and eventually reaches to zero at $t = t_1$. Now, the shortages occur during this time and it will be completely backlogged. The available stock dimension $I(t)$ at any time t is administered by the accompanying differential conditions

$$\frac{\mathrm{d}I(t)}{\mathrm{d}t} + \theta I(t) = -U, \text{ for } 0 \le t \le t_1 \tag{1}$$

$$\frac{\mathrm{d}I(t)}{\mathrm{d}t} = -U_2 \text{ for } t_1 \le t \le T \tag{2}$$

with

$$I(0) = Q, \text{ and } I(t_1) = 0. \tag{3}$$

Now, by solving (1) and (2) using the conditions in (3), the final solution is given by

$$I(t) = U\left[\mathrm{e}^{(t_1 - t)} - 1\right], \quad \text{for } 0 \le t \le t_1 \tag{4}$$

and

$$I(t) = U[t_1 - t], \quad \text{for } t_1 \le t \le T \tag{5}$$

Since $0 < \theta \ll 1$, the exponential series expansion is applied in (4); so, ignoring higher power term of θ, we obtain

$$\begin{aligned} I(t) &= \frac{U}{\theta}\left[1 + (t_1 - t)\theta + (t_1 - t)^2\theta^2 - 1\right], \quad \text{for } 0 \le t \le t_1 \\ &= U\left(t_1 - t + \theta\, t_1^2 + \theta\, t^2 - 2\theta\, t_1 t\right), \text{ for } 0 \le t \le t_1 \end{aligned} \tag{6}$$

Using the initial condition: $I(0) = Q$, in (6), we obtain

$$Q = U\left(t_1 + \theta t_1^2\right), \tag{7}$$

Total average number of holding units (I_H) during the period $[0, T]$ is given by

$$I_H = \int_0^{t_1} I(t)\mathrm{d}t = U\left(\frac{1}{2}t_1^2 + \frac{1}{3}\theta\, t_1^3\right) \tag{8}$$

All out number of decayed units (I_D) amid the period [0, T] is given by

$$\begin{aligned} I_D &= Q - \text{ Total Demand} \\ &= Q - \int_0^{t_1} U\mathrm{d}t = U\theta\, t_1^2, \end{aligned} \tag{9}$$

All out normal number of deficiency units (I_S) the period [0, T] is given by

$$I_S = -\int_{t_1}^{T} I(t)\mathrm{d}t = \frac{-U}{2}\left(2Tt_1 - T^2 - t_1^2\right) = \frac{U}{2}(T - t_1)^2 \tag{10}$$

Total shortage cost per unit time

$$C_{TS} = \frac{1}{T}[C_\mathrm{h} I_S] = \frac{(T - t_1)^2}{2T} UC_S \tag{11}$$

Then the function ATC (t_1, T) is given by

$$\begin{aligned} \mathrm{ATC}(t_1, T) &= \frac{1}{T}[C_\mathrm{o} + C_\mathrm{h} I_H + C_\mathrm{h} I_D + C_\mathrm{h} I_S] \\ &= \frac{1}{T}\left[C_\mathrm{o} + UC_\mathrm{h}\left(\frac{1}{2}t_1^2 + \frac{1}{3}\theta\, t_1^3\right) + UC_\mathrm{h}\theta\, t_1^2 + \frac{U}{2}C_\mathrm{h}(T - t_1)^2\right] \end{aligned} \tag{12}$$

In our model, the demand rate U follows the uniform probability distribution function with expectation as $\frac{a+b}{2}$, $a > 0$, $b > 0$ and $a < b$. Then the Eq. (12) can be rewritten as

$$\begin{aligned} \mathrm{ATC}(t_1, T) = \frac{1}{T}\bigg[C_\mathrm{o} &+ \left(\frac{a+b}{2}\right)\left(\frac{1}{2}t_1^2 + \frac{1}{3}\theta t_1^3\right)C_\mathrm{h} \\ &+ \left(\frac{a+b}{2}\right)C_\mathrm{h}\theta t_1^2 + \left(\frac{a+b}{4}\right)C_\mathrm{h}(T - t_1)^2\bigg] \end{aligned} \tag{13}$$

To minimize the function ATC (t_1, T), the minimum values of t_1 and T are determined by solving the following:

$$\frac{\partial \mathrm{ATC}(t_1, T)}{\partial t_1} = 0, \text{ and } \frac{\partial \mathrm{ATC}(t_1, T)}{\partial T} = 0$$

$$\frac{\partial \text{ATC}(t_1, T)}{\partial t_1} = \frac{1}{T}\left[\left(\frac{a+b}{2}\right)(t_1 + \theta t_1^2)C_h + 2\left(\frac{a+b}{2}\right)C_h\theta t_1 - \left(\frac{a+b}{2}\right)C_h(T - t_1)\right] = 0$$

$$0 \Rightarrow C_h\left(t_1 + \theta t_1^2\right) + 2C_h\theta t_1 - C_h(T - t_1) = 0 \quad (14)$$

and

$$\frac{\partial \text{ATC}(t_1, T)}{\partial T} = -\frac{1}{T^2}\left[C_o + \left(\frac{a+b}{2}\right)\left(\frac{1}{2}t_1^2 + \frac{1}{3}\theta t_1^3\right)C_h + \left(\frac{a+b}{2}\right)C_h\theta t_1^2 + \left(\frac{a+b}{4}\right)C_h(T - t_1)^2\right] + \frac{1}{T}\left(\frac{a+b}{2}\right)C_h(T - t_1) = 0$$

$$\Rightarrow C_o + \left(\frac{a+b}{2}\right)\left(\frac{1}{2}t_1^2 + \frac{1}{3}\theta t_1^3\right)C_h + \left(\frac{a+b}{2}\right)C_h\theta t_1^2 + \left(\frac{a+b}{4}\right)C_h(T - t_1)^2 - T\left(\frac{a+b}{2}\right)C_h(T - t_1) = 0$$

$$\Rightarrow C_o + \left(\frac{a+b}{2}\right)C_h\left(\frac{1}{2}t_1^2 + \frac{1}{3}\theta t_1^3\right) + \left(\frac{a+b}{2}\right)C_h\theta t_1^2 + \left(\frac{a+b}{2}\right)C_h(T - t_1)\left(\frac{\tau - t_1}{2} - T\right) = 0$$

$$\Rightarrow C_o + \left(\frac{a+b}{2}\right)\left[C_h\left(\frac{1}{2}t_1^2 + \frac{1}{3}\theta t_1^3\right) + C_h\theta t_1^2 - \frac{1}{2}C_h\left(T^2 - t_1^2\right)\right] = 0 \quad (15)$$

The nonlinear Eqs. (14)–(15) can be solved by using a suitable computer software. However, the calculation of second derivatives of the function ATC (t_1, T) is very complicated. Therefore, with the help of a graph, the convexity of function ATC (t_1, T) can be examined.

4 Algorithm

Step 1: Solve the nonlinear Eqs. (14)–(15) for t_1 and T.
Step 2: Assume $t_1 = t_1^*$, and $T = T^*$ be the optimal solution as obtained in step 1. Calculate the value
Step 3: Determine the optimal value of ATC for $t_1 = t_1^*$, and $T = T^*$ using Eq. (13).
Step 5: Assume ATC = ATC* be the optimal solution as obtained in Step 3.
Step 6: Final optimal solution is ATC = ATC*, when $t_1 = t_1^*$, and $T = T^*$.
Step 7: The convexity of function ATC can be seen from the plot of a 3D graph for ATC = ATC*, $t_1 = t_1^*$, and $T = T^*$.

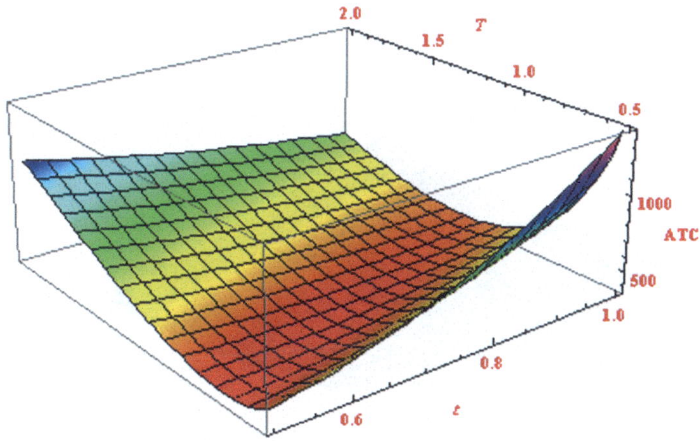

Fig. 1 Convexity of ATC function (Example 1)

5 Numerical Examples

To illustrate the model, two examples are given below:

Example 1 Let us consider the input data: $C_o = \$150$, $C_h = \$5$, $C_s = \$15$, $C_p = \$20$, $\theta = 0.03$, $a = 80$, $b = 140$. We have the optimum solution as $t_1^* = 0.5975$, $T^* = 0.7228$ and ATC* = 413.42. The function ATC shows convex in Fig. 1.

Example 2 Let us consider the input data: $C_o = \$1650$, $C_h = \$12$, $C_s = \$25$, $C_p = \$35$, $\theta = 0.08$, $a = 50$, $b = 350$. The optimum solution is $t_1^* = 0.7192$, $T^* = 1.2453$ and ATC* = 2630.80. The function ATC shows convex in Fig. 2.

6 Post-Optimal Analysis

The post-optimal analysis is applied to evaluate the effect of changing inventory parameters on model conditions and measuring the flexibility of model. We change only one parameter at a time and the outputs are shown in Table 1. The comparing bends of the all-out expenses are shown in Fig. 3a–g, respectively.

7 Observations

The accompanying perceptions are recorded from Table 1:

(i) For the increase of θ, the optimum values of t_1 and T decline. By this impact, the function ATC increases.

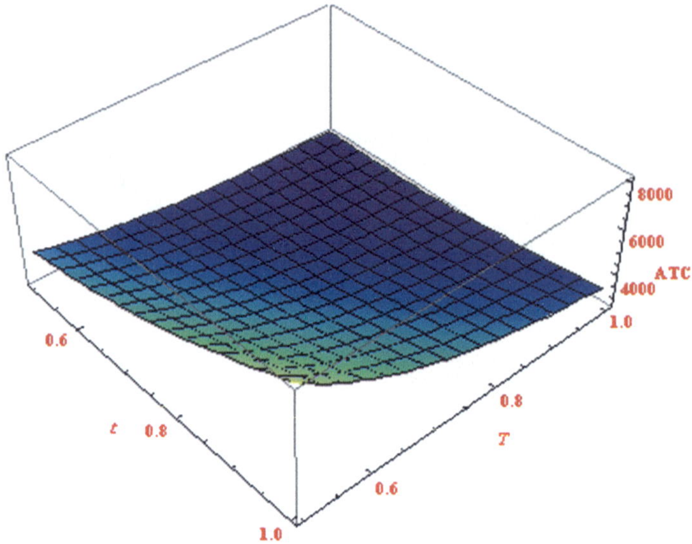

Fig. 2 Convexity of ATC function (Example 2)

(ii) For the increase of C_o, the optimum value of t_1 increases and T decline. By this impact, the function ATC increases.

(iii) For the increase of C_h, the optimum values of both t_1 and T decline. By this impact, the function ATC increases.

(iv) For the increase of C_s, the optimum value of t_1 increases and T decline. By this impact, the function ATC increases.

(v) For the increase of C_h, the optimum values of both t_1 and T decline. By this impact, the function ATC increases.

(vi) For the increase of parameters a and b, the optimum values of both t_1 and T decline. By this impact, the function ATC increases.

(vii) The function ATC is more sensitive to $C_{h,}$ a and b. On the other side, it is less sensitive to C_p and θ.

8 Conclusion

A deteriorating inventory model with uniformly distributed random demand considering the complete backlog has been discussed. Deterioration and shortages cannot be avoided in any inventory model. In this paper, a constant deterioration was considered. The proposed model is developed as an optimization of average total cost function. Post-optimal analysis has been performed. It shows that the function ATC is more sensitive to holding cost. Thus, the decision maker (DM) can take the managerial decision to control over optimal average total cost, and for other related

Table 1 Post-optimal analysis for Example 1

Changing parameters	Change in parameters	t_1	T	ATC
θ	0.01	0.6515	0.7693	389.27
	0.02	0.6229	0.7446	401.69
	0.03	*0.5975*	*0.7228*	*413.42*
	0.04	0.5747	0.7034	424.53
	0.05	0.5541	0.6859	435.09
C_o	130	0.5567	0.6733	384.77
	140	0.5775	0.6985	399.35
	150	*0.5975*	*0.7228*	*413.42*
	160	0.6170	0.7463	427.03
	170	0.6357	0.7691	440.23
C_h	3	0.7463	0.8525	350.31
	4	0.6614	0.7777	384.09
	5	*0.5975*	*0.7228*	*413.42*
	6	0.5472	0.6804	439.34
	7	0.5062	0.6464	462.55
C_s	11	0.5798	0.7455	400.98
	13	0.5898	0.7325	408.01
	15	*0.5975*	*0.7228*	*413.42*
	17	0.6036	0.7153	417.70
	19	0.6086	0.7094	421.17
C_p	16	0.6111	0.7344	406.80
	18	0.6042	0.7285	410.13
	20	*0.5975*	*0.7228*	*413.42*
	22	0.5910	0.7173	416.65
	24	0.5847	0.7120	419.84
a	60	0.6264	0.7578	394.25
	70	0.6115	0.7397	403.95
	80	*0.5975*	*0.7228*	*413.42*
	90	0.5845	0.7070	422.70
	100	0.5723	0.6923	431.73
b	120	0.6264	0.7578	394.25
	130	0.6114	0.7397	403.95
	140	*0.5975*	*0.7228*	*413.42*
	150	0.5845	0.7070	422.67
	160	0.5723	0.6923	431.72

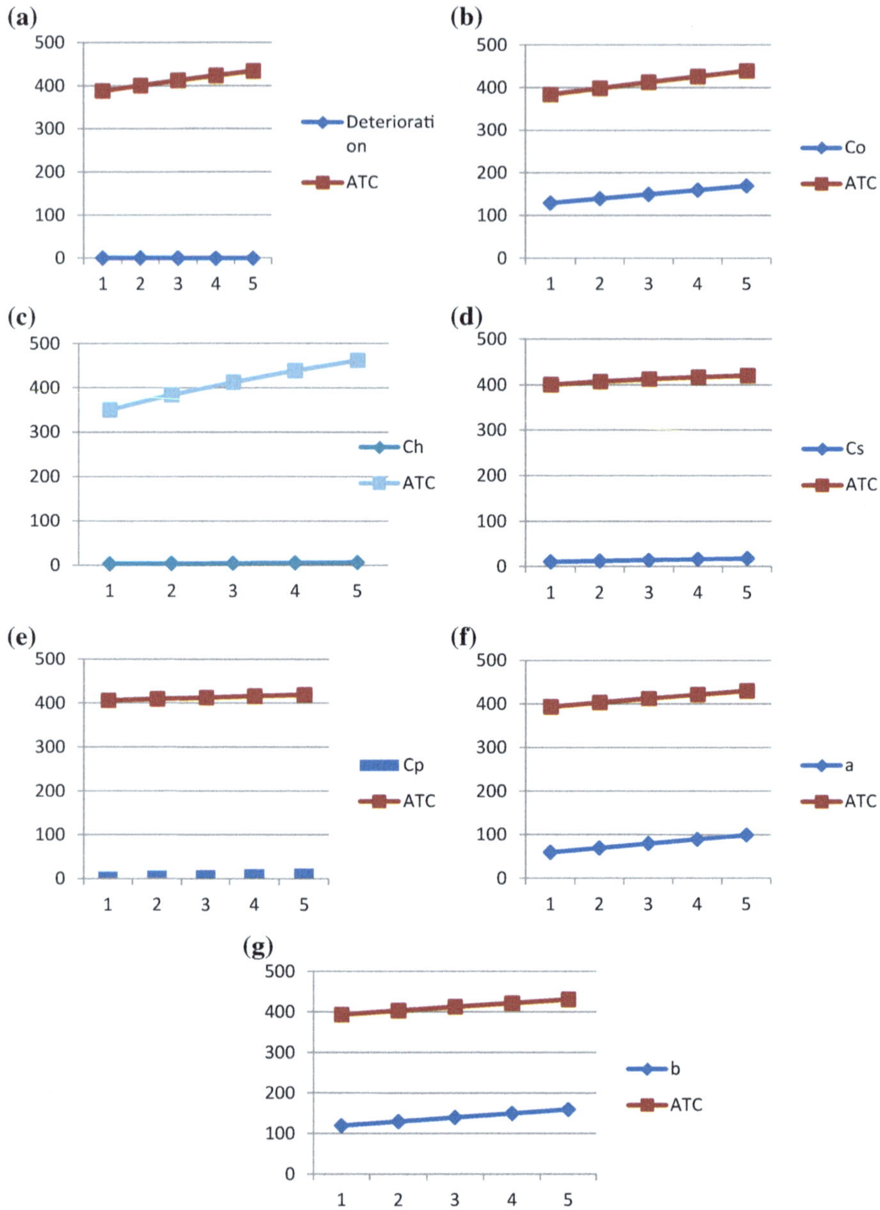

Fig. 3 Graphical representation of post-optimal analysis

parameters, as per this model situations, in case of seasonal items such as bathing suits, winter coats, etc.

References

1. Kao, C.K., Hsu, W.K.H.: A single-period inventory model with fuzzy demand. Comput. Math. Appl. **43**(6–7), 841–848 (2002)
2. Samanta, G.P., Roy, A.: A production inventory model with deteriorating items and shortages. Yugosl. J. Oper. Res. **14**(2), 219–230 (2004)
3. Alfares, H.K.: Inventory model with stock-level dependent demand rate and variable holding cost. Int. J. Prod. Econ. **108**(1–2), 259–265 (2007)
4. Pan, J.C.H., Yang, M.F.: Integrated inventory models with fuzzy annual demand and fuzzy production rate in a supply chain. Int. J. Prod. Res. **46**(3), 753–770 (2008)
5. Zhou, Y.W., Min, J., Goyal, S.K.: Supply-chain coordination under an inventory-level-dependent demand rate. Int. J. Prod. Econ. **113**(2), 518–527 (2008)
6. Tripathy, C.K., Mishra, U.: An inventory model for WEIBULL deteriorating items with price dependent demand and time-varying holding cost. Appl. Math. Sci. **4**(41–44), 2171–2179 (2010)
7. Patra, S.K.: An order level inventory model for deteriorating items with partial backlog and partial lost sales. Int. J. Adv. Oper. Manag. **2**(3), 185–200 (2010)
8. Singh, C., Singh, S.R.: An integrated supply chain model for the perishable items with fuzzy production rate and fuzzy demand rate. Yugosl. J. Oper. Res. **21**(1), 47–64 (2011)
9. Jaggi, C.K., Pareek, S., Sharma, A., Nidhi, A.: Fuzzy inventory model for deteriorating items with time-varying demand and shortages. Am. J. Oper. Res. **2**(6), 81–92 (2012)
10. Maihami, R., Kamalabadi, I.N.: Joint pricing and inventory control for non-instantaneous deteriorating items with partial backlogging and time and price dependent demand. Int. J. Prod. Econ. **136**(1), 116–122 (2012)
11. Giri, B.C., Maiti, T.: Supply chain model for a deteriorating product with time-varying demand and production rate. J. Oper. Res. Soc. **63**(5), 665–673 (2012)
12. Dutta, D., Kumar, P.: Fuzzy inventory model for deteriorating items with shortages under fully backlogged condition. Int. J. Soft Comput. Eng. **3**(2), 393–398 (2013)
13. Dutta, D., Kumar, P.: A partial backlogging inventory model for deteriorating items with time-varying demand and holding cost: an interval number approach. Croat. Oper. Res. Rev. **6**(2), 321–334 (2015)
14. Dutta, D., Kumar, P.: A partial backlogging inventory model for deteriorating items with time-varying demand and holding cost. Int. J. Math. Oper. Res. **7**(3), 281–296 (2015)
15. Manna, P., Manna, S.K., Giri, B.C.: An economic order quantity model with ramp type demand rate, constant deterioration rate and unit production cost. Yugosl. J. Oper. Res. **26**(3), 305–316 (2016)
16. Zhao, S.T., Wu, K., Yuan, X.M.: Optimal production-inventory policy for an integrated multi-stage supply chain with time-varying demand. Eur. J. Oper. Res. **255**(2), 364–379 (2016)
17. Shabani, S., Mirzazadeh, A., Sharifi, E.: A two-warehouse inventory model with fuzzy deterioration rate and fuzzy demand rate under conditionally permissible delay in payment. Int. J. Prod. Eng. **33**(2), 134–142 (2016)
18. Palanivel, M., Uthayakumar, R.: A production-inventory model with promotional effort, variable production cost and probabilistic deterioration. Int. J. Syst. Assur. Eng. Manag. **8**(1), 290–300 (2017)
19. Saha, S.: An EPQ model for deteriorating items with probabilistic demand and variable production rate. Am. J. Eng. Res. **7**(6), 153–159 (2018)

Analysis of $M/E_K/1$ Queue Model in Bulk Service Environment

Manish Kumar Pandey and D. K. Gangeshwer

Abstract In this paper, we consider bulk arrival queue with bulk services to analyse the proposed $M/E_K/1$ queuing model. A single server including bulk services queue with bulk arrival Poisson input and service time distribution is exponential. A steady-state probability generating function for queue length of a bulk arrival using ambulance service as a server has been derived. The mean queue length and other operational behaviours have been obtained. Finally, some special cases with respect capacity of an ambulance services have derived.

Keywords $M/E_K/1$ queue model · Bulk queue and bulk services · Steady-state probability generating function

1 Introduction

The pioneer work in the field of queuing theory was done by Erlang [1], an engineer at the Copenhagen telephone exchange, who worked on the theory of probabilities and telephone conversations. After Erlang [1], Molina [2] succeeded queuing theory and published his theory under the title of application of the theory of probability to telephone trunking problems. Fry [3] in this book on probability and its engineering uses throws further light on the above problem. In the literature, Jakson [4] defined the idea of queuing process with phase type service in which he discussed about service, be it human or mechanical, and is referred to as a customer who may be finite or infinite. Balachandran [5] deals with two parametric policies for a single services facility when the arrival is distributed as passion and the service durations have a general distribution. Choudhury and Madan [6] considered a batch arrival queue with a Bernoulli vacation schedule under a restricted admissibility policy. In which they derive the steady-state queue size distribution at a random point of time as well as at departure epoch. Zadeh [7] investigated about a batch arrival queue system with

M. K. Pandey (✉) · D. K. Gangeshwer
Department of Applied Mathematics, Bhilai Institute of Technology Durg, Durg, CG 491001, India

D. K. Gangeshwer
e-mail: dk.gangeshwar@bitdurg.ac.in

D. Dutta and B. Mahanty (eds.), *Numerical Optimization in Engineering and Sciences*, Advances in Intelligent Systems and Computing 979,
https://doi.org/10.1007/978-981-15-3215-3_21

coxian-2 server vacation and admissibility restricted, and he derived the probability generating function of the system; by using them, the performance measures are obtained. Bhagat and Jain [8] found $M_x/G/1$ retrial queue with unreliable server and general retrial times i which they examined the effect of several parameters on the system performance.

Bharathidass et al. [9] analysed a single-server Erlangian bulk service queue with certain concept like vacation break down and repair. Jiang et al. [10] explained about an N-policy GI/M/I queue in multi-phase service environment with disaster, and they derived the stationary queue length distribution with numerical examples. Pandey and Gangeshwer [11] studied that the application of the diffusion approximation to hospital using $G_{\infty}/G_M/1$ double-ended queue model and considered no limit for patients arrival, but available M-service are limited. Mohamed and Karthikeyen [12] showed that on bulk queuing models, wherein they have provided sufficient information to analysis managers and industry people who are interested in using queuing theory to model congestion problem and want to locate the detail to relevant models. Queuing systems with $M/E_k/1$ queue models have attracted much attention from numerous researchers, such as Jaiswal [13], Restrepo [14], Heyman [15], Kambo Chaudhary [16], Madan [17], Ke [18], Jain and Agrawal [19], Janssen and Leeuwaarden Van [20], Griffiths et al. [21], Yu et al. [22] and Miaomiao et al. [23].

In this paper, we considered bulk arrival queue with bulk services to analyse the queuing model. A single server including bulk services queue with bulk arrival Poisson input and service time distribution is exponential. In section two, $M/E_k/1$ queue model in bulk service environment using ambulance service has been constructed. Section three explains the generating function of the state probabilities based on ambulance capacity. The mean queue length and other operational behaviours have been obtained. Some special cases with respect to capability of ambulance services to hospital have derived.

2 $M/E_k/1$ Model in Bulk Service Environment

The queuing system may be described on the line of Tuteja and Arora [24], wherein an infinite capacity queuing facility where patients arrive at a service facility according to a Poisson process. To avoid higher cost of the system, relaxing the assumption that server start service as soon as. Patients arrive at service facility in Erlangian process having 'k' services with rate $k\alpha$ for each service. The server provides one phase's service to each patient. The service discipline is assumed to be first come first served (FCFS) basis. The server follows general bulk service rule that is the server does not accept less than 'a' and more than 'b' patients at a time. The service time distribution is exponential with rate β. Ambulance follows random idle policy that means if there are less than 'a' unit in the queue then server goes to idleness and further assumed that if there is no unit in the queue, the server again goes to the idle state for the same distributed random period. In real-life situation, if any accidents

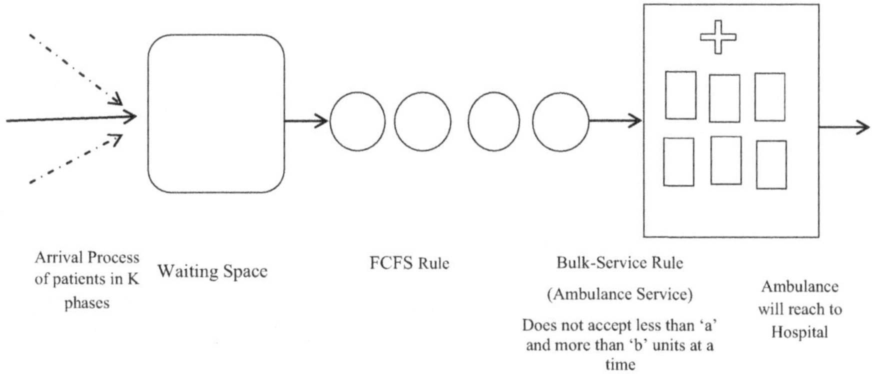

Fig. 1 M/E$_k$/1 queue model in bulk service environment using ambulance service

occur at certain place, then ambulance comes and there is limited number of seats available in ambulance to carry hospital for the treatment.

The following notations and definitions have been considered to describe the proposed model when ambulance is idle state:

(i) $P_{n,r}\ (t)$ = the probability that at time t there are n patients in the queue, when arrival is in the rth phase. That is $(0 \leq n < \infty,\ 1 \leq r < k)$ and $P_{n,r} = \lim_{n\to\infty} P_n, r(t)$

(ii) When ambulances in operative state
$Q_{n,\ r}\ (t)$ = the probability that at time t there are n patients in the queue, when arrival is in the rth phase. That is $(0 \leq n < \infty, 1 \leq r < k)$ $Q_{n,r} = \lim_{n\to\infty} Q_n, r(t)$.

Steady-State Equations

Considering the aforesaid queue model, the following formulation of the steady-state equations on the line Tuteja and Arora [24, p. 123] (Fig. 1):

$$K\alpha P_{0,1} = \beta Q_{0,r} + K\alpha P_{0r-1} \tag{1}$$

$$K\alpha P_{0,1} = \beta Q_{0,1} \tag{2}$$

$$K\alpha P_{n,r} = \beta Q_{n,r} + K\alpha P_{n,r-1}(n < a) \tag{3}$$

$$K\alpha P_{n,1} = \beta Q_{n,1} + K\alpha P_{n-1,k}(n < a) \tag{4}$$

$$(K\alpha + \theta)P_{n,r} = K\alpha P_{n,r-1} \quad (n \geq a) \tag{5}$$

$$(K\alpha + \theta)P_{n,1} = K\alpha P_{n-1,k} \quad (n \geq a) \tag{6}$$

$$(K\alpha + \beta)Q_{0,1} = \beta \sum_{j=a}^{b} Q_{j,1} + \theta \sum_{j=a}^{b} P_{j,1} \tag{7}$$

$$(K\alpha + \beta)Q_{0,r} = \beta \sum_{j=a}^{b} Q_{j,r} + \theta \sum_{j=a}^{b} P_{j,r} + K\alpha\, Q_{0,r-1} \tag{8}$$

$$(K\alpha + \beta)Q_{n,r} = \beta Q_{n+b,1} + \theta P_{n+b,1} + K\alpha\, Q_{n+1,k}(n \geq 1) \tag{9}$$

$$(K\alpha + \beta)Q_{n,r} = \beta Q_{n+b,r} + \theta P_{n+b,r} + K\alpha\, Q_{n,r-1}(n \geq 1) \tag{10}$$

Partial Probability Generating Function

In this section, the different partial probability generating functions are defined when the server is on working:

$$\begin{aligned}
Q_r(z) &= \sum_{n=0}^{\infty} Q_{n,r} z^n, \\
Q(x, z) &= \sum_{r=1}^{k} Q_r(z) x^r \\
Q_n(x) &= \sum_{r=1}^{k} Q_{n,r} x^r\, P_r(z) = \sum_{n=0}^{\infty} P_{n,r} z^n \\
P(x, z) &= \sum_{r=1}^{k} P_r(z) x^r\, P_n(x) = \sum_{r=1}^{k} P_{n,r} x^n
\end{aligned} \tag{11}$$

Further, $P(x, z)$ and $Q(x, z)$, the generating function of the state, is derived based on Tuteja and Arora [24]:

$$P(x, z) = \frac{\theta \sum_{n=0}^{a-1} P_n(x) z^n + \beta \sum_{n=0}^{a-1} Q_n(x) z^n + K\alpha P_k(z)\left(z - x^k\right)}{[k\alpha(1 - x) + \theta]} \tag{12}$$

Similarly, we get

$$Q(x, z) = \frac{\begin{array}{c} [k(1 - x) + \theta]\left[\sum_{j=a}^{b-1} \left(z^b - z^j\right)\left\{\beta Q_j(x) + \theta P_j(x)\right\}\right] \\ - \sum_{j=0}^{a-1} z^j \left[\beta Q_j(x) + \theta P_j(x)\right][K\alpha(1 - x)] \\ + K_\alpha P_k(z) x\left(z - x^k\right) + [K\alpha(1 - x) + \theta]\left[K\alpha z^b x Q_k(z)\left(z - x^k\right)\right] \end{array}}{\left[K\alpha z^b(1 - x) - \beta\left(1 - z^b\right)\right][K\alpha(1 - x) + \theta]} \tag{13}$$

3 Generating Function of the State Probabilities Based on Ambulance Capacity

In this section, immediate service in health sector is required when any accident occurs such as under-constructed flyover collapsed, train accident, stampede. Then, the ambulance service is available immediately in which the service capacity is maximum two or three. Tuteja and Arora [24] derived the functions for $a = 1$ and $b = 1$. Subsequently, the following generating functions of the state probabilities have been derived:

When the ambulance service capacity is minimum capacity $a = 1$ and maximum capacity $b = 2$, then,

$$P(x, z) = \frac{\theta P_0(x) + \beta Q_0(x) + K\alpha P_k(z)\left(z - x^k\right)}{K\alpha(1 - x) + \theta}$$

$$Q(x, z) = \frac{\begin{array}{c}\{k\alpha(1-x)+\theta\}\left(z^2 - z\right)\{\beta Q_1(x) + \theta P_1(x)\} - \{\beta Q_0(x) + \theta P_0(x)\}\{k\alpha(1-x)\} \\ +\theta k\alpha P_k(z)x\left(z - x^k\right) + \{k\alpha(1-x)+\theta\}\left\{k\alpha z^2 x Q_k(z)\left(z - x^k\right)\right\}\end{array}}{\left\{k\alpha z^2(1-x) - \beta\left(1 - z^2\right)\right\}\{k\alpha(1-x)+\theta\}}$$

$$L_{q_1} = \frac{k\alpha p_k(1)}{\theta}$$

$$L_{q_1} = \frac{\begin{array}{c}\left[2\theta\{\beta Q_1(1) + \theta P_1(1)\} + k\alpha\left\{2P_k'(1) + 4Q_k'(1) + 6Q_k(1)\right\}\right]2\theta\beta \\ -2\theta\beta[\theta\{\beta Q_1(1) + \theta P_1(1)\} + \theta K\alpha\{P_k(1) + 2Q_k(1)\} + \theta\{K\theta\alpha P_k(1) + 2Q_k(1)\}]\end{array}}{2(2\theta\beta)^2}$$

$$E_1 = \frac{\theta P_0(1) + \beta Q_0(1)}{\theta}$$

$$E_2 = \frac{[\beta Q_1(1) + \theta P_1(1)] + K_\alpha[2Q_k(1) + P_k(1)]}{2\beta}$$

$$L_{P_1} = \frac{\theta\left[\theta P_0'(1) + \beta Q_0'(1) - K_\alpha^2 P_k(1)\right] + K_\alpha[\theta P_0(1) + \beta Q_0(1)]}{\theta}$$

$$L_{P_2} = \frac{\begin{array}{c}\left[2K_\alpha\left\{\beta Q_0'(1) + \theta P_0'(1) + K_\alpha^2 Q_k(1)\right\} - K_\alpha^2\theta(k+1)\{P_k(1) + Q_k(1)\}(-k\theta\alpha)\right] \\ -2K_{\alpha^2}^2[K_\alpha\{\beta Q_0(1) + \theta P_0(1)\}] - K_\alpha^2\theta\{P_k(1) + Q_k(1)\}\end{array}}{2[-K\theta\alpha]^2}$$

Similarly, when the ambulance service capacity occurs with minimum capacity $a = 1$ and maximum capacity $b = 3$, then,

$$Q(x, z) = \frac{\begin{array}{c}[k\alpha(1-x)+\theta]z(z-1)[(z+1)\{\beta Q_1(x) + \theta P_1(x)\} \\ +z\{\beta Q_2(x) + \theta P_2(x)\}] - [\beta Q_0(x) - \theta P_0(x)] \\ \left[k\alpha(1-x) + \theta k\alpha P_k(z)x\left(z - x^k\right)\right] \\ [k\alpha(1-x)+\theta]\left[k\alpha z^3 x Q_k(z)\left(z - x^k\right)\right]\end{array}}{\left[k\alpha z^3(1-x) - \beta\left(1 - z^3\right)\right][k\alpha(1-x)+\theta]}$$

$$L_{q_2} = \frac{\beta[4Q_1(1) + 3Q_2(1)] - 2\theta[P_1(1) + P_2(1)] + k\alpha\left[\{2P_k'(1) - P_k(1)\} + 3\{2Q_k'(1) + 3Q_k(1)\}\right]}{3\beta}$$

$$E_2 = \frac{\beta[2Q_1(1) + Q_2(1)] + \theta[2P_1(1) + P_2(1)] + k\alpha[3Q_k(1) + P_k(1)]}{3\beta}$$

$P(x, z)$, L_{q_1}, E_1, L_{p_1} and L_{p_2} are same as in $a = 1$ and $b = 2$.

4 Conclusion

Present paper established analysis of M/E$_k$/1 queue model in bulk service environment considering the ambulance in which steady-state probability generating function for queue length of a bulk arrival queue with bulk service, mean queue length with condition and other operational behaviours with different think have been obtained. The ambulance capacity when $a = 1$ (minimum) and $b = 2$ (maximum) and $a = 1$ (minimum) and $b = 3$ (maximum) have been derived. The proposed development would be a useful for many emergency (or accident) situations where optimal numbers of ambulances are needed to carry all the patients to hospital.

References

1. Erlang, A.K.: The theory of probabilities and telephone conversations. Matematika, **20**(A), 33–39 (1909)
2. Molina, E.C.: Application of the theory of probability to telephone trunking problems. Bell Syst. Tech. J. **7**, 461–494 (1927)
3. Fry, T.C.: Probability and its Engineering Uses. Princeton, Van Nostrand, NJ (1928)
4. Jackson, R.R.P.: Queuing systems with phase type service. J. Oper. Res. Soc. **5**(4), 109–120 (1954)
5. Balachandran, K.R.: Control policies for a single server system. Manage. Sci. **19**(9), 1013–1018 (1973)
6. Choudhury, G., Madan, K.C.: A batch arrival queue with Bernoulli vacation schedule under multiple vacation policy. Manag. Sci. Financ. Eng. **12**(2), 1–18 (2006)
7. Zadeh, A.B.: A batch arrival queue system with Coxian-2 server vacations and admissibility restricted. Am. J. Ind. Bus. Manag. **2**(2), 47–54 (2012)
8. Bhagat, A., Jain, M.: Unreliable MX/G/1 retrials queue with multi-optional services and impatient customers. Int. J. Oper. Res. **17**(2), 248–273 (2013)
9. Bharathidass, S., Arivukkarasu, V., Ganesan, V.: Bulk service queue with server breakdown and repairs. Int. J. Stat. Appl. Math. **3**(1), 136–142 (2018)
10. Jiang, T., Ammar, S.I., Chang, B., Liu, L.: Analysis of an N-Policy GI/M/1 queue in a multi-phase service environment with disasters. Int. J. Appl. Math. Comput. Sci. **28**(2), 375–386 (2018)

11. Pandey, M.K., Gangeshwer, D.K.: Applications of the Diffusion Approximation to Hospital Sector Using G∞/GM/1 Double Ended Queue Model. J. Comput. Math. Sci. **9**(4), 302–308 (2018)
12. Mohamed, S.Y., Karthikeyen, N.: Analysis of bulk arrivals in queuing models. Int. Res. J. Multidiscip. Stud. **4**(4), 1–6 (2018)
13. Jaiswal, Nk: Bulk service queuing system. Oper. Res. **8**, 139–143 (1960)
14. Restrepo, R.A.: A queue with simultaneous arrival and Erlangian service distribution. Oper. Res **13**, 375–388 (1965)
15. Heyman, D.P.: Optimal operating policies for M/G/1 queuing systems. Oper. Res. **16**(2), 362–382 (1968)
16. Kambo, N.S., Chaudhry, M.L.: Distribution of busy period for the bulk-service queuing system Ek/Ma, b/1. Comput. Oper. Res. **11**(3), 267–274 (1984)
17. Madan, K.C.: An M/G/1 queue with second optional service. Queuing Syst. **34**, 37–46 (2000)
18. Ke, J.C.: The analysis of a general input queue with N policy and exponential vacations. Queuing Syst. **45**(2), 135–160 (2003)
19. Jain, M., Agrawal, P.K.: M/Ek/1 queuing system with working vacation. Qual. Technol. Quant. Manag. **4**(4), 455–470 (2007)
20. Janssen, A.J.E.M., Van Leeuwaarden, J.S.H.: Analytic computation schemes for the discrete-time bulk service queue. Queuing Syst. **50**, 141–163 (2005)
21. Griffiths, J.D., Leonenko, G.M., Williams, J.E.: The transient solution to M/Ek/1 queue. Oper. Res. Lett. **34**(3), 349–354 (2006)
22. Yu, Z., Liu, M., Ma, Y.: Steady-state queue length analysis of a batch arrival queue under N-policy with single vacation and setup times. Intell. Inf. Manag. **2**(6), 365–374 (2010)
23. Miaomiao, Yu., Tang, Yinghui, Fuc, Yonghong, Pan, Lemeng: Queuing system with no damage service interruptions. Math. Comput. Model. **54**(5–6), 1262–1272 (2011)
24. Tuteja, R.K., Arora, R.T.: A single server control policy having an Erlangian arrival. Int. Res. J. Multidiscip. Stud. **6**(2), 122–127 (1990)

Role of Consistency and Random Index in Analytic Hierarchy Process—A New Measure

V. Shyamprasad and P. Kousalya

Abstract This paper explains about the role of consistency and Random Index in analytic hierarchy process [2], which is one of the popular and most effectively used decision-making techniques developed by professor Saaty in the 1970s. The techniques of AHP have been applied in many areas and several decision-making issues. Consistency plays a vital role in AHP [2], Saaty introduced consistency test to accept or reject the decision-maker's pairwise comparison matrix, i.e., Consistency Ratio $= \frac{\text{Consistency Index}}{\text{Random Index}}$ should less than 0.1 [3]. Here Consistency Index (C.I) $= \frac{\lambda_{\max}-x}{x-1}$ and Random Index (R.I) decide the consistency of comparison matrix obtained by decision-maker. Various authors have proposed different R.I values based on the size of the comparison matrix. This paper explains a statistical technique to get R.I values for various sizes of pairwise comparison matrices by using R.I $(x) = \frac{\bar{\lambda}_{\max}-x}{x-1}$. The effectiveness of the new technique was also verified with illustrations, and the results were analyzed [1].

Keywords Analytic hierarchy process · Consistency · Random Index · Pairwise comparison

1 Introduction

Analytic hierarchy process is the most commonly applied decision-making technique. It has been applied in several decision-making areas and real-time applications during the last four decades. AHP uses a reciprocal pairwise comparison matrix; hence, the data will be given in linguistic manner [4]. Pairwise comparison matrix obtained by the decision-maker should satisfy the consistency condition if not he/she has to redefine the judgment matrix. Consistency Index (C.I) $= \frac{\lambda_{\max}-x}{x-1}$ and Random

V. Shyamprasad (✉)
Guru Nanak Institutions Technical Campus, Hyderabad, Telangana 501506, India
e-mail: shyamnow4u@gmail.com

P. Kousalya
Visiting Faculty, IIITDM, Jagannathagattu, Dinnedevarapadu, Kurnool, Andhra Pradesh 518007, India

D. Dutta and B. Mahanty (eds.), *Numerical Optimization in Engineering and Sciences*, Advances in Intelligent Systems and Computing 979,
https://doi.org/10.1007/978-981-15-3215-3_22

Index (R.I) decide the nature of the comparison matrix. If the pairwise comparison matrix is perfectly consistent, then $\lambda_{\max} = x$ and C.I = 0. Random Index is the average of C.I values of various sizes of comparison matrices. Evaluation of R.I values for higher-order comparison matrices is complicated. This paper explains a new statistical technique $\mathrm{R.I}(x) = \frac{\bar{\lambda}_{\max}-x}{x-1}$ to get R.I(x) values for higher-order comparison matrices [1].

2 Study of Random Index Values

The values of Random Index for various sizes of comparison matrices can be obtained by the following steps [1]:

(i) Generating comparison matrix based on people's opinions and Saaty's ratio scale.
(ii) Evaluating the Consistency Index for each size of the comparison matrix.
(iii) Obtaining the average value of Consistency Index.

Random Index values for various sizes of the comparison matrices depend on the Consistency Index values.

2.1 First Attempt to Estimate Random Index Values Using Cubic Function

In the first attempt, R.I(x) values for various sizes of the comparison matrices were obtained by using a cubic function. $\boldsymbol{I}(\boldsymbol{x}) = 0.0015\boldsymbol{x}^3 - 0.0512\boldsymbol{x}^2 + 0.5915\boldsymbol{x} - 0.7912$. This was obtained by the least squares method [1].

2.2 Least Squares Cubic Function for 'x' and R.I(x)

$$\text{Assume } \mathrm{R.I}(x) = a_0 + a_1 x + a_2 x + a_3 x^3 \tag{1}$$

be the cubic function to be constructed by using least squares method.

The four normal equations are:

$$m a_0 + a_1 \sum x + a_2 \sum x^2 + a_3 \sum x^3 = \sum \mathrm{R.I}(x) \tag{2}$$

$$a_0 \sum x + a_1 \sum x^2 + a_2 \sum x^3 + a_3 \sum x^4 = \sum x.\mathrm{R.I}(x) \tag{3}$$

$$a_0 \sum x^2 + a_1 \sum x^3 + a_2 \sum x^4 + a_3 \sum x^5 = \sum x^2.\mathrm{R.I}(x) \tag{4}$$

$$a_0 \sum x^3 + a_1 \sum x^4 + a_2 \sum x^5 + a_3 \sum x^6 = \sum x^3.\text{R.I}(x) \quad (5)$$

$$\Rightarrow 8a_0 + 52a_1 + 380a_2 + 3016a_3 = 9.49 \quad (6)$$

$$52a_0 + 380a_1 + 3016a_2 + 25{,}316a_3 = 67.5 \quad (7)$$

$$380a_0 + 3016a_1 + 25{,}316a_2 + 220{,}792a_3 = 512.31 \quad (8)$$

$$3016a_0 + 25{,}316a_1 + 220{,}792a_2 + 1{,}978{,}340a_3 = 4248.67 \quad (9)$$

from Eqs. (6), (7), (8), and (9), we get

$\text{R.I}(n) = 0.0015x^3 - 0.0512x^2 + 0.5915x - 0.7912$, which is the required cubic function obtained by the method of least squares.

Table 1 explains the variations between existing and new estimated R.I(x) values obtained by cubic function.

Figure 1 shows the variations between existing (Saaty's) R.I values and new estimated R.I values obtained by cubic function.

Table 1 Estimated and existing Saaty's R.I(x) values

x	iii	iv	v	vi	vii	viii	ix	x
Existing R.I values	0.52	0.90	1.12	1.24	1.32	1.41	1.45	1.49
Estimated R.I values	0.56	0.85	1.07	1.23	1.35	1.43	1.48	1.50

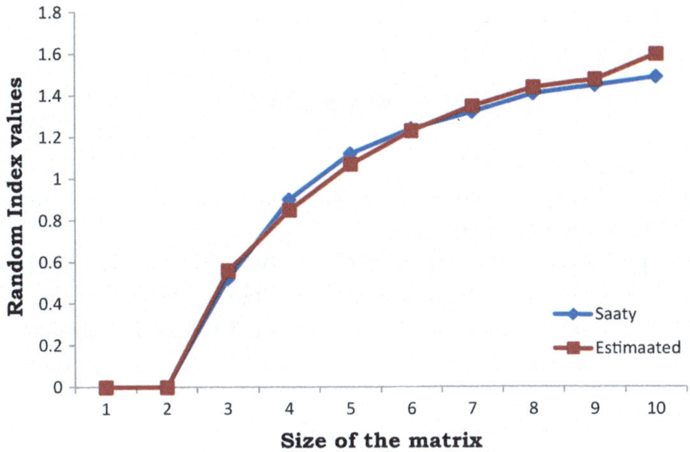

Fig. 1 Comparison between existing (Saaty's) and new estimated R.I values

2.3 Second Attempt to Evaluate Random Index Values by a New Measure

In the second attempt, R.I(x) values for different sizes of higher-order comparison matrices were estimated with the help of a statistical measure R.I$(x) = \frac{\bar{\lambda}_{max} - x}{x-1}$ instead of taking the average of C.I values. $\bar{\lambda}_{max}$ values for various PCMs of order 3–10 were obtained by randomly generated comparison matrices, and for the comparison matrices of sizes, more than 10 were obtained by the following function $\bar{\lambda}_{\mathbf{max}}(\boldsymbol{x}) = \mathbf{2.7698}(\boldsymbol{x}) - \mathbf{4.3512}$; this was obtained by the method of least squares for order of the matrix 'x' and '$\bar{\lambda}_{max}$' values [1].

2.4 Least Squares Straight Line for 'x' and $\bar{\lambda}_{max}$

$$\text{Assume } \bar{\lambda}_{max}(x) = a_0 + a_1(x) \tag{10}$$

be the equation of straight line to be constructed by the method of least squares.

The two normal equations are:

$$ma_0 + a_1 \sum x = \sum \bar{\lambda}_{max} \tag{11}$$

$$a_0 \sum x + a_1 \sum x^2 = \bar{\lambda}_{max}\, x \tag{12}$$

$$\Rightarrow 8a_0 + 52a_1 = 109.22 \tag{13}$$

$$52a_0 + 380a_1 = 826.2616 \tag{14}$$

from Eqs. (13) and (14), we get

$\bar{\lambda}_{max}(x) = 2.7698(x) - 4.3512$, which is the required straight line equation obtained by the method of least squares.

Table 2 explains the variations between existing (Saaty's) R.I values and new estimated R.I values obtained by new measure R.I$(x) = \frac{\bar{\lambda}_{max} - x}{x-1}$.

Figure 2 shows the variations between Saaty's R.I values and new estimated R.I values obtained by new measure R.I$(x) = \frac{\bar{\lambda}_{max} - x}{x-1}$; from Fig. 2, it can be observed that

Table 2 Estimated and existing (Saaty's) R.I(x) values

x	iii	iv	v	vi	Vii	viii	Ix	x
Existing R.I values	0.52	0.90	1.12	1.24	1.32	1.41	1.45	1.49
Estimated R.I values	0.52	0.89	1.11	1.25	1.34	1.40	1.45	1.48

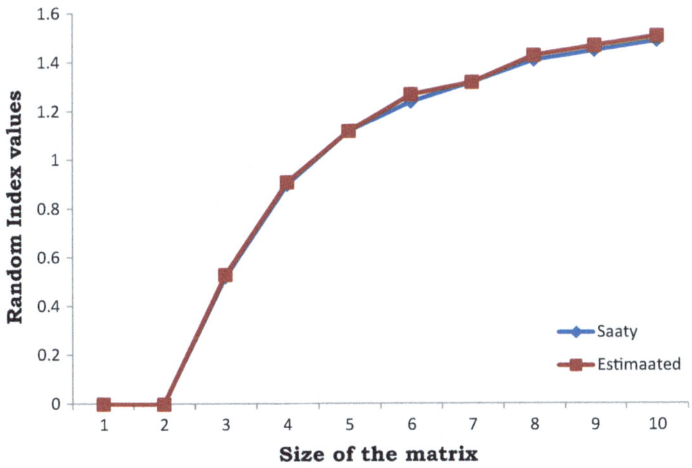

Fig. 2 Comparison between existing (Saaty's) and new estimated R.I values

both Saaty's and estimated R.I values are almost the same and it is an easy method to obtain R.I values even for higher-order comparison matrices.

3 Illustrations

The correctness of the evaluated R.I(x) values has been checked with a randomly generated comparison matrices of order '4' and '5'; the results were analyzed.

3.1 *Comparison Matrix of Dimension '4'*

$$A = \begin{pmatrix} 1 & 5 & 3 & 7 \\ 1/5 & 1 & 1/3 & 5 \\ 1/3 & 3 & 1 & 6 \\ 1/7 & 1/5 & 1/6 & 1 \end{pmatrix}$$

Consistency ratio using existing (Saaty's) R.I(x) values: 0.079.
Consistency ratio using estimated R.I(x) values: 0.078.
Difference between existing and estimated R.I(x) values: 0.001.

3.2 Comparison Matrix of Dimension '5'

$$= \begin{pmatrix} 1 & 1 & 3 & 1 & 1/3 \\ 1 & 1 & 3 & 3 & 5 \\ 1/3 & 1/3 & 1 & 1/3 & 1 \\ 1 & 1/3 & 3 & 1 & 3 \\ 3 & 1/5 & 1 & 1/3 & 1 \end{pmatrix}$$

Consistency ratio using existing (Saaty's) R.I(x) values: 0.202.
Consistency ratio using estimated R.I(x) values: 0.202.
Difference between existing and estimated R.I(x) values: 0.

4 Limitations

(i) Various authors proposed different R.I(x) values to check the consistency of pairwise comparison matrices, decision-maker may opt any suitable R.I(x) values which are similar to Saaty's values [1].
(ii) The differences in R.I(x) values of various authors is mainly due to the number of comparison matrices used in their experiment is very small [1].
(iii) The proposed R.I(x) values are only to check the consistency of comparison matrices in analytic hierarchy process introduced by Thomas L. Saaty [3].

5 Conclusion

Consistency plays a key role in analytic hierarchy process [2]. Saaty introduced consistency ratio test, i.e., $\text{C.R} = \frac{\text{C.I}}{\text{R.I}}$ should be less than 0.1 [3]. Calculation of R.I values for higher-order comparison matrices is not an easy task. In the first attempt, R.I(x) values for different sizes of comparison matrices were obtained by using a cubic function $\text{R.I}(x) = 0.0015x^3 - 0.0512x^2 + 0.5915x - 0.7912$. It was observed that R.I(x) values obtained by cubic function were slightly deviated with existing values, and there was an error between existing and estimated R.I(x) values [1].

In the second attempt, R.I(x) values for different sizes of comparison matrices were obtained by using a new statistical measure $\text{R.I}(x) = \frac{\bar{\lambda}_{\max} - x}{x-1}$. R.I($x$) values obtained by new measure were almost the same with existing values. The effectiveness of the estimated R.I(x) values has been verified with comparison matrices of sizes '4' and '5'. It has been observed that the consistency ratio using existing and estimated R.I(x) values was the same. It is an easy approach to obtain Random Index values [1].

References

1. Alonso, A.: Consistency in the AHP: a new approach. Int. J. Uncertain., Knowl. Based Syst. **14**, 445–459 (2016)
2. Ergu, D., Kou, G., Peng, Y., Shi, Y.: An easy method to develop the consistency ratio of pairwise comparison matrices in AHP. Eur. J. Oper. Res. **213**, 246–259 (2011)
3. Saaty, T.L.: The analytic hierarchy process-what it is and how it is used. J. Math. Model. **9**, 161–176 (1987)
4. Saaty, T.L.: How to make a decision: the analytic hierarchy process. Eur. J. Oper. Res. **48**, 9–26 (1990)

Sensitivity Analysis Through RSAWM—A Case Study

S. Supraja and P. Kousalya

Abstract Decision making plays a very vital role in today's lives. Though the solutions are acquired by applying available multi-criteria decision making (MCDM) methods, solutions thus obtained may be consistent or inconsistent. T. L. Saaty has already developed a method to check the consistency. However, Triantaphyllou suggests that performing sensitivity analysis improves the consistency of the solution. Thus, based on their suggestions, in this paper a MCDM method named revised simple additive weighting method (RSAWM) is applied initially to choose a best computer system out of three systems (*A*, *B* and *C*) corresponding to four criteria, whose solution is already acquired by analytical hierarchy process (AHP). Later, sensitivity analysis is utilized to look at the developments in the decisions by modifying the weight of criteria in three different cases. It is clear that the results remained the same in all the cases. Hence, selecting the system *A* is the suitable decision.

Keywords MCDM methods · RSAWM · AHP · Sensitivity analysis · Consistency

1 Introduction

For the long past, most of the professionals of various domains happen to face the issue of multi-criteria to single out a best decision amongst the alternatives. In 1980, T. L. Saaty has come up with the hierarchical approach to arrive at answers. Thenceforth, AHP has been utilized for problems with more than one criterion in varied spheres. AHP is a much simpler, adaptable mechanism and is one among the MCDM Methods. It, being a hierarchical model, generates a hierarchical structure between objective, alternatives and criteria. The basic statistical information is made up by relying on Saaty's [1] comparison table. The outcomes are taken as consistent banking on the ratio elucidated by him [2, 3]. Harker [4] has described the significance

S. Supraja (✉)
VITS, Deshmukhi, Hyderabad, India
e-mail: sprj10@gmail.com

P. Kousalya
Visiting Faculty, IIITDM, Jagannathagattu, Dinnedevarapadu, Kurnool, Andhra Pradesh 518007, India

D. Dutta and B. Mahanty (eds.), *Numerical Optimization in Engineering and Sciences*, Advances in Intelligent Systems and Computing 979,
https://doi.org/10.1007/978-981-15-3215-3_23

of AHP in the scientific area. According to Triantaphyllou [5, 6], sensitivity analysis is an important tool to improve consistency. In the present day, the application of sensitivity analysis has come to the limelight. The key role of sensitivity analysis in deriving the best decision for complex situations was discussed by Triantaphyllou [6]. He also discussed several methods under sensitivity analysis with suitable examples [6]. Sensitivity analysis is the process of observing the changes in the outcomes or results by giving some changes in the inputs. In this regard, Memariani [7] presented his work with a case study. The data and the ideas used in this paper are taken from [6–8].

2 Methodology

The study in this paper is subdivided into two parts, firstly applying RSAW method to acquire solution (Sect. 2.1) then, secondly performing sensitivity analysis by changing the weight of criteria (Sect. 2.2) to note the changes in the solution. However, the sample data (i.e., pairwise comparison matrices) and weights of the criteria and alternatives obtained by AHP [8] are considered in this paper.

2.1 Algorithm of RSAW Method

Step-1: Normalize the matrix by

$$\eta_{yz} = \frac{q_y^{\min}}{e_{yz}} \times H_{\Omega i} \quad y = 1, 2, \ldots u \text{ and } z = 1, 2, \ldots v \tag{1}$$

Since $\sum \Omega_k = 1$.

Step-2: Assess the priorities of each alternative by

$$J_y = \sum_{z=1}^{v} \eta_{yz} \Omega_z \tag{2}$$

2.2 Sensitivity Analysis

Step-1: Consider Ω'_k is the altered weight of kth criterion. Then

$$\Omega'_k = \frac{(n+1)\Omega_k}{n} \tag{3}$$

Step-2: Accordingly, the remaining attributes change their weights by

$$\Omega'_z = \left[1 - \frac{\Omega_k}{n(1-\Omega_k)}\right] \cdot \Omega_z,\ z = 1, 2, \ldots v, z \neq k \tag{4}$$

where $n \neq 1, n = 2, 3, \ldots$ such that $\sum \Omega'_k = 1$.
Step-3: Repeat in step-2 of Sect. 2.1 with new weights. The alternative with the highest weight is ranked first.

3 Illustration

For the sake of illustrative application, let us take a case in which one wants to improve a computer system in a computer manufacturing facility. Several configurations of a system may be available to make a choice. Our decision to make a choice may take into account some other aspects like price, performance, software, expandability, maintenance, etc. Then, these aspects also will be the criteria to make a decision. In this particular case, we need to identify the best alternative. However, in a different case one may want to know the relative importance of all the available choices that is when financing a set of contending projects is concerned knowing their relative importance becomes inevitable. In the case of selecting the best computer, let there be the configurations *A*, *B*, *C* that are available. In addition to that hardware, expandability is also one of the decision criteria, then it is taken as, *A* is better than *B*, which in turn is better than *C* (Fig. 1).

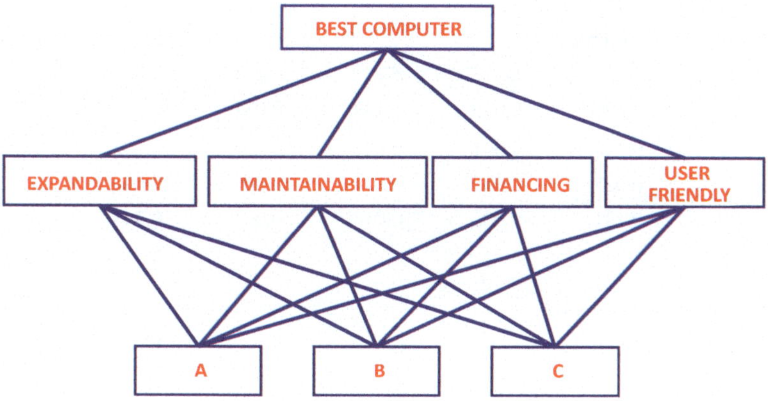

Fig. 1 Hierarchical structure

Table 1 Rank of the alternatives

Alternatives	Criteria				PV	Rank
	G_1	G_2	G_3	G_4		
	0.553	0.131	0.271	0.045		
A	0.116	0.089	0.058	0.011	0.273	1
B	0.022	0.017	0.030	0.003	0.072	3
C	0.105	0.002	0.018	0.002	0.128	2

4 Ranks of the Alternatives by RSAWM

The ranks of the alternatives given in Table 1 are obtained by applying the relation (2).

5 Changing the Weight of Criteria

Some researchers may choose the highest ranked criteria as critical, and others may pickup criteria with lowest rank as critical. In this section, the highest ranked criterion weight G_1, criteria taken at random G_2 and criterion that is least ranked G_4 are changed by using relation (3). This gives a relative variation in the remaining criteria by relation (4) of Sect. 2.2. Then, by repeating steps 1 and 2 of Sect. 2.1, the rankings of the alternatives will be obtained. The priority values of the alternatives acquired in the above discussed three cases are given in Tables 2, 3 and 4, respectively.

Table 2 Rank of the alternatives

Alternatives	Criteria				PV	Rank
	G_1	G_2	G_3	G_4		
	0.691	0.090	0.187	0.031		
A	0.145	0.061	0.040	0.007	0.253	1
B	0.027	0.012	0.021	0.002	0.062	3
C	0.131	0.002	0.013	0.002	0.147	2

Table 3 Rank of the alternatives

Alternatives	Criteria				PV	Rank
	G_1	G_2	G_3	G_4		
	0.532	0.163	0.260	0.043		
A	0.112	0.111	0.055	0.010	0.288	1
B	0.021	0.021	0.029	0.003	0.074	3
C	0.101	0.003	0.018	0.002	0.124	2

Table 4 Rank of the alternatives

Alternatives	Criteria				PV	Rank
	G_1	G_2	G_3	G_4		
	0.546	0.129	0.267	0.056		
A	0.115	0.088	0.057	0.013	0.272	1
B	0.022	0.017	0.029	0.004	0.072	3
C	0.104	0.002	0.018	0.003	0.127	2

5.1 Highest Ranked Criteria

See Table 2.

5.2 Criteria at Random

See Table 3.

5.3 Least Ranked Criteria

See Table 4.

6 Conclusion

In this paper, the alternatives order achieved from RSAW method is akin to the alternatives order arrived by varying the weights of criterion with the highest rank, the criterion with least rank and criterion chosen at random as the opinion matrices considered are consistent (i.e., their consistency ratio is less than 0.1 [1, 2]). Here, it is observed that the result obtained in all the cases does not show any change in its decisions. Hence, the solution obtained is said to be consistent by Triantaphyllou [6]. Finally, choosing alternative *A* is the best decision for the situation considered.

References

1. Saaty, T.L.: Inconsistency and rank preservation. J. Math. Psychol. **28**(2) (1984)
2. Saaty, T.L.: The Analytical Hierarchy Process. Tata McGraw Hill, New York (1980)

3. Saaty, T.L.: Fundamentals of Decision Making and Priority Theory with the AHP. RWS Publications, Pittsburgh (1994)
4. Harker, P.T.: The Art of Science and Decision-Making: The Analytic Hierarchy Process. Springer (1989)
5. Triantaphyllou, E., Mann, S.H.: A computational evaluation of the AHP and the revised AHP when the eigen value method is used under a continuity assumption. Comput. Ind. Eng. **26**(3), 609–618 (1994)
6. Triantaphyllou, E., Sanchez, A.: A sensitivity analysis approach for some deterministic multi-criteria decision making methods. Decis. Sci. **28**(1), 151–194 (1997)
7. Memariani, A., et al.: Sensitivity analysis of a simple additive weight method: the results of change in the weight of one attribute on the final ranking of the alternatives. J. Ind. Eng. **4**, 13–18 (2009)
8. Kousalya, P., Supraja, S.: A membership function solution approach to AHP. Am. J. Math. Sci. **2**(1) (2013). ISSN No: 2250-3102

RSAWM for the Selection of All Round Excellence Award—An Illustration

S. Supraja and P. Kousalya

Abstract Triantaphyllou strongly suggests that sensitivity analysis is the best technique which develops the consistency of the solution. This paper is aimed to check what happens when sensitivity analysis is applied to an inconsistent problem. In this paper, a MCDM method, Revised Simple Additive Weighting Method (RSAWM), is applied initially to select a branch of students (out of five branches) for All Round Excellence Award depending on seven criteria (few criteria have subcriteria), whose solution is already acquired by AHP. Later, sensitivity analysis is made use of to have hands on the variations in the decisions by altering the weight of criterion whose rank is highest, least and taken at random. The outcome obtained in all the cases is similar. Hence, the branch of ECE students received the award.

Keywords MCDM methods · RSAWM · AHP · Sensitivity analysis · Inconsistent

1 Introduction

AHP is a well-known and most widely used MCDM method to obtain the decisions for complex problems involving more number of criteria. The evolution, applications, methodology and history are explained in detail by Saaty [1–3]. The importance of AHP in the field of science is given by Harker [4]. Sensitivity analysis has been a facet of discussion in MCDM area. Sensitivity analysis is the process of observing the changes in the outcomes by giving the changes in its inputs. A large number of researchers have investigated greatly in this regard for about a decade. Memariani [5] presented his work regarding sensitivity analysis through SAW method. According to Triantaphyllou [6, 7], sensitivity analysis is an important tool to improve the consistency. He also discussed several methods under sensitivity analysis with suitable

S. Supraja (✉)
VITS, Deshmukhi, Hyderabad, India
e-mail: sprj10@gmail.com

P. Kousalya
Visiting Faculty, IIITDM, Jagannathagattu,Dinnedevarapadu, Kurnool, Andhra Pradesh 518007, India

D. Dutta and B. Mahanty (eds.), *Numerical Optimization in Engineering and Sciences*, Advances in Intelligent Systems and Computing 979,
https://doi.org/10.1007/978-981-15-3215-3_24

examples to arrive best decision [7]. The data and the ideas used in this paper are taken from [5, 7, 8].

2 Methodology

After getting the solution by AHP [8] (the pairwise comparison matrices and the rankings of the alternatives refer Appendix 3), this study is categorized into two parts: 1. applying RSAW method (refer Appendix 1) to attain the result and 2. performing sensitivity analysis to observe the variations in the result achieved by changing the weight of criteria (refer Appendix 2).

2.1 *Algorithm of RSAW Method*

Step 1: Normalize the matrix by

$$\eta_{yz} = \frac{q_y^{\min}}{e_{yz}} \times H_{\Omega i} \quad y = 1, 2, \ldots u \text{ and } z = 1, 2, \ldots v \tag{1}$$

Since $\sum \Omega_k = 1$.

Step 2: Assess the priorities of each alternative by

$$J_y = \sum_{z=1}^{v} \eta_{yz} \Omega_z \tag{2}$$

2.2 *Sensitivity Analysis*

Step 1: Consider Ω_k' is the altered weight of kth criterion. Then

$$\Omega_k' = \frac{(n+1)\Omega_k}{n} \tag{3}$$

Step 2: Accordingly the remaining attributes change their weights by

$$\Omega_z' = \left[1 - \frac{\Omega_k}{n(1-\Omega_k)}\right] \cdot \Omega_z, z = 1, 2, \ldots v, z \neq k \tag{4}$$

where $n \neq 1, n = 2, 3, \ldots$ such that $\sum \Omega_k' = 1$.
Step 3: Repeat step 2 of Sect. 2.1 with new weights.

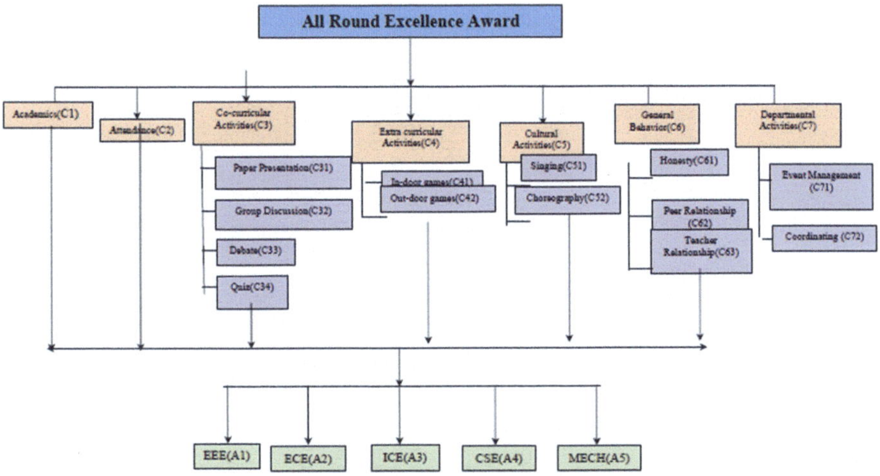

Fig. 1 Hierarchical decomposition of criteria, subcriteria and alternatives

3 Illustration

Consider a case of selecting a branch of student for All Round Excellence Award in an engineering college. The different branches available are taken as alternatives. There are many qualities as criteria to choose from, namely academics, attendance, curricular, cultural activities, etc. These criteria play a part in arriving at a decision. Also, suppose that under seven main criteria, there are subcriteria—for example, if we consider cultural activities criteria in which we have two subcriteria like singing and choreography. Suppose that ECE branch student more active in academics and in curricular activities than the student of CSE branch. Also EIE branch student is very poor in attendance and sports than the student of mechanical branch (Fig. 1).

The ranking of the alternatives obtained by using relations (1) and (2) of Sect. 2.1 is given in Table 1 of Sect. 4.

4 Ranks of the Alternatives by RSAWM

See Table 1.

5 Changing the Weight of Criteria

The ranking of the alternatives obtained by changing the weight of criteria in three cases is given in Tables 2, 3 and 4, respectively.

Table 1 Ranks of the alternatives

Alt.	Criteria															Sum	Rank
	G_1	G_2	G_3	G_4	G_5	G_6	G_7	G_8	G_9	G_{10}	G_{11}	G_{12}	G_{13}	G_{14}	G_{15}		
	0.32	0.04	0.20	0.20	0.20	0.20	0.50	0.50	0.50	0.50	0.33	0.33	0.33	0.50	0.50		
H_1	0.03	0.01	0.02	0.02	0.02	0.03	0.08	0.06	0.04	0.06	0.03	0.04	0.03	0.05	0.04	0.55	2
H_2	0.01	0.00	0.01	0.01	0.01	0.01	0.11	0.12	0.08	0.05	0.02	0.03	0.02	0.03	0.05	0.58	1
H_3	0.01	0.00	0.00	0.00	0.00	0.00	0.00	0.00	0.01	0.01	0.00	0.00	0.00	0.01	0.01	0.06	5
H_4	0.01	0.01	0.00	0.00	0.00	0.00	0.01	0.01	0.01	0.01	0.01	0.01	0.01	0.01	0.01	0.11	3
H_5	0.01	0.01	0.00	0.00	0.00	0.00	0.00	0.00	0.00	0.00	0.01	0.00	0.01	0.01	0.01	0.08	4

Table 2 Rank of the alternatives

Alt.	Criteria															Sum	Rank
	G_1	G_2	G_3	G_4	G_5	G_6	G_7	G_8	G_9	G_{10}	G_{11}	G_{12}	G_{13}	G_{14}	G_{15}		
	0.47	0.03	0.15	0.15	0.15	0.15	0.38	0.38	0.38	0.38	0.25	0.25	0.25	0.38	0.38		
H_1	0.04	0.01	0.02	0.01	0.02	0.02	0.06	0.05	0.03	0.05	0.02	0.03	0.02	0.04	0.03	0.44	2
H_2	0.02	0.00	0.01	0.01	0.01	0.01	0.08	0.09	0.07	0.04	0.02	0.02	0.02	0.03	0.04	0.45	1
H_3	0.01	0.00	0.00	0.00	0.00	0.00	0.00	0.00	0.01	0.01	0.00	0.00	0.00	0.00	0.00	0.05	5
H_4	0.01	0.01	0.00	0.00	0.00	0.00	0.01	0.01	0.01	0.01	0.00	0.01	0.00	0.01	0.01	0.09	3
H_5	0.02	0.01	0.00	0.00	0.00	0.00	0.00	0.00	0.00	0.00	0.01	0.00	0.01	0.00	0.00	0.07	4

Table 3 Rank of the alternative

Alt.	Criteria															Sum	Rank
	G_1	G_2	G_3	G_4	G_5	G_6	G_7	G_8	G_9	G_{10}	G_{11}	G_{12}	G_{13}	G_{14}	G_{15}		
	0.16	0.02	0.10	0.10	0.10	0.10	0.75	0.25	0.25	0.25	0.17	0.17	0.17	0.25	0.25		
H_1	0.01	0.00	0.01	0.01	0.01	0.01	0.12	0.03	0.02	0.03	0.01	0.02	0.01	0.02	0.02	0.35	2
H_2	0.01	0.00	0.00	0.01	0.00	0.00	0.16	0.06	0.04	0.03	0.01	0.02	0.01	0.02	0.02	0.39	1
H_3	0.00	0.00	0.00	0.00	0.00	0.00	0.01	0.00	0.00	0.00	0.00	0.00	0.00	0.00	0.00	0.04	5
H_4	0.00	0.00	0.00	0.00	0.00	0.00	0.01	0.01	0.00	0.00	0.00	0.00	0.00	0.01	0.01	0.07	3
H_5	0.01	0.00	0.00	0.00	0.00	0.00	0.00	0.00	0.00	0.00	0.00	0.00	0.00	0.00	0.00	0.04	4

Table 4 Rank of the alternatives

Alt.	Criteria															Sum	Rank
	G_1	G_2	G_3	G_4	G_5	G_6	G_7	G_8	G_9	G_{10}	G_{11}	G_{12}	G_{13}	G_{14}	G_{15}		
	0.31	0.06	0.20	0.20	0.20	0.20	0.49	0.49	0.49	0.49	0.32	0.32	0.32	0.49	0.49		
H_1	0.02	0.01	0.02	0.02	0.02	0.02	0.08	0.06	0.04	0.06	0.03	0.04	0.03	0.05	0.04	0.54	2
H_2	0.01	0.00	0.01	0.01	0.01	0.01	0.10	0.12	0.08	0.05	0.02	0.03	0.02	0.03	0.05	0.57	1
H_3	0.01	0.01	0.00	0.00	0.00	0.00	0.00	0.00	0.01	0.01	0.00	0.00	0.00	0.01	0.01	0.06	5
H_4	0.01	0.01	0.00	0.00	0.00	0.00	0.01	0.01	0.01	0.01	0.01	0.01	0.01	0.01	0.01	0.12	3
H_5	0.01	0.01	0.00	0.00	0.00	0.00	0.00	0.00	0.00	0.00	0.01	0.00	0.01	0.01	0.01	0.08	4

5.1 Highest Ranked Criteria

See Table 2.

5.2 Criteria at Random

See Table 3.

5.3 Least Ranked Criteria

See Table 4.

6 Conclusion

The order of ranks secured by RSAW method is indistinguishable from the order of ranks attained by differing the weights of criterion with highest rank, the criterion with least rank and criterion chosen at random, though the opinion matrices taken are inconsistent (i.e., their consistency ratio is more than 0.1 by Saaty). Here, it is observed that the result obtained in all the cases does not show any changes in its decisions. This is true for the condition that the sum of the weights of all criteria is unity (the results may vary by relaxing the condition). Hence, the solution obtained is said to be consistent by Triantaphyllou [7]. Finally, ECE branch students are eligible to receive the award for the taken situation.

Appendix 1

Let $E = \left[e_{yz}\right]_{u \times v}$ be the decision matrix formed by the vectors obtained from the pairwise comparison matrices of alternative with respect to each criterion, as its column elements as shown in Table 5.
where H_y—alternatives, G_z—criteria Ω_z—weight of criteria.

$H_{\Omega y}$—weight of the alternative obtained by AHP for $y = 1, 2, 3$ and $z = 1, 2, 3$.

Now take minimum in each row

Table 5 Decision matrix

	Ω_1	Ω_2	Ω_3	
	G_1	G_2	G_3	PV
H_1	e_{11}	e_{12}	e_{13}	$H_{\Omega 1}$
H_2	e_{21}	e_{22}	e_{23}	$H_{\Omega 2}$
H_3	e_{31}	e_{32}	e_{33}	$H_{\Omega 3}$

	Ω_1	Ω_2	Ω_3	
	G_1	G_2	G_3	Row min
H_1	e_{11}	e_{12}	e_{13}	$q_1^{\min}$
H_2	e_{21}	e_{22}	e_{23}	$q_2^{\min}$
H_3	e_{31}	e_{32}	e_{33}	$q_3^{\min}$

Now divide $q_y^{\min}$ by each corresponding row elements respectively in the decision matrix. Then

	Ω_1	Ω_2	Ω_3	
	G_1	G_2	G_3	PV
H_1	$\frac{q_1^{\min}}{e_{11}}$	$\frac{q_1^{\min}}{e_{11}}$	$\frac{q_1^{\min}}{e_{11}}$	$H_{\Omega 1}$
H_2	$\frac{q_2^{\min}}{e_{21}}$	$\frac{q_2^{\min}}{e_{22}}$	$\frac{q_2^{\min}}{e_{23}}$	$H_{\Omega 2}$
H_3	$\frac{q_3^{\min}}{e_{31}}$	$\frac{q_3^{\min}}{e_{32}}$	$\frac{q_3^{\min}}{e_{33}}$	$H_{\Omega 3}$

Let $\frac{q_y^{\min}}{e_{yz}} = Q_{yz}$

	Ω_1	Ω_2	Ω_3	
	G_1	G_2	G_3	PV
H_1	Q_{11}	Q_{12}	Q_{13}	$H_{\Omega 1}$
H_2	Q_{21}	Q_{22}	Q_{23}	$H_{\Omega 2}$
H_3	Q_{31}	Q_{32}	Q_{33}	$H_{\Omega 3}$

Table 6 Normalized decision matrix

	Ω_1	Ω_2	Ω_3
	G_1	G_2	G_3
H_1	η_{11}	η_{12}	η_{13}
H_2	η_{21}	η_{22}	η_{23}
H_3	η_{31}	η_{32}	η_{33}

Now multiply each row element with the corresponding alternative weight $H_{\Omega y}$

	Ω_1	Ω_2	Ω_3
	G_1	G_2	G_3
H_1	$Q_{11} \times H_{\Omega 1}$	$Q_{12} \times H_{\Omega 1}$	$Q_{13} \times H_{\Omega 1}$
H_2	$Q_{21} \times H_{\Omega 2}$	$Q_{22} \times H_{\Omega 2}$	$Q_{23} \times H_{\Omega 2}$
H_3	$Q_{31} \times H_{\Omega 3}$	$Q_{32} \times H_{\Omega 3}$	$Q_{33} \times H_{\Omega 3}$

Assume $Q_{yz} \times H_{\Omega y} = \eta_{yz}$.

The elements in Table 6 represent the normalized values of the values present in Table 5. To get preference values, each column elements are multiplied with the corresponding weight of the criteria and take the sum of each row. Then

$$PV = \eta_{11} \cdot \Omega_1 + \eta_{12} \cdot \Omega_2 + \eta_{13} \cdot \Omega_3 + \eta_{21} \cdot \Omega_1 + \eta_{22} \cdot \Omega_2 + \eta_{23} \cdot \Omega_3 + \eta_{31} \cdot \Omega_1 + \eta_{32} \cdot \Omega_2 + \eta_{33} \cdot \Omega_3$$

Generalizing the above relation we get

$$J_y = \sum_{z=1}^{v} \eta_{yz}\Omega_z \tag{5}$$

Then the alternative with highest preference value is the best alternative.

Appendix 2

The result of change in the weight of criterion on the other criteria

Memariani [5] derived the following theorem to change the weight of the criteria

Theorem *Consider the change of weight of* k*th attribute is* σ_k*, then the weight of other criteria alters as*

$$\frac{1 - \Omega_k'}{1 - \Omega_k} \cdot \Omega_z, z = 1, 2, \ldots v, z \neq k \tag{6}$$

Taking the above theorem into consideration, the following theory is developed and applied to find the changes in the weights of criteria in this study.

Theorem to change the weight of criteria 1/*n* times of its original weight

Suppose the weight of the kth attribute is added by $\frac{1}{n}$th part of the weight of Ω_k.

Proof If

$$\Omega_k' = \Omega_k + \sigma_k = \Omega_k + \frac{\Omega_k}{n}$$

By adding $\frac{1}{n}$th part of the weight of Ω_k as σ_k. Then

$$\Omega_k' = \Omega_k + \frac{\Omega_k}{n} = \frac{(n+1)\Omega_k}{n} \tag{7}$$

Then by relation (6) the weights of the other attributes vary as

$$\begin{aligned}\Omega_z' &= \frac{1-\Omega_k'}{1-\Omega_k} \cdot \Omega_z = \frac{1-\frac{(n+1)\Omega_k}{n}}{1-\Omega_k} \cdot \Omega_z \\ &= \frac{n - n\Omega_k - \Omega_k}{n(1-\Omega_k)} \cdot \Omega_z \\ &= \left[1 - \frac{\Omega_k}{n(1-\Omega_k)}\right]\Omega_z, z = 1, 2, \ldots v, z \neq k \end{aligned} \tag{8}$$

is the equation used to change the weight of other attributes.
where $n \neq 1, n = 2, 3 \ldots$.

Appendix 3

Tables 7, 8 and 9 are the opinion matrices/pairwise comparison matrices formed by comparing the alternatives with respect to alternatives corresponding to each criterion and subcriteria. Table 10 gives the comparison matrix of criteria with respect to criteria. The priority values obtained in this table are the weights of the criteria (Table 11).

Table 7 Pairwise comparison matrices for level 2

G_1							G_2						
	H_1	H_2	H_3	H_4	H_5	PV		H_1	H_2	H_3	H_4	H_5	PV
H_1	1	0.33	5	3	5	0.28	H_1	1	0.33	0.2	0.33	5	0.1
H_2	3	1	5	3	5	0.44	H_2	3	1	5	3	7	0.51
H_3	0.2	0.2	1	0.33	1	0.06	H_3	5	0.2	1	1	1	0.16
H_4	0.33	0.33	3	1	3	0.15	H_4	3	0.33	1	1	3	0.2
H_5	0.2	0.2	1	0.3	1	0.06	H_5	0.2	0.14	1	0.33	1	0.06
$\lambda_{max} = 5.1892$, C.I = 0.0473, C.R = 0.0426							$\lambda_{max} = 5.9761$, C.I = 0.244, C.R = 0.2198						
G_{31}							G_{33}						
	H_1	H_2	H_3	H_4	H_5	PV		H_1	H_2	H_3	H_4	H_5	PV
H_1	1	0.33	5	1	5	0.23	H_1	1	1	3	1	3	0.28
H_2	3	1	5	3	5	0.45	H_2	1	1	1	3	3	0.28
H_3	0.2	0.2	1	0.33	0.33	0.05	H_3	0.33	1	1	0.33	0.33	0.09
H_4	1	0.33	3	1	3	0.19	H_4	1	0.33	3	1	3	0.23
H_5	0.2	0.2	3	0.33	1	0.08	H_5	0.33	0.33	3	0.33	1	0.12
$\lambda_{max} = 5.2914$, C.I = 0.0729, C.R = 0.0656							$\lambda_{max} = 5.6874$, C.I = 0.1719, C.R = 0.1548						
G_{32}							G_{34}						
	H_1	H_2	H_3	H_4	H_5	PV		H_1	H_2	H_3	H_4	H_5	PV
H_1	1	0.33	3	1	3	0.19	H_1	1	1	3	1	0.33	0.18
H_2	3	1	5	3	5	0.46	H_2	1	1	3	3	5	0.38
H_3	0.33	0.2	1	0.33	0.33	0.06	H_3	0.33	0.33	1	0.33	1	0.09
H_4	1	0.33	3	1	3	0.19	H_4	1	0.33	3	1	3	0.22

(continued)

Table 7 (continued)

G_1							G_2						
	H_1	H_2	H_3	H_4	H_5	PV		H_1	H_2	H_3	H_4	H_5	PV
H_5	0.33	0.2	3	0.33	1	0.09	H_5	3	0.2	1	0.33	1	0.13
$\lambda_{max} = 5.1952$, C.I $= 0.0488$, C.R $= 0.044$							$\lambda_{max} = 5.9064$, C.I $= 0.2266$, C.R $= 0.2041$						
G_{41}							G_{42}						
	H_1	H_2	H_3	H_4	H_5	PV		H_1	H_2	H_3	H_4	H_5	PV
H_1	1	1	3	0.33	0.33	0.14	H_1	1	3	1	1	0.33	0.18
H_2	1	1	0.33	0.33	0.33	0.09	H_2	0.33	1	0.33	0.33	0.33	0.08
H_3	0.33	3	1	0.33	0.33	0.11	H_3	1	3	1	1	1	0.23
H_4	3	3	3	1	0.33	0.26	H_4	1	3	1	1	1	0.23
H_5	3	3	3	3	1	0.41	H_5	3	3	1	1	1	0.28
$\lambda_{max} = 5.547$, C.I $= 0.1368$, C.R $= 0.1232$							$\lambda_{max} = 5.1416$, C.I $= 0.0354$, C.R $= 0.0319$						

Table 8 Pairwise comparison matrices for level 2

G_{51}							G_{52}						
	H_1	H_2	H_3	H_4	H_5	Priority vector		H_1	H_2	H_3	H_4	H_5	Priority vector
H_1	1.00	3.00	3.00	1.00	1.00	0.27	H_1	1.00	0.33	3.00	1.00	1.00	0.18
H_2	0.33	1.00	3.00	0.33	0.33	0.11	H_2	3.00	1.00	3.00	0.33	0.33	0.18
H_3	0.33	0.33	1.00	0.33	0.33	0.07	H_3	0.33	0.33	1.00	0.33	0.33	0.07
H_4	1.00	3.00	3.00	1.00	1.00	0.27	H_4	1.00	3.00	3.00	1.00	1.00	0.28
H_5	1.00	3.00	3.00	1.00	1.00	0.27	H_5	1.00	3.00	3.00	1.00	1.00	0.28
$\lambda_{max} = 5.1372$, C.I $= 0.0343$, C.R $= 0.0309$							$\lambda_{max} = 5.5737$, C.I $= 0.1434$, C.R $= 0.1292$						
G_{61}							G_{62}						
	H_1	H_2	H_3	H_4	H_5	Priority vector		H_1	H_2	H_3	H_4	H_5	Priority vector
H_1	1.00	1.00	3.00	1.00	3.00	0.27	H_1	1.00	1.00	1.00	1.00	1.00	0.20
H_2	1.00	1.00	3.00	1.00	3.00	0.27	H_2	1.00	1.00	1.00	1.00	1.00	0.20
H_3	0.33	0.33	1.00	0.33	1.00	0.09	H_3	1.00	1.00	1.00	1.00	1.00	0.20
H_4	1.00	1.00	3.00	1.00	3.00	0.27	H_4	1.00	1.00	1.00	1.00	1.00	0.20
H_5	0.33	0.33	1.00	0.33	1.00	0.09	H_5	1.00	1.00	1.00	1.00	1.00	0.20
$\lambda_{max} = 4.988$, C.I $= -0.003$, C.R $= -0.0027$							$\lambda_{max} = 5.000$, C.I $= 0.0$, C.R $= 0.0$						
G_{63}							G_{71}						
	H_1	H_2	H_3	H_4	H_5	Priority vector		H_1	H_2	H_3	H_4	H_5	Priority vector
H_1	1.00	1.00	3.00	1.00	3.00	0.27	H_1	1.00	1.00	3.00	1.00	1.00	0.23
H_2	1.00	1.00	3.00	1.00	3.00	0.27	H_2	1.00	1.00	3.00	1.00	3.00	0.28
H_3	0.33	0.33	1.00	0.33	1.00	0.09	H_3	0.33	0.33	1.00	0.33	0.33	0.08
H_4	1.00	1.00	3.00	1.00	3.00	0.27	H_4	1.00	1.00	3.00	1.00	1.00	0.23
H_5	0.33	0.33	1.00	0.33	1.00	0.09	H_5	1.00	0.33	3.00	1.00	1.00	0.18
$\lambda_{max} = 4.988$, C.I $= -0.003$, C.R $= -0.0027$							$\lambda_{max} = 5.1416$, C.I $= 0.0354$, C.R $= 0.0319$						

Table 9 Pairwise comparison matrix with respect to subcriteria G_{72}

G_{72}						
	H_1	H_2	H_3	H_4	H_5	Priority vector
H_1	1	3	3	1	1	0.29227
H_2	0.33	1	3	1	1	0.18796
H_3	0.33	0.33	1	0.33	1	0.09684
H_4	1	1	3	1	1	0.23461
H_5	1	1	1	1	1	0.18833
$\lambda_{max} = 5.2931$, C.I $= 0.0733$, $C.R = 0.066$						

Table 10 Comparison matrix of criteria with respect to criteria

	G_1	G_2	G_3	G_4	G_5	G_6	G_7	
G_1	1	5	3	5	5	1	3	0.316851
G_2	0.2	1	0.33	0.33	0.33	0.2	1	0.045595
G_3	0.33	3	1	3	3	1	1	0.158735
G_4	0.2	3	0.33	1	1	0.33	0.33	0.067133
G_5	0.2	3	0.33	1	1	0.33	1	0.078653
G_6	1	5	1	3	3	1	3	0.234051
G_7	0.33	1	1	3	1	0.33	1	0.098985

$\lambda_{\max} = 7.4524$, C.I $= 0.0754$, $C.R = 0.0559$

Table 11 Global and priority vector matrix by AHP

Alt.	Criteria															PV	Rank
	G_1	G_2	G_3	G_4	G_5	G_6	G_7	G_8	G_9	G_{10}	G_{11}	G_{12}	G_{13}	G_{14}	G_{15}		
	0.31	0.04	0.20	0.20	0.20	0.20	0.50	0.50	0.50	0.50	0.33	0.33	0.33	0.50	0.50		
H_1	0.28	0.10	0.23	0.28	0.19	0.18	0.14	0.18	0.27	0.18	0.27	0.20	0.27	0.23	0.29	0.23	3
H_2	0.44	0.51	0.45	0.28	0.46	0.38	0.09	0.08	0.11	0.18	0.27	0.20	0.27	0.28	0.19	0.24	1
H_3	0.06	0.01	0.05	0.09	0.06	0.09	0.11	0.23	0.07	0.08	0.09	0.20	0.09	0.08	0.10	0.10	5
H_4	0.15	0.02	0.19	0.22	0.19	0.22	0.26	0.23	0.27	0.28	0.27	0.20	0.27	0.23	0.23	0.23	2
H_5	0.06	0.01	0.00	0.11	0.09	0.10	0.40	0.20	0.27	0.28	0.09	0.20	0.09	0.18	0.10	0.20	4

References

1. Saaty, T.L.: Inconsistency and rank preservation. J. math. Psychol. **28**(2) (1984)
2. Saaty, T.L.: The Analytical Hierarchy Process. Tata McGraw Hill, New York (1980)
3. Saaty, T.L.: Fundamentals of Decision Making and Priority Theory with the AHP. RWS Publications, Pittsburgh (1994)
4. Harker, P.T.: The Art of Science and Decision-Making: The Analytic Hierarchy Process. Springer (1989)
5. Memariani, A., et al.: Sensitivity analysis of a simple additive weight method: the results of change in the weight of one attribute on the final ranking of the alternatives. J. Ind. Eng. **4**, 13–18 (2009)
6. Triantaphyllou, E., Mann, S.H.: A computational evaluation of the AHP and the revised AHP when the eigen value method is used under a continuity assumption. Comput. Ind. Eng. **26**(3), 609–618 (1994)
7. Triantaphyllou, E., Sanchez, A.: A sensitivity analysis approach for some deterministic multi-criteria decision making methods. Decis. Sci. **28**(1), 151–194 (1997)
8. Kousalya, P., Mahender Reddy, G.: Selection of a student for all round excellence award using fuzzy AHP and TOPSIS methods. Int. J. Eng. Res. Appl. (IJERA) 1(4), 1993–2002 (2011)

Solving Bi-Level Linear Fractional Programming Problem with Interval Coefficients

Suvasis Nayak and Akshay Kumar Ojha

Abstract In this paper, a method is developed to determine a compromise solution of bi-level linear fractional programming problem (BL-LFPP) which comprises intervals as coefficients of decision variables and constants involved both in the objective and constraint functions. Non-dominated solution at each level is separately determined by transforming fractional objectives into interval-valued linear functions using variable transformation and Taylor series approximation. Goal programming method ignoring the over deviational variables for maximization goals is applied to determine the compromise solution and range of optimal objective values. A numerical example is illustrated to demonstrate the proposed method.

Keywords Bi-level optimization · Linear fractional programming · Non-dominated solution · Interval coefficients

1 Introduction

In hierarchical organizations, bi-level programming is considered as an important problem which comprises two sequential optimization problems as upper and lower level. Upper and lower level decision makers (ULDM and LLDM) control set of decision variables independently. As decisions of both levels get influenced by each other, preference is given to each decision in a cooperative environment. Linear fractional programming (LFP) [13] comprises the objective function as fraction of affine (linear plus constant) functions which has wide range of applications in science, engineering, economics, finance, business, management, production, infor-

S. Nayak (✉)
School of Applied Sciences, KIIT Deemed to be University, Bhubaneswar, India
e-mail: nksuvasis@gmail.com

A. K. Ojha
School of Basic Sciences, Indian Institute of Technology Bhubaneswar, Bhubaneswar, India
e-mail: akojha@iitbbs.ac.in

D. Dutta and B. Mahanty (eds.), *Numerical Optimization in Engineering and Sciences*, Advances in Intelligent Systems and Computing 979,
https://doi.org/10.1007/978-981-15-3215-3_25

mation theory and many more. Some common instances of fractional objectives are profit/cost, cost/volume, inventory/sale, debt/equity, output/employee and so on. Mishra [7] developed a method to solve BL-LFPP using analytic hierarchy process and weighting sum method. Emam [5] proposed interactive approach to solve bi-level integer multi-objective FPP. Baky [2] used fuzzy goal programming to solve a decentralized bi-level problem. A bi-level nonlinear decision-making problem under fuzziness was studied by Abo-Sinna [1], whereas Baky and Abo-Sinna [3] solved a bi-level problem with multiple nonlinear objectives at each level. Sakawa et al. [11] developed a fuzzy approach to solve BL-LFPP with fuzzy parameters. Stanojević and Stanojević [14] solved a BL-LFPP with fuzzy coefficients in the objective function only by using the concepts of α-cut and fuzzy. Singh and Chakrobarty [12] developed a method to solve a bi-level linear programming problem (BL-LPP) in a fuzzy environment, whereas Ren [10] proposed another solving approach in which all the coefficients are fuzzy random variables.

This paper studies BL-LFPP with interval coefficients considering the situation that DM does not have always precise information of all the required data while formulating practical problems into suitable mathematical models. In such cases, use of a range of values in form of intervals is more appropriate instead of fixed values. The proposed method derives a compromise solution and a range of optimal objective values of the problem.

The organization of the paper is as follows: Sect. 2 interprets arithmetic operations on intervals, and variable transformation method. Section 3 formulates BL-LFPP with interval coefficients. The proposed solution approach is incorporated in Sect. 4 to determine the compromise solution. A numerical example is discussed in Sect. 5, and finally, Sect. 6 includes some conclusions.

2 Preliminaries

2.1 Arithmetic Operations on Intervals

Consider $p, q \in I \subset \mathbb{R}$ where I denotes the set of closed intervals in the real line. If $p = [p^L, p^U]$ and $q = [q^L, q^U]$, the details of arithmetic operations on intervals can be found in [8, 9] from which the basic operations can be defined as follows.

- $p + q = [p^L + q^L, p^U + q^U]$
- $p - q = [p^L - q^U, p^U - q^L]$
- $pq = [p^L, p^U][q^L, q^U] = [min S_1, max S_1]$, where $S_1 = \{p^L q^L, p^L q^U, p^U q^L, p^U q^U\}$
- $p/q = [p^L, p^U]/[q^L, q^U] = [min S_2, max S_2]$, where $S_2 = [\frac{p^L}{q^L}, \frac{p^L}{q^U}, \frac{p^U}{q^L}, \frac{p^U}{q^U}]$ and $0 \notin$ q.
- for $\alpha \in \mathbb{R}, \alpha p = \begin{cases} [\alpha p^L, \alpha p^U], \alpha \geq 0 \\ [\alpha p^U, \alpha p^L], \alpha < 0 \end{cases}$

As 'I' is not totally ordered, $\preccurlyeq$ is defined as a partial ordering relation on it. $p \preccurlyeq q$, i.e., p is inferior to q or q is superior to p iff $p^L \leq q^L$ and $p^U \leq q^U$. $p \prec q$ if one of the following condition holds [15].

$$\{p^L < q^L, p^U < q^U\} \text{ or } \{p^L < q^L, p^U \leq q^U\} \text{ or } \{p^L \leq q^L, p^U < q^U\}.$$

2.2 *Variable Transformation Method*

Charnes and Cooper [4] developed variable transformation method to determine the optimal solution of a LFPP using an extra variable and an additional constraint. Consider the following two optimization models M_1 and M_2.

$$(M_1)\text{: max } f(x) = \frac{cx + \alpha}{dx + \beta} \quad \text{subject to} \quad \Omega_1 = \{Ax \leq b, x \geq 0\}$$

$$(M_2)\text{: max } g(y, z) = cy + \alpha t \quad \text{subject to} \quad \Omega_2 = \{dy + \beta t = 1, Ay - bt \leq 0,\ \ y \geq 0, t > 0\}$$

where the transformations $t = \frac{1}{dx+\beta}$ and $y = xt$ derive (M_2) from (M_1).

Lemma 1 [13] *For any feasible solution $(y, t) \in \Omega_2$, t is positive.*

Theorem 1 [13] *If (y^*, t^*) is an optimal solution of (M_2), then $x^* = \frac{y^*}{t^*}$ is the optimal solution of (M_1).*

Consider the following two problems where $f : \mathbb{R}^n \to I$ and $g : \mathbb{R}^n \to \mathbb{R}$.

$$(P_1) : \text{max } f(x) = [f^L(x), f^U(x)] \quad \text{subject to } x \in \Omega_1$$

$$(P_2) : \text{max } g(x) = f^L(x) + f^U(x) \quad \text{subject to } x \in \Omega_1$$

Definition 1 [15] x^* is a non-dominated solution of the problem: max $f(x) = [f^L(x), f^U(x)]$ subject to $x \in \Omega_1$ if there exists no $\hat{x} \in \Omega_1$ such that $f(x^*) \prec f(\hat{x})$.

Theorem 2 [15] *If x^* is an optimal solution of (P_2), then x^* is a non-dominated solution of (P_1).*

3 Problem Formulation

The mathematical formulation of the bi-level LFPP (BL-LFPP) with interval coefficients can be defined in a cooperative environment as follows.

$$\text{Upper level (DM}_1) : \max_{X_1} \ f_1(x) = \frac{c_1 x + \alpha_1}{d_1 x + \beta_1}$$

$$\text{Lower level (DM}_2\text{)} : \max_{X_2} \; f_2(x) = \frac{c_2x + \alpha_2}{d_2x + \beta_2}$$

subject to

$$x \in \Omega = \{A_1X_1 + A_2X_2 \leq \; b; X_1, X_2 \geq 0\}$$

where DM_i controls the set of decision variables $X_i, (i = 1, 2)$ independently. $X_i \in \mathbb{R}^{n_i}$ and $x = (X_1, X_2) = (x_1, x_2, \ldots, x_n) \in \mathbb{R}^n \geq 0, n = n_1 + n_2$. $A_i = (a_{kj}) \in \mathbb{R}^{m \times n_i}, \; i = 1, 2$ and $b = (b_k) \in \mathbb{R}^m, \; k = 1, 2, \ldots, m$. $c_i = (c_{ij}), \; d_i = (d_{ij}) \in \mathbb{R}^n$ and $\alpha_i, \beta_i \in \mathbb{R}$ for $i = 1, 2; \; j = 1, 2, \ldots, n$.

Assume that $c_{ij}, d_{ij}, \alpha_i, \beta_i \in I^+, a_{kj}, b_j \in I$ and $d_ix + \beta_i > 0 \, \forall x \in \Omega$ such that, $c_{ij} = [c_{ij}^L, c_{ij}^U], d_{ij} = [d_{ij}^L, d_{ij}^U], \alpha_i = [\alpha_i^L, \alpha_i^U], \beta_i = [\beta_i^L, \beta_i^U], a_{kj} = [a_{kj}^L, a_{kj}^U], b_j = [b_j^L, b_j^U]$.

4 Proposed Method of Solution

The objective functions at upper and lower level of the BL-LFPP can be stated as follows:

$$f_i(x) = \frac{\sum_{j=1}^{n}[c_{ij}^L, c_{ij}^U]x_j + [\alpha_i^L, \alpha_i^U]}{\sum_{j=1}^{n}[d_{ij}^L, d_{ij}^U]x_j + [\beta_i^L, \beta_i^U]}, i = 1, 2$$

Using arithmetic operations on intervals [8], the objectives at upper and lower level can be transformed into interval-valued functions as:

$$f_i(x) = \left[\frac{\sum_{j=1}^{n} c_{ij}^L x_j + \alpha_i^L}{\sum_{j=1}^{n} d_{ij}^U x_j + \beta_i^U}, \frac{\sum_{j=1}^{n} c_{ij}^U x_j + \alpha_i^U}{\sum_{j=1}^{n} d_{ij}^L x_j + \beta_i^L}\right] = [f_i^L(x), f_i^U(x)], i = 1, 2$$

The constraints of the problem BL-LFPP can be stated in the following form:

$$\sum_{j=1}^{n}[a_{kj}^L, a_{kj}^U]x_j \preccurlyeq [b_k^L, b_k^U], k = 1, 2, \ldots, m$$

It can be further simplified as follows.

$$\sum_{j=1}^{n} a_{kj}^L x_j \leq b_k^L, \sum_{j=1}^{n} a_{kj}^U x_j \leq b_k^U, k = 1, 2, \ldots, m$$

As the BL-LFPP is considered of maximization type, the individual maximal solutions of the lower bounds $f_i^L(x)$ and upper bounds $f_i^U(x)$ of the interval-valued functions $f_i(x)$ are derived at both levels. The mathematical formulation to determine the maximal solution of $f_i^L(x)$ using variable transformation method can be stated as follows.

$$\max \sum_{j=1}^{n} c_{ij}^{L} y_j + t\alpha_i^{L} \tag{1}$$

subject to

$$\sum_{j=1}^{n} d_{ij}^{U} y_j + t\beta_i^{U} = 1,\ \sum_{j=1}^{n} a_{kj}^{L} y_j \le t b_k^{L},\ \sum_{j=1}^{n} a_{kj}^{U} y_j \le t b_k^{U},\ k = 1, 2, \ldots, m$$
$$y_j \ge 0, t > 0$$

Using Theorem 1, the individual maximal solution of $f_i^L(x)$ can be obtained by solving problem (1). Similarly, mathematical formulation for $f_i^U(x)$ can be modeled to determine its maximal solution. Let $x_i^{L*} = (x_{ik}^{L*}, k = 1, 2, \ldots, n)$ and $x_i^{U*} = (x_{ik}^{U*}, k = 1, 2, \ldots, n)$ be the individual maximal solutions obtained for $f_i^L(x)$ and $f_i^U(x)$, respectively. The fractional lower and upper bound functions $f_i^L(x)$ and $f_i^U(x)$ at both levels ($i = 1, 2$) can be approximated by linear functions using their Taylor series expansion about own individual maximal solutions.

These approximations can be formulated as follows.

$$f_i^L(x) \approx \tilde{f}_i^L(x) = f_i^L(x_i^{L*}) + \sum_{k=1}^{n} (x_k - x_{ik}^{L*}) \frac{\partial f_i^L(x_i^{L*})}{\partial x_k},\ i = 1, 2$$

$$f_i^U(x) \approx \tilde{f}_i^U(x) = f_i^U(x_i^{U*}) + \sum_{k=1}^{n} (x_k - x_{ik}^{U*}) \frac{\partial f_i^U(x_i^{U*})}{\partial x_k},\ i = 1, 2$$

The objective functions at upper and lower levels of the BL-LFPP can be stated as follows.

$$\tilde{f}_i(x) = [\tilde{f}_i^L(x), \tilde{f}_i^U(x)],\ i = 1, 2$$

By Theorem 2, DM$_i$ solve the following problems separately for ($i = 1, 2$) to determine the non-dominated solutions of level-i.

$$\max\ \tilde{f}_i^L(x) + \tilde{f}_i^U(x) \tag{2}$$

subject to

$$\sum_{j=1}^{n} a_{kj}^{L} x_j \le b_k^{L},\ \sum_{j=1}^{n} a_{kj}^{U} x_j \le b_k^{U}, x_j \ge 0,\ k = 1, 2, \ldots, m$$

Let x^{1*} and x^{2*} be the non-dominated solutions of upper and lower level problems, respectively. As ULDM controls the set of decision variables X_1, the aspiration values X_1^* of X_1 are determined from the corresponding coordinates of the non-dominated solution x^{1*}. The aspiration values of $\tilde{f}_i^L(x)$ and $\tilde{f}_i^U(x)$ are obtained as:

At level-1: $\tilde{f}_1^L(x^{1*}) = \tilde{f}_1^{L*}$ and $\tilde{f}_1^U(x^{1*}) = \tilde{f}_1^{U*}$

At level-2: $\tilde{f}_2^L(x^{2*}) = \tilde{f}_2^{L*}$ and $\tilde{f}_2^U(x^{2*}) = \tilde{f}_2^{U*}$

Finally, the compromise solution of the BL-LFPP with interval coefficients can be obtained on solving the following problem which is formulated due to goal programming method [6] as follows.

$$\min \sum_{i=1}^{2} d_{iL}^{-} + \sum_{i=1}^{2} d_{iL}^{+} + \sum_{i=1}^{2} d_{iU}^{-} + \sum_{i=1}^{2} d_{iU}^{+} + (e^{-} + e^{+}) \tag{3}$$

subject to

$$\begin{gathered}
\tilde{f}_i^L(x) + d_{iL}^{-} - d_{iL}^{+} = \tilde{f}_i^{L*},\ \tilde{f}_i^U(x) + d_{iU}^{-} - d_{iU}^{+} = \tilde{f}_i^{U*},\ i = 1, 2 \\
X_1 + e^{-} - e^{+} = X_1^{*} \\
\sum_{j=1}^{n} a_{kj}^L x_j \le b_k^L,\ \sum_{j=1}^{n} a_{kj}^U x_j \le b_k^U,\ k = 1, 2, \ldots, m \\
x_j \ge 0, d_{iL}^{-}, d_{iL}^{+}, d_{iU}^{-}, d_{iU}^{+}, e^{-}, e^{+} \ge 0 \\
d_{iL}^{-}.d_{iL}^{+} = 0,\ d_{iU}^{-}.d_{iU}^{+} = 0,\ e^{-}.e^{+} = 0
\end{gathered}$$

Since over deviation from the aspiration level shows the state of complete achievement, the over deviational variables d_{iL}^{+} and d_{iU}^{+} can be ignored for the objective functions of maximization types. Therefore, the above formulation (3) can be simplified as follows:

$$\min \sum_{i=1}^{2} d_{iL}^{-} + \sum_{i=1}^{2} d_{iU}^{-} + (e^{-} + e^{+}) \tag{4}$$

subject to

$$\begin{gathered}
\tilde{f}_i^L(x) + d_{iL}^{-} \ge \tilde{f}_i^{L*},\ \tilde{f}_i^U(x) + d_{iU}^{-} \ge \tilde{f}_i^{U*},\ i = 1, 2 \\
X_1 + e^{-} - e^{+} = X_1^{*} \\
\sum_{j=1}^{n} a_{kj}^L x_j \le b_k^L,\ \sum_{j=1}^{n} a_{kj}^U x_j \le b_k^U,\ k = 1, 2, \ldots, m \\
x_j \ge 0, d_{iL}^{-}, d_{iU}^{-}, e^{-}, e^{+} \ge 0,\ e^{-}.e^{+} = 0
\end{gathered}$$

5 Numerical Example

To illustrate the solution procedure, consider the following bi-level LFPP in a cooperative environment.

Upper level: $\max_{x_1} \ f_1(x) = \dfrac{[2, 3]x_1 + [5, 7]x_2 + [1, 2]x_3 + [1, 2]}{[3, 5]x_1 + [2, 6]x_2 + [2, 3]x_3 + [2, 4]}$

Lower level: $\max_{x_2, x_3} \ f_2(x) = \dfrac{[2, 5]x_1 + [4, 7]x_2 + [3, 5]x_3 + [3, 4]}{[1, 2]x_1 + [3, 5]x_2 + [5, 7]x_3 + [4, 5]}$

subject to

$[-1, 1]x_1 + [1, 1]x_2 + [-1, 1]x_3 \preccurlyeq [-1, 5]$

$[-2,3]x_1 + [-1,-1]x_2 + [1,2]x_3 \preccurlyeq [-1,7],\ x_1, x_2, x_3 \geq 0$

Using arithmetic operations on intervals, the objective at both levels are formulated as follows.

$$\text{Upper level: } \max_{x_1}\ f_1(x) = \left[\frac{2x_1 + 5x_2 + x_3 + 1}{5x_1 + 6x_2 + 3x_3 + 4}, \frac{3x_1 + 7x_2 + 2x_3 + 2}{3x_1 + 2x_2 + 2x_3 + 2}\right]$$

$$\text{Lower level: } \max_{x_2, x_3}\ f_2(x) = \left[\frac{2x_1 + 4x_2 + 3x_3 + 3}{2x_1 + 5x_2 + 7x_3 + 5}, \frac{5x_1 + 7x_2 + 5x_3 + 4}{x_1 + 3x_2 + 5x_3 + 4}\right]$$

subject to

$$\Omega = \begin{cases} x_1 + x_2 + x_3 \leq 5,\ x_1 - x_2 + x_3 \geq 1, \\ 3x_1 - x_2 + 2x_3 \leq 7,\ 2x_1 + x_2 - x_3 \geq 1,\ x_1, x_2, x_3 \geq 0 \end{cases}$$

According to the proposed method, the objectives are approximated by linear functions as $f_i(x) = [\tilde{f}_i^L(x), \tilde{f}_i^U(x)]$ at level $i = 1, 2$.

At upper level ($i = 1$):

$$\tilde{f}_1^L(x) = -0.0298x_1 + 0.0630x_2 - 0.0255x_3 + 0.5104$$
$$\tilde{f}_1^U(x) = -0.1870x_1 + 0.2701x_2 - 0.1246x_3 + 1.6647$$

At lower level ($i = 2$):

$$\tilde{f}_2^L(x) = 0.0181x_1 - 0.0023x_2 - 0.127x_3 + 0.7598$$
$$\tilde{f}_2^U(x) = 0.1893x_1 - 0.0473x_2 - 0.5917x_3 + 2.0652$$

The non-dominated solution at upper level problem is obtained as $x^{1*} = (0.6667, 2, 2.3333)$ where $\tilde{f}_1^{L*} = 0.5570$ and $\tilde{f}_1^{U*} = 1.7895$. The aspiration value for x_1 controlled by ULDM is ascertained as $x_1^* = 0.6667$. The non-dominated solution at lower level problem is obtained as $x^{2*} = (3, 2, 0)$ where $\tilde{f}_2^{L*} = 0.8095$ and $\tilde{f}_2^{U*} = 2.5385$. Using goal programming method, the problem is formulated as follows.

$$\min\ \sum_{i=1}^{2} d_{iL}^- + \sum_{i=1}^{2} d_{iU}^- + (e^- + e^+)$$
$$\text{subject to}$$
$$(-0.0298x_1 + 0.0630x_2 - 0.0255x_3 + 0.5104) + d_{1L}^- \geq 0.5570$$
$$(0.0181x_1 - 0.0023x_2 - 0.127x_3 + 0.7598) + d_{2L}^- \geq 0.8095$$
$$(-0.1870x_1 + 0.2701x_2 - 0.1246x_3 + 1.6647) + d_{1U}^- \geq 1.7895$$
$$(0.1893x_1 - 0.0473x_2 - 0.5917x_3 + 2.0652) + d_{2U}^- \geq 2.5385$$
$$x_1 + e^- - e^+ = 0.6667$$
$$x_1 + x_2 + x_3 \leq 5,\ x_1 - x_2 + x_3 \geq 1,\ 3x_1 - x_2 + 2x_3 \leq 7,\ 2x_1 + x_2 - x_3 \geq 1$$
$$x_1, x_2, x_3 \geq 0$$
$$x_1, x_2, x_3 \geq 0, d_{1L}^-, d_{2L}^-, d_{1U}^-, d_{2U}^-, e^-, e^+ \geq 0, e^-.e^+ = 0$$

Table 1 Objective values at the compromise solution

$Level : i$	$f_i^L(x^*)$	$\tilde{f}_i^L(x^*)$	$f_i^U(x^*)$	$\tilde{f}_i^U(x^*)$
$Level : i = 1$	0.3200	0.4820	1	1.4985
$Level : i = 2$	0.6154	0.7295	1.4211	1.9942

On solving the above problem, the compromise solution is obtained as $x^* = (0.6667, 0, 0.3333)$ where the values of the lower and upper bound functions of the fractional and approximated linear objectives are shown in Table 1 and bar graph.

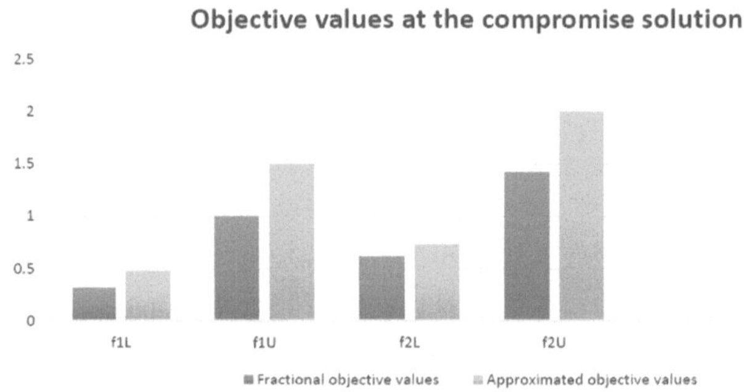

The values of the fractional objectives $[f_i^L(x^*), f_i^U(x^*)], i = 1, 2$ are considered as the range of optimal objective values.

6 Conclusion

In hierarchical organizations, if DMs have no precise information about the resources, then intervals are more appropriate to be considered instead of fixed values. In this view, this paper develops a method to solve a BL-LFPP with interval coefficients, and optimal range of objective values is determined for the objectives of both levels as the solution. This method is also useful for BL-LFPP with fuzzy coefficients (linear, triangular, trapezoidal, etc.) as fuzzy numbers can be transformed into interval forms using fuzzy α-cuts. Numerical example is discussed to illustrate the solution procedure.

References

1. Abo-Sinna, M.A.: A bi-level non linear multi-objective decision-making under fuzziness. J. Oper. Res. Soc. India **38**(5), 484–495 (2001)
2. Baky, I.A.: Fuzzy goal programming algorithm for solving decentralized bi-level multi-objective programming problems. Fuzzy Sets Syst. **160**(18), 2701–2713 (2009)
3. Baky, I.A., Abo-Sinna, M.A.: TOPSIS for bi-level MODM problems. Appl. Math. Model. **37**(3), 1004–1015 (2013)
4. Charnes, A., Cooper, W.W.: Management models and industrial applications of linear programming. Manage. Sci. **4**(1), 38–91 (1957)
5. Emam, O.E.: Interactive approach to bi-level integer multi-objective fractional programming problem. Appl. Math. Comput. **223**, 17–24 (2013)
6. Miettinen, K.: Nonlinear Multiobjective Optimization, 12th edn. Springer Science & Business Media, Berlin (2012)
7. Mishra, S.: Weighting method for bi-level linear fractional programming problems. Euro. J. Oper. Res. **183**(1), 296–302 (2007)
8. Moore, R.E.: Interval Analysis. Prince-Hall, Englewood Cliffs, NJ (1996)
9. Nayak, S., Ojha, A.K.: Generating pareto optimal solutions of multi-objective LFPP with interval coefficients using-constraint method. Math. Modell. Anal. **20**(3), 329–345 (2015)
10. Ren, A.: Solving the general fuzzy random bilevel programming problem through Me measure-based approach. IEEE Access. **6**, 25610–25620 (2018)
11. Sakawa, M., Nishizaki, I., Uemura, Y.: Interactive fuzzy programming for two-level linear fractional programming problems with fuzzy parameters. Fuzzy Sets Syst. **115**(1), 93–103 (2000)
12. Singh, V.P., Chakraborty, D.: Solving bi-level programming problem with fuzzy random variable coefficients. J. Intell. Fuzzy Syst. **32**(1), 521–528 (2017)
13. Stancu-Minasian, I.M.: Fractional programming : Theory, Methods and Applications. Kluwer Academic Publishers, Berlin (1997)
14. Stanojević, B., Stanojević, M.: Solving method for linear fractional optimization problem with fuzzy coefficients in the objective function. Int. J. Comput. Commun. Control **8**(1), 146–152 (2013)
15. Wu, H.-C.: On interval-valued nonlinear programming problems. J. Math. Anal. Appl. **338**(1), 299–316 (2008)

RBF-FD Based Method of Lines with an Optimal Constant Shape Parameter for Unsteady PDEs

Chirala Satyanarayana

Abstract It is known that, the RBF-FD scheme combined with standard Runge-Kutta methods in Method of Lines (MOL) provides robust and efficient numerical solutions for unsteady Partial Differential Equations (PDE). If the infinitely smooth RBF like multiquadric is used as basic function in RBF-FD scheme, then the scaling (shape) parameter present in them plays a substantial role in obtaining the accurate numerical solutions. In this work, an optimization process is proposed to determine a constant optimal scaling parameter of basis function in RBF-FD scheme, through method of lines. In this process, the error function is written in terms of local truncation error and then an (near) optimal shape parameter is found by minimizing the error function. The proposed optimization process is validated for one dimensional heat and wave equations.

Keywords RBF-FD scheme · Optimal shape parameter · Multiquadric

1 Introduction

The Finite Difference (FD) numerical schemes derived using Radial Basis Function (RBF) [2, 8] are more appealing alternative for solving Partial Differential Equations (PDE). These schemes are truely meshfree, and can be applied over the scattered nodes. The resulting global matrix is better conditioned and flexible to solve nonlinear PDEs. In the RBF-FD Method Of Lines (MOL), the spatial derivatives in PDE are discretized using RBF-FD scheme. Then the PDE reduces to the system of ordinary differential equations (ODE). Finally, these initial value ODEs are solved using Runge-Kutta methods.

If the very smooth basis functions, such as Multiquadric (MQ) and Gaussian (GA) are used as interpolants, then the accuracy of corresponding scheme depends on the scaling parameter (ϵ) present in the basis function. Many researchers [1, 3, 4, 6]

C. Satyanarayana (✉)
Department of Mathematics, Mahindra École Centrale, Hyderabad, Telangana, India
e-mail: Satyanarayana.Chirala@mechyd.ac.in

D. Dutta and B. Mahanty (eds.), *Numerical Optimization in Engineering and Sciences*, Advances in Intelligent Systems and Computing 979,
https://doi.org/10.1007/978-981-15-3215-3_26

have proposed variety of algorithms to find an (near) optimal shape parameter in RBF interpolation, global and local collocation to obtain better accurate numerical solutions. Roque [5] have tested the global optimization algorithm [3] to find a constant optimal shape parameter in RBF-PS method to solve unsteady problems. However, due to global nature of this algorithm the condition number of matrix increases exponentially, if the points increases. Sanyasiraju [7] have proposed a localized Rippa's algorithm [6] to find a better value of shape parameter in local RBF grid-free schemes and then tested for the steady convection diffusion equations. Bayona [1] have proposed an optimization method for RBF-FD scheme based on the analytical expressions of local truncation error in spatial derivatives, then validated on the Poission equations. Marjan [4] have tested the Rippa's algorithm [6] to find a good value of shape parameter to solve the unsteady PDEs in global collocation method. Then concluded that Rippa's algorithm gives better accurate results for interpolation problem, however, the same algorithm may fail for unsteady PDEs.

In the present work, the optimization algorithm proposed for RBF-FD scheme [1] to solve Poission equation is continued to the unsteady PDEs using method of lines approach. In this process, the objective function depends on the local truncation error of the discretization process. Then the optimal scaling factor of basis function is obtained after minimizing this function over a fixed interval of ϵ. The RBF-FD MOL along with optimization process is validated on one dimensional heat and wave equations. The present article is organized as the following.

The RBF-FD based method of lines for solving unsteady PDEs is presented in Sect. 2. In Sect. 3, the optimization method to find an optimal scaling parameter for unsteady PDEs is described. In Sect. 4 the validation of proposed optimization for one dimensional heat and wave equations is presented. Some conclusions are made at the end.

2 RBF-FD Based MOL for Unsteady PDEs

The general form of unsteady PDE is described in the following equations.

$$\frac{\partial^\alpha u(\underline{x}, t)}{\partial t^\alpha} + \mathscr{L}u(\underline{x}, t) = f(\underline{x}, t), \ \underline{x} \in \Omega \subseteq \mathbb{R}^d, \ t \in (0, T], \ T > 0, \ \alpha \in \mathbb{N} \quad (1)$$

$$\mathscr{B}u(\underline{x}, t) = g(\underline{x}, t), \ \underline{x} \in \partial\Omega \quad (2)$$

$$u(\underline{x}, 0) = u_0(\underline{x}), \ \underline{x} \in \Omega \quad (3)$$

where $\mathscr{L}$ and $\mathscr{B}$ are the linear differential operators. The functions f, g and u_0 are chosen so that the PDE (1) and (2) is well-posed. In the present work, the spatial derivatives in Eqs. (1) and (2) are approximated using RBF-FD method [2, 8].

Let M and N be the total number of points in the computational domain and on its boundary, respectively. Consider $\Omega_i = \{\underline{x}_1, \underline{x}_2, \ldots, \underline{x}_{m_i}\}$ be the local support at the center $\underline{x}_i$, having $m_i (\ll M)$ centers, including $\underline{x}_i$. Using RBF-FD scheme, the spatial derivatives in Eqs. (1) and (2) are discretized as,

$$\frac{d^\alpha u(\underline{x}_i, t)}{dt^\alpha} + \sum_{j=1}^{m_i} \beta_j u(\underline{x}_j, t) = f(\underline{x}_i, t), \ t \in (0, T], \ i = 1, 2, \ldots, M \tag{4}$$

$$\sum_{k=1}^{m_k} \gamma_k u(\underline{x}_k, t) = g(\underline{x}_b, t), \ t \in (0, T], \ b = 1, 2, \ldots, N \tag{5}$$

The initial condition for (4) and (5) is $u(0) = u_0(\underline{x}_i)$, $i = 1, 2, \ldots, M$. The higher order ODEs in Eq. (4) are transformed into the system of first order ODEs and then solved using fourth order explicit Runge-Kutta method. The β's in Eq. (4) are calculated by solving the following local system,

$$A[\beta/\bar{\gamma}] = (\mathscr{L} B(\underline{x}_i))^T, \quad A = \begin{bmatrix} \boldsymbol{\Xi} & \mathbf{p} \\ \mathbf{p}^T & \mathbf{0} \end{bmatrix}, \quad \text{for each } \underline{x}_i \in \Omega \tag{6}$$

where $\boldsymbol{\Xi} := \phi(\|\underline{x}_i - \underline{x}_j\|_2, \varepsilon)$, $i, j = 1, 2, \ldots, m_i$, $B(\underline{x}) = [\phi(\|\underline{x} - \underline{x}_1\|_2, \varepsilon), \ldots, \phi(\|\underline{x} - \underline{x}_{m_i}\|_2, \varepsilon), p_1(\underline{x}), \ldots, p_l(\underline{x})]$. The polynomial basis, $\mathbf{p} := p_j(\underline{x}_i)$, $j = 1, 2, \ldots, l$, $i = 1, 2, \ldots, m_i$, may be required to ensure the nonsingularity of A, defined in the Eq. (6). A similar procedure is applied to find γ's in Eq. (5). It is known that RBF-FD scheme works on a random set of nodes. However in the present work, the uniform distribution of the nodes, with $m_i = 3$ and time step $k = 0.001$, are chosen to compare the proposed RBF-FD MOL with finite difference scheme. All numerical results are obtained with respect to multiquadric RBF, $\phi(r) = \sqrt{(1 + (\varepsilon r)^2}$, where the Euclidean distance (r) is defined as, $r = \|\underline{x} - \underline{x}_i\|_2$.

3 Optimal Shape Parameter

In this section, an optimization process to determine the (near) optimal constant shape parameter for RBF-FD scheme is presented to solve unsteady PDEs. The error function is defined in terms of local truncation error of the discretization process. Then the optimal (scaling) shape parameter ($\tilde{\epsilon}$) is obtained after minimizing the error function. The error function (LTE) is defined as the following:

(a) If $\alpha = 1$ in Eq. (4), then the error function is defined as,

$$\text{LTE}(\tilde{\epsilon}) = \left\| \frac{du_i}{dt} + \sum_{j=1}^{m_i} \beta_j u_j - f_i \right\|_\infty + \left\| \sum_{k=1}^{m_k} \gamma_k u_k - g_b \right\|_\infty \tag{7}$$

Then the analytical expression of the local truncation error in the time discretization is considered as,

$$\mathrm{LTE}_1 = \frac{k^5}{5!}\frac{\partial^5 u}{\partial t^5} + O(k^6) \tag{8}$$

(b) If $\alpha = 2$ in Eq. (4), using $\dfrac{du}{dt} = v$, then (4) transforms into the system of first order ODEs. Define the error function as,

$$\mathrm{LTE}(\tilde{\epsilon}) = \left\| \frac{du_i}{dt} - v_i \right\|_\infty + \left\| \frac{dv_i}{dt} + \sum_{j=1}^{m_i} \beta_j u_j - f_i \right\|_\infty + \left\| \sum_{k=1}^{m_k} \gamma_k u_k - g_b \right\|_\infty \tag{9}$$

The local truncation error with respect to time discretization is,

$$\mathrm{LTE}_2 = \frac{k^5}{5!}\frac{\partial^5 U}{\partial t^5} + O(k^6), \text{ where } U = [u, v]^T \tag{10}$$

In the discretization of second order spatial derivative, the analytical expression of local truncation error at (x_i, t_j), with respect to MQ-RBF [1], is obtained as,

$$\mathrm{LTE}_{MQ} = \frac{h^2}{12}\frac{\partial^4 u}{\partial x^4} + h^2\epsilon^2 \frac{\partial^2 u}{\partial x^2} - \frac{3}{4} h^2 \epsilon^4 u(x, t) \tag{11}$$

To approximate the partial derivatives in Eqs. (8), (10) and (11), first, PDE (1) and (2) is solved using RBF-FD with a fixed value of shape parameter. Then the *fminbnd* of MATLAB function handle is used to find the (near) optimal constant shape parameter ($\tilde{\epsilon}$), with $\tilde{\epsilon}_{\min} = 0.001$ and $\tilde{\epsilon}_{\max} = 10$. Further, the accuracy of RBF-FD is improved with the (near) optimal constant shape parameter.

4 Validation

In this section, the RBF-FD based method of lines with an optimal constant parameter is validated on one dimensional heat and wave equations.

Problem 1 1D Heat equation [5]

$$\frac{\partial u}{\partial t} = \frac{\partial^2 u}{\partial x^2}, \quad 0 < x < 1, \ t > 0 \tag{12}$$

$$u_x(0, t) = 0, \ u_x(1, t) = 0, \ t \geq 0 \tag{13}$$

the initial condition $u(x, 0)$ is obtained from the analytical solution

$$u(x, t) = 9 + 3\, e^{-\pi^2 t} \cos(\pi x) + 5\, e^{-16\pi^2 t} \cos(4\pi x) \tag{14}$$

Problem 2 1D Wave equation

$$\frac{\partial^2 u}{\partial t^2} = \frac{\partial^2 u}{\partial x^2}, 0 < x < 1, t > 0 \tag{15}$$

$$u_x(0,t) = 0,\ u_x(1,t) = 0,\ u(x,0) = \frac{1}{8}\cos(\pi x),\ u_t(x,0) = 0,\ 0 \le x \le 1 \tag{16}$$

with analytical solution $u(x,t) = \frac{1}{8}\cos(\pi x)\cos(\pi t)$.

First, the one dimensional heat and wave equations are solved using MQ RBF-FD on a fixed interval of shape parameter for various time levels. Then the infinite norm error and the corresponding local truncation error functions are compared in Fig. 1 at $M = 41, T = 0.25$. It is observed from these figures that, the shape parameter depends on the basis, and unknown function in PDE and its derivatives. Further, it is independent of grid size and time. This suggests that, if the shape parameter is known for a particular values of M and T, then the same value can be used for any grid size and time.

It is also observed from these figures that, the error function perfectly imitates the infinite norm error curve of RBF-FD scheme. They have a global minima in the same interval of shape parameter. In Fig. 1, the horizontal line represents the infinite norm errors obtained with Finite Difference (FD) scheme. Therefore, it is clear from these curves that, there exists an interval of shape parameter in which the RBF-FD based MOL is more accurate than the finite difference scheme. In Table 1, the infinite norm errors obtained using multiquadric RBF-FD based MOL with actual (ϵ) and approximate ($\tilde{\epsilon}$) optimal constant shape parameter are compared with finite difference (FD) scheme. It is observed from Table 1 that, RBF-FD is at least one decimal place accurate than FD scheme. Therefore, the RBF-FD with an optimal constant parameter produces better accurate numerical solution than finite difference scheme for solving unsteady partial differential equations.

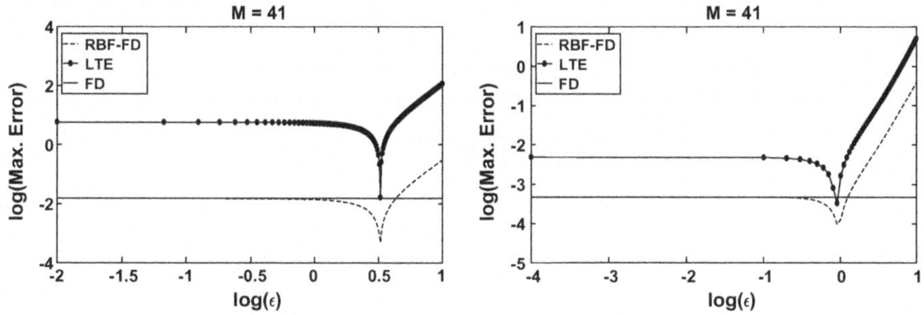

Fig. 1 Comparison of maximum errors for MQ RBF-FD, FD and Local Truncation Error (LTE) function for Problems 1 and 2 (Left and Right) at $M = 41, T = 0.25$

Table 1 The infinite norm errors calculated using RBF-FD and FD scheme for Problems 1 and 2 at $T = 0.01$

		$M = 11$	$M = 21$	$M = 41$	$M = 81$
Problem 1	FD	1.646e−01	5.732e−02	1.518e−02	3.264e−03
	ϵ	3.2255	3.2255	3.2255	3.2255
	RBF−FD with ϵ	4.423e−02	1.562e−03	4.202e−04	4.078e−05
	$\tilde{\epsilon}$	3.2825	3.2825	3.2825	3.2825
	RBF−FD with $\tilde{\epsilon}$	6.352e−02	3.962e−03	4.934e−04	9.503e−05
	$\mid \epsilon - \tilde{\epsilon} \mid$	0.0570	0.0570	0.0570	0.0570
Problem 2	FD	3.697e−06	1.736e−06	3.518e−07	1.726e−07
	ϵ	0.8173	0.8173	0.8173	0.8173
	RBF−FD with ϵ	6.073e−07	1.998e−07	2.446e−08	3.090e−09
	$\tilde{\epsilon}$	0.8814	0.8813	0.8813	0.8813
	RBF−FD with $\tilde{\epsilon}$	4.459e−07	1.091e−07	7.684e−08	6.556e−09
	$\mid \epsilon - \tilde{\epsilon} \mid$	0.0641	0.0640	0.0640	0.0640

5 Conclusion

In this work, the RBF-FD based method of lines (MOL) is presented for solving unsteady partial differential equations. The computations are carried out using multiquadirc radial basis function. Also, an algorithm to find the (near) optimal shape parameter for RBF-FD MOL is proposed based on the local truncation error of the discretization process. Then the following observations are made in the present work.

There exists an interval for shape parameter in which RBF-FD based method of lines is more accurate than finite difference method. The shape parameter depends on the basis, unknown function in PDE and its derivatives. The error function defined in terms of local truncation errors perfectly imitates the infinite norm error curve of RBF-FD scheme. The shape parameter is independent of grid size and time. Therefore, if the shape parameter is known at a fixed number of centers and time, then it can be used for rest of the computations.

References

1. Bayona, V., Moscoso, M., Carretero, M., Kindein, M.: RBF-FD formulas and convergence properties. J. Comput. Phys. **229**, 8281–8295 (2010). https://doi.org/10.1016/j.jcp.2010.07.008
2. Chandini, G., Sanyasiraju, Y.V.S.S.: Local radial basis function based gridfree scheme for unsteady incompressible viscous flows. J. Comput. Phys. **227**, 8922–8948 (2008). https://doi.org/10.1016/j.jcp.2008.07.004

3. Fasshauer, G.E., Zhang, J.: On choosing optimal shape parameters for RBF approximation. Numerical Algorithms **45**, 345–368 (2007). https://doi.org/10.1007/s11075-007-9072-8
4. Marjan, U.: On the selection of a good value of shape parameter in solving time-dependent partial differential equations using RBF approximation method. Appl. Math. Model. **38**, 135–144 (2014). https://doi.org/10.1016/j.apm.2013.05.060
5. Neves, A.M.A., Roque, C.M.C., Ferreira, A.J.M.: Solving time-dependent problems by an RBF-PS method with an optimal shape parameter. J. Phys: Conf. Ser. **181**, 012053 (2009). https://doi.org/10.1088/1742-6596/181/1/012053
6. Rippa, S.: An algorithm for selecting a good value for the parameter c in radial basis function interpolation. Adv. Comput. Math. **11**, 193–210 (1999). https://doi.org/10.1023/A:1018975909870
7. Sanyasiraju, Y.V.S.S., Satyanarayana, C.: On optimization of the RBF shape parameter in a grid-free local scheme for convection dominated problems over non-uniform centers. Appl. Math. Model. **37**, 7245–7272 (2013). https://doi.org/10.1016/j.apm.2013.01.054
8. Wright, G.B., Fornberg, B.: Scattered node compact finite difference-type formulas generated from radial basis functions. J. Comput. Phys. **212**(1), 99–123 (2006). https://doi.org/10.1016/j.jcp.2005.05.030

Parametric Accelerated Over Relaxation (PAOR) Method

V. B. Kumar Vatti, G. Chinna Rao and Srinesh S. Pai

Abstract By introducing a parameter α, a method named parametric accelerated over relaxation (PAOR) is considered and the choices of the parameters involved in PAOR method are given in terms of the eigenvalues of the Jacobi matrix. It is shown through some numerical examples that PAOR method surpasses the other methods considered in this paper.

Keywords AOR · SOR · Gauss–Seidel · Jacobi

1 Introduction

Many iterative methods play a major role in solving the linear system of equations

$$AX = b \tag{1.1}$$

(or)

$$(I - L - U)X = b \tag{1.2}$$

where I, L, U are unit, a strictly lower and upper triangular parts of A of order $n \times n$ and X & b are unknown and known vectors of order $n \times 1$, respectively.

In this paper, we consider PAOR method in Sect. 2 and in Sect. 3, the choices of the parameters associated in PAOR method are estimated. In Sect. 4, we consider some numerical examples to distinguish PAOR method over the other methods and concluded in Sect. 5.

V. B. Kumar Vatti (✉) · G. Chinna Rao
Department of Engineering Mathematics, Andhra University, Visakhapatnam, Andhra Pradesh, India
e-mail: drvattivbk@yahoo.co.in

S. S. Pai
CMT Laboratory, South Western Railways, Indian Railways, Hubli, India

D. Dutta and B. Mahanty (eds.), *Numerical Optimization in Engineering and Sciences*, Advances in Intelligent Systems and Computing 979,
https://doi.org/10.1007/978-981-15-3215-3_27

2 PAOR Method

The well-known AOR method for solving (1.2) is given by

$$(I-\omega L)X^{(n+1)} = \{(1-r)I + (r-\omega)L + rU\}X^{(n)} + rb$$
$$(n = 0, 1, 2, 3, \ldots) \tag{2.1}$$

By introducing a parameter $\alpha \neq -1$ in (2.1), we have

$$[(1+\alpha)I - \omega L]X^{(n+1)} = \{(1+\alpha-r)I + (r-\omega)L + rU\}X^{(n)} + rb$$
$$(n = 0, 1, 2, 3, \ldots) \tag{2.2}$$

This method (2.2) can be called as parametric AOR (PAOR) method and the methods such as AOR, SOR, Gauss–Seidel and Jacobi methods [3–5] can be realized for the choices of α, r and ω as

$$(\alpha, r, \omega) = (0, r, \omega), (0, \omega, \omega), (0, 1, 1), (0, 1, 0) \tag{2.3}$$

respectively.

The iteration matrix of PAOR method is

$$P_{\alpha,r,\omega} = [(1+\alpha)I - \omega L]^{-1}\{(1+\alpha-r)I + (r-\omega)L + rU\} \tag{2.4}$$

Now, $P_{0,r,\omega}$, $P_{0,\omega,\omega}$, $P_{0,1,1}$, $P_{0,1,0}$ become the iteration matrices of AOR, SOR, Gauss–Seidel and Jacobi methods [3–5] and the spectral radii of these methods are known to be

$$S(P_{0,r,w}) = \begin{cases} \dfrac{\overline{\mu}^2}{\left(1+\sqrt{1-\overline{\mu}^2}\right)^2} & \text{when } \underline{\mu} = 0 \text{ (or)} \sqrt{1-\overline{\mu}^2} \leq 1-\underline{\mu}^2 \text{(or) } 0 < \underline{\mu} < \overline{\mu} \\ 0 & \text{when } \underline{\mu} = \overline{\mu} \\ \dfrac{\underline{\mu}\sqrt{\overline{\mu}^2-\underline{\mu}^2}}{\sqrt{1-\underline{\mu}^2}\left(1+\sqrt{1-\overline{\mu}^2}\right)} & \text{when } 0 < \underline{\mu} < \overline{\mu} \;\&\; 1-\underline{\mu}^2 < \sqrt{1-\overline{\mu}^2} \end{cases} \tag{2.5}$$

$$S(P_{0,\omega,\omega}) = \frac{\overline{\mu}^2}{\left(1+\sqrt{1-\overline{\mu}^2}\right)^2} \tag{2.6}$$

$$S(P_{0,1,1}) = \overline{\mu}^2 \tag{2.7}$$

$$S(P_{0,1,0}) = \overline{\mu} \tag{2.8}$$

where $\underline{\mu}$ and $\overline{\mu}$ are the smallest and largest eigenvalues of Jacobi matrix $P_{0,1,0}$.

Theorem 2.1 *If* μ *be the eigenvalue of Jacobi matrix* $P_{0,1,0}$ *and '*λ*' be the eigenvalue of PAOR matrix* $P_{\alpha,r,\omega}$*, then* μ *and* λ *are connected by the relation*

$$[\lambda(1+\alpha)-(1+\alpha-r)]^2 = r\mu^2(\lambda\omega + r - \omega).$$

Proof Let

$$\begin{aligned}
&\left|P_{\alpha,r,\omega}-\lambda I\right| = 0\,(\text{or})\ \left|\lambda I - P_{\alpha,r,\omega}\right| = 0\\
&\left|\lambda I - [(1+\alpha)I-\omega L]^{-1}\{(1+\alpha-r)I+(r-\omega)L+rU\}\right| = 0\\
&|[\lambda(1+\alpha)-(1+\alpha-r)]I - [(\lambda\omega+r-\omega)L+rU]| = 0\\
&\Rightarrow \left|\left(\frac{\lambda(1+\alpha)-(1+\alpha-r)}{(\lambda\omega+r-\omega)^{\frac{1}{2}}r^{\frac{1}{2}}}\right)I-(L+U)\right| = 0
\end{aligned}$$

If 'μ' be the eigenvalue of $(L+U)$, we have

$$\Rightarrow \frac{\lambda(1+\alpha)-(1+\alpha-r)}{(\lambda\omega+r-\omega)^{\frac{1}{2}}r^{\frac{1}{2}}} = \mu$$

$$\Rightarrow [\lambda(1+\alpha)-(1+\alpha-r)]^2 = r\mu^2(\lambda\omega+r-\omega) \tag{2.1.1}$$

Theorem 2.2 *For* $\omega = \frac{2(1+\alpha)}{1+\sqrt{1-\mu^2}}(\alpha \neq -1)$*, the eigenvalues of the matrix* $P_{\alpha,r,\omega}$ *are* $\lambda = \frac{r\omega\mu^2}{2(1+\alpha)^2} - \frac{r}{1+\alpha} + 1$.

Proof From (2.1.1), we have

$$\begin{aligned}
&(1+\alpha)^2\lambda^2 - \left[2(1+\alpha)^2 - 2r(1+\alpha)+\mu^2 r\omega\right]\lambda\\
&\quad + \left[(1+\alpha)^2 - 2r(1+\alpha)+\mu^2 r\omega - \mu^2 r^2 + r^2\right] = 0
\end{aligned}$$

$$\Rightarrow \lambda = \frac{\left[2(1+\alpha)^2-2r(1+\alpha)+\mu^2 r\omega\right]\pm\sqrt{\Delta}}{2(1+\alpha)^2} \tag{2.2.1}$$

where

$$\begin{aligned}
\Delta &= \left[2(1+\alpha)^2 - 2r(1+\alpha)+\mu^2 r\omega\right]^2\\
&\quad - 4(1+\alpha)^2\left[(1+\alpha)^2 - 2r(1+\alpha)+\mu^2 r\omega - \mu^2 r^2 + r^2\right]\\
&= \mu^2 r^2\left[\mu^2\omega^2 - 4(1+\alpha)\omega + 4(1+\alpha)^2\right]
\end{aligned}$$

which will be zero if $\mu^2\omega^2 - 4(1+\alpha)\omega + 4(1+\alpha)^2 = 0$
(Or)
if $\omega = \frac{2(1+\alpha)}{1+\sqrt{1-\mu^2}}$ for any μ.
Therefore, λ of (2.2.1) becomes

$$\lambda = \frac{r\omega\mu^2}{2(1+\alpha)^2} - \frac{r}{1+\alpha} + 1.$$

3 Choice of the Parameters α, r and ω

The eigenvalues of PAOR matrix are obtained in Theorem 2.2 as

$$\lambda = \left(\frac{r\omega\mu^2}{2(1+\alpha)^2}\right) - \left(\frac{r}{1+\alpha} - 1\right) \tag{3.1}$$

If the two terms of RHS in (3.1) are connected by the relation

$$\frac{r\omega\mu^2}{2(1+\alpha)^2} = k\left(\frac{r}{1+\alpha} - 1\right) \tag{3.2}$$

where k is any constant, then rewriting r in terms k and k in terms of r from (3.2), we obtain

$$r = \frac{(1+\alpha)k}{k - \frac{\omega\mu^2}{2(1+\alpha)}} \tag{3.3}$$

and

$$k = \frac{r\omega\mu^2}{(1+\alpha)} \cdot \frac{1}{2[r-(1+\alpha)]} \tag{3.4}$$

Equating the values of $k(1+\alpha)$ obtained from (3.3) and (3.4), we get

$$[r-(1+\alpha)]\left[k - \frac{\omega\mu^2}{2(1+\alpha)}\right] = \frac{\omega\mu^2}{2} \tag{3.5}$$

Now, fixing ω in (3.5) as

$$\omega = \frac{2(1+\alpha)}{1+\sqrt{1-\overline{\mu}^2}} \tag{3.6}$$

and multiplying and dividing the RHS term of (3.5) by $\omega + \frac{\overline{\mu}^2 - \underline{\mu}^2}{2}$,

$$[r-(1+\alpha)]\left[k - \frac{\omega\mu^2}{2(1+\alpha)}\right] = \left[\omega + \frac{\overline{\mu}^2 - \underline{\mu}^2}{2}\right]\left[\frac{\frac{\omega\mu^2}{2}}{\omega + \frac{\overline{\mu}^2 - \underline{\mu}^2}{2}}\right] \tag{3.7}$$

Equating Eq. (3.7) of the first term in LHS with first term in RHS and similarly the second term in LHS with second term in RHS, we get

$$r = 1 + \alpha + \omega + \frac{\overline{\mu}^2 - \underline{\mu}^2}{2} \tag{3.8}$$

$$k = 1 - \sqrt{1 - \overline{\mu}^2} + \frac{\frac{\omega \underline{\mu}^2}{2}}{\omega + \frac{\overline{\mu}^2 - \underline{\mu}^2}{2}} \tag{3.9}$$

It is observed and verified that if $k > 1$, then r should be taken as r in (3.8) and if $k < 1$, then r should be taken as $\frac{r}{2}$.

We summarize above results by giving the choices of ω and r for various values of α:

Type 1: when $\underline{\mu} = \overline{\mu}$ and $k = 1$

$$\omega = \frac{2(1+\alpha)}{1 + \sqrt{1 - \overline{\mu}^2}} \quad \& \quad r = \frac{(1+\alpha)}{\sqrt{1 - \overline{\mu}^2}} \tag{3.10}$$

Type 2: when $\underline{\mu} \neq \overline{\mu}$ and $k > 1$

$$\omega = \frac{2(1+\alpha)}{1 + \sqrt{1 - \overline{\mu}^2}} \quad \& \quad r = 1 + \alpha + \omega + \frac{\overline{\mu}^2 - \underline{\mu}^2}{2} \tag{3.11}$$

Type 3: when $\underline{\mu} \neq \overline{\mu}$ and $k < 1$

$$\omega = \frac{2(1+\alpha)}{1 + \sqrt{1 - \overline{\mu}^2}} \quad \& \quad r = \left(1 + \alpha + \omega + \frac{\overline{\mu}^2 - \underline{\mu}^2}{2}\right)/2 \tag{3.12}$$

4 Numerical Examples

Example 4.1 For the matrix $\begin{bmatrix} 3 & -4 \\ 2 & -3 \end{bmatrix}$ considered by Hadjimos [1],

$$\underline{\mu} = \frac{2\sqrt{2}}{3} = \overline{\mu}.$$

It is estimated that

$$S(P_{-2,-3,-1.5}) = S(P_{0,3,1.5}) = S(P_{1,6,3}) = 0 < S(P_{0,1.5,1.5})$$

$$= 0.5 < S(P_{0,1,1}) = 0.88889 < S(P_{0,1,0}) = 0.94281$$

Example 4.2 For the matrix $A = \begin{bmatrix} 1 & 0 & \frac{1}{5} & \frac{1}{5} \\ 0 & 1 & -\frac{71}{10} & \frac{113}{10} \\ \frac{16}{5} & \frac{1}{5} & 1 & 0 \\ 2 & \frac{1}{5} & 0 & 1 \end{bmatrix}$ considered by Avdelas and Hadjimos [2], $\underline{\mu} = \frac{\sqrt{23}}{5}$, $\overline{\mu} = \frac{\sqrt{24}}{5}$. It is calculated from the iteration matrices that

$$S\left(P_{0,2.68666667,\frac{5}{3}}\right) = 0.56893 < S\left(P_{0,\frac{5}{3},\frac{5}{3}}\right) = \frac{2}{3} < S(P_{0,1,1}) = 0.96 < S(P_{0,1,0}) = 0.9798$$

It is interesting to note that $S\left(P_{0,-\frac{5}{4},\frac{5}{3}}\right)$ happens to be 1.30703262 but not $\frac{\sqrt{46}}{12}$ as mentioned by Avdelas and Hadjimos [2].

5 Conclusion

It is observed in many examples including the above two that spectral radius of PAOR method lesser than the other methods for any $\alpha \neq -1$ and hence this method plays a major role in the rate of convergence over the other methods considered in this paper.

References

1. Hadjidimos, A.: Accelerated over relaxation method. Math. Comput. **32**(141), 149–157 (1978). https://doi.org/10.1090/S0025-5718-1978-0483340-6
2. Avdelas, G., Hadjidimos, A.: Optimum accelerated overrelaxation method in a special case. Math. Comput. **36**(153), 183–187 (1981). https://doi.org/10.1090/S0025-5718-1981-0595050-5
3. Varga, R.S.: Iterative Analysis. Prentice Hall, Englewood Cliffs, NJ (1962)
4. Varga, R.S.: Iterative solution of elliptic systems and applications to the neutron diffusion equations of reactor physics (Eugene L. Wachspress). SIAM Rev. **9**(4), 756–758 (1967)
5. Youssef, I.K., Taha, A.A.: On the modified successive overrelaxation method. Appl. Math. Comput. **219**(9), 4601–4613 (2013). https://doi.org/10.1016/j.amc.2012.10.071

Solving Multi-choice Fractional Stochastic Transportation Problem Involving Newton's Divided Difference Interpolation

Prachi Agrawal and Talari Ganesh

Abstract There are limited numbers of methods to choose the optimal choice among the multiple choices. A numerical technique named Newton's Divided Difference Interpolation is used to find out the solution of multi-choice fractional stochastic transportation problem. Because of the uncertainty, the parameters of the problem supplies and demands are considered as multi-choice random parameters which are treated as independent random variables follows Logistic distribution. Also, the coefficients of the decision variables in the fractional objective function are taken as multi-choice type. To get the deterministic model, chance constrained programming is applied to the probabilistic constraints, and the transformed mathematical model is presented. An illustration illustrates the methodology and also solved using Lagrange's Interpolation.

Keywords Fractional programming · Multi-choice random parameter · Stochastic programming · Transportation problem

1 Introduction

Business and industry are practically faced with economic optimization such as cost minimization, profit maximization. The transportation problem is concerned with the optimal way in which a homogeneous product is produced at different production houses (supplies locations) and distributed to a number of stores (destinations).

In day to day, the decision maker has to face many difficulties/challenges to find the optimal solution of the problem which occurs because of the vagueness and multiple choices. To solve the multi-choice programming, Chang [1] introduced the concept of binary variables and revised the model in which the binary variables has

P. Agrawal (✉) · T. Ganesh
Department of Mathematics and Scientific Computing,
National Institute of Technology Hamirpur, Hamirpur,
Himachal Pradesh 177005, India
e-mail: prachiagrawal202@gmail.com

D. Dutta and B. Mahanty (eds.), *Numerical Optimization in Engineering and Sciences*,
Advances in Intelligent Systems and Computing 979,
https://doi.org/10.1007/978-981-15-3215-3_28

replaced by continuous variables. Mahapatra et al. [2] developed a model for multi-choice stochastic transportation problem involving extreme value distribution using binary variables. To optimize one or more ratios of the function, the fractional programming problem has been widely used. The extension of fractional programming in transportation problem is proposed by Swarup [3], which plays an important role in logistics and supply management.

Most of the studies considered with deterministic cases, in which the parameters are precisely known to the decision maker. But in real life applications, it is not always possible to find the exact value of the parameters. To tackle with the situation of uncertainty and impreciseness in making decisions, Zadeh [4] introduced the fuzzy theory. Liu [5] solved the fractional transportation problem in which all the parameters are considered as triangular fuzzy numbers. Also, fuzzy programming is used to convert the multi-objective function into a single objective function by Mahapatra et al. [6].

A computation method explored by Acharya et al. [7] for solving the fuzzy stochastic transportation problem. A technique which includes normal randomness is used to convert the stochastic transportation problem into a deterministic. To handle the situation of the multi-choice parameter in the transportation problem, Roy [8] proposed Lagrange's interpolating polynomial to convert the multi-choice parameter into a single choice. Pradhan et al. [9] presented a linear programming model in which each alternative of multi-choice parameters considered as a random variable and used the different type of optimization models (V-model, fractile criterion model, probability maximization model, and E-model).

In this paper, a fractional stochastic transportation problem (MCFSTP) is considered in which the objective function is in the ratio such that the cost and the corresponding profit are in multi-choice type. The supplies and demands are taken as multi-choice random parameters and each alternative of multi-choice parameter considered as an independent random variable which follows Logistic distribution. The detailed problem statement is presented in Sect. 2 and the solution procedure of the presented problem is shown in Sects. 3 and 4 contains the concluding remarks.

2 Problem Statement

A transportation company wants to ship its product from different production houses to various retail stores. Assuming there are p number of production houses and q number of retail stores and a homogeneous product is transferred from kth production house to tth retail store. Let x_{kt} denotes the number of units of the product. Due to the uncertainty and the different choices in the environment, the values of the parameters are not always fixed therefore, the supplies and the demands are considered as multi-choice random parameters. Hence, the constraints of the formulated problem are in a probabilistic manner with their aspiration level. The objective function is in the ratio form, which represents the transportation cost associated with the obtained profit per unit of the product from kth production house to tth retail store. Because of the various choices for the transportation and other factors, the transportation cost and their profit are assumed as multi-choice type, so that the mathematical formulation of the said problem stated as

$$\min\ Z = \left\{ \frac{\sum_{k=1}^{p} \sum_{t=1}^{q} (c_{kt}^{1}, c_{kt}^{2}, \ldots, c_{kt}^{v})\, x_{kt} + \xi}{\sum_{k=1}^{p} \sum_{t=1}^{q} (r_{kt}^{1}, r_{kt}^{2}, \ldots, r_{kt}^{v})\, x_{kt} + \eta} \right\} \tag{1}$$

$$s.t.\ P\left(\sum_{t=1}^{q} x_{kt} \leq (S_k^1, S_k^2, \ldots, S_k^m) \right) \geq 1 - \theta_k \quad k = 1, 2, \ldots, p \tag{2}$$

$$P\left(\sum_{k=1}^{p} x_{kt} \geq (R_t^1, R_t^2, \ldots, R_t^n) \right) \geq 1 - \sigma_t \quad t = 1, 2, \ldots, q \tag{3}$$

$$x_{kt} \geq 0 \quad \text{for every } k,\ t \tag{4}$$

where, $(S_k^1, S_k^2, \ldots, S_k^m)$ are the multi-choice random parameters for the total availability S_k at the kth production house, considered as independent random variable. $(R_t^1, R_t^2, \ldots, R_t^n)$ are the multi-choice random parameters for the total requirement R_t of the product at the tth retail store which considered as independent random variable. θ_k and σ_t represent the probability of meeting the constraints and η and ξ are the fixed costs in the objective function.

3 Solution Methodology

This section presents the solution methodology to solve the multi choice fractional stochastic transportation problem. Firstly, the Newton's Divided Difference Interpolation is used to convert the multi-choice parameter and then the probabilistic constraints converted into its deterministic form using chance constrained technique.

3.1 Newton's Divided Difference Interpolating Polynomial for Multi-choice Parameters

To transform the multi-choice parameter into its optimal choice, the numerical approximation technique Newton's Divided Difference Interpolation is used. For each alternative of multi-choice parameter, introducing an integer variable such that the interpolating polynomial is formulated. Since there are v number of alternatives to the cost and profit parameters in the above problem, the integer variables $w_{c_{kt}}^{i}$ $(i = 0, 1, \ldots, v-1)$, $w_{r_{kt}}^{i}$ $(i = 0, 1, \ldots, v-1)$ introduced respectively.

Being supplies and demands are as multi-choice random parameters, the integer variables $w_{S_k}^{j}$ $(j = 0, 1, \ldots, m-1)$, $w_{R_t}^{u}$ $(u = 0, 1, \ldots, n-1)$ to each alternative

Table 1 Divided difference (DD) table

$w^i_{c_{kt}}$	$F_{c_{kt}}(w^i_{c_{kt}})$	First DD	Second DD	Third DD
0	c^1_{kt}			
		$f[w^0_{c_{kt}}, w^1_{c_{kt}}]$		
1	c^2_{kt}		$f[w^0_{c_{kt}}, w^1_{c_{kt}}, w^2_{c_{kt}}]$	
		$f[w^1_{c_{kt}}, w^2_{c_{kt}}]$		$f[w^0_{c_{kt}}, w^1_{c_{kt}}, w^2_{c_{kt}}, w^3_{c_{kt}}]$
2	c^3_{kt}		$f[w^1_{c_{kt}}, w^2_{c_{kt}}, w^3_{c_{kt}}]$	
		$f[w^2_{c_{kt}}, w^3_{c_{kt}}]$		
3	c^4_{kt}			

is introduced. Different order of divided difference calculated according to alternatives for each multi-choice parameter. Table 1 shows the different order of divided difference therefore using Table 1, the Newton's Divided Difference interpolating polynomial is formulated for the cost parameter in Eq. (5).

$$\begin{aligned} F_{c_{kt}}(w_{c_{kt}}) = f[w^0_{c_{kt}}] + (w_{c_{kt}} - w^0_{c_{kt}})\, f[w^0_{c_{kt}}, w^1_{c_{kt}}] + (w_{c_{kt}} - w^0_{c_{kt}}) \\ (w_{c_{kt}} - w^1_{c_{kt}})\, f[w^0_{c_{kt}}, w^1_{c_{kt}}, w^2_{c_{kt}}] + \cdots + (w_{c_{kt}} - w^0_{c_{kt}}) \\ (w_{c_{kt}} - w^1_{c_{kt}}) \ldots (w_{c_{kt}} - w^{v-1}_{c_{kt}})\, f[w^0_{c_{kt}}, w^1_{c_{kt}}, \ldots, w^{v-1}_{c_{kt}}] \end{aligned} \tag{5}$$

$$\begin{aligned} F_{c_{kt}}(w_{c_{kt}}) = c^1_{kt} + (w_{c_{kt}} - w^0_{c_{kt}})\,(c^2_{kt} - c^1_{kt}) + (w_{c_{kt}} - w^0_{c_{kt}})(w_{c_{kt}} - w^1_{c_{kt}}) \\ \left(\frac{c^3_{kt} - 2c^2_{kt} + c^1_{kt}}{w^2_{c_{kt}} - w^0_{c_{kt}}}\right) + \cdots + \left(\sum_{i=1}^{v} \frac{c^{(i)}_{kt}}{\prod_{i \neq j+1, j=0}^{v-1}(w^{i-1}_{c_{kt}} - w^j_{c_{kt}})}\right) \end{aligned} \tag{6}$$

In the similar way, the interpolating polynomials formulated for the profit, supply and demand which denoted by $F_{r_{kt}}(w_{r_{kt}})$, $F_{S_k}(w_{S_k})$, $F_{R_t}(w_{R_t})$ and the multi-choice parameters replaced by its interpolating polynomial in the problem, then the mathematical model can be formulated as

$$\min\ Z = \left\{ \frac{\sum_{k=1}^{p} \sum_{t=1}^{q} F_{c_{kt}}(w_{c_{kt}})\, x_{kt} + \xi}{\sum_{k=1}^{p} \sum_{t=1}^{q} F_{r_{kt}}(w_{r_{kt}})\, x_{kt} + \eta} \right\} \tag{7}$$

$$\text{s.t. } P\left(\sum_{t=1}^{q} x_{kt} \leq F_{S_k}(w_{S_k})\right) \geq 1 - \theta_k \quad k = 1, 2, \ldots, p \tag{8}$$

$$P\left(\sum_{k=1}^{p} x_{kt} \geq F_{R_t}(w_{R_t})\right) \geq 1 - \sigma_t \quad t = 1, 2, \ldots, q \tag{9}$$

$$x_{kt} \geq 0 \quad \text{for every } k,\ t \tag{10}$$

3.2 Conversion of Probabilistic Constraints

The multi-choice parameters converted into their interpolating polynomial so that the obtained probabilistic constraints will convert into its deterministic form. We consider the supplies probabilistic constraints to covert its deterministic one.

Consider the constraint (8) for every $k = 1, 2, \ldots, p$,

$$P\left(\sum_{t=1}^{q} x_{kt} \leq F_{S_k}(w_{S_k})\right) \geq 1 - \theta_k \tag{11}$$

$$\Longrightarrow 1 - P\left(F_{S_k}(w_{S_k}) \leq \sum_{t=1}^{q} x_{kt}\right) \geq 1 - \theta_k \tag{12}$$

Applying Chance constrained technique, this implies

$$\Longrightarrow P\left(\frac{F_{S_k}(w_{S_k}) - E(F_{S_k}(w_{S_k}))}{\sqrt{V(F_{S_k}(w_{S_k}))}} \leq \frac{\sum_{t=1}^{q} x_{kt} - E(F_{S_k}(w_{S_k}))}{\sqrt{V(F_{S_k}(w_{S_k}))}}\right) \leq \theta_k \tag{13}$$

$$\Longrightarrow P\left(\zeta_{S_k} \leq \frac{\sum_{t=1}^{q} x_{kt} - E(F_{S_k}(w_{S_k}))}{\sqrt{V(F_{S_k}(w_{S_k}))}}\right) \leq \theta_k \tag{14}$$

$$\Longrightarrow \phi\left(\frac{\sum_{t=1}^{q} x_{kt} - E(F_{S_k}(w_{S_k}))}{\sqrt{V(F_{S_k}(w_{S_k}))}}\right) \leq \phi(-g_{\theta_k}) \tag{15}$$

$$\Longrightarrow \frac{\sum_{t=1}^{q} x_{kt} - E(F_{S_k}(w_{S_k}))}{\sqrt{V(F_{S_k}(w_{S_k}))}} \leq -g_{\theta_k} \tag{16}$$

$$\Longrightarrow \sum_{t=1}^{q} x_{kt} \leq E(F_{S_k}(w_{S_k})) - g_{\theta_k} \sqrt{V(F_{S_k}(w_{S_k}))} \tag{17}$$

where, $E(F_{S_k}(w_{S_k}))$, $V(F_{S_k}(w_{S_k}))$ denote the mean and the variance of interpolating polynomial $F_{S_k}(w_{S_k})$ respectively. Also, ϕ be the cumulative distribution function of Standard Normal distribution and g_{θ_k} denotes the value of standard normal variable. Hence Eq. (17) represents the deterministic constraint of the probabilistic constraint (8).

Likewise, applying the same procedure to the demands constraints (9), the equivalent deterministic constraint for every $t = 1, 2, \ldots, q$ is as

$$\sum_{k=1}^{p} x_{kt} \leq E(F_{R_t}(w_{R_t})) + g_{\sigma_t} \sqrt{V(F_{R_t}(w_{R_t}))} \tag{18}$$

where, $E(F_{R_t}(w_{R_t}))$, $V(F_{R_t}(w_{R_t}))$ denote the mean and the variance of the $F_{R_t}(w_{R_t})$ respectively and g_{σ_t} denotes the value of standard normal variable. We calculate mean and the variance of the random interpolating polynomial as

$$E(F_{S_k}(w_{S_k})) = E\Bigg\{S_k^1 + (w_{S_k} - w_{S_k}^0)\,(S_k^2 - S_k^1) + (w_{S_k} - w_{S_k}^0)(w_{S_k} - w_{S_k}^1)\left(\frac{S_k^3 - 2S_k^2 + S_k^1}{w_{S_k}^2 - w_{S_k}^0}\right) + \ldots + \left(\sum_{i=1}^{v} \frac{S_k^{(i)}}{\prod_{i \neq j+1, j=0}^{v-1} (w_{S_k}^{i-1} - w_{S_k}^{j})}\right)\Bigg\} \tag{19}$$

$$= E(S_k^1) + (w_{S_k} - w_{S_k}^0)\,(E(S_k^2) - E(S_k^1)) + (w_{S_k} - w_{S_k}^0)(w_{S_k} - w_{S_k}^1)\left(\frac{E(S_k^3) - 2E(S_k^2) + E(S_k^1)}{w_{S_k}^2 - w_{S_k}^0}\right) + \ldots + \left(\sum_{i=1}^{v} \frac{E(S_k^{(i)})}{\prod_{i \neq j+1, j=0}^{v-1} (w_{S_k}^{i-1} - w_{S_k}^{j})}\right)\Bigg\} \tag{20}$$

$$V(F_{S_k}(w_{S_k})) = V\Bigg\{S_k^1 + (w_{S_k} - w_{S_k}^0)\,(S_k^2 - S_k^1) + (w_{S_k} - w_{S_k}^0)(w_{S_k} - w_{S_k}^1)\left(\frac{S_k^3 - 2S_k^2 + S_k^1}{w_{S_k}^2 - w_{S_k}^0}\right) + \ldots + \left(\sum_{i=1}^{v} \frac{S_k^{(i)}}{\prod_{i \neq j+1, j=0}^{v-1} (w_{S_k}^{i-1} - w_{S_k}^{j})}\right)\Bigg\} \tag{21}$$

$$= V(S_k^1) + (w_{S_k} - w_{S_k}^0)^2\,(V(S_k^2) + V(S_k^1)) + (w_{S_k} - w_{S_k}^0)^2(w_{S_k} - w_{S_k}^1)^2\left(\frac{V(S_k^3) + 4E(S_k^2) + V(S_k^1)}{(w_{S_k}^2 - w_{S_k}^0)^2}\right) + \ldots + \left(\sum_{i=1}^{v} \frac{V(S_k^{(i)})}{\prod_{i \neq j+1, j=0}^{v-1} (w_{S_k}^{i-1} - w_{S_k}^{j})^2}\right)\Bigg\} \tag{22}$$

Equations (20) and (22) represents the mean and the variance of the $F_{S_k}(w_{S_k})$. In the similar way, the mean and the variance of the interpolating polynomial for the demands can be calculated using Eqs. (20) and (22).

Deterministic Model: Using Newton's Divided Difference Interpolation and chance constrained technique, the deterministic model is

$$\min\ Z = \left\{\frac{\sum_{k=1}^{p}\sum_{t=1}^{q} F_{c_{kt}}(w_{c_{kt}})\, x_{kt} + \xi}{\sum_{k=1}^{p}\sum_{t=1}^{q} F_{r_{kt}}(w_{r_{kt}})\, x_{kt} + \eta}\right\} \tag{23}$$

$$\text{s.t.} \sum_{t=1}^{q} x_{kt} \le E(F_{S_k}(w_{S_k})) - g_{\theta_k}\sqrt{V(F_{S_k}(w_{S_k}))} \quad k = 1, 2, \ldots, p \tag{24}$$

$$\sum_{k=1}^{p} x_{kt} \le E(F_{R_t}(w_{R_t})) + g_{\sigma_t}\sqrt{V(F_{R_t}(w_{R_t}))} \quad t = 1, 2, \ldots, q \tag{25}$$

$$x_{kt} \ge 0 \quad \text{for every } k,\ t \tag{26}$$

4 Numerical Example

To better understand the example, we present the hypothetical example which illustrates the methodology. We modeled the problem and solved it using solution procedure.

There are three types of almonds available in the three different origins. Origin 1, 2 and 3 contains Mamra, Gurbandi and Californian respectively. The amount of content (fat, protein and cost production) for different types of almonds is different. We assume that, the cost of raw material, fat content and protein content of each type of almonds are random variables which follow normal distribution. The almonds should be transported from each origin to each destination such that, the total requirement must be fulfilled. Also, the requirement of the each type of almonds is considered as random variable due to fluctuations in the sale of each type of almonds. The main aim is to fulfill the requirement of the demand locations according to the availability of the product and minimize the objective function.The following matrix represents the cost matrix in which first element [16, 19, 20] indicates that x_{11} route either 16 or 19 or 20 required cost.

$$\begin{array}{c} \\ V_1 \\ V_2 \\ V_3 \end{array} \begin{array}{c} \begin{array}{ccc} M_1 & M_2 & M_3 \end{array} \\ \begin{pmatrix} [16, 19, 20] & [18, 19, 21] & [12, 16, 18, 19] \\ [20, 21, 23, 26, 29] & [13, 16, 18] & [13, 17] \\ [15, 16] & [20, 21, 22, 25] & [14, 16, 17] \end{pmatrix} \end{array}$$

The following matrix presents the profit matrix from first vendor to first destination.

$$\begin{array}{c} \\ V_1 \\ V_2 \\ V_3 \end{array} \begin{array}{c} \begin{array}{ccc} M_1 & M_2 & M_3 \end{array} \\ \begin{pmatrix} [10, 12, 13] & [14, 16, 17] & [20, 21, 22, 25] \\ [13, 16, 18] & [20, 21, 23, 26, 29] & [13, 17] \\ [14, 16] & [14, 17, 19, 22] & [18, 19, 21] \end{pmatrix} \end{array}$$

The data which is required for the model is presented in Tables 2 and 3.

Using the methodology and data, the deterministic model can be expressed as follows:

Table 2 Data for almonds

Almonds	Avg. Production cost	Variance of cost	Mean of fat	Variance of fat	Mean of protein	Variance of protein	Significance level
Mamra	13.673	0.7712	11.28	0.194	5.2	0.25	0.89
Gurbandi	13.64764	0.489293	10.1	0.17	0.24	0.24	0.97
Californian	13.07041	0.74055	0.4	0.15	0.2	0.2	0.96

Table 3 For demand: mean, variance and their significance level

Random variable	Mean	Variance	Probabilities
b_1	10	3	0.15
b_2	4	2	0.20
b_3	5	1	0.25

$$\min Z = (16 + 3w_{11} - w_{11}(w_{11} - 1))x_{11} + (18 + w_{12} + 0.5w_{12}(w_{12} - 1))x_{12} + (12+ 4w_{13} - w_{13}(w_{13} - 1) + (1/6)w_{13}(w_{13} - 1)(w_{13} - 2))x_{13} + (20 + w_{21} + 0.5w_{21} (w_{21} - 1) - (1/24)w_{21}(w_{21} - 1)(w_{21} - 2)(w_{21} - 3))x_{21} + (13 + 3w_{22} - 0.5w_{22} (w_{22} - 1))x_{22} + (13 + 4w_{23})x_{23} + (15 + w_{31})x_{31} + (20 + w_{32} + (1/3)w_{32}(w_{32} -1)(w_{32} - 2))x_{32} + (14 + 2w_{33} - 0.5w_{33}(w_{33} - 1)x_{33} + 700)/((10 + 2y_{11}- 0.5y_{11}(y_{11} - 1))x_{11} + (14 + 2y_{12} - 0.5y_{12}(y_{12} - 1))x_{12} + (20 + y_{13} + (1/3)y_{13} (y_{13} - 1)(y_{13} - 2))x_{13} + (13 + 3y_{21} - 0.5y_{21}(y_{21} - 1))x_{21} + (20 + y_{22} + 0.5y_{22} (y_{22} - 1) - (1/24)y_{22}(y_{22} - 1)(y_{22} - 2)(y_{22} - 3))x_{22} + (13 + 4y_{23})x_{23} + (14+ 2y_{31})x_{31} + (14 + 3y_{32} - 0.5y_{32}(y_{32} - 1)(y_{32} - 2))x_{32} + (18 + y_{33} + 0.5y_{33} (y_{33} - 1))x_{33} + 10)$$

s.to.

$$z_{11} + z_{12} + z_{13} <= 11.28 + v_1(5.2 - 11.28) + 0.5v_1(v_1 - 1)(13.673 - 2 \times 5.2 + 11.28) + \left(\sqrt{\frac{0.194 + v_1^2(0.25 + 0.194) + (1/4)(v_1^2)(v_1 - 1)^2}{(0.7712 + 4 \times 0.25 + 0.194)}}\right) \phi^{-1}(0.89)$$

$$z_{21} + z_{22} + z_{23} <= 10.1 + v_2(5 - 10.1) + 0.5v_2(v_2 - 1)(13.64764 - 2 \times 5 + 10.1) + \left(\sqrt{\frac{0.17 + v_2^2(0.24 + 0.17) + (1/4)(v_2^2)(v_2 - 1)^2}{(0.489293 + 4 \times 0.24 + 0.17)}}\right) \phi^{-1}(0.97)$$

$$z_{31} + z_{32} + z_{33} <= 9.4 + v_3(4.74 - 9.4) + 0.5v_3(v_3 - 1)(13.07041 - 2 \times 4.74 + 9.4) + \left(\sqrt{\frac{0.15 + v_3^2(0.2 + 0.15) + (1/4)v_3^2(v_3 - 1)^2}{(0.74055 + 4 \times 0.2 + 0.15)}}\right) \phi^{-1}(0.96)$$

$$z_{11} + z_{21} + z_{31} >= 10 + \sqrt{3}\ \phi^{-1}(1 - 0.15),\ \ z_{12} + z_{22} + z_{32} >= 4 + \sqrt{2}\ \phi^{-1}(1 - 0.20)$$

$$z_{13} + z_{23} + z_{33} >= 5 + \sqrt{1}\ \phi^{-1}(1 - 0.25),\quad 0 \le v_k \le 2,\ \ k = 1, 2, 3$$

$$0 \le w_{11} \le 2,\ 0 \le w_{12} \le 2,\ 0 \le w_{13} \le 3,\ 0 \le w_{21} \le 4,\ 0 \le w_{22} \le 2$$

$$0 \le w_{23} \le 1,\ 0 \le w_{31} \le 1,\ 0 \le w_{32} \le 3,\ 0 \le w_{33} \le 2$$

$$0 \leq y_{11} \leq 2,\ 0 \leq y_{12} \leq 2,\ 0 \leq y_{13} \leq 3,\ 0 \leq y_{21} \leq 2,\ 0 \leq y_{22} \leq 4$$

$$0 \leq y_{23} \leq 1,\ 0 \leq y_{31} \leq 1,\ 0 \leq y_{32} \leq 3,\ 0 \leq y_{33} \leq 2,$$

$$z_{kt} \geq 0\ k = 1, 2, 3,\ \ t = 1, 2, 3,\ \ v_k,\ w_{kt}, y_{kt} \in Z^+$$

5 Results and Discussion

The numerical examples presents the objective function in the fractional form and the constraints in which supplies are considered as multi-choice random parameter, demands is taken as random variable which follow Normal Distribution. The solution of aforementioned model is $x_{11} = 2,\ x_{13} = 10,\ x_{22} = 10,\ x_{31} = 10$, the remaining decision variables are zero and the value of objective function is $Z = 1.370460$, which are obtained using LINGO 11.0 software.

6 Conclusion

In this paper, the solution methodology of fractional stochastic transportation problem with multi-choice parameters is presented. With the help of Newton's Divided Difference Interpolation, the multi-choice parameters converted into its single choice so that the obtained solution should be optimal. Also, in the real-life situations, the decision maker may not get the exact information about the supplies and demands, therefore, it considered as a multi-choice random parameter in which each alternative is treated as independent random variables, which follow Logistic distribution. The chance constrained programming is applied to the probabilistic constraints to get the deterministic constraints. An illustration presents the methodology and the solution procedure for the proposed problem and solved using Lagrange's Interpolation [8]. The value obtained using Lagrange's Interpolation is $Z = 0.1758028$ and the values of decision variables are $x_{11} = 2,\ x_{13} = 6,\ x_{22} = 10,\ x_{31} = 10$, rest of the decision variables are zero. This fractional stochastic transportation problem plays an important role in various fields such as logistics and supply management to reduce cost and improving service.

References

1. Chang, C.T.: Multi-choice goal programming. Omega **35**(4), 389–396 (2007)
2. Mahapatra, D.R., Roy, S.K., Biswal, M.P.: Multi-choice stochastic transportation problem involving extreme value distribution. Appl. Math. Model. **37**(4), 2230–2240 (2013)
3. Swarup, K.: Transportation technique in linear fractional functional programming. J. R. Naval Sci. Ser. **21**(5), 256–260 (1966)

4. Zadeh, L.A.: Information and control. Fuzzy Sets **8**(3), 338–353 (1965)
5. Liu, S.T.: Fractional transportation problem with fuzzy parameters. Soft. Comput. **20**(9), 3629–3636 (2016)
6. Mahapatra, D.R., Roy, S.K., Biswal, M.P.: Stochastic based on multi-objective transportation problems involving normal randomness. Adv. Model. Optim. **12**(2), 205–223 (2010)
7. Acharya, S., Ranarahu, N., Dash, J.K., Acharya, M.M.: Computation of a multi-objective fuzzy stochastic transportation problem. Int. J. Fuzzy Comput. Model. **1**(2), 212–233 (2014)
8. Roy, S.K.: Lagranges interpolating polynomial approach to solve multi-choice transportation problem. Int. J. Appl. Comput. Math. **1**(4), 639–649 (2015)
9. Pradhan, A., Biswal, M.P.: Multi-choice probabilistic linear programming problem. Opsearch **54**(1), 122–142 (2017)

On Stability of Multi-quadric-Based RBF-FD Method for a Second-Order Linear Diffusion Filter

Mahipal Jetta and Satyanarayana Chirala

Abstract The usual three node approximations for the first-order spatial derivatives in the radial basis functions based finite difference (RBF-FD) scheme allow relatively smaller time step size than the standard scheme (forward difference in time and central difference in space) in obtaining stable results. In this paper, to better mimic the continuous model, we use nine nodal points in RBF-FD approximation of first-order spatial derivatives. We then employ these approximations to solve a two-dimensional linear diffusion equation. The constructed numerical scheme is shown to be stable using the maximum-minimum principle. We also show that the scheme allows four times bigger time step size than the standard scheme.

Keywords Multi-quadric RBF · Stability · Linear diffusion filter · Finite difference method

1 Introduction

The method of radial basis functions (RBF) is an important tool for the interpolation of multidimensional scattered data. Franke [4] has tested the performance of various basis functions (multi-quadric, Gaussian and Splines) for the interpolation of scattered data and concluded that the multi-quadric (MQ) basis produces the most accurate results when compared with other basis functions. This observation leads to development of accurate numerical schemes for the partial differential equations [3, 9].

In most of the existing works on RBF-FD-based numerical schemes, the spatial derivatives have been approximated with values of the underlying function at grid points on one principal direction, for example, u_x has been approximated with values of u at grid points on the X-axis. This enables the user to employ efficient

---

M. Jetta (✉) · S. Chirala
Mahindra École Centrale, Hyderabad, India
e-mail: mahipal.jetta@mechyd.ac.in

D. Dutta and B. Mahanty (eds.), *Numerical Optimization in Engineering and Sciences*, Advances in Intelligent Systems and Computing 979,
https://doi.org/10.1007/978-981-15-3215-3_29

algorithms for solving the resulting system of equations. However, these three node-based schemes suffer from heavy numerical dissipation.

To address this issue, in this paper, an RBF-FD-based scheme is employed to obtain the 3×3 stencil approximations for the first-order spatial derivatives. These stencils have been used to approximate the spatial derivatives in the linear parabolic partial differential equation, and the time derivative is discretized using the forward difference approximation. Further, we analyze the stability of the proposed scheme and show that it is stable under some restriction on the time step size. This restriction (bound) on the time step size is a function of the grid size and shape parameter of multi-quadric basis functions.

The rest of the paper is organized as follows. In Sect. 2, the RBF-FD scheme for solving unsteady partial differential equations is presented. In Sect. 3, the stability analysis of the proposed scheme for two-dimensional linear diffusion equation is carried out. Finally, Sect. 4 constitutes the concluding remarks.

2 An RBF-FD Scheme for Unsteady Problems

Consider a well-posed linear unsteady partial differential equation:

$$\frac{\partial u}{\partial t}(\mathbf{x}, t) = \mathscr{L}u(\mathbf{x}, t) \ \text{ in } \ \Omega, \tag{1}$$

$$u_0(\mathbf{x}) = u(\mathbf{x}, t = 0), \tag{2}$$

$$\frac{\partial u}{\partial n} = 0 \ \text{ on } \partial\Omega \tag{3}$$

where u is an unknown function, $\mathbf{x} \in \mathbb{R}^d$, $\mathscr{L}$ is a linear partial differential operator, and $\frac{\partial u}{\partial n}$ is a directional derivative of u in the outward normal direction n. To solve this equation, we employ RBF-FD scheme [9], which is briefly described as follows.

Let N be the total number of centers in the computational domain, and let

$$\mathbf{C}_i = \{\mathbf{x}_1, \mathbf{x}_2, \ldots, \mathbf{x}_{n_i}\}$$

be the local support of the center $\mathbf{x}_i$ having n_i neighboring centers. Then, $\mathscr{L}u(\mathbf{x}_i)$ is expressed approximately as a linear combination of u values at $n_i (\ll N)$ centers of the local support $\mathbf{C}_i$. Symbolically, this is given by

$$\mathscr{L}u(\mathbf{x}_i) \approx \sum_{j=1}^{n_i} c_j u(\mathbf{x}_j), \quad \text{for each } \ \mathbf{x}_i \in \Omega \tag{4}$$

The weights, $\mathbf{c} = \{c_j\}_{j=1}^{n_i}$ in (4), are obtained by solving the linear system,

$$A[\mathbf{c}/\gamma] = (\mathscr{L}B(\mathbf{x}_i))^T, \tag{5}$$

where $A = \begin{bmatrix} \Xi & \mathbf{1} \\ \mathbf{1}^T & 0 \end{bmatrix}$ with $\mathbf{1}^T = [1, 1, \dots 1]_{1\times n_i}$, γ is a dummy variable, and $B(\mathbf{x}) = [\phi(\|\mathbf{x}-\mathbf{x}_1\|, \varepsilon), \dots, \phi(\|\mathbf{x}-\mathbf{x}_{n_i}\|, \varepsilon), \ 0]$.

The elements of the sub matrix Ξ are given by

$$\Xi_{i,j} := \phi(\|\mathbf{x}_i - \mathbf{x}_j\|, \varepsilon), \ \ i, j = 1, 2, \dots, n_i. \tag{6}$$

Michelli [7] showed that the choice of multi-quadric basis produces a non-singular interpolation matrix (A) in Eq. (5). The last row in the matrix A is used to have the zero sum of the coefficients.

Now, the discretized form of Eq. (1), with the obtained weights c_j, reads as

$$\frac{u(\mathbf{x}_i, t_{k+1}) - u(\mathbf{x}_i, t_k)}{\Delta t} = \sum_{j=1}^{n_i} c_j u(\mathbf{x}_j, t_k) \tag{7}$$

The accuracy of this approximation depends on the choice of the underlying basis functions. It is known that the multi-quadric RBF gives better accurate results than with that of Gaussian and Splines [6]. Therefore in this work, all the computations are carried out with multi-quadric basis function, $\phi = \sqrt{1+\epsilon^2 r^2}$, where r is the Euclidean distance between the center $\mathbf{x}$ and $\mathbf{x}_i$, and ϵ is the (scaling) shape parameter of basis function. It has been realized that this parameter plays an important role in the accuracy of numerical solution [1]. Note that throughout this article we take uniorm distribution of the nodes.

3 Linear Diffusion Filter

In this section, we compare the stability of finite difference and RBF-FD schemes for a two-dimensional heat equation.

Consider a partial differential equation of the form

$$\frac{\partial u}{\partial t}(\mathbf{x}, t) = div(\nabla u(\mathbf{x}, t)), \ \ \mathbf{x} \in \Omega \subset \mathbb{R}^2 \tag{8}$$

with initial condition $u_0(\mathbf{x}) = u(\mathbf{x}, t = 0)$ and homogeneous Neumann boundary condition: $\dfrac{\partial u}{\partial n} = 0$ on $\partial\Omega$. This model has been applied to remove noise from an image [8]. The approach here is to consider the noisy image as an initial condition and evolve it according to (8). As this equation allows equal amount of diffusion in all the directions, this model is considered as an isotropic diffusion filter.

3.1 Stability of RBF-FD Scheme

We now analyze stability of multi-quadric-based RBF-FD scheme for the considered linear diffusion model (8).

Five Node Approximation Approximating the time derivative with forward difference and spatial derivatives with central differences using three nodal points in each direction, we obtain the following scheme for (8)

$$\frac{U_{i,j}^{n+1} - U_{i,j}^{n}}{\Delta t} = d_1 U_{i,j}^{n} + d_2 U_{i-1,j}^{n} + d_3 U_{i+1,j}^{n} + d_4 U_{i,j+1}^{n} + d_5 U_{i,j-1}^{n} \tag{9}$$

where Δt is the time step size and $U_{i,j}^{n}$ denotes the approximation of $u(x, y; t)$ in the pixel (i, j) at time $n\Delta t$. Here, the coefficients in Eq. (9) can be obtained by solving the local system (5) with $n_i = 5$. It can be observed that $d_2 = d_3 = d_4 = d_5 = \frac{1}{h^2} + \frac{5}{6}\epsilon^2$ and $d_1 = -\frac{4}{h^2} - \frac{10}{3}\epsilon^2$ [1]. Equation (9) can be rewritten as

$$U_{i,j}^{n+1} = d_2 \Delta t \left(U_{i-1,j}^{n} + U_{i+1,j}^{n} + U_{i,j-1}^{n} + U_{i,j+1}^{n}\right) + (1 + d_1 \Delta t) U_{i,j}^{n} \tag{10}$$

The maximum-minimum principle will hold for (10) if each coefficient of U is non-negative and sum of all the coefficients of U on the right hand side of (10) is equal to 1. These conditions enforce the restriction on time step size which is given by

$$\Delta t \le min \left\{ \frac{-1}{d_1}, \frac{-1}{d_1 + d_2} \right\}.$$

In Fig. 1, the time step size bound for (10) is plotted against the shape parameter ϵ. One can observe from this figure that the time step size bound for finite difference scheme dominates the RBF-FD scheme bound for all the values of the

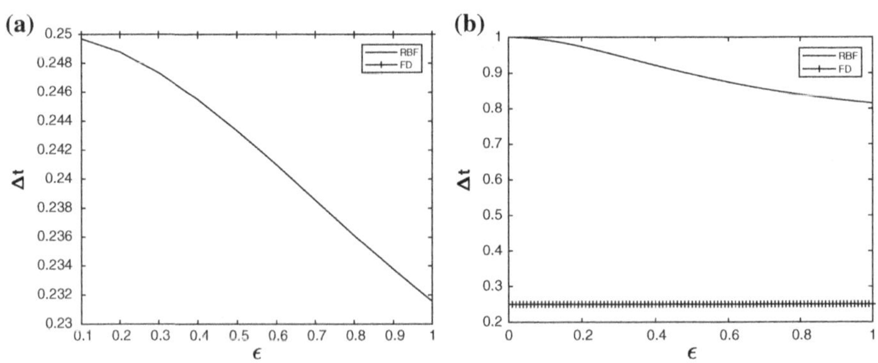

Fig. 1 Stability bound for FD and RBF-FD schemes for two-dimensional heat equation **a** Five node approximation **b** Nine node approximation

shape parameter. Therefore, this RBF-FD scheme is relatively less efficient than the standard finite difference scheme.

Nine Node Approximation To improve the efficiency of RBF-FD scheme, we consider nine nodes in the approximation of first-order spatial derivatives. Considering $n_i = 9$ in (4) and solving the resulting system give the following approximations, in stencil form, for the first-order spatial derivatives

$$u_x = \begin{pmatrix} -\alpha & 0 & \alpha \\ -\beta & 0 & \beta \\ -\alpha & 0 & \alpha \end{pmatrix}, \text{ and}$$
$$u_y = \begin{pmatrix} \alpha & \beta & \alpha \\ 0 & 0 & 0 \\ -\alpha & -\beta & -\alpha \end{pmatrix}, \tag{11}$$

where α and β are functions of the shape parameter ϵ. The approximation of second-order spatial derivatives can be obtained by consecutive application of first-order stencils. For example, the approximating stencil for u_{xx} is

$$L = \begin{pmatrix} \alpha^2 & 0 & -2\alpha^2 & 0 & \alpha^2 \\ 2\alpha\beta & 0 & -4\alpha\beta & 0 & 2\alpha\beta \\ 2\alpha^2+\beta^2 & 0 & -4\alpha^2-2\beta^2 & 0 & 2\alpha^2+\beta^2 \\ 2\alpha\beta & 0 & -4\alpha\beta & 0 & 2\alpha\beta \\ \alpha^2 & 0 & -2\alpha^2 & 0 & \alpha^2 \end{pmatrix}. \tag{12}$$

For $u_t = u_{xx}$ in the two-dimensional domain, the resulting explicit scheme will take the following form

$$\frac{U_{i,j}^{n+1} - U_{i,j}^{n}}{\Delta t} = L_{i,j} * U_{i,j}^{n} \tag{13}$$

Here, the convolution $L_{i,j} * U_{i,j}^{n}$ is a discretization of u_{xx} at the n-th time level.

The column wise arrangement of U, which can be obtained using single index $l = (i-1)N + j$ with N being the number of grid points in a principal axis direction, enables us to rewrite (13) as

$$U_l^{n+1} = (I + S)U_l^n \tag{14}$$

where I is the identity matrix and S is the sparse matrix representing the convolution of U with small convolution kernel of $\Delta t \times L$. The stability of (14) is guaranteed if spectral radius of the iteration matrix $(I + S)$ is less than 1. One may note that this condition also ensures the stability of

$$U_l^{n+1} = -(I + S)U_l^n \tag{15}$$

We rewrite (15) as

$$U_l^{n+1} + ((2I + S) - I)U_l^n = 0 \tag{16}$$

The stability of (16) is guaranteed if $(2I + S)$ is diagonally dominant [2, 5]. Considering the approximations in the inner part of the computational domain (obtained by removing two boundary rows and two boundary columns), the non-zero matrix elements of S at each row are given in the following order

$$\Delta t \times [\alpha^2 \quad -2\alpha^2 \quad \alpha^2 \quad 2\alpha\beta \quad -4\alpha\beta \quad 2\alpha\beta \quad 2\alpha^2+\beta^2 \quad -4\alpha^2-2\beta^2 \\ 2\alpha^2+\beta^2 \quad 2\alpha\beta \quad -4\alpha\beta \quad 2\alpha\beta \quad \alpha^2 \quad -2\alpha^2 \quad \alpha^2].$$

Here, the diagonal entry of S is $\Delta t(-4\alpha^2 - 2\beta^2)$. Therefore, the diagonal dominant condition will be satisfied if

$$|2 + \Delta t(-4\alpha^2 - 2\beta^2)| > \Delta t(12\alpha^2 + 2\beta^2 + 16\alpha\beta). \tag{17}$$

By simplifying this inequality, one can see that the non-negativite bound on Δt as

$$\Delta t < \frac{1}{8\alpha^2 + 2\beta^2 + 8\alpha\beta}. \tag{18}$$

This bound on time step size involves ϵ as the implicit parameter. Therefore, one can find an optimal time step size by maximizing the bound over the shape parameter ϵ.

Note that, in general, the bound on Δt is bigger for the rows corresponding to boundary nodal points than the internal nodal points [5].

It is easy to extend these ideas to two-dimensional linear diffusion equation, and one can obtain the time step size bound on Δt as

$$\Delta t < \frac{1}{16\alpha^2 + 4\beta^2 + 16\alpha\beta}. \tag{19}$$

4 Conclusion

In this paper, we have analyzed multi-quadric-based RBF-FD scheme for two-dimensional linear diffusion equation. It has been shown that the stability of RBF-FD scheme, derived using three node approximation for first-order spatial derivatives, is guaranteed for a relatively smaller time step size than that of the standard finite difference scheme.

We have improved the time step size bound for RBF-FD scheme by considering the nine nodes in the approximation of first-order spatial derivatives. We then analyzed the stability of resulting 5×5 stencil scheme using the maximum-minimum principle for a two-dimensional parabolic equation and shown that this scheme allows four to five times larger time step size than the standard finite difference scheme. In our future work, we wish to find an optimal shape parameter in the RBF-FD method for the time-dependent nonlinear partial differential equations, appearing in image processing.

References

1. Bayona, V., Moscoso, M., Carretero, M., Kindein, M.: RBF-FD formulas and convergence properties. J. Comput. Phys. **229**, 8281–8295 (2010). https://doi.org/10.1016/j.jcp.2010.07.008
2. Bulirsch, R., Stoer, J.: Introduction to Numerical Analysis. Springer, Heidelberg (2002)
3. Chandini, G., Sanyasiraju, Y.V.S.S.: Local RBF-FD solutions for steady convection-diffusion problems. Int. J. Numer. Meth. Eng. **72**, 352–378 (2007). https://doi.org/10.1002/nme.2024
4. Franke, R.: Scattered data interpolation: test of some methods. Math. Comput. **48**, 181–200 (1982)
5. Hajiaboli, M.R.: An anisotropic fourth-order diffusion filter for image noise removal. Int. J. Comput. Vision **92**(2), 177–191 (2011)
6. Kansa, E.J., Carlson, R.: Improved accuracy of multiquadric interpolation using variable shape parameters. Comput. Math. Appl. **24**(12), 99–120 (1992). https://doi.org/10.1016/0898-1221(92)90174-G
7. Michelli, C.A.: Interpolation of scattered data: disctance matrices and conditionally positive definite functions. Constr. Approximation **2**, 11–22 (1986)
8. Weickert, J.: Anisotropic Diffusion in Image Processing, vol. 1. Teubner, Stuttgart (1998)
9. Wright, G., Fornberg, B.: Scattered node compact finite difference-type formulas generated from radial basis functions. J. Comput. Phys. **212**, 99–123 (2006). https://doi.org/10.1016/j.jcp.2005.05.030

Portfolio Optimization Using Particle Swarm Optimization and Invasive Weed Optimization

Pulak Swain and Akshay Kumar Ojha

Abstract The problem of efficient asset allocation is an important part of financial decision-making, and it can be done by the use of portfolio optimization. In this study, we considered mean–variance and mean–semivariance portfolio models. Our aim is to find an efficient portfolio with minimum risk for a fixed lower bound of return. We solved those problems by particle swarm optimization (PSO) and invasive weed optimization (IWO). The results show that IWO produces lower risks and higher return values as compared to PSO.

Keywords Portfolio optimization · Particle swarm optimization · Invasive weed optimization

1 Introduction

Portfolio optimization is based on the process of the best allocation of wealth among several assets. This has been one of the top research areas since Harry Markowitz developed the modern portfolio theory [1, 2]. Portfolio models are mainly concerned with an investment with two criteria: expected return and risk. An investor always prefers to make the investment in such a way that he gets the maximum return from the portfolio with a minimum risk. For this, Markowitz proposed a mean–variance optimization model by considering the expected return as the reward and variance of return as the risk measure. Over the years, the study has been extended into multiperiod portfolio problems, and the introduction of uncertainty factor has taken the portfolio theory into another level [3, 4]. Many heuristic methods such as genetic algorithm [5], tabu search, PSO have been used to solve portfolio problems [6–8]. IWO is a new heuristic method in optimization, and nowadays, it is being used to

P. Swain (✉) · A. K. Ojha
Indian Institute of Technology Bhubaneswar, Bhubaneswar, India
e-mail: ps28@iitbbs.ac.in

A. K. Ojha
e-mail: akojha@iitbbs.ac.in

D. Dutta and B. Mahanty (eds.), *Numerical Optimization in Engineering and Sciences*, Advances in Intelligent Systems and Computing 979,
https://doi.org/10.1007/978-981-15-3215-3_30

solve various optimization problems. This motivates us to see that how it works for portfolio problems.

The main contributions of the study are given by:

(i) Mean–variance and mean–semivariance portfolio models are formulated from the stock market return data.
(ii) The portfolio models have been solved by two heuristic methods—PSO and IWO. A comparative analysis has been done on the results obtained from the two methods to know the better-suited method for such portfolio problems.

The rest of the paper is organized as follows. In Sect. 2, we have discussed on portfolio optimization and the two methods—PSO and IWO. Section 3 presents the results of portfolio problems by using these methods, and a comparison is presented between the results of these two methods. Section 4 incorporates the concluding remarks and the future scope of research.

2 Preliminaries

2.1 *Risk–Return Portfolio Analysis*

Consider n assets $S_j (j = 1, 2, \ldots n)$ with returns $R_j (j = 1, 2, \ldots n)$, respectively. Let μ_i denotes the expected return of asset S_i and σ_{ij} is the covariance of returns between assets S_i and S_j. A portfolio is represented by an n-dimensional vector $\boldsymbol{x} = [x_1 \ldots x_n]^T$, where x_i is the proportion of the total fund invested in asset S_i. Let $\boldsymbol{\mu}$ be the vector containing all μ_i's and $\boldsymbol{\Sigma}$ be the $n \times n$ covariance matrix containing σ_{ij} values. Then the expected return and the variance of resulting portfolio are, respectively, given by,

$$E[\boldsymbol{x}] = \sum_i x_i \mu_i = \boldsymbol{\mu}^\top \boldsymbol{x}, \qquad Var[\boldsymbol{x}] = \sum_{i,j} x_i x_j \sigma_{ij} = \boldsymbol{x}^\top \boldsymbol{\Sigma} \boldsymbol{x} \tag{1}$$

The Markowitz mean–variance model can be formulated as follows [9]:

$$\min \quad \frac{1}{2} \sum_{i=1}^{n} \sum_{j=1}^{n} x_i x_j \sigma_{ij} \tag{2}$$

$$\text{s.t}: \quad \sum_{i=1}^{n} x_i \mu_i \geq t, \quad \sum_{i=1}^{n} x_i = 1, \quad x_i \geq 0 \tag{3}$$

The above problem is a quadratic optimization problem. Here the constant $\frac{1}{2}$ is multiplied in the objective function for convenience in optimality conditions. The objective function minimizes the total risk associated with the portfolio (variance of portfolio return) subject to the expected return of the portfolio being above some target return value t. However for non-normal return data, some other risk measures, such as *lower semiabsolute deviation, lower semivariance, value at risk, conditional value at risk*, can be used instead of variance [10, 11].

Now let $R(\boldsymbol{x})$ be the rate of return of the portfolio given by $\sum_{j=1}^{n} R_j x_j$ and $\mu(\boldsymbol{x}) = \sum_{j=1}^{n} \mu_j x_j$ be the expected value of $R(\boldsymbol{x})$. Then the lower semivariance of the portfolio is defined as:

$$V_{-}(\boldsymbol{x}) = E(|R(\boldsymbol{x}) - \mu(\boldsymbol{x})|_{-}^{2}) \tag{4}$$

where $|R(\boldsymbol{x}) - \mu(\boldsymbol{x})|_{-} = -\min\{0, [R(\boldsymbol{x}) - \mu(\boldsymbol{x})]\}$. That means the lower semivariance considers only those return values which are below the expected return.

2.2 *Particle Swarm Optimization*

Particle swarm optimization (PSO) [12, 13] is based on the behavior of group of birds while searching for food. Here, a randomly distributed set of particles (potential solution) is chosen, and the PSO algorithm tries to improve the solution in each iteration.

1. Suppose there are N number of particles $i = 1, 2, \ldots, N$ moving in the n-dimensional search space. Let the position vector and the velocity vector of the ith particle are $x_i = (x_{i1}, x_{i2}, \ldots, x_{in})$ and $v_i = (v_{i1}, v_{i2}, \ldots, v_{in})$, respectively.
2. Each particle i in the swarm has a personal best $p_{best_i} = (p_{i1}, p_{i2}, \ldots, p_{in})$, which is the best position; the particle i has visited since the first time step. And the whole swarm has a global best $g_{best} = (g_1, g_2, \ldots, g_n)$ which is the best position found by the swarm.
3. In each iteration for $i = 1, 2, \ldots, N$ and $j = 1, 2, \ldots, n$ the velocity of each particle is updated by the formula:

$$v_{ij}(t+1) = wv_{ij}(t) + c_1 r_1 [p_{ij}(t) - x_{ij}(t)] + c_2 r_2 [g_j(t) - x_{ij}(t)] \tag{5}$$

   where w is the weight vector for the velocity component, c_1 and c_2 are the learning factors, r_1 and r_2 are random numbers whose values lie in between 0 and 1, t is the iteration number.
4. And the particle's position for the next iteration is changed by the formula:

$$x_{ij}(t+1) = x_{ij}(t) + v_{ij}(t+1) \tag{6}$$

5. The process is repeated until the terminating condition is achieved, and we get the optimal solution of the problem.

2.3 Invasive Weed Optimization

Invasive weed optimization (IWO) [13, 14] is based on the behavior of weeds. The IWO method is described as follows:

1. *Initialization*: An initial population is randomly generated with n number of weeds in search space.
2. *Reproduction*: Every weed produces a number of seeds depending on its fitness value. Let $S_{\min}$ and $S_{\max}$ be the number of seeds produced by the weakest and strongest weed, respectively.
3. *Dispersion of seeds*: The produced seeds are randomly distributed over the search area with zero mean and standard deviation given by,

$$\sigma_{\text{iter}} = \left(\frac{\text{iter}_{\max} - \text{iter}}{\text{iter}_{\max}} \right)^{n} (\sigma_{\text{initial}} - \sigma_{\text{final}}) + \sigma_{\text{final}} \tag{7}$$

4. *Elimination*: When the number of plants exceeds a maximum limit ($P_{\max}$), the weed with worst fitness value is removed from the colony.
5. The process continues till the maximum number of iterations is reached.

3 Results and Discussions

Here we have considered three portfolio problems. In *Problem-1*, the closing price data of three stocks IBM (IBM), WMT (Walmart), and SEHI (Southern Electric) has been taken from November 2000 to November 2001. Our aim is to build a portfolio with these assets which will give a return of at least 0.05 with a minimum risk. Let the weights associated with these three stocks be w_1, w_2, and w_3, respectively. After finding the expected return vector and the covariance matrix of these assets, we formulated the mean–variance portfolio model as:

$$\begin{aligned} \min \quad & \frac{1}{2}\{0.017092333w_1^2 + 0.005878743w_2^2 + 0.062948688w_3^2 + \\ & 2(0.0033025w_1w_2) + 2(0.004493854w_1w_3) + 2(0.004493854w_2w_3)\} \\ \text{s.t}: \quad & 0.026w_1 + 0.008083333w_2 + 0.07375w_3 \geq 0.05, \quad \sum_{i=1}^{3} w_i = 1, \quad w_i \geq 0 \end{aligned}$$

In *Problem-2*, we have taken the daily return data of three stocks AAPL (Apple Inc.), BAC (Bank of America Corp), TEVA (Teva Pharmaceutical Industries) for the year 2018 from www.kaggle.com. Let their weights in the portfolio be x_1, x_2, and x_3, respectively. Here the lower bound of the expected portfolio return is 0.001. So the mean–variance model for this problem is given by:

$$\min \quad \frac{1}{2}\{0.000209147x_1^2 + 2.14E - 04x_2^2 + 6.57E - 04x_3^2 + 2(9.73E - 05x_1x_2) + 2(8.63E - 05x_1x_3) + 2(1.54E - 04x_2x_3)\}$$

$$\text{s.t}: \quad 0.00145648x_1 + 0.00018374x_2 + 0.00068512x_3 \geq 0.001, \sum_{i=1}^{3} x_i = 1, x_i \geq 0$$

In *Problem-3*, we formulated the mean–semivariance model by using the same return data which was used in *Problem-2*. The problem is given by:

$$\min \quad \frac{1}{2}\{6.97E - 05x_1^2 + 9.37E - 05x_2^2 + 0.000255371x_3^2 + 2(3.43E - 05x_1x_2) + 2(3.22E - 05x_1x_3) + 2(6.21E - 05x_2x_3)\}$$

$$\text{s.t}: \quad 0.00145648x_1 + 0.00018374x_2 + 0.00068512x_3 \geq 0.001, \sum_{i=1}^{3} x_i = 1, x_i \geq 0$$

We have solved these three problems using PSO and IWO (taking the initial solution as $w_1 = 0, w_2 = 0, w_3 = 1$ for *Problem-1* and $x_1 = 1, x_2 = 0, x_3 = 0$ for *Problems-2 and 3*). The parameter settings for the two methods were as follows: For PSO: particle number $(N) = 20$, maximum iterations $(\text{iter}_{\max}) = 300$, cognition learning factor $(c_1) = 2$, social learning factor $(c_2) = 2$, inertia weight $(w) = 0.6$. For IWO: initial population size $(n) = 10$, maximum population size $(P_{\max}) = 25$, maximum iterations $(\text{iter}_{\max}) = 300$, minimum seeds $(S_{\min}) = 0$, maximum seeds $(S_{\max}) = 5$, initial standard deviation $(\sigma_{\text{initial}}) = 0.5$, final standard deviation $(\sigma_{\text{final}}) = 0.001$.
We made a comparison of the results in Tables 1, 2, and 3. The convergence characteristic curves of the two methods are shown in Figs. 1, 2, and 3.

Table 1 Comparison of results for Problem-1

	PSO	IWO
Optimal portfolio weights	[0 0.35816 0.63696]	[0.50057 0.0054544 0.49869]
Portfolio risk	0.014183	0.0103018
Portfolio return	0.0500000	0.0499615
No. of iterations for convergence	34	245

Table 2 Comparison of results for Problem-2

	PSO	IWO
Optimal portfolio weights	[0.642273 0.352727 0]	[0.613040 0.305270 0.078025]
Portfolio risk	7.851610E−5	7.728261E−05
Portfolio return	0.001000265	0.001002425
No. of iterations for convergence	112	30

Table 3 Comparison of results for Problem-3

	PSO	IWO
Optimal portfolio weights	[0.621166 0.3064053 0.067430]	[0.62671 0.30367 0.061954]
Portfolio risk	2.75858393E−05	2.74445599E−05
Portfolio return	0.00100721	0.00101103
No. of iterations for convergence	21	174

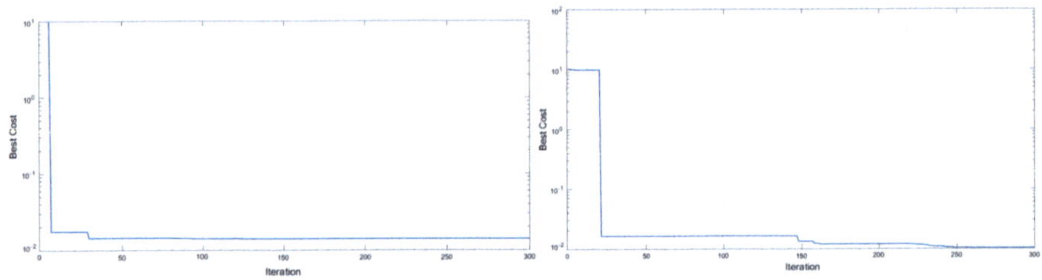

Fig. 1 Convergence characteristic curves of PSO (left) and IWO (right) for Problem-1

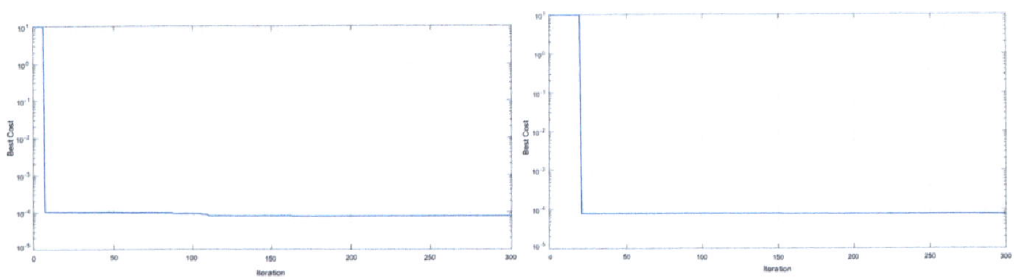

Fig. 2 Convergence characteristic curves of PSO (left) and IWO (right) for Problem-2

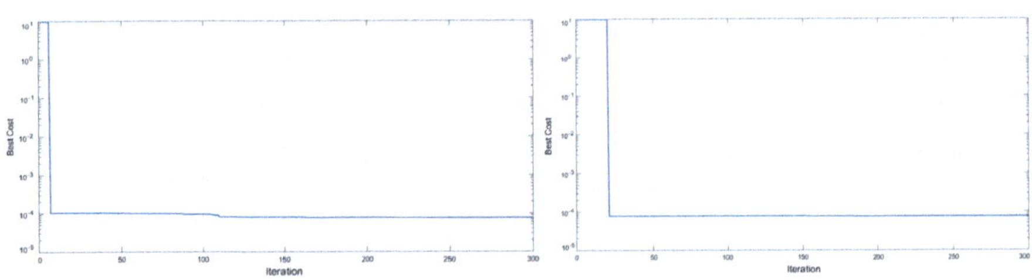

Fig. 3 Convergence characteristic curves of PSO (left) and IWO (right) for Problem-3

4 Concluding Remarks

On solving the problems, we obtained the optimal portfolio weights from which we calculated the expected return and the values of variance–semivariance return. The results show that the optimal weights obtained from IWO method give rise to higher expected portfolio return values and lower risk values (variance and semivariance) as compared to PSO method. It justifies that the results were better in IWO as compared to PSO. But in *Problems-1 and 3*, we got faster convergence in PSO method, which shows that none of the discussed methods is more efficient than the other in all the aspects. However, since in portfolio problems our main aim is to get the best portfolio and not to get the faster convergence, so IWO may be considered as a better method. There is a scope of future research in solving the multiperiod and uncertain portfolio problems by IWO method.

Acknowledgements Authors are grateful to the editors and the reviewers for their useful comments and suggestions for the improvement of the quality of our paper.

References

1. Markowitz, H.: Portfolio selection. J. Finan. **7**, 77–91 (1952)
2. Markowitz, H.: Portfolio Selection: Efficient Diversification of Investments. Wiley, New York (1959)
3. Kar, M.B., Majumder, S., Kar, S., Pal, T.: Cross-entropy based multi-objective uncertain portfolio selection problem. J. Intell. Fuzzy Syst. **32**(6), 4467–4483 (2017)
4. Guo, S., Yu, L., Li, X., Kar, S.: Fuzzy multi-period portfolio selection with different investment horizons. Euro. J. Oper. Res. **254**(3), 1026–1035 (2016)
5. Majumder, S., Saha, B., Anand, P., Kar, S., Pal, T.: Uncertainty based genetic algorithm with varying population for random fuzzy maximum flow problem. Expert Syst. **35**(4), e12264 (2018)
6. Chang, T.J., Yang, S.C., Chang, K.J.: Portfolio optimization problems in different risk measures using genetic algorithm. Expert Syst. Appl. **36**(7), 10529–10537 (2009)
7. Golmakani, H.R., Fazel, M.: Constrained portfolio selection using particle swarm optimization. Expert Syst. Appl. **38**(7), 8327–8335 (2011)
8. Kar, M.B., Kar, S., Guo, S., Li, X., Majumder, S.: A new bi-objective fuzzy portfolio selection model and its solution through evolutionary algorithms. Soft. Comput. **23**(12), 4367–4381 (2019)
9. Cornuejols, G., Tutuncu, R.: Optimization Methods in Finance, vol. 5, pp. 141–169. Cambridge University Press, Cambridge (2006)
10. Konno, H., Waki, H., Yuuki, A.: Portfolio optimization under lower partial risk measures. Asia-Pac. Finan. Markets. **9**(2), 127–140 (2002)
11. Qin, Z., Kar, S., Zheng, H.: Uncertain portfolio adjusting model using semiabsolute deviation. Soft. Comput. **20**(2), 717–725 (2016)
12. Parsopoulos, K.E., Vrahatis, M.N.: Particle swarm optimization method for constrained optimization problems. Intell. Technol. Theory Appl.: New Trends Intell. Technol. **76**(1), 214–220 (2002)

13. Shahriar, M.S., et. al.: Comparison of Invasive Weed Optimization (IWO) and Particle Swarm Optimization (PSO) in improving power system stability by UPFC controller employing a multi-objective approach. In: 1st International Conference on Advanced Information and Communication Technology 2016 (ICAICT 2016), pp. 1–7 (2016)
14. Naidu, Y.R., Ojha, A.K.: Solving nonlinear constrained optimization problems using invasive weed optimization. In: Intelligent Computing, Communication and Devices. Springer, New Delhi, pp. 127–133 (2015)

The Influence of Lewis Number on Natural Convective Nanofluid Flows in an Enclosure: Buongiorno's Mathematical Model: A Numerical Study

C. Venkata Lakshmi, A. Shobha, K. Venkatadri and K. R. Sekhar

Abstract A numerical simulation is performed on transport phenomena in a square cavity with the Lewis number influence. The enclosure filled with water-based nanofluid and the square cavity enclosed by thermal walls with various thermal conditions, vertical walls of the cavity having different temperatures comparatively one is higher than the other. The effective tool for implemented to eliminate the pressure term, named as vorticity—stream function formulation. The nonlinear coupled governing partial differential equations are employed with finite difference method along with central difference approximation. The visualization of fluid flow, temperature distribution and nanoparticles distribution is depicted in the form of streamlines, isotherms and isoconcentration with the influence of various governing parameters.

Keywords Natural convection · Laminar · Enclosure · Nanofluid · Heat transfer

1 Introduction

Transport phenomena have been playing a vital role in industries, science and engineering, etc. Due to this importance to enhance the interest of the scientists and researchers on transport phenomena in fluid flows with various geometries. Importance of heat transfer in fluids in industries, whereas a suitable heat transfer fluid is employed for heat transportation in heat exchangers. The phenomena thus considered notable attention for various applications in engineering [1–6]. Cooling technique plays challenges in several industries, like vapor deposition instrument of chemicals, energy collectors, energy system buildings and non-Newtonian chemical processes. The controlling heat transfer by utilizing water-based nanoparticles is passionate tool in recent years. The flows driven in an enclosure with the influence of Buoyancy force is an interesting problem in fluid mechanics. In recent years, increase

C. Venkata Lakshmi · A. Shobha
Department of Applied Mathematics, Sri Padmavathi Mahila
Visva Vidyalayam, Tirupati, A.P., India

K. Venkatadri (✉) · K. R. Sekhar
Department of Mathematics, Vemu Institute of Technology, P.kothakota, Chitoor, A.P., India
e-mail: venkatadri.venki@gmail.com

D. Dutta and B. Mahanty (eds.), *Numerical Optimization in Engineering and Sciences*,
Advances in Intelligent Systems and Computing 979,
https://doi.org/10.1007/978-981-15-3215-3_31

interest to study the natural convection in an enclosure with numerically as well as experimentally [7–16]. Nanofluid convective flows within the enclosures are examined by two-phase and single-phase model. Thermophoresis and Brownian motion parameter-based study are carried out by the researchers [10, 17, 18].

CVFEM-based simulation of convective flow in an L-shaped enclosure with account of Brownian motion and thermophoresis effects are reported by [19]. In addition to the influence of magnetic field on convective flow in a C-shaped enclosure filled with CuO-water nanofluid proposed by [20]. The convective flow of CuO nanofluid in triangular enclosure is examined under the influence of aspect ratio is reported [21]. Natural convection of domain of interest with a nanofluid and is simulated affected by magnetic field is conducted by [22]. Numerical simulation of sedimentation, thermophoresis and Brownian motion effects on a natural convection of Al_2O_3/water nanofluid is examined by [23]. Numerical investigation of an unsteady heat and mass transfer of a nanofluid thin film flow caused by stretching sheet is conducted by [24]. The temperature-dependent properties are studied in an enclosure along with two different nanofluid models that are carried out by [25]. A similar investigation has been reported by [26, 27]. The effects of Brownian motion and thermophoresis are considered by [27] and studied natural convection in an enclosure. Pakravan and Yaghoubi [26] reported the role of Dufour effects on heat transfer analytically. Their result reveals that the heat transfer rate is reduced by the Dufour effect. They noticed our computational results that give good agreement when comparing to Nusselt number values with the results of various experimental data. They are not taken Brownian motion and thermophoresis effects. Sheikholeslami [28] reported on heat transfer in porous enclosure by the influence of nanoparticles shapes.

The present numerical simulation is examined on convective heat transfer in nanofluid. Both influence of Brownian motion and thermophoresis are included. The numerical simulation is conducted for various governing key parameters, which include, buoyancy ratio (N), thermal Rayleigh number (Ra) and Lewis number (Le).

2 Mathematical Governing Equations

The working fluid in an enclosure taken as nanofluid, the enclosure vertical walls temperature is T_h and T_c (i.e., $T_c < T_h$), respectively. Except for the density, all properties of fluid flow are considered to be constant. In addition that buoyancy force is included through the Boussinesq approximation. The effects like magnetic field, viscous, radiation and Joule heating were ignored in present computation (Fig. 1).

Under these assumptions, we can write the governing equations of the present study. The conservation of mass, momentum and thermal energy can be considered by [29, 30].

$$\nabla . \boldsymbol{V} = 0 \tag{1}$$

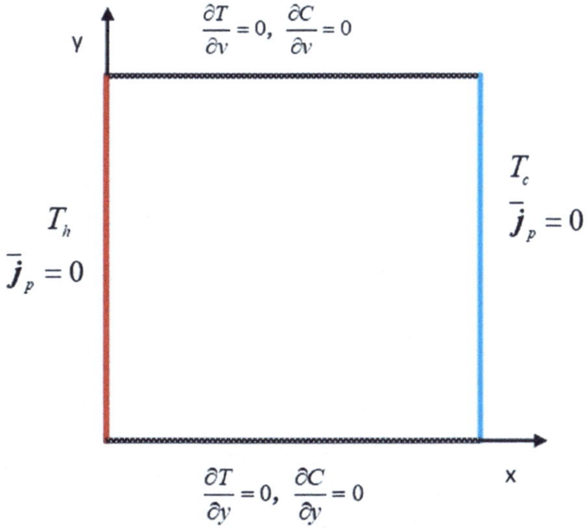

Fig. 1 Schematic physical model

$$\rho_f\left[\frac{\partial \boldsymbol{V}}{\partial t}+(\boldsymbol{V}.\nabla)\boldsymbol{V}\right]=-\nabla p+\mu\nabla^2\boldsymbol{V}+\left[C_{\rho p}+(1-C)\rho_{f_0}(1-\beta(T-T_c))\right]\boldsymbol{g} \tag{2}$$

$$\frac{\partial T}{\partial t}+(\boldsymbol{V}.\nabla)T=\alpha\,\nabla^2 T+\delta[D_B\nabla C.\nabla T+(D_T/T_C)\nabla T.\nabla T] \tag{3}$$

$$\rho_p\left[\frac{\partial C}{\partial t}+(\boldsymbol{V}.\nabla)C\right]=-\nabla.\boldsymbol{j}_p \tag{4}$$

where $\boldsymbol{j}_p=-\rho_p[D_B\nabla C+(D_T/T_C)\nabla T]$.

Equations’ (1)–(4) are solved in vorticity and stream function for these criteria, we have

$u=\frac{\partial\overline{\psi}}{\partial y}$, $v=-\frac{\partial\overline{\psi}}{\partial x}$ and $\overline{\omega}=\frac{\partial v}{\partial x}-\frac{\partial u}{\partial y}$.

By using the below dimensionless quantities:

$$\tau=\frac{\alpha t}{L^2},X=\frac{x}{L},Y=\frac{y}{L}\psi=\frac{\overline{\psi}}{\alpha},\omega=\frac{\overline{\omega}L^2}{\alpha},\theta=\frac{(T-T_c)}{(T_h-T_c)},\phi=\frac{C}{C_0}$$

The non-dimensional governing equations of natural convective flow in an enclosure in form of vorticity and stream function become (see [31, 32]):

$$\omega=-\frac{\partial^2\psi}{\partial X^2}-\frac{\partial^2\psi}{\partial Y^2} \tag{5}$$

$$\frac{\partial\omega}{\partial\tau}+\frac{\partial\psi}{\partial Y}\frac{\partial\omega}{\partial X}-\frac{\partial\psi}{\partial X}\frac{\partial\omega}{\partial Y}=\mathrm{Pr}\ \nabla^2\omega+\mathrm{Ra.\,Pr}\left(\frac{\partial\theta}{\partial X}-N\frac{\partial\phi}{\partial X}\right) \tag{6}$$

$$\frac{\partial \theta}{\partial \tau}+\frac{\partial \psi}{\partial Y}\frac{\partial \theta}{\partial X}-\frac{\partial \psi}{\partial X}\frac{\partial \theta}{\partial Y}=\nabla^2\theta+\mathrm{Nb}\left(\frac{\partial \phi}{\partial X}\frac{\partial \theta}{\partial X}+\frac{\partial \phi}{\partial Y}\frac{\partial \theta}{\partial Y}\right)+\mathrm{Nt}\left[\left(\frac{\partial \theta}{\partial X}\right)^2+\left(\frac{\partial \theta}{\partial Y}\right)^2\right] \tag{7}$$

$$\frac{\partial \phi}{\partial \tau}+\frac{\partial \psi}{\partial Y}\frac{\partial \phi}{\partial X}-\frac{\partial \psi}{\partial X}\frac{\partial \phi}{\partial Y}=\frac{1}{\mathrm{Le}}\nabla^2\phi+\frac{\mathrm{Nt}}{\mathrm{Le\,Nb}}\nabla^2\theta \tag{8}$$

The current investigation of square enclosure wall boundary conditions [27] defined as follows:

$$\psi=0,\ \ \theta=1,\ \ \bar{\boldsymbol{j}}_p=0\left(\mathrm{or\,Nb}\frac{\partial \phi}{\partial X}+\mathrm{Nt}\frac{\partial \theta}{\partial X}=0\right)\quad \text{on } X=0$$

$$\psi=0,\ \ \theta=0,\ \ \bar{\boldsymbol{j}}_p=0\left(\mathrm{or\,Nb}\frac{\partial \phi}{\partial X}+\mathrm{Nt}\frac{\partial \theta}{\partial X}=0\right)\quad \text{on } X=1$$

$$\psi=0,\ \frac{\partial \theta}{\partial Y}=0,\ \frac{\partial \phi}{\partial Y}=0\quad \text{on } Y=0,1$$

Buoyancy ratio $N=\frac{(C_h-C_c)(\rho_p-\rho_{f_0})}{\beta(1-C_c)\rho_{f_0}(T_h-T_c)}$, Brownian motion parameter $\mathrm{Nb}=D_B\frac{(\rho c)_p}{(\rho c)_f}\frac{(C_h-C_c)}{\alpha_f}$, Prandtl number $\mathrm{Pr}=\frac{\nu_f}{\alpha_f}$, thermophoresis parameter $\mathrm{Nt}=\frac{D_T}{T_c}\frac{(\rho c)_p}{(\rho c)_f}\frac{(T_h-T_c)}{\alpha_f}$, Rayleigh number $\mathrm{Ra}=\frac{g\beta(T_h-T_c)(1-C_c)L^3}{\nu_f^2}$ and Lewis number $\mathrm{Le}=\frac{\alpha_f}{D_B}$.

The heat transfer rate is calculated with local Nusselt number *Nu* and the average Nusselt Nu. These parameters are defined as $\mathrm{Nu}=-\frac{\partial\theta}{\partial n}$, $\overline{\mathrm{Nu}}=\int_0^L \mathrm{Nu}\,\mathrm{d}Y$.

3 Numerical Method and Validation

The governing non-dimensional vorticity-stream function formulated partial differential equations (5)–(8) with appropriate boundary conditions were solved by second-order accurate approximation finite difference method. The computation is carried with uniform grid system, i.e., the step length of both *X*- and *Y*-directions are constant. The approximation of the convective terms and diffusion term was conducted by the help of central difference finite scheme. The Poisson equation of stream function (5) is solved by the effective iterative method named as successive over-relaxation (SOR).

$$\psi_{i,j}=\frac{(\psi_{i+1,j}+\psi_{i-1,j})+(\psi_{i,j+1}+\psi_{i,j-1})+h^2\omega_{i,j}}{4\beta^{-1}}+(1-\beta)\psi_{i,j}$$

$$\frac{\omega_{i+1,j}^{n+1}-\omega_{i-1,j}^{n}}{\mathrm{d}t}=-\frac{\partial \psi}{\partial Y}\frac{\partial \omega}{\partial X}+\frac{\partial \psi}{\partial X}\frac{\partial \omega}{\partial Y}+\mathrm{Pr}\,\nabla^2\omega+\mathrm{Ra.Pr}\left(\frac{\partial \theta}{\partial X}-N\frac{\partial \phi}{\partial X}\right)$$

$$\frac{\phi_{i+1,j}^{n+1}-\phi_{i-1,j}^{n}}{\mathrm{d}t}=\left\{-\frac{\partial\psi}{\partial Y}\frac{\partial\phi}{\partial X}+\frac{\partial\psi}{\partial X}\frac{\partial\phi}{\partial Y}+\frac{1}{\mathrm{Le}}\nabla^2\phi+\frac{Nt}{\mathrm{Le\,Nb}}\nabla^2\theta\right\}$$

$$\frac{\theta_{i+1,j}^{n+1}-\theta_{i-1,j}^{n}}{\mathrm{d}t}=\left\{\begin{array}{l}-\frac{\partial\psi}{\partial Y}\frac{\partial\theta}{\partial X}+\frac{\partial\psi}{\partial X}\frac{\partial\theta}{\partial Y}+\nabla^2\theta\\+\mathrm{Nb}\left(\frac{\partial\phi}{\partial X}\frac{\partial\theta}{\partial X}+\frac{\partial\phi}{\partial Y}\frac{\partial\theta}{\partial Y}\right)+\mathrm{Nt}\left[\left(\frac{\partial\theta}{\partial X}\right)^2+\left(\frac{\partial\theta}{\partial Y}\right)^2\right]\end{array}\right\}$$

The explicit time integration scheme starts as following with appropriate initial values of all key variables

$$\omega_{i+1,j}^{n+1}=\omega_{i-1,j}^{n}+\mathrm{d}t\left\{-\frac{\partial\psi}{\partial Y}\frac{\partial\omega}{\partial X}+\frac{\partial\psi}{\partial X}\frac{\partial\omega}{\partial Y}+\mathrm{Pr}\,\nabla^2\omega+\mathrm{Ra.Pr}\left(\frac{\partial\theta}{\partial X}-N\frac{\partial\phi}{\partial X}\right)\right\}^n$$

$$\theta_{i+1,j}^{n+1}=\theta_{i-1,j}^{n}+\mathrm{d}t\left\{\begin{array}{l}-\frac{\partial\psi}{\partial Y}\frac{\partial\theta}{\partial X}+\frac{\partial\psi}{\partial X}\frac{\partial\theta}{\partial Y}+\nabla^2\theta\\+\mathrm{Nb}\left(\frac{\partial\phi}{\partial X}\frac{\partial\theta}{\partial X}+\frac{\partial\phi}{\partial Y}\frac{\partial\theta}{\partial Y}\right)+\mathrm{Nt}\left[\left(\frac{\partial\theta}{\partial X}\right)^2+\left(\frac{\partial\theta}{\partial Y}\right)^2\right]\end{array}\right\}^n$$

$$\phi_{i+1,j}^{n+1}=\phi_{i-1,j}^{n}+\mathrm{d}t\left\{-\frac{\partial\psi}{\partial Y}\frac{\partial\phi}{\partial X}+\frac{\partial\psi}{\partial X}\frac{\partial\phi}{\partial Y}+\frac{1}{\mathrm{Le}}\nabla^2\phi+\frac{Nt}{\mathrm{Le\,Nb}}\nabla^2\theta\right\}^n$$

The time integration is applied for all governing coupled nonlinear partial differential equations, the time integration is terminated when they reach the following convergence criteria.

$$\frac{\left|\lambda_{i,j}^{k+1}-\lambda_{i,j}^{k}\right|}{\left|\lambda_{i,j}^{k+1}\right|}\leq 10^{-8}$$

In the above expression, λ indicates the generic variable, which represents the ω, θ, ϕ. The detail computational procedure is expressed in the given flowchart Fig. 2. The above-mentioned convergence criteria are adopted in implemented MATLAB code as the mentioned numerical procedure flowchart.

The developed house-computational MATLAB code has been validated with [33] buoyancy-driven flow in an enclosure for $\mathrm{Ra} = 10^4$ and $\mathrm{Pr} = 0.71$. Figure 3 reveals a justified agreement between the results of obtained contour plots of streamlines, isotherms and vorticity for $\mathrm{Ra} = 10^4$ and $\mathrm{Pr} = 0.71$ and numerical results of Saitoh et al. [33]. These comparisons give the confidence for further investigation.

4 Results and Discussion

The current simulation is conducted heat transfer analysis within an enclosure filled with water-based nanofluid with utilizing Buongiorno model for $\mathrm{Ra} = 10^4$ to 10^6, $\mathrm{Pr} = 6.2$ and $\mathrm{Le} = 1$–100. The enclosure of the horizontal wall is kept at adiabatic and side wall are maintained different uniform temperatures (i.e., $T_h > T_c$). The

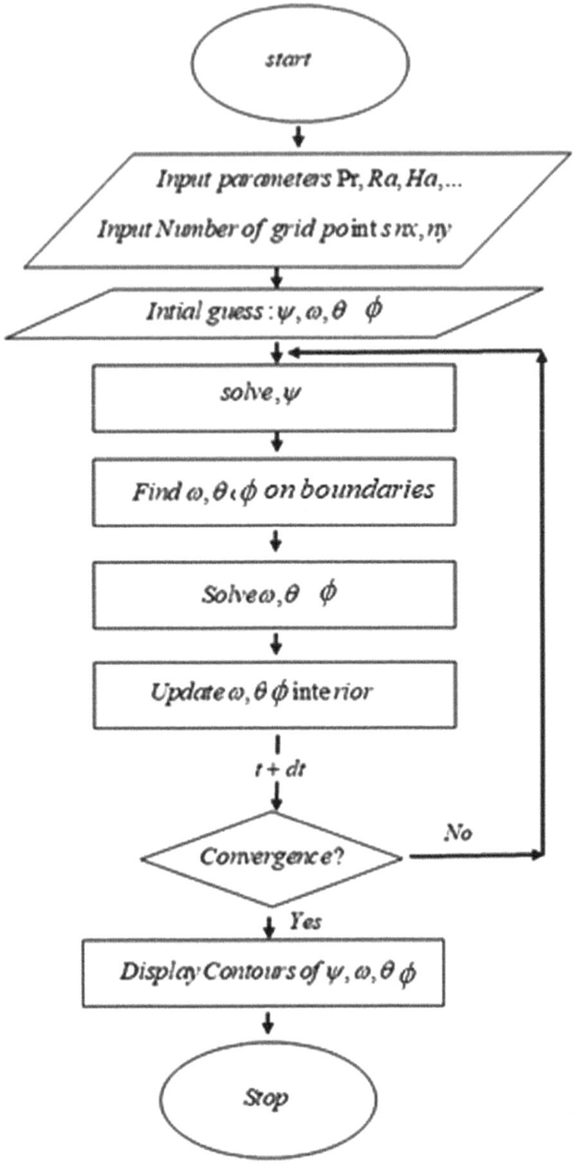

Fig. 2 Flowchart for the solution procedure of present computation

influence of thermal Rayleigh number, Brownian motion parameter, Lewis number and thermophoretic parameter on convective flow in a square cavity is examined numerically.

Figure 4 reveals the significant effect of different values of Rayleigh number $Ra = 10^4$ to 10^6, on the flow patterns, temperature contours and isoconcentration contours for $Le = 1$, $Nt = 0.1$, $Nb = 0.1$, $N = 0.1$ and $Pr = 6.2$. When the left side wall is heated

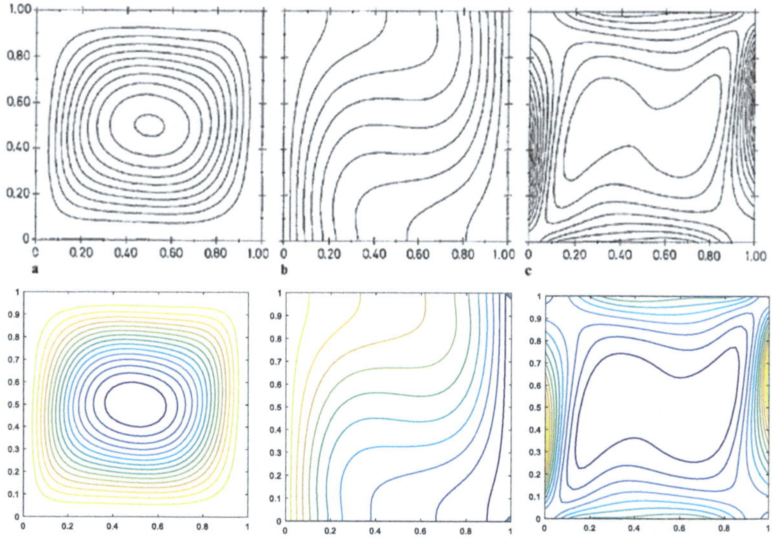

Fig. 3 Comparison of present and Saitoh et al. [28] (first row) for $Pr = 0.71$ and $Ra = 10^4$

uniformly as usually, due to the right cold side wall the nanofluid particles rise up from the bottom of hot wall and fall down to along the cold right wall develops the circular mono eddy within the center of the enclosure at $Ra = 10^4$. As increasing of Rayleigh number from $Ra = 10^4$ to 10^6, the streamline patterns are registered significant changes in circular flow. At $Ra = 10^5$, the mono-circular clockwise cell shifted to horizontal elliptic circulation, whereas at $Ra = 10^6$, the horizontal elliptic circulation magnitudes are very strong in addition to the two minor eddies are appeared near to the side walls. At $Ra = 10^4$, the inclined isotherms occupy whole cavity, in addition to the isotherms are clustered at the top corner of cold wall and bottom portion of hot wall. When an increase to $Ra = 10^5$, the convection enhanced and the thermal boundary layer developed along the side walls. The thermal boundary layer is more strengthen along the isothermal thermal walls due to the Rayleigh number is enhanced to $Ra = 10^6$. Nanoparticles distribution is shown in Fig. 4 for various values of Ra when $Ra \leq 10^5$, the degree of non-uniformity in the particle distribution exists.

The influence of Lewis number on stream lines, isotherms and nanoparticles distribution is depicted in Fig. 5. For $Ra = 10^5$, $Nt = 0.1$, $Nb = 0.1$, $N = 0.1$ and $Pr = 6.2$. Regardless of Le the fluid flow formed a single enlarged circulation within the enclosure and the corresponding isotherms are the considerable changes are registered due to strong thermal plume is developed along the side walls. The significant changes are noticed on isoconcentrations for increasing of Lewis number. The non-uniform nanoparticles distribution is exhibited for low value of Lewis number, i.e., $Le = 5$. The reduction of non-uniform nanoparticles distribution is observed at $Le = 20$ as well as $Le = 30$. When increases to $Le = 50$, high isoconcentration boundary

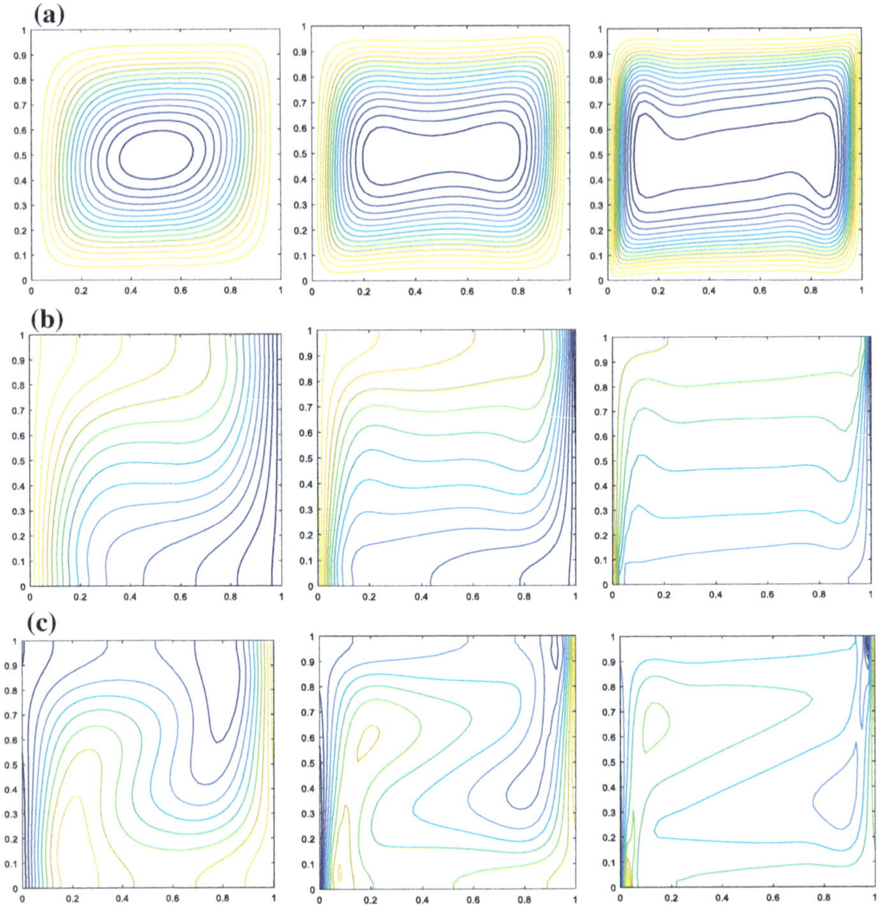

Fig. 4 (**a**) Streamlines, (**b**) isotherms, and (**c**) isoconcentrations for Le = 1, Nt = 0.1, Nb = 0.1, N = 0.1 and Pr = 6.2—first column Ra = 10^4, second column Ra = 10^5, third column Ra = 10^6

layer is formed along the isothermal side walls due to the nanoparticles distribution is in uniform more.

The effect of Lewis number on local Nusselt number of hot vertical wall with different Rayleigh number is presented in Fig. 6a, b. The Lewis number is proportional to the thermal conductivity, in addition to that in nanoparticles momentum equation, the inverse of the Lewis number accelerates the nanoparticles due to this influence when the augment of Lewis number with de-accelerate in local Nusselt number is seen in Fig. 6a, b. The variation of local Nusselt number with the effect of Rayleigh number is described in Fig. 7. The heat transfer rate in water-based nanofluid is gradually increased when increasing of Rayleigh number. The constant rate of heat transfer is observed at low value of Ra = 10^3. The higher heat transfer rate is noticed at Ra = 10^6.

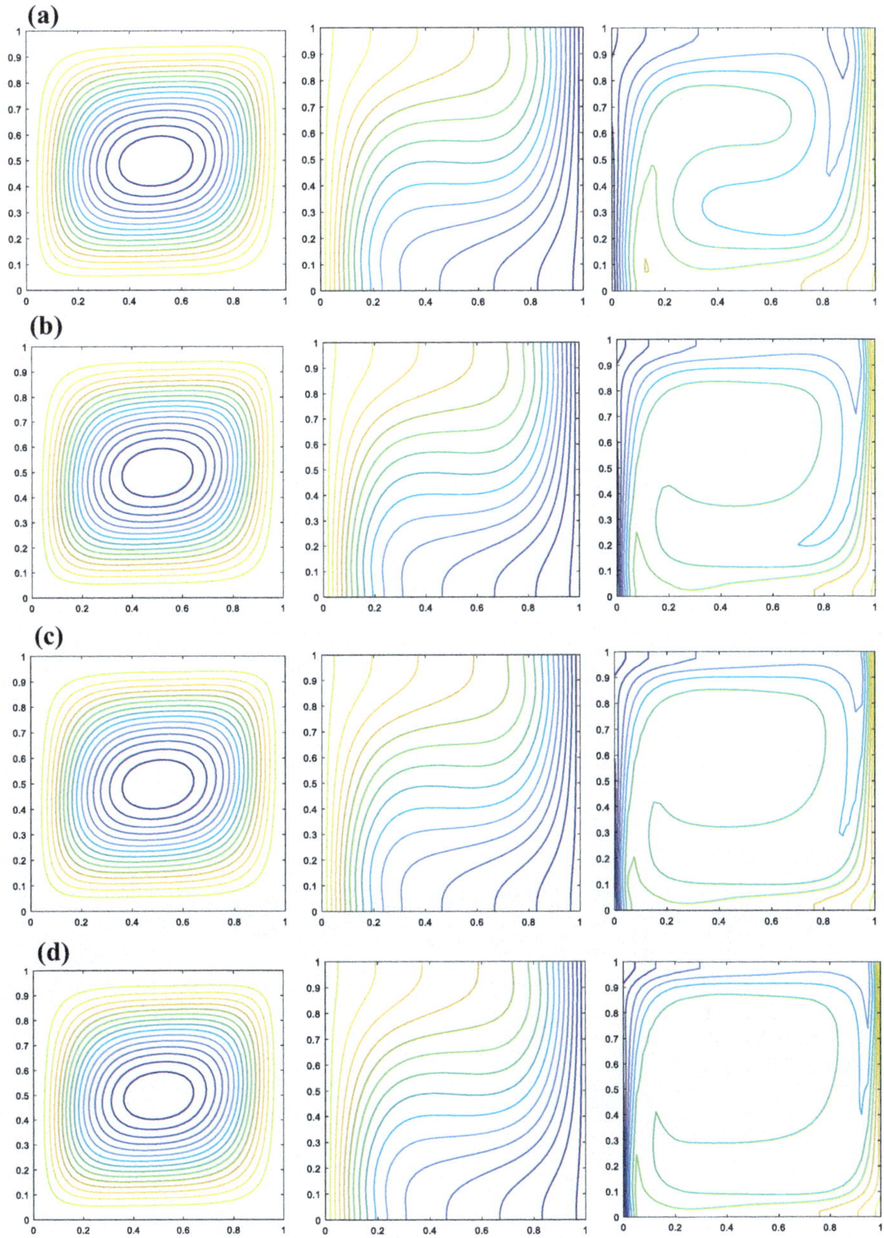

Fig. 5 Streamlines, isotherms and isoconcentrations for Ra = 10^5, Nt = 0.1, Nb = 0.1, N = 0.1, Pr = 6.2 and Le = 5—(**a**), Le = 20—(**b**), Le = 30—(**c**), Le = 50—(**d**)

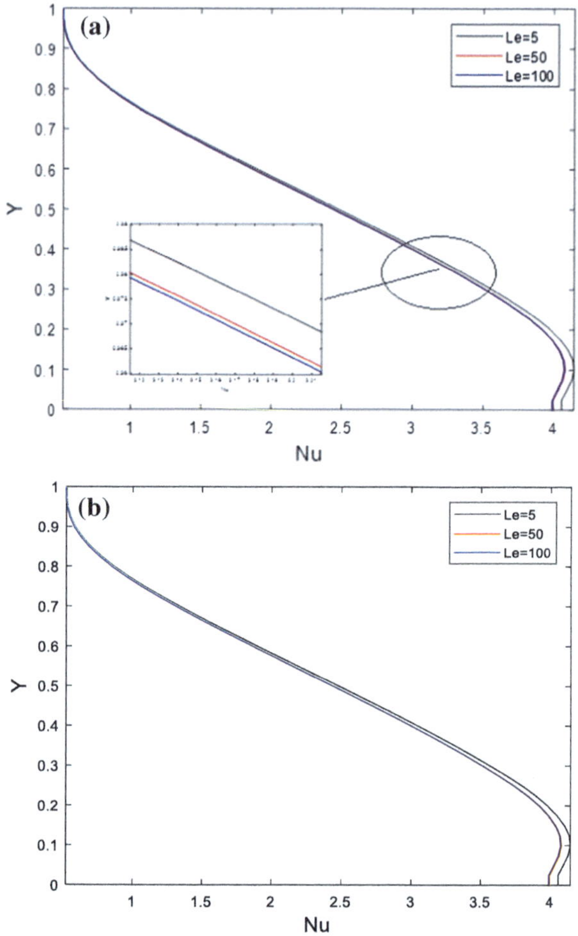

Fig. 6 Variation of the local Nu at hot surface with Le for Nt = 0.1, Nb = 0.1, N = 0.1, Pr = 6.2; Ra = 10^4—(**a**) and Ra = 10^5—(**b**)

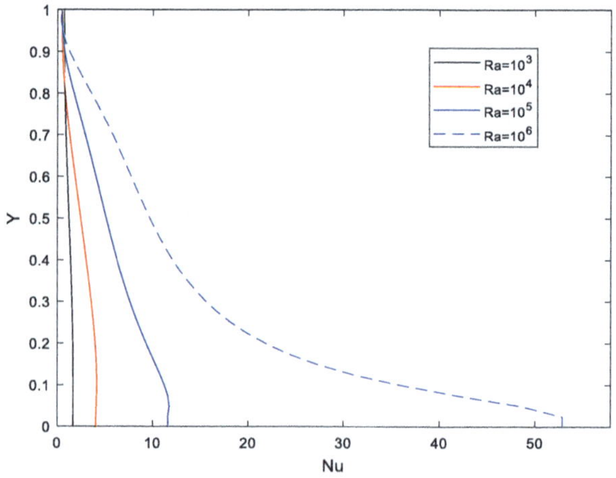

Fig. 7 Variation of the local Nu of hot wall with Ra for Le = 5, Nt = 0.1, Nb = 0.1, N = 0.1 and Pr = 6.2

5 Conclusion

The transport phenomenon in a water-based nanofluid within an enclosure was examined numerically by using vorticity and stream function. The two-phase nanofluid model and the numerical solution are conducted on the basis of a second-order accurate difference scheme. The domain of interest was a square enclosure filled with water-based nanofluid with the isothermal side walls and insulated horizontal walls. The wide range of key parameters influence on distribution of flow patterns, temperature contours and isoconcentration has been obtained. We conclude the following findings:

1. An increasing of thermal Rayleigh number Ra leads to an increase in convective flow and heat transfer.
2. Regardless of Lewis number Le the flow and heat transfer rate is enhanced when increasing of Rayleigh number Ra.
3. The non-uniform nanoparticles distribution rate is reduced gradually when increasing of Lewis number Le.
4. Regardless of Rayleigh number Ra an increasing of the reduction rate of local Nusselt number when increasing of Lewis number Le.

References

1. Lai, F.S., Yang, Y.T.: Lattice Boltzmann simulation of natural convection heat transfer of Al_2O_3/water nanofluids in a square cavity. Int. J. Therm. Sci. **50**, 1930–1941 (2011). https://doi.org/10.1016/j.ijthermalsci.2011.04.015
2. Kakac, S., Pramuanjaroenkij, A.: Review of convective heat transfer enhancement with nanofluids. Int. J. Heat Mass Transf. **52**, 3187–3196 (2009). https://doi.org/10.1016/j.ijheatmasstransfer.2009.02.006
3. Trisaksri, V., Wongwises, S.: Critical review of heat transfer characteristics of nanofluids. Renew. Sustain. Energy Rev. **11**(3), 512–523 (2007). https://doi.org/10.1016/j.rser.2005.01.010
4. Wang, X.Q., Mujumdar, A.S.: Heat transfer characteristics of nanofluids. Int. J. Therm. Sci. **46**, 1–19 (2007). https://doi.org/10.1016/j.ijthermalsci.2006.06.010
5. Daungthongsuk, W., Wongwises, S.: A critical review of convective heat transfers nanofluids. Renew. Sustain. Energ. Rev. **11**, 797–817 (2007). https://doi.org/10.1016/j.rser.2005.06.005
6. Gupta, M., Arora, N., Kumar, R., Kumar, S., Dilbaghi, N.: A comprehensive review of experimental investigations of forced convective heat transfer characteristics for various nanofluids. Int. J. Mech. Mater. Eng. **9**(11), 1–21 (2014). https://doi.org/10.1186/s40712-014-0011-x
7. Jourabian, M., Farhadi, M., Rabienataj Darzi, A.: Outward melting of ice enhanced by Cu nanoparticles inside cylindrical horizontal annulus: Lattice Boltzmann approach. Appl. Math. Model. 8813–8825 (2013). https://doi.org/10.1007/s12206-015-0828-0
8. Hatami, M., Sheikholeslami, M., Ganji, D.D.: Laminar flow and heat transfer of nanofluid between contracting and rotating disks by least square method. Powder Technol. **253**, 769–779 (2014). https://doi.org/10.1016/j.powtec.2013.12.053
9. Sheikholeslami, M., Hatami, M., Ganji, D.D.: Analytical investigation of MHD nanofluid flow in a semi-porous channel. Powder Technol. **246**, 327–336 (2013). https://doi.org/10.1016/j.powtec.2013.05.030

10. Hatami, M., Song, D., Jing, D.: Optimization of a circular-wavy cavity filled by nanofluid under the natural convection heat transfer condition. Int. J. Heat Mass Transf. **98**, 758–767 (2016). https://doi.org/10.1016/j.ijheatmasstransfer.2016.03.063
11. Song, D., Jing, D., Luo, B., Geng, J., Ren, Y.: Modeling of anisotropic flow and thermodynamic properties of magnetic nanofluids induced by external magnetic field with varied imposing directions. J. Appl. Phys. **118** (2015). https://doi.org/10.1063/1.4927043
12. Abouei Mehrizi, A., Farhadi, M., Hassanzade Afroozi, H., Sedighi, H., Rabienataj Darzi, A.A.: Mixed convection heat transfer in a ventilated cavity with hot obstacle: effect of nanofluid and outlet port location. Int. Commun. Heat Mass Transf. **39**, 1000–1008 (2012). https://doi.org/10.1016/j.icheatmasstransfer.2012.04.002
13. Khazayinejad, M., Hatami, M., Jing, D., Khaki, M., Domairry, G.: Boundary layer flow analysis of a nanofluid past a porous moving semi-infinite flat plate by optimal collocation method. Powder Technol. **301**, 34–43 (2016). https://doi.org/10.1016/j.powtec.2016.05.053
14. Hatami, M., Khazayinejad, M., Jing, D.: Forced convection of Al_2O_3-water nanofluid flow over a porous plate under the variable magnetic field effect. Int. J. Heat Mass Transf. **102**, 622–630 (2016). https://doi.org/10.1016/j.ijheatmasstransfer.2016.06.075
15. M'hamed, B., Che Sidik, N.A., Muhammad Yazid, M.N.A.W., Mamat, R., Najafi, G., Kefayati, G.: A review on why researchers apply external magnetic field on nanofluids. Int. Commun. Heat Mass Transf. **78**, 60–67 (2016). https://doi.org/10.1016/j.icheatmasstransfer.2016.08.023
16. Che Sidik, N.A., Adamu, I.M., Mahmud Jamil, M., Kefayati, G., Mamat, R., Najafi, G.: Recent progress on hybrid nanofluids in heat transfer applications: a comprehensive review. Int. Commun. Heat Mass Transf. **78**, 68–79 (2016). https://doi.org/10.1016/j.icheatmasstransfer.2016.08.019
17. Haddad, Z., Abu-Nada, E., Oztop, H.F., Mataoui, A.: Natural convection in nanofluids: are the thermophoresis and Brownian motion effects significant in nanofluid heat transfer enhancement? Int. J. Therm. Sci. **57**, 152–162 (2012). https://doi.org/10.1016/j.ijthermalsci.2012.01.016
18. Sheikholeslami, M., Gorji-Bandpy, M., Soleimani, S.: Two phase simulation of nanofluid flow and heat transfer using heatline analysis. Int. Commun. Heat Mass Transf. **47**, 73–81 (2013). https://doi.org/10.1016/j.icheatmasstransfer.2013.07.006
19. Sheikholeslami, M., Chamkha, A.J., Rana, P., Moradi, R.: Combined thermophoresis and Brownian motion effects on nanofluid free convection heat transfer in an L-shaped enclosure. Chin. J. Phys. (2017). https://doi.org/10.1016/j.cjph.2017.09.011
20. Mliki, B., Abbassi, M.A., Omri, A., Belkacem, Z.: Lattice Boltzmann analysis of MHD natural convection of CuO water nanofluid in inclined C-shaped enclosures under the effect of nanoparticles Brownian motion. Powder Technol. (2016). https://doi.org/10.1016/j.powtec.2016.11.054
21. Ghasemi, B., Aminossadati, S.M.: Brownian motion of nanoparticles in a triangular enclosure with natural convection. Int. J. Therm. Sci. **49**, 931–940 (2010). https://doi.org/10.1016/j.ijthermalsci.2009.12.017
22. Mliki, B., Abbassi, M.A., Omri, A., Belkacem, Z.: Effects of nanoparticles Brownian motion in a linearly/sinusoidally heated cavity with MHD natural convection in the presence of uniform heat generation/absorption. Powder Technol. **295**, 69–83 (2016). https://doi.org/10.1016/j.powtec.2016.03.038
23. Ho, C.J., Chen, D.-S., Yan, W.-M., Mahian, O.: Rayleigh-Benard convection of Al_2O_3/water nanofluids in a cavity considering sedimentation, thermophoresis and Brownian motion. Int. Commun. Heat Mass Transfer **57**, 22–26 (2014). https://doi.org/10.1016/j.icheatmasstransfer.2014.07.014
24. Qasim, M., Khan, Z.H., Lopez, R.J., Khan, W.A.: Heat and mass transfer in nanofluid thin film over an unsteady stretching sheet using Buongiorno's model. Eur. Phys. J. Plus **131**, 1–16 (2016). https://doi.org/10.1140/epjp/i2016-16016-8
25. Corcione, M., Cianfrini, M., Quintino, A.: Two-phase mixture modeling of natural convection of nanofluids with temperature-dependent properties. Int. J. Therm. Sci. **71**, 182–195 (2013). https://doi.org/10.1016/j.ijthermalsci.2013.04.005

26. Pakravan, H.A., Yaghoubi, M.: Analysis of nanoparticles migration on natural convective heat transfer of nanofluids. Int. J. Therm. Sci. **68**, 79–93 (2013). https://doi.org/10.1016/j.ijthermalsci.2012.12.012
27. Sheikhzadeh, G.A., Dastmalchi, M., Khorasanizadeh, H.: Effects of nanoparticles transport mechanisms on Al_2O_3 water nanofluid natural convection in a square enclosure. Int. J. Therm. Sci. **66**, 51–62 (2013). https://doi.org/10.1016/j.ijthermalsci.2012.12.001
28. Sheikholeslami, M.: Magnetic field influence on CuO–H_2O nanofluid convective flow in a permeable cavity considering various shapes for nanoparticles. Int. J. Hydrogen Energy **42**, 19611–19621 (2017)
29. Hossain, M.S., Alim, M.A.: MHD free convection within trapezoidal cavity with non-uniformly heated bottom wall. Int. J. Heat Mass Transf. **69**, 327–336 (2014). https://doi.org/10.1016/j.ijheatmasstransfer.2013.10.035
30. Buongiorno, J.: Convective transport in nanofluids. ASME J Heat Transfer **128**, 240–250 (2006). https://doi.org/10.1115/1.2150834
31. Grosan, T., Revnic, C., Pop, I., Ingham, D.B.: Magnetic field and internal heat generation effects on the free convection in a rectangular cavity filled with a porous medium. Int. J. Heat Mass Transf. **52**, 1525–1533 (2009). https://doi.org/10.1016/j.ijheatmasstransfer.2008.08.011
32. Sheremet, M.A., Pop, I.: Natural convection in a wavy porous cavity with sinusoidal temperature distributions on both side walls filled with a nanofluid: Buongiorno's mathematical model. ASME J. Heat Transf. **137**, 072601 (2015). https://doi.org/10.1115/1.4029816
33. Saitoh, T., Hirose, K.: High-accuracy bench mark solutions to natural convection in a square cavity. Comput. Mech. **4**, 417–427 (1989). https://doi.org/10.1007/BF00293047

Reliability Model for 4-Modular and 5-Modular Redundancy System by Using Markov Technique

G. Saritha, M. Tirumala Devi and T. Sumathi Uma Maheswari

Abstract Redundancy is the provision of alternative means or parallel paths in a system for accomplishing a task such that all means must fail before causing a system failure. There are many models that have been developed on the reliability analysis using Markov technique. Markov models are very useful in finding the system reliability in various life situations. In this paper, reliability has been modelled for 4-modular and 5-modular redundancy system by using Markov technique by considering at least two modules must operate, three modules must operate and four modules must operate for successful operation of the system.

Keywords Reliability · Fault tolerance · Modular redundancy · Markov technique

Notations

λ	Failure rate of one module
S_i	The state which has i no. of modules fail
$P_i(t)$	Probability of i no. of modules good at time t
R_{mn}	Reliability of n-modular redundancy system out of which at least m-modules should operate
$P_F(t)$	Failure probability of the system at time t

G. Saritha (✉) · M. Tirumala Devi · T. Sumathi Uma Maheswari
Department of Mathematics, Kakatiya University, Warangal, Telangana, India
e-mail: gouravenisaritha66@gmail.com

D. Dutta and B. Mahanty (eds.), *Numerical Optimization in Engineering and Sciences*, Advances in Intelligent Systems and Computing 979,
https://doi.org/10.1007/978-981-15-3215-3_32

1 Introduction

A Markov model is a stochastic method for randomly changing systems where it is assumed that the future states depend only on present states, not depend on past states. In Markov model, it is assumed that more than one transition in a small time is negligible. These models show all possible states as well as the transition, rate of transitions and probabilities between them. Markov models are used to model the probabilities of different states and the rate of transition among them.

Modularity is a technique to organize and simplify a complex system from elementary to complex. If the system has n modules, there is a chance that the system may work if at least some modules are good. It is assumed that the failure rate of the module is constant.

Reliability is defined as the probability that a system will perform satisfactorily during a specified period of time under the given conditions [1]. Kalairasi et al. [2] studied the analysis of system reliability using Markov technique. Reni sagayaraj et al. [3] have described the Markov models in system reliability with applications. Koteswarrao et al. [4] studied the design and implementation of 4-bit multiplier using fault-tolerant self-repairing full adder. Sawant [5] estimated the applications of Laplace transform in engineering fields. Sadi et al. [6] studied towards tolerating faults by triple modular redundancy. Jansi Rani et al. [7] have described multiple error recovery in TMR systems. Rahman et al. [8] analysed a fault-tolerant voter circuit for triple modular redundant system.

2 Reliability Modelling of a 4-Modular Redundancy System

2.1 *At Least Two Modules Must Operate for Functioning of the System*

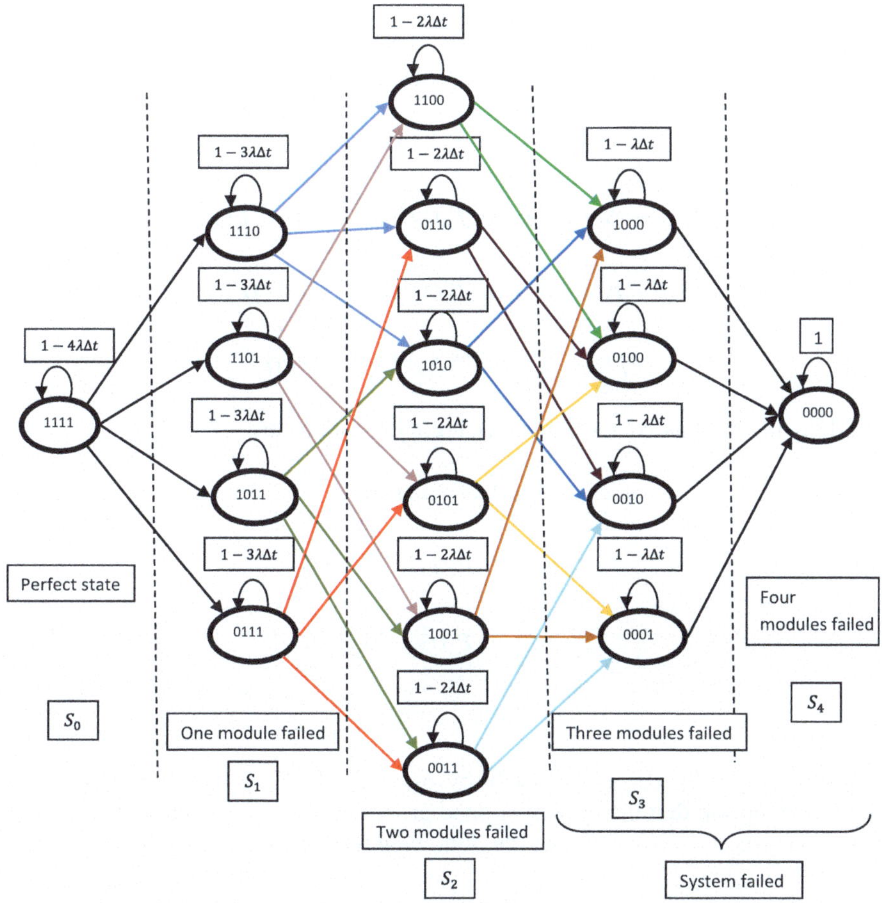

The 4-modular redundancy system can have five states, i.e. S_0, S_1, S_2, S_3 and S_4. The 4-modular redundancy system has 16 distinct states. The system state diagram is partitioned into four major parts. The perfect state function (1111), the one module failed states (1110), (1101), (1011) and (0111), the two modules failed states (1100), (0110), (1010), (0101), (1001) and (0011), and the failed states are (1000), (0100), (0010), (0001) and (0000).

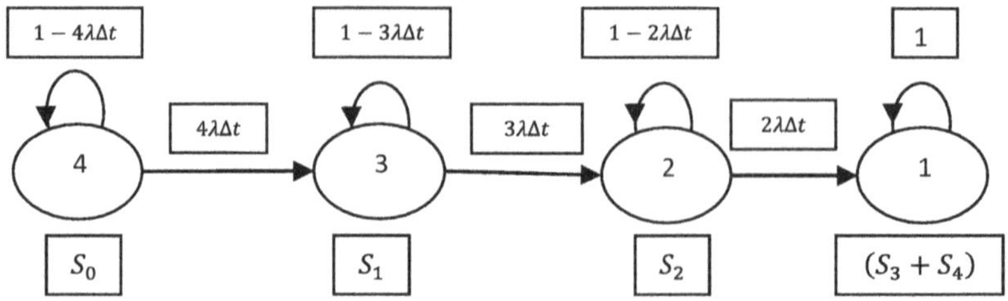

The state with constant hazard model has a probability transition that is approximated by $\lambda\Delta t$ within a specified time Δt.

The system's probability in four states at $(t + \Delta t)$ time

$$P_4(t + \Delta t) = (1 - 4\lambda\Delta t)P_4(t) \tag{1}$$

$$P_3(t + \Delta t) = (4\lambda\Delta t)P_4(t) + (1 - 3\lambda\Delta t)P_3(t) \tag{2}$$

$$P_2(t + \Delta t) = (3\lambda\Delta t)P_3(t) + (1 - 2\lambda\Delta t)P_2(t) \tag{3}$$

$$P_F(t + \Delta t) = (2\lambda\Delta t)P_2(t) + P_F(t) \tag{4}$$

Taking the limit as $\Delta t \to 0$, the differential equations are

$$P_4'(t) = -4\lambda P_4(t) \tag{5}$$

$$P_3'(t) = 4\lambda P_4(t) - 3\lambda P_3(t) \tag{6}$$

$$P_2'(t) = 3\lambda P_3(t) - 2\lambda P_2(\mathrm{t}) \tag{7}$$

$$P_F'(t) = 2\lambda P_2(t) \tag{8}$$

Solving Eqs. (5), (6), (7) and (8), by using the Laplace transformations, and taking assumption that the system is good at initial time $t = 0$, hence, $P_4(0) = 1$, $P_3(0) = P_2(0) = P_F(0) = 0$, then the following results are obtained

$$P_4(t) = \mathrm{e}^{-4\lambda t} \tag{9}$$

$$P_3(t) = 4\mathrm{e}^{-3\lambda t} - 4\mathrm{e}^{-4\lambda t} \tag{10}$$

$$P_2(t) = 6\mathrm{e}^{-2\lambda t} - 12\mathrm{e}^{-3\lambda t} + 6\mathrm{e}^{-4\lambda t} \tag{11}$$

$$P_F(t) = 1 - 6\mathrm{e}^{-2\lambda t} + 8\mathrm{e}^{-3\lambda t} - 3\mathrm{e}^{-4\lambda t} \tag{12}$$

The 4-modular redundancy system reliability, if at least two modules should operate

$$R_{24} = P_2(t) + P_3(t) + P_4(t) \tag{13}$$

The failure probability of the system is

$$P_F(t) = 1 - R_{24}. \tag{14}$$

2.2 At Least Three Modules Must Operate for Functioning of the System

The 4-modular redundancy system has 16 distinct states. The system state diagram is partitioned into three major parts. The perfect state function (1111), the one module failed states (1110), (1101), (1011) and (0111), and the failed states are (1100), (0110), (1010), (0101), (1001), (0011), (1000), (0100), (0010), (0001) and (0000).

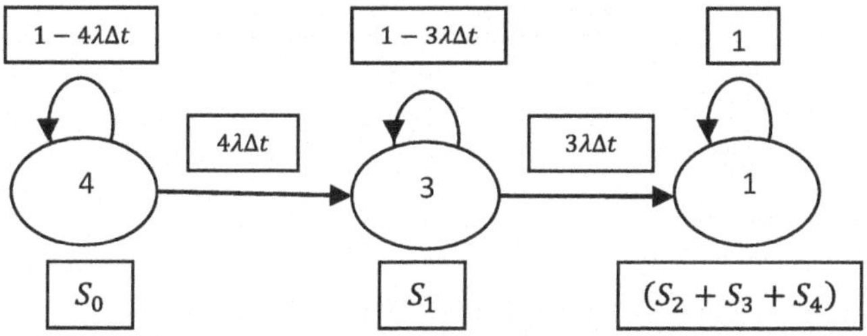

Using procedure of case (i), we get

The 4-modular redundancy system reliability, if at least three modules must operate

$$R_{34} = P_3(t) + P_4(t) \tag{15}$$

The system's failure probability is

$$P_F(t) = 1 - R_{34}. \tag{16}$$

3 Reliability Modelling of a 5-Modular Redundancy System

3.1 At Least Two Modules Must Operate for Functioning of the System

In a similar method of 4-modular redundancy system, 5-modular redundancy system has been developed. In this case, the system has 6 states which are, S_0, S_1, S_2, S_3, S_4 and S_5. The 5-modular redundancy system has 32 distinct states that can operate. The system state diagram is partitioned into five major parts. The perfect state function (11111), the one module failed function states (11110), (11101), (11011), (10111), (01111), the two modules failed states (10110), (11100), (10011), (11001), (00111), (01110), (11010), (10101), (01011), (01101), the three modules failed states (11000), (10001), (00011), (00110), (00101), (10010), (01010), (10100), (01001), (01100), and the failed states are (10000), (01000), (00100), (00010), (00001) and (00000).

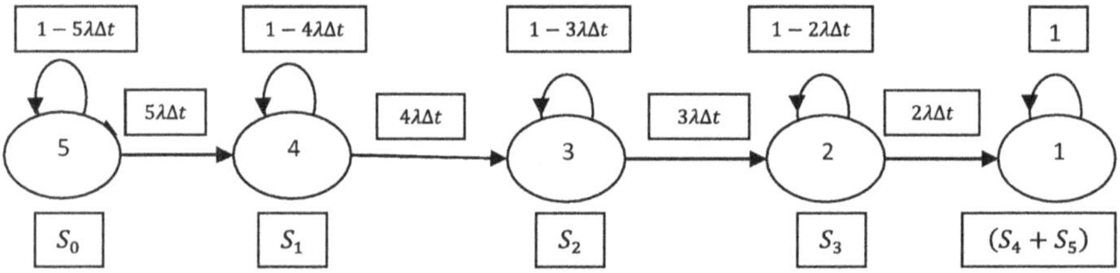

The system's probability for five states at $(t + \Delta t)$ time

$$P_5(t + \Delta t) = (1 - 5\lambda\Delta t)P_5(t) \tag{17}$$

$$P_4(t + \Delta t) = (5\lambda\Delta t)P_5(t) + (1 - 4\lambda\Delta t)P_4(t) \tag{18}$$

$$P_3(t + \Delta t) = (4\lambda\Delta t)P_4(t) + (1 - 3\lambda\Delta t)P_3(t) \tag{19}$$

$$P_2(t + \Delta t) = (3\lambda\Delta t)P_3(t) + (1 - 2\lambda\Delta t)P_2(t) \tag{20}$$

$$P_F(t + \Delta t) = (2\lambda\Delta t)P_2(t) + P_F(t).1 \tag{21}$$

The differential equations are

$$P_5'(t) = -5\lambda P_5(t) \tag{22}$$

$$P_4'(\mathrm{t}) = 5\lambda P_5(\mathrm{t}) - 4\lambda P_4(\mathrm{t}) \tag{23}$$

$$P_3'(t) = 4\lambda P_4(t) - 3\lambda P_3(\mathrm{t}) \tag{24}$$

$$P_2'(t) = 3\lambda P_3(t) - 2\lambda P_2(\mathrm{t})$$

$$P_F'(t) = 2\lambda P_2(t) \tag{25}$$

Solving Eqs. (22), (23), (24) and (25), by using the Laplace transformations, and taking the assumption that the system is in perfect condition at initial time $t = 0$, hence $P_5(0) = 1$, $P_4(0) = P_3(0) = P_2(0) = 0$ and $P_F(0) = 0$, and then the following results are obtained

$$P_5(t) = \mathrm{e}^{-5\lambda t} \tag{26}$$

$$P_4(t) = 5\mathrm{e}^{-4\lambda t} - 5\mathrm{e}^{-5\lambda t} \tag{27}$$

$$P_3(t) = 10\mathrm{e}^{-3\lambda t} - 20\mathrm{e}^{-4\lambda t} + 10\mathrm{e}^{-5\lambda t} \tag{28}$$

$$P_2(t) = 10\mathrm{e}^{-2\lambda t} - 30\mathrm{e}^{-3\lambda t} + 30\mathrm{e}^{-4\lambda t} - 10\mathrm{e}^{-5\lambda t} \tag{29}$$

$$P_F(t) = 1 - 10\mathrm{e}^{-2\lambda t} + 20\mathrm{e}^{-3\lambda t} - 15\mathrm{e}^{-4\lambda t} + 4\mathrm{e}^{-5\lambda t} \tag{30}$$

The 5-modular redundancy system reliability, if at least two modules should operate

$$R_{25} = P_2(t) + P_3(t) + P_4(t) + P_5(t) \tag{31}$$

The failure probability of the system is

$$P_F(t) = 1 - R_{25}. \tag{32}$$

3.2 *At Least Three Modules Must Operate for Functioning of the System*

The 5-modular redundancy system has 32 distinct states. The system state diagram is partitioned into four major parts. The perfect state function (11111), the one module failed function states (11110), (11101), (11011), (10111), (01111), the two modules failed states (10110), (11100), (10011), (11001), (00111), (01110), (11010), (10101), (01011), (01101), and the failed states are (11000), (10001), (00011),

(00110), (00101), (10010), (01010), (10100), (01001), (01100), (10000), (01000), (00100), (00010), (00001) and (00000).

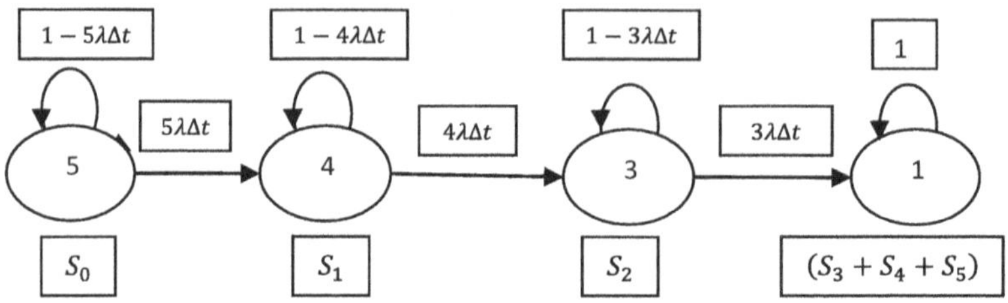

Using procedure of case (i), we get

The reliability of the 5-modular redundancy system, if at least three modules must operate

$$R_{35} = P_3(t) + P_4(t) + P_5(t) \tag{33}$$

The failure probability of the system is

$$P_F(t) = 1 - R_{35} \tag{34}$$

and similarly it can be derived that the 5-modular redundancy system reliability, if at least four modules must operate

$$R_{45} = P_5(t) + P_4(t) \tag{35}$$

and the system's failure probability is

$$P_F(t) = 1 - R_{45}. \tag{36}$$

4 Numerical Results

Case (i): Reliabilities of 4-modular redundancy system, if at least two modules should operate (R_{24}) and three modules should operate (R_{34}) for various values of λ.

t	λ	R_{24}	R_{34}
2	0.01	0.999969	0.997709
2	0.02	0.999766	0.99125
2	0.03	0.999245	0.981197
2	0.04	0.998287	0.968064
2	0.05	0.996799	0.952313
2	0.06	0.994707	0.934355
2	0.07	0.991955	0.91456
2	0.08	0.988504	0.893256
2	0.09	0.984329	0.870736

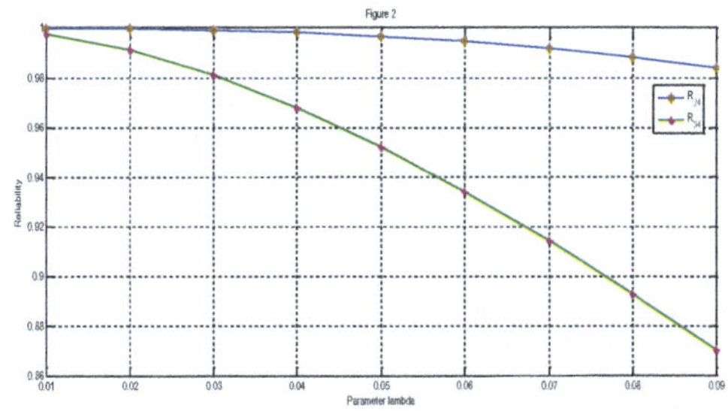

Case (ii): Reliabilities of 5-modular redundancy system, if at least two modules should operate (R_{25}), three modules should operate (R_{35}) and four modules should operate (R_{45}) for various values of λ.

t	λ	R_{25}	R_{35}	R_{45}
2	0.01	0.999999	0.999925	0.996232
2	0.02	0.999989	0.999432	0.985796
2	0.03	0.999945	0.998194	0.969866
2	0.04	0.999836	0.995963	0.949465
2	0.05	0.999621	0.992565	0.925478
2	0.06	0.999256	0.987882	0.89867
2	0.07	0.998696	0.981844	0.869704
2	0.08	0.997893	0.974421	0.839146
2	0.09	0.996803	0.965617	0.807483

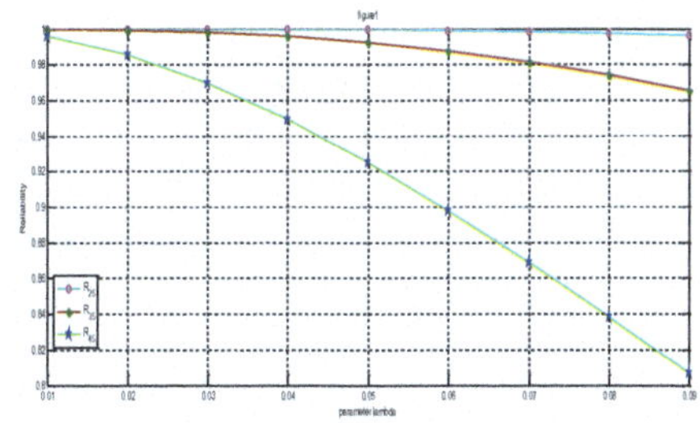

Case (iii): Reliabilities of 4-modular redundancy system, if at least two modules should operate (R_{24}) and three modules should operate (R_{34}) for various values of *t*.

t	λ	R_{24}	R_{34}
5	0.01	0.999553	0.98664
10	0.01	0.996799	0.952313
15	0.01	0.990319	0.904078
20	0.01	0.979414	0.84726
25	0.01	0.96389	0.785828
30	0.01	0.943895	0.722696
35	0.01	0.919801	0.65996
40	0.01	0.89211	0.599087
45	0.01	0.861393	0.541064

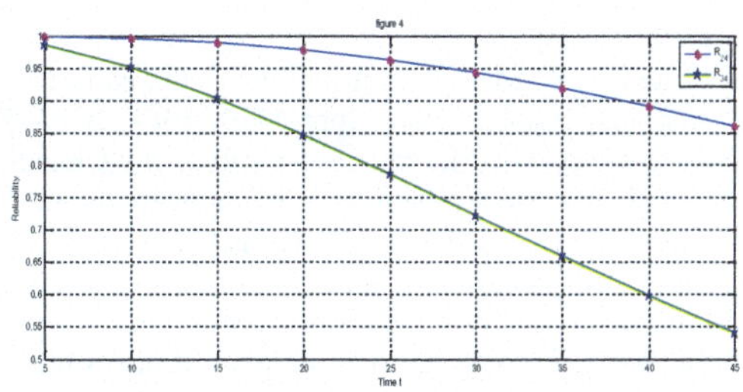

Case (iv): Reliabilities of 5-modular redundancy system, if at least two modules should operate (R_{25}), three modules should operate (R_{35}) and four modules must operate (R_{45}) for various values of t.

t	λ	R_{25}	R_{35}	R_{45}
5	0.01	0.999973	0.998923	0.978451
10	0.01	0.999621	0.992565	0.925478
15	0.01	0.998328	0.978306	0.854592
20	0.01	0.995384	0.955459	0.775127
25	0.01	0.990148	0.924503	0.693378
30	0.01	0.982116	0.886564	0.61345
35	0.01	0.970957	0.843067	0.537889
40	0.01	0.956512	0.795506	0.468141
45	0.01	0.938778	0.745315	0.404898

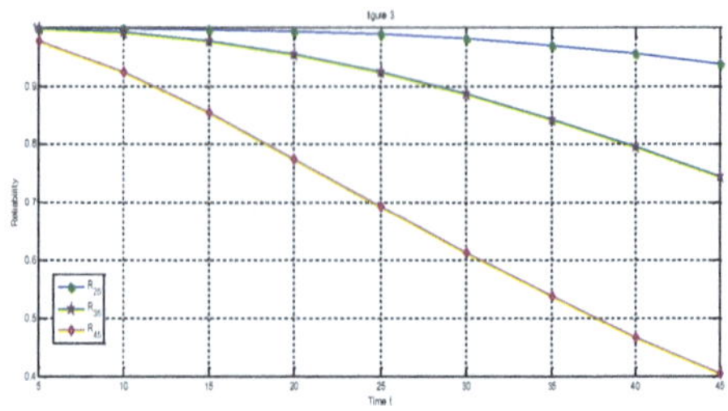

5 Conclusion

Reliability has been derived for the 4-modular redundancy system for at least two modules and three modules should operate and also has been derived that the reliability of 5-modular redundancy system, if at least two, three and four modules should operate. It is observed from the computations that the reliability decreases as λ & t increase for the 4-modular and 5-modular redundancy systems. It is also observed that when at least two modules operate, then the 4-modular and 5-modular redundancy system is more reliable.

References

1. Srinath, L.S.: Reliability Engineering, 4th edn. Affiliated East West Press Private Limited, New Delhi (2005)
2. Kalairasi, S., Merceline Anita, A., Gethanjalii, R.: Analysis of system reliability using markov technique. Glob. J. Pure Appl. Math. **13**(9), 5265–5275 (2017)
3. Reni sagayaraj, M., Merceline Anita, A., Chandrababu, A., Gowtham Prakash, S.: Markov models in system reliability with applications. Int. J. Innov. Res. Dev. (IJIRD) **3**(11) (2014)
4. Koteswarrao, B., Mahesh, D., Hima bindhu, S.K.: Design and Implementation of 4-bit multiplier using fault-tolerant self repairing full adder. Int. J. Recent Innov. Eng. Res. **120**(6), 2017–2041 (2018)
5. Sawant, L.S.: Applications of laplace transform in engineering fields. Int. Res. J. Eng. Technol. (IRJET) **05**(05) (2018)
6. Sadi, M.S., Sultana, S., Moin Al Rashid, Md., Lateef, K.F.: Towards tolerating faults by triple modular redundancy. J. Softw. Eng. Intell. Syst. **2**(2) (2017)
7. Jansi Rani, K., Vinitha, P., Prasanna Venkatesan, G.K.: Multiple error recovery in TMR systems. Int. J. Eng. Trends Technol. **11** (2014)
8. Rahman, M.H., Rafique, S., Alam, M.S.: A fault tolerant voter circuit for triple modular redundant system. J. Electr. Electron. Eng. **5**(5), 156–166 (2017)

An Improved Secant-Like Method and Its Convergence for Univariate Unconstrained Optimization

R. Bhavani and P. Paramanathan

Abstract We describe an improved Secant-like method based on divided difference approximation of third derivative for optimization of univariate unconstrained function. The convergence rate of the method is discussed with sufficient number of numerical examples to show the applications of our method. The collation of our consequence with subsisting models are given.

Keywords Unconstrained optimization · Univariate optimization · Iterative method · Secant-like method

1 Introduction

In the optimization of univariate unceasingly differentiable function $f(x)$, the minimization operation can still be executed just using function evaluation when the derivatives are not available at the evaluating time. Many optimization algorithms developed using various quasi newton approaches, aim to refining an initial rough approximation [1–4]. For rectifying the disadvantages of these methods, different versions of secant methods are introduced by various authors for optimizing univariate minimization problem

$$\min f(x) \tag{1}$$

where f is described on a real space $\mathbb{R}$.

In that, modification in the classical Secant method stationed on the development of cubic approximation gave markable improved results than the existing one [5]. Also replacing the second derivative of Newton's method using taylor series expansion we can get an variant Secant method,

R. Bhavani · P. Paramanathan (✉)
Department of Mathematics,
Amrita School of Engineering - Coimbatore Amrita Vishwa Vidyapeetham,
Coimbatore, India
e-mail: p_paramanathan@cb.amrita.edu

D. Dutta and B. Mahanty (eds.), *Numerical Optimization in Engineering and Sciences*,
Advances in Intelligent Systems and Computing 979,
https://doi.org/10.1007/978-981-15-3215-3_33

$$x_{d+1} = x_d - \frac{f'(x_d)(x_d - x_{d-1})}{2f'(x_d) - \frac{f(x_d)-f(x_{t-d})}{(x_d - x_{d-1})}} \tag{2}$$

In order to overcome the drawbacks of variant of Secant method we again use the taylor series expansion and second derivative of variant secant method to replace the third derivative and get modified Secant method.

$$x_{d+1} = x_d - \frac{1}{f'''(x_d)}[f''(x_d) + \sqrt{f''(x_d)^2 - 2f'''(x_d)f'(x_d)}] \tag{3}$$

This method converges faster compared to Secant method and variant of Secant method. But this technique can also be improved in a better way using divided difference property.

A unified convergence analysis for Wang's method at $\theta = -2$ with quadratical convergence is found for variant of Secant method using Divided difference property is excellent compared Secant method and variant of Secant method [6]

$$x_{d+1} = x_d - \frac{f'(x_d)(x_d - x_{d-1})}{4f'(x_d) - 6f[x_d, x_{d-1}] + 2f'(x_{d-1})} \tag{4}$$

But we can get faster convergence if the divided difference property is applied on f'''.

For solving nonlinear equations, a new secant-like method is found where no derivative is required and survey of the confluence manifest that the asymptotical order of convergence is $1 + \sqrt{3}$ [7]

$$y_{d+1} = x_{d+1} - \frac{(x_d - y_d)f(x_{d+1})}{f(x_d) - f(y_d)}$$

It is efficient only when the evaluation cost of the derivative is high.

A modification of Secant method, helps to upgrade the felicitous of the Secant method that changes the aspiration of a linear system in every step. They does not utilize derivatives and has R-order of convergence $\frac{1+\sqrt{5}}{2}$ [8]. Only the quantitative firmness of the Secant method is enhanced in [9].

A modification Secant-like method is demonstrated to have confluence order of 2.732 [9] and it is effective than the order of convergence 2.414 [10]. But increasing the order of derivative can give a faster and better convergence rate.

A unaccustomed technique for minimisation of univariate functions which evince an vitally quadratic convergence and whose confluence interim is only restricted by the actuality of closeby maxima is found [11]. Entailing only the first and second-order derivatives may help to find minimum but increasing the order helps to converge smoothly.

This paper introduces an improved secant-like method for finding the solution of univariate minimization problem by replacing third derivative f''' using divided difference property. Since the logic of any algorithm relatively depends on one-dimensional optimization [12] and they are absolutely necessary, we discuss one-dimensional optimization [5]. In most problems, certain degree of smoothness is possessed by functions. To utilize this smoothness, execution on polynomial approximation is designed. Order of convergence is greater than unity in this method [5]. Newton's method is prepossessing with regard to its convergence amid the techniques of polynomial approximation. Here using divided difference property in the modified Secant method, an improved Secant-like method is derived. Using x_d and x_{d-1} as our initial values for $d = 1, 2, 3, \ldots$ we will find x_{d+1} and until it satisfies the terminating condition (i.e) $|f'(x_{d+1})| < \epsilon$ we will iterate by replacing this x_{d+1} value with x_d if $f'(x_{d+1}) > 0$ else replace x_{d-1} if $f'(x_{d+1}) < 0$.

The execution of improved Secant-like method is compared and concluded based on its faster convergence using six standard test functions and its convergence rate is found. The best part of this method is for each value of θ, a new class is obtained and its convergence is found and compared. Since this method is used to solve non-linear functions we have taken functions of degree 2 or more and transcendental functions like exponential and cosine functions.

In this paper, we entrust an improved Secant-like method (9). Section 2 describes the details of the proposed method and its convergence properties. Section 3 manifest the numerical outcome of the proposed technique and its comparison with Secant method, variant of Secant method, Modified Secant method and Wang's method.

2 An Improved Secant-Like Method

Modified secant method [5] is given below as follows:

$$x_{d+1} = x_d - \frac{1}{f'''(x_d)}[f''(x_d) + \sqrt{f''(x_d)^2 - 2f'''(x_d)f'(x_d)}]$$

It has an order of convergence $p = 1.618\ldots$.

Using the divided difference property (see [6]), we have

$$\begin{aligned} f'''(x_d) &\approx \frac{f''(x_d) - f''(x_{d-1})}{x_d - x_{d-1}} \\ &= \frac{M - f[x_d, x_d, x_{d-1}] + f[x_d, x_d, x_{d-1}] - f[x_d, x_{d-1}, x_{d-1}] + f[x_d, x_{d-1}, x_{d-1}]}{x_d - x_{d-1}} \\ &= f[x_d, x_d, x_d, x_{d-1}] + f[x_d, x_d, x_{d-1}, x_{d-1}] + f[x_d, x_{d-1}, x_{d-1}, x_{d-1}] \end{aligned} \quad (5)$$

where,

$$M = \frac{f[x_d, x_d, x_d] - f[x_{d-1}, x_{d-1}, x_{d-1}]}{x_d - x_{d-1}}$$

Generally, in the backward difference of $f'''(x_d)$ the possibility of getting an error is high so in order to reduce the error we are implementing divided difference property.
and

$$f'''(x_d) \approx \frac{3f''(x_d) - h_d}{x_d - x_{d-1}} \tag{6}$$

This $f'''(x_d)$ is proposed in [5].
where,

$$f''(x_d) = f[x_d, x_d, x_{d-1}] + f[x_d, x_{d-1}, x_{d-1}]$$

and

$$\begin{aligned} h_d &= \frac{2f'(x_d) - [\frac{f(x_d)-f(x_{d-1})}{x_d-x_{d-1}}]}{x_d - x_{d-1}} \\ &= 2\frac{f[x_d, x_d] - f[x_d, x_{d-1}]}{x_d, x_{d-1}} \\ &= 2f[x_d, x_d, x_{d-1}] \end{aligned}$$

Therefore,

$$\begin{aligned} f'''(x_d) &= 3\frac{f[x_d, x_d, x_{d-1}] + f[x_d, x_{d-1}, x_{d-1}] - 2f[x_d, x_d, x_{d-1}]}{x_d - x_{d-1}} \\ &= -3f[x_d, x_d, x_{d-1}, x_{d-1}] \end{aligned} \tag{7}$$

where $f\underbrace{\ldots}_{d+1}$ are divided difference of order d for d=1,2,...[6].

From [6], suppose $\theta \in R$ is a persistent. By linear fusion of (4) and (6), we acquire

$$\begin{aligned} f'''(x_d) &\approx \theta(f[x_d, x_d, x_d, x_{d-1}] + f[x_d, x_d, x_{d-1}, x_{d-1}] + f[x_d, x_{d-1}, x_{d-1}, x_{d-1}]) - \\ &\quad - 3(1-\theta)(f[x_d, x_d, x_{d-1}, x_{d-1}]) \\ &= \frac{\theta f''(x_d) - 3(1+\theta)(f[x_d, x_d, x_{d-1}] - f[x_d, x_{d-1}, x_{d-1}]) - \theta f''(x_{d-1})}{x_d - x_{d+1}} \end{aligned} \tag{8}$$

Hence substituting $f'''(x_d)$ in Eq. (2), we get an improved Secant-like method as follows,

$$x_{d+1} = x_d - \frac{1}{f'''(x_d)}[f''(d_t)^2 + N] \tag{9}$$

where,

$$f'''(x_d) = \frac{\theta f''(x_d) - 3(1+\theta)(f[x_d, x_d, x_{d-1}] - f[x_d, x_{d-1}, x_{d-1}]) - \theta f''(x_{d-1})}{x_d - x_{d-1}},$$

$$N = \sqrt{f''(x_d)^2 - 2(\frac{\theta f''(x_d) - 3(1+\theta)(f[x_d, x_d, x_{d-1}] - f[x_d, x_{d-1}, x_{d-1}]) - \theta f''(x_{d-1})}{x_d - x_{d-1}})f'(x_d)}$$

For each value of θ we will get a new class and each class is compared with secant method, variant of Secant method and Modified Secant method based on its faster convergence.

This method has an order of convergence atleast $p = 1.618\ldots$.

Algorithm 1 This is the algorithm helps to calculate the optimal minimum point x^*.

1. Choose the value of θ, $\theta \in R$
2. Choose two points u and v such that $f'(u) < 0$ and $f'(v) > 0$ and also choose a small integer ϵ
   set $x_{d-1} = u$ for $d = 1, 2, \ldots$
   set $x_d = v$ for $d = 1, 2, \ldots$
3. Calculate the new point x_{d+1} using Eq.(2) and evaluate $f'(x_{d+1})$
4. If $|f'(x_{d+1})| < \epsilon$, Terminate
   elseif $f'(x_{d+1}) < 0$, set $x_{d-1} = x_{d+1}$ and go to Step 3.
   elseif $f'(x_{d+1}) > 0$, set $x_d = x_{d+1}$ and go to Step 3.

3 Numerical Test

We will compare the execution of improved Secant-like method with Secant method, Variant of Secant method, Modified Secant method, Wang's method based on its number of iterations using six Test functions which were used in [5, 6, 10, 12]. In Table 1, the test function and its minimum point x^* is given. Tables 2 and 3, will give the improved Secant-like method results. Table 4 gives the final results by comparing the following Eqs. (10), (2), (3), (4), (9) using their initial points.

Secant method:

$$x_{d+1} = x_d - \frac{f'(x_d)(x_d - x_{d-1})}{f'(x_d) - f'(x_{d-1})} \tag{10}$$

Table 2 represents the first four test functions with its number of iterations for θ values (5, 10, 15, 20, 25), they converges and at $\theta = 10$ the number of iterations are less compared to other θ values. For θ values (0, −5, −10, −15, −20) these four functions are increasing function and they does not converge.

Table 1 Standard optimization problems

No.	Function	Minimum point
1	$x^4 - 8.5 * x^3 - 31.0625 * x^2 - 7.50 * x + 45$	8.2784
2	$(x+2)^2(x+4)(x+5)(x+8)(x-16)$	12.6791
3	$e^x - 3 * x^2$	2.8331
4	$Cosx + (x-2)^2$	2.3542
5	$3774.522(1/x) + 2.27x - 181.529, x > 0$	40.7772
6	$10.2(1/x) + 6.2 * x^3, x > 0$	0.8605

Table 2 Secant-like method results (1)

No.	$\theta = 5$	$\theta = 10$	$\theta = 15$	$\theta = 20$	$\theta = 25$
1	4	3	4	3	4
2	12	6	8	6	8
3	9	5	9	6	6
4	6	4	5	4	5

Table 3 Secant-like method results (2)

No.	$\theta = -5$	$\theta = -10$	$\theta = -15$	$\theta = -20$	$\theta = -25$
5	11	5	6	5	6
6	10	5	8	6	12

Table 4 The results

Functions	x_0	x_{-1}	(10)	(2)	(3)	(4)	(9)
1	8.0	7.8	6	6	4	6	3
2	12.0	12.1	7	7	4	5	6
3	2.6	2.5	4	4	3	6	5
4	2.0	1.8	6	5	4	4	4
5	32.0	30.0	7	7	5	5	5
6	0.5	0.4	7	7	4	5	5

Table 3 represents the last two test functions with its number of iterations for θ values $(-5, -10, -15, -20, -25)$, they converges and at $\theta = -10$ the number of iterations are less compared to other θ values. For θ values (0, 5, 10, 15, 20) these four functions are increasing function and they does not converge.

4 Conclusion

We have developed an improved Secant-like method to solve univariate unconstrained optimization problems, based on modified Secant method from [5, 6]. The Proposed method examines using divided difference property and shares the same convergence properties with modified secant method [5]. The improved Secant-like method is excellent compared to proposed methods.

References

1. Ortega, J.M., Rheinbolt, W.C.: Iterative Solution of Nonlinear Equations in Several Variables. Academic Press, New York (1970)
2. Mamta, V.K., Kukreja, V.K., Singh, S.: On some third-order iterative methods for solving nonlinear equations. Appl. Math. Comput. **171**, 272–280 (2005)
3. Kou, J., Li, Y., Wang, X.: On modified Newton methods with cubic convergence. Appl. Math. Comput. **176**, 123–127 (2006)
4. Rardin, R.L.: Optimization in Operations Research. Prentice-Hall Inc., New York, NJ (1998)
5. Kahya, E., Chen, J.H.: A modified Secant method for unconstrained optimization. Appl. Math. Comput. **186**, 1000–1004 (2007)
6. Ren, Hongmin, Qingbiao, Wu: A class of modified Secant methods for unconstrained optimization. Appl. Math. Comput. **206**, 716–720 (2008)
7. Wang, X., Kou, J., Gu, C.: A new modified secant-like method for solving nonlinear equations. Comput. Math. Appl. **60**, 1633–1638 (2010)
8. Amata, S., Hernndez-Vern, M.A., Rubio, M.J.: Improving the applicability of the Secant method to solve nonlinear systems of equations. Appl. Math. Comput. **247**, 741–752 (2014)
9. Ren, H.: A second-derivative-free modified Secant-like method with order 2.732... for unconstrained optimization. Appl. Math. Comput. **202**, 688–692 (2008)
10. Kahya, E.: Modified Secant-type methods for unconstrained optimization. Appl. Math. Comput. **181**, 1349–1356 (2006)
11. Salgueiro da Silva, M.A.: A novel method for robust minimisation of univariate functions with quadratic convergence. Appl. Math. Comput. **200**, 168–177 (2007)
12. Kahya, E.: A new unidimensional search method for optimization: linear interpolation method. Appl. Math. Comput. **171**, 912–926 (2005)

Integrability Aspects of Deformed Fourth-Order Nonlinear Schrödinger Equation

S. Suresh Kumar

Abstract A systematic investigation on the integrability of deformed fourth-order nonlinear Schrödinger (D4oNLS) equation is presented. It is shown that the D4oNLS equation admits a Lax representation satisfying certain differential constrains on the deforming functions. From our results, we observed that the D4oNLS equation and the deformed second-order nonlinear Schrödinger (NLS) equation admit the same sets of differential constraints on the deforming functions. It is also shown that the D4oNLS equation possesses one-soliton, two-soliton and three-soliton solutions. These soliton solutions have been derived by using Hirota's method.

Keywords Integrability · Deformed fourth-order nonlinear Schrödinger equation · Soliton solution · Hirota's method

1 Introduction

Many nonlinear physical phenomena can be modeled in the form of NLS equations, which are arising in the field of science and engineering such as optical fibers, fluid dynamics etc, [1–8, 12, 15, 16, 19–23]. The standard second-order NLS (2oNLS) equation (or simply NLS equation) is an important basic model for the propagation, dynamics of waves and magnetostatic spin waves [1, 3, 7, 8, 20]. However, upon increasing the wave amplitude, many researchers have considered higher order effects (that is, by adding various higher order terms to the standard 2oNLS equation), which are not included in the standard 2oNLS equation [4–6, 21–23]. The integrability of standard 2oNLS equation may be lost when adding various higher order terms to standard 2oNLS equation. It is well known that if a given nonlinear partial differential equation (NPDE) with two independent variables t and x admits a Lax pair, then it is integrable in the sense of Lax [1, 12, 13]. It is also well known that a given NPDE

S. Suresh Kumar (✉)
PG and Research Department of Mathematics, C. Abdul Hakeem College (Autonomous), Melvisharam, Vellore Dt, Tamilnadu 632509, India
e-mail: kds_sureshmurali@yahoo.com

Ramanujan Institute for Advanced Study in Mathematics, University of Madras, Chepauk, Chennai 600005, Tamilnadu, India

D. Dutta and B. Mahanty (eds.), *Numerical Optimization in Engineering and Sciences*, Advances in Intelligent Systems and Computing 979,
https://doi.org/10.1007/978-981-15-3215-3_34

is said to be integrable if it possesses multi-soliton solutions [10]. In this paper, we consider integrable 4oNLS equation [6] given by,

$$iq_t + q_{xx} + 2q^2q^* + \delta(q_{xxxx} + 8qq^*q_{xx} + 2q^2q^*_{xx} + 4qq_xq^*_x + 6q_x^2q^* + 6q^3q^{*2}) = 0 \tag{1}$$

along with its Lax pair M, N satisfying the Lax equation

$$M_t - N_x + [M, N] = 0 \quad \text{(or)} \quad M_t - N_x + MN - NM = 0, \tag{2}$$

where M and N are 2×2 matrices given by

$$\begin{aligned} M(q, q^*, \lambda) &= J_1 - iJ_0\lambda, \\ N(q, q^*, \lambda) &= -(i + 3i\delta qq^*)J_2 + i\delta J_4 + (2J_1 - 2\delta J_3)\lambda \\ &\quad + (-2iJ_0 + 4i\delta J_2)\lambda^2 - 8\delta J_1\lambda^3 + 8i\delta J_0\lambda^4, \end{aligned} \tag{3}$$

λ is a spectral parameter and

$$\begin{aligned} J_0 &= \begin{pmatrix} 1 & 0 \\ 0 & -1 \end{pmatrix}, J_1 = \begin{pmatrix} 0 & q \\ -q^* & 0 \end{pmatrix}, J_2 = \begin{pmatrix} -qq^* & -q_x \\ -q^*_x & qq^* \end{pmatrix}, \\ J_3 &= \begin{pmatrix} -(q^*q_x - qq^*_x) & -(q_{xx} + 2q^2q^*) \\ q^*_{xx} + 2q^{*2}q & q^*q_x - qq^*_x \end{pmatrix}, \\ J_4 &= \begin{pmatrix} qq^*_{xx} - q_xq^*_x + q^*q_{xx} & q_{xxx} + 3qq^*q_x \\ q^*_{xxx} + 3qq^*q^*_x & -qq^*_{xx} + q_xq^*_x - q^*q_{xx} \end{pmatrix}. \end{aligned}$$

If an integrable NPDE gets perturbed or deformed, then it is of interest to investigate whether it preserves the integrability of un-perturbed NPDE. However, recent investigations shown that the perturbed or deformed NPDEs preserve its integrability [11, 14, 17, 18]. In Ref. [17, 18], we have showed that the deformed NLS equation and deformed Hirota equation admit Lax pair which indicates the integrability of both deformed equations. In the present work, we have shown that the D4oNLS equation given by

$$iq_t + q_{xx} + 2q^2q^* + \delta(q_{xxxx} + 8qq^*q_{xx} + 2q^2q^*_{xx} + 4qq_xq^*_x + 6q_x^2q^* + 6q^3q^{*2}) = g, \tag{4a}$$

with the differential constraints on the deforming functions g and h are given by,

$$g_x = -2iqh, \tag{4b}$$

$$h_x = iqg^* - iq^*g \tag{4c}$$

admits a Lax representation satisfying certain differential constrains on the deforming functions. Also, we derive its one-soliton, two-soliton and three-soliton solutions by using Hirota's method.

2 Lax Pair and Soliton Solutions of D4oNLS Equation

2.1 *Lax Pair*

To construct Lax matrices for the D4oNLS equation (4), we have introduced the negative power of the spectral parameter λ (that is, λ^{-1}) in the Lax matrix N given in (3) and keeping the Lax matrix M unchanged. Then, the Lax Eq. (2) leads to the D4oNLS equation (4) provided

$$M(q, q^*, \lambda) = J_1 - iJ_0\lambda,$$
$$N(q, q^*, \lambda) = -\frac{i}{2\lambda} J_{\text{def}} - (i + 3i\delta q q^*) J_2 + i\delta J_4 + (2J_1 - 2\delta J_3)\lambda$$
$$+ (-2i J_0 + 4i\delta J_2)\lambda^2 - 8\delta J_1 \lambda^3 + 8i\delta J_0 \lambda^4, \text{ where } J_{\text{def}} = \begin{pmatrix} h & -ig \\ ig^* & -h \end{pmatrix}.$$

2.2 *Soliton Solutions*

To obtain soliton solutions of the D4oNLS equation (4), first, we rewrite Eq. (4) as

$$iq_t + q_{xx} + 2q^2 q^* + \delta(q_{xxxx} + 8qq^* q_{xx} + 2q^2 q^*_{xx}$$
$$+ 4qq_x q^*_x + 6q_x^2 q^* + 6q^3 q^{*2}) = v_x, \tag{5a}$$

$$v_{xx} = -2iqw_x, \tag{5b}$$

$$w_{xx} = iqv^*_x - iq^* v_x, \tag{5c}$$

where $g = \frac{\partial v}{\partial x}$ and $h = \frac{\partial w}{\partial x}$. According to the Hirota's method [9], the solution of (5) is assumed to be

$$q(x,t) = \frac{Q(x,t)}{S(x,t)}, v(x,t) = \frac{V(x,t)}{S(x,t)}, w(x,t) = \frac{W(x,t)}{S(x,t)}, \tag{6}$$

where Q, V are complex functions and W, S are real functions. Substituting the above transformations (6) into (5) and rearranging the terms, we find that the Eq. (5) can be written into the following trilinear form

$$(iD_t + D_x^2 + \delta D_x^4)Q.S = D_x V.S, \tag{7a}$$

$$D_x^2 S.S = 2QQ^*, \tag{7b}$$

$$S\big(D_x^2 V.S\big) - V\big(D_x^2 S.S\big) + 2iQ(D_x W.S) = 0, \tag{7c}$$

$$S\big(D_x^2 W.S\big) - W\big(D_x^2 S.S\big) + iQ^*(D_x V.S) - iQ\big(D_x V^*.S\big) = 0, \tag{7d}$$

where D is the Hirota operator [9] defined by

$$D_t^n D_x^m g.h = \left(\frac{\partial}{\partial t} - \frac{\partial}{\partial t'}\right)^n \left(\frac{\partial}{\partial x} - \frac{\partial}{\partial x'}\right)^m g(x,t)h\big(x',t'\big)|_{(x',t')=(x,t)}.$$

Expanding the functions Q, V, W, S as power series in ε, that is,

$$Q = \sum_{n=1}^{\infty} \varepsilon^{2n-1} Q_{2n-1}, \quad V = \sum_{n=1}^{\infty} \varepsilon^{2n-1} V_{2n-1},$$

$$W = W_0(x,t) + \sum_{n=1}^{\infty} \varepsilon^{2n} W_{2n}, \quad S = 1 + \sum_{n=1}^{\infty} \varepsilon^{2n} S_{2n}$$

and using them in (7), one can construct the N-soliton solutions in the usual way.

2.2.1 One-Soliton Solution

To compute one-soliton solution of (5), we now assume that

$$Q = \varepsilon Q_1,\ V = \varepsilon V_1,\ W = W_0 + \varepsilon^2 W_2,\ S = 1 + \varepsilon^2 S_2, \tag{8}$$

where ε is an arbitrary small parameter. Substituting the above expressions (8) in Eq. (7) and then equating the different powers of ε to zero yield an overdetermined system of linear partial differential equations (LPDEs). Solving them consistently yields

$$Q_1 = e^{\eta},\ V_1 = -8i\, m(t)e^{\eta+\mu},$$

$$S_2 = e^{\eta+\eta^*+\mu},\ W_0 = m(t)x,\ W_2 = \frac{m(t)(\alpha x - 4)}{\alpha} e^{\eta+\eta^*+\mu},$$

where $\eta = \alpha x + i\alpha^2 t + i\delta\alpha^4 t - \frac{2}{\alpha}\int m(t)\mathrm{d}t + \beta + i\gamma$, $e^{\mu} = \frac{1}{4\alpha^2}$ and α, β, γ are real constants and $m(t)$ is an arbitrary function. So, the one-soliton solution of Eq. (5) is

$$q = \frac{Q}{S} = \frac{e^{\eta}}{1 + e^{\eta+\eta^*+\mu}}, \tag{9a}$$

$$v = \frac{V}{S} = -\frac{8im(t)e^{\eta+\mu}}{1 + e^{\eta+\eta^*+\mu}}, \tag{9b}$$

$$w = \frac{W}{S} = \frac{m(t)\left[x + \frac{\alpha x-4}{\alpha}e^{\eta+\eta^*+\mu}\right]}{1 + e^{\eta+\eta^*+\mu}}. \tag{9c}$$

Hence, the one-soliton solution of the D4oNLS equation (4) is

$$q = \frac{e^{\eta}}{1 + e^{\eta+\eta^*+\mu}}, \tag{10a}$$

$$g = -\frac{2im(t)e^{\eta}\left(1 - e^{\eta+\eta^*+\mu}\right)}{\alpha(1 + e^{\eta+\eta^*+\mu})^2}, \tag{10b}$$

$$h = \frac{m(t)\left[1 - 6e^{\eta+\eta^*+\mu} + e^{2\eta+2\eta^*+2\mu}\right]}{(1 + e^{\eta+\eta^*+\mu})^2}. \tag{10c}$$

Note that when $m(t) = 0$ (that is, the deformed functions $g = h = 0$), the solution (10) is the one-soliton solution of fourth-order NLS Eq. (1). Also, note that when $\delta = 0$, the solution (10) is the one-soliton solution of deformed NLS equation. The above one-soliton solution (9) depends on the arbitrary function $m(t)$. The evolution of one-soliton solution (9) with different choices of $m(t)$ and some different parameters are given in Figs. 1, 2, 3. In Fig. 2, we can see that the soliton represents the parabolic-type propagation with intensity $|q|^2$. In Fig. 3, the soliton represents the periodicity oscillated-type propagation with intensity $|q|^2$.

2.2.2 Two-Soliton Solution

To compute two-soliton solution of (5), we now assume that

$$Q = \varepsilon Q_1 + \varepsilon^3 Q_3, \quad V = \varepsilon V_1 + \varepsilon^3 V_3,$$
$$W = W_0 + \varepsilon^2 W_2 + \varepsilon^4 W_4, \quad S = 1 + \varepsilon^2 S_2 + \varepsilon^4 S_4.$$

Using them in Eq. (7) and then equating the different powers of ε to zero yield an overdetermined system of LPDEs. Solving them consistently yields

$$Q_1 = e^{\eta_1} + e^{\eta_2}, \quad Q_3 = e^{\eta_1+\eta_1^*+\eta_2+\mu_1} + e^{\eta_1+\eta_2+\eta_2^*+\mu_2},$$
$$V_1 = -8i\, m(t)\left(e^{\eta_1+\mu_3} + e^{\eta_2+\mu_4}\right),$$

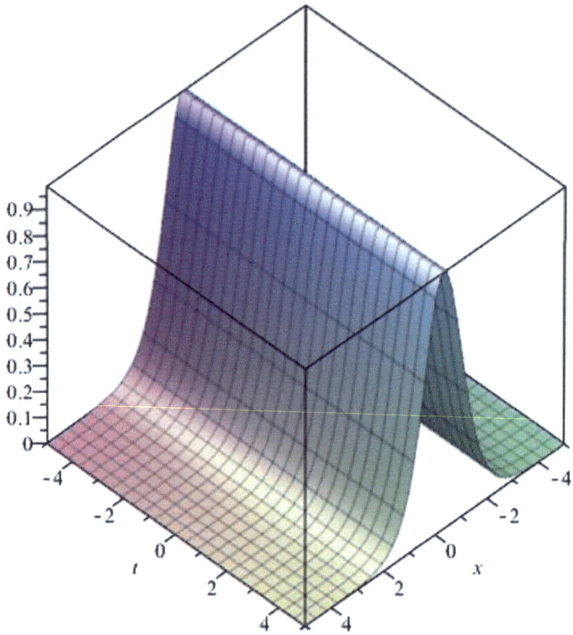

Fig. 1 One-soliton solution (9) with $\delta = 1, \alpha = 1, \beta = 1, \gamma = 0, m(t) = 0$

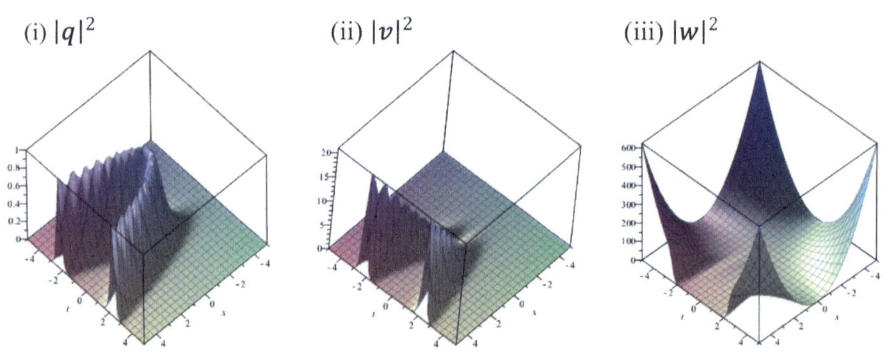

Fig. 2 One-soliton solution (9) with $\delta = 1, \alpha = 1, \beta = 1, \gamma = 0, m(t) = t$

$$
\begin{aligned}
V_3 &= -8i\, m(t)\big(e^{\eta_1+\eta_1^*+\eta_2+\mu_1+\mu_4} + e^{\eta_1+\eta_2+\eta_2^*+\mu_2+\mu_3}\big),\\
S_2 &= e^{\eta_1+\eta_1^*+\mu_3} + e^{\eta_2+\eta_2^*+\mu_4} + e^{\eta_1+\eta_2^*+\mu_5} + e^{\eta_1^*+\eta_2+\mu_5},\\
S_4 &= e^{\eta_1+\eta_1^*+\eta_2+\eta_2^*+\mu_1+\mu_2},\ W_0 = m(t)x,\\
W_2 &= m(t)\big[(\alpha_1 x - 4)e^{\eta_1+\eta_1^*+\kappa_1} + (\alpha_2 x - 4)e^{\eta_2+\eta_2^*+\kappa_2}\\
&\quad +(\alpha_1\alpha_2 x - 2\alpha_1 - 2\alpha_2)\big(e^{\eta_1+\eta_2^*+\kappa_3} + e^{\eta_1^*+\eta_2+\kappa_3}\big)\big],\\
W_4 &= m(t)(\alpha_1\alpha_2 x - 4\alpha_1 - 4\alpha_2)e^{\eta_1+\eta_1^*+\eta_2+\eta_2^*+\kappa_4},
\end{aligned}
$$

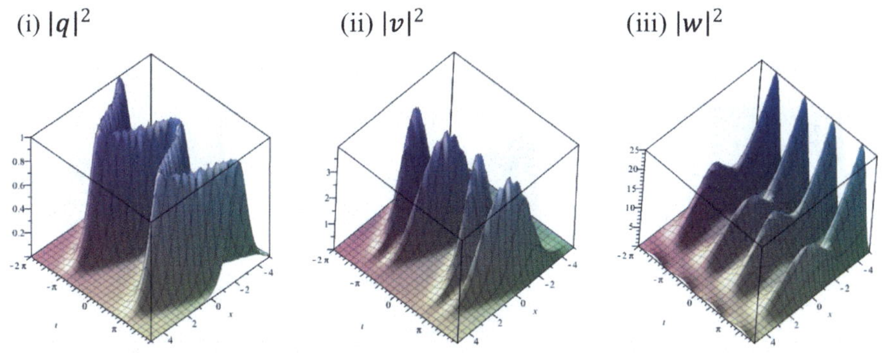

Fig. 3 One-soliton solution (9) with $\delta = 1, \alpha = 1, \beta = 1, \gamma = 0, m(t) = \sin t$

where

$$\eta_l = \alpha_l + i\alpha_l^2 t + i\delta\alpha_l^4 t - \frac{2}{\alpha_l}\int m(t)\mathrm{d}t + \beta_l + i\gamma_1, l = 1, 2,$$

$$e^{\mu_1} = \frac{(\alpha_1 - \alpha_2)^2}{4\alpha_1^2(\alpha_1 + \alpha_2)^2}, e^{\mu_2} = \frac{(\alpha_1 - \alpha_2)^2}{4\alpha_2^2(\alpha_1 + \alpha_2)^2}, e^{\mu_3} = \frac{1}{4\alpha_1^2},$$

$$e^{\mu_4} = \frac{1}{4\alpha_2^2}, e^{\mu_5} = \frac{1}{(\alpha_1 + \alpha_2)^2}, e^{\kappa_1} = \frac{1}{4\alpha_1^3}, e^{\kappa_2} = \frac{1}{4\alpha_2^3},$$

$$e^{\kappa_3} = \frac{1}{\alpha_1\alpha_2(\alpha_1 + \alpha_2)^2}, e^{\kappa_4} = \frac{(\alpha_1 - \alpha_2)^4}{16\alpha_1^3\alpha_2^3(\alpha_1 + \alpha_2)^4}$$

and $\alpha_1, \alpha_2, \beta_1, \beta_2, \gamma_1$ are real constants and $m(t)$ is an arbitrary function with $\alpha_1^2 t + \alpha_1^4 t = \alpha_2^2 t + \alpha_2^4 t + 2k\pi$, k is a nonzero integer. So, the two-soliton solution of Eq. (5) is

$$q = \frac{e^{\eta_1} + e^{\eta_2} + e^{\eta_1+\eta_1^*+\eta_2+\mu_1} + e^{\eta_1+\eta_2+\eta_2^*+\mu_2}}{S(x,t)},$$

$$v = \frac{-8im(t)\left[e^{\eta_1+\mu_3} + e^{\eta_2+\mu_4} + e^{\eta_1+\eta_1^*+\eta_2+\mu_1+\mu_4} + e^{\eta_1+\eta_2+\eta_2^*+\mu_2+\mu_3}\right]}{S(x,t)},$$

$$\begin{aligned} w = \frac{m(t)}{S(x,t)}\big[&x + (\alpha_1 x - 4)e^{\eta_1+\eta_1^*+\kappa_1} + (\alpha_2 x - 4)e^{\eta_2+\eta_2^*+\kappa_2} \\ &+ (\alpha_1\alpha_2 x - 2\alpha_1 - 2\alpha_2)\left(e^{\eta_1+\eta_2^*+\kappa_3} + e^{\eta_1^*+\eta_2+\kappa_3}\right) \\ &+ (\alpha_1\alpha_2 x - 4\alpha_1 - 4\alpha_2)e^{\eta_1+\eta_1^*+\eta_2+\eta_2^*+\kappa_4}\big], \end{aligned}$$

where

$$\begin{aligned} S(x,t) = 1 &+ e^{\eta_1+\eta_1^*+\mu_3} + e^{\eta_2+\eta_2^*+\mu_4} + e^{\eta_1+\eta_2^*+\mu_5} + e^{\eta_1^*+\eta_2+\mu_5} \\ &+ e^{\eta_1+\eta_1^*+\eta_2+\eta_2^*+\mu_1+\mu_2}. \end{aligned}$$

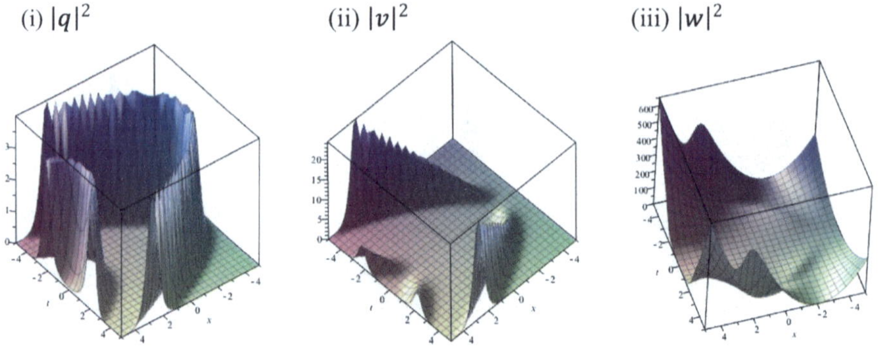

Fig. 4 Two-soliton solution with $\delta = 1, \alpha_1 = 20, \alpha_2 = 20.1, \beta_1 = \beta_2 = 1, \gamma_1 = 0, m(t) = t$

The propagation of two-soliton solution with $\delta = 1, \alpha_1 = 20, \alpha_2 = 20.1, \beta_1 = \beta_2 = 1, \gamma_1 = 0, m(t) = t$ is given in Fig. 4.

2.2.3 Three-Soliton Solution

Proceeding in a similar manner explained in previous subsections, we checked that Eq. (5) admits a three-soliton solution given by

$$q(x,t) = \frac{Q}{S} = \frac{Q_1 + Q_3 + Q_5}{1 + S_2 + S_4 + S_6},$$

where $Q_1 = e^{\eta_1} + e^{\eta_2} + e^{\eta_3}$,

$$\begin{aligned}
Q_3 &= \sum_{l=1}^{3} B_l e^{\eta_1+\eta_2+\eta_l^*} + \sum_{l=1}^{3} B_{l+3} e^{\eta_1+\eta_3+\eta_l^*} + \sum_{l=1}^{3} B_{l+6} e^{\eta_2+\eta_3+\eta_l^*},\\
Q_5 &= C_1 e^{\eta_1+\eta_2+\eta_3+\eta_1^*+\eta_2^*} + C_2 e^{\eta_1+\eta_2+\eta_3+\eta_1^*+\eta_3^*} + C_3 e^{\eta_1+\eta_2+\eta_3+\eta_2^*+\eta_3^*},\\
S_2 &= \sum_{l=1}^{3} D_l e^{\eta_1+\eta_l^*} + \sum_{l=1}^{3} D_{l+3} e^{\eta_2+\eta_l^*} + \sum_{l=1}^{3} D_{l+6} e^{\eta_3+\eta_l^*},\\
S_4 &= A_1 e^{\eta_1+\eta_2+\eta_1^*+\eta_2^*} + A_2 e^{\eta_1+\eta_2+\eta_1^*+\eta_3^*} + A_3 e^{\eta_1+\eta_2+\eta_2^*+\eta_3^*}\\
&\quad + A_4 e^{\eta_1+\eta_3+\eta_1^*+\eta_2^*} + A_5 e^{\eta_1+\eta_3+\eta_1^*+\eta_3^*} + A_6 e^{\eta_1+\eta_3+\eta_2^*+\eta_3^*}\\
&\quad + A_7 e^{\eta_2+\eta_3+\eta_1^*+\eta_2^*} + A_8 e^{\eta_2+\eta_3+\eta_1^*+\eta_3^*} + A_9 e^{\eta_2+\eta_3+\eta_2^*+\eta_3^*},\\
S_6 &= Y e^{\eta_1+\eta_2+\eta_3+\eta_1^*+\eta_2^*+\eta_3^*}
\end{aligned}$$

and

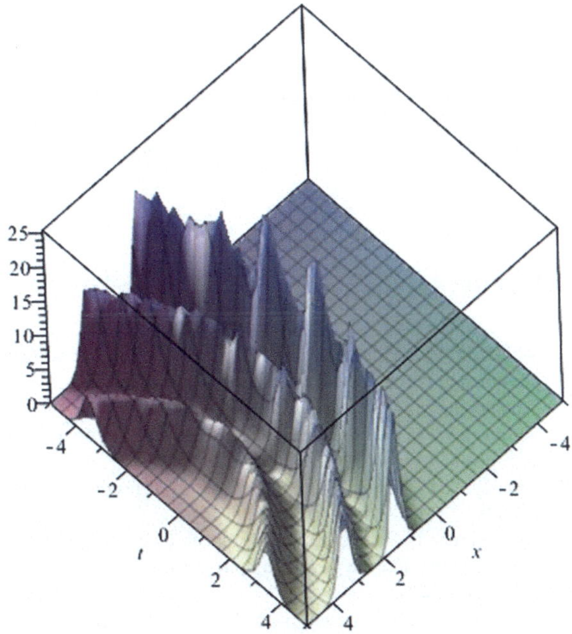

Fig. 5 Three-soliton solution with $\delta = 1, \alpha_1 = 4, \alpha_2 = 3.5, \alpha_3 = 2.5, \beta_1 = \beta_2 = \beta_3 = 1, \gamma_1 = 0, m(t) = t$

$$\eta_l = \alpha_l + i\alpha_l^2 t + i\delta\alpha_l^4 t - \frac{2}{\alpha_l} \int m(t)\mathrm{d}t + \beta_l + i\gamma_1, l = 1, 2, 3,$$

$\alpha_1, \alpha_2, \alpha_3, \beta_1, \beta_2, \beta_3, \gamma_1$ are real constants and $m(t)$ is an arbitrary function. The expressions of $v(x, t)$, $w(x, t)$ and the real parameters $A_l, B_l, D_l, l = 1, \ldots, 9,\ C_1, C_2, C_3, Y$ involving lengthy expressions, therefore we omit from presenting here. The propagation of three-soliton solution with $\delta = 1, \alpha_1 = 4, \alpha_2 = 3.5, \alpha_3 = 2.5, \beta_1 = \beta_2 = \beta_3 = 1, \gamma_1 = 0, m(t) = t$ is given in Fig. 5.

3 Conclusion

A systematic investigation on the integrability of D4oNLS equation is presented. It is shown that the D4oNLS equation admits a Lax representation satisfying certain differential constrains on the deforming functions. From our results, we observed that the D4oNLS equation and the deformed second-order nonlinear Schrödinger equation admit the same sets of differential constraints on the deforming functions. It is also shown that the D4oNLS equation possesses one-soliton, two-soliton and three-soliton solutions. These soliton solutions have been derived by using Hirota's method.

Acknowledgements The author is thankful to the anonymous referees for their constructive suggestions.

The author gratefully acknowledges the support of the Professor R. Sahadevan (RIASM, University of Madras, Chennai) for his valuable suggestions which have improved the paper. The author would like to thank the Management and Principal of C.Abdul Hakeem College (Autonomous), Melvisharam, for their support and encouragement.

References

1. Ablowitz, M.J., Clarkson, P.A.: Solitons, Nonlinear Evolution Equations and Inverse Scattering. Cambridge University Press, Cambridge (1992)
2. Ahmed, I., Seadawy, A.R., Lu, D.: Kinky breathers, W-shaped and multi-peak solitons interaction in (2 + 1)-dimensional nonlinear Schrödinger equation with Kerr law of nonlinearity. Eur. Phys. J. Plus **134**(3), 120 (2019)
3. Benney, D.J., Newell, A.C.: Propagation of nonlinear wave envelopes. J. Math. Phys. **46**, 133 (1967)
4. Brugarinoa, T., Sciacca, M.: Singularity analysis and integrability for a higher order nonlinear Schrödinger equation governing pulse propagation in a generic fiber optic. Opt. Commun. **262**, 250–256 (2006)
5. Chowdury, A., Kedziora, D.J., Ankiewicz, A., Akhmediev, N.: Soliton solutions of an integrable nonlinear Schrödinger equation with quintic terms. Phys. Rev. E **90**, 032922 (2014)
6. Guo, R., Hao, H.Q.: Breathers and multi-soliton solutions for the higher-order generalized nonlinear Schrödinger equation. Communi. Nonlinear Sci. Numer. Simulat. **18**(9), 2426–2435 (2013)
7. Hasegawa, A., Kodama, Y.: Solitons in Optical Communications. Oxford University Press, Oxford (1995)
8. Hasegawa, A., Tappert, F.: Transmission of stationary nonlinear optical pulses in dispersive dielectric fibers. Appl. Phys. Lett. **23** (1973)
9. Hietarinta, J.: Introduction to the Hirota bilinear method. Lect. Notes. Phys. **495** (1997)
10. Hietarinta, J.: Hirota's Bilinear method and its connection with integrability. Lect. Notes. Phys. **767**, 279–314 (2009)
11. Kundu, A.: Nonlinearizing linear equations to integrable systems including new hierarchies with nonholonomic deformations. J. Math. Phys. **50**, 102702 (2009)
12. Lakshmanan, M., Rajasekar, S.: Nonlinear Dynamics, Integrability, Chaos and Patterns. Springer, Berlin (2003)
13. Lax, P.D.: Integrals of nonlinear equations of evolution and solitary waves. Commun. Pure Appl. Math. **21**, 467–490 (1968)
14. Sahadevan, R., Nalinidevi, L.: Integrability of certain deformed nonlinear partial differential equations. J. Nonlinear Math. Phys. **17**, 379–396 (2010)
15. Sahadevan, R., Radhakrishnan, R., Lakshmanan, M.: Integrability and singularity structure of coupled nonlinear Schrödinger equations. Chaos Solitons Fractals **5**(12), 2315–2327 (1995)
16. Shao, Y., Zeng, Y.: The solutions of the NLS equations with self-consistent sources. J. Phys. A **38**, 2441 (2005)
17. Suresh Kumar, S., Balakrishnan, S., Sahadevan, R.: Integrability and lie symmetry analysis of deformed N-coupled nonlinear Schrödinger equations. Nonlinear Dyn. **90**, 2783–2795 (2017)
18. Suresh Kumar, S., Sahadevan, R.: Integrability and group theoretical aspects of deformed N-coupled Hirota equations. Int. J. Appl. Comput. Math **5**, 1–32 (2019)
19. Yong, X., Fany, Y., Huangz, Y., Ma, W.X., Tian, J.: Darboux transformation and solitons for an integrable nonautonomous nonlinear integro-differential Schrödinger equation. Mod. Phys. Lett. B **31**(30), 1750276 (2017)

20. Zakharov, V.E.: Stability of periodic waves of finite amplitude on a surface of a deep fluid. J. Appl. Mech. Tech. Phys. **9**, 190 (1968)
21. Zhang, Z., Tian, B., Liu, L., Sun, Y., Zhong, Du: Lax pair, breather-to-soliton conversions, localized and periodic waves for a coupled higher-order nonlinear Schrödinger system in a birefringent optical fiber. Eur. Phys. J. Plus **134**, 129 (2019)
22. Zhang, H.Q., Yuan, S.S.: General *N*-dark vector soliton solution for multi-component defocusing Hirota system in the optical fiber media. Commun. Nonlinear Sci. Numer. Simulat. **51**, 124–132 (2017)
23. Zhou, C., He, X.T., Chen, S.: Basic dynamic properties of the high-order nonlinear Schrödinger equation. Phys. Rev. A **46**, 2277 (1992)

A New Approach for Finding a Better Initial Feasible Solution to Balanced or Unbalanced Transportation Problems

B. S. Surya Prabhavati and V. Ravindranath

Abstract In this paper, we propose a method to find a better initial feasible solution to a balanced or unbalanced transportation problem. Unlike other methods, the entering nonbasic variable in this method is chosen based on the quantities of supply or demand corresponding in addition to cost. The method is applied over several numerical examples studied in the literature and solution performance compared with the competing methods JHM, TOM, etc., in terms of minimizing transportation cost and computational efficiency. Solutions obtained by the proposed method are found to be optimal in many examples and near optimal in the other cases in terms of taking lesser number of iterations to reach optimality.

Keywords Transportation problem · Linear programming (LP) · Optimization initial basic solution · Vogel's approximation method · Unbalanced transportation problem

1 Introduction

One important application of linear programming (LP) is in the area of physical distribution (transportation) of goods and services from several supply centers to many demand centers. The structure of transportation problem involves large number of shipping routes from several supply origins to several demand destinations. The objective is to determine the number of units of an item that should be shipped from an origin to a destination in order to satisfy the requirement of goods or services at each destination. This should be done within the limited quantity of goods or services available at each source with minimum transportation cost and/or time. Mathematically, it is easy to express a transportation problem in terms of an LP model which can be solved by using simplex method. Since LP model involves large number of variables and constraints, it generally takes longer time to solve by the conventional simplex method. Several heuristic methods have been developed to

B. S. Surya Prabhavati · V. Ravindranath (✉)
Department of Mathematics, JNTUK, Kakinada 533003, India
e-mail: nath_vr@yahoo.com

D. Dutta and B. Mahanty (eds.), *Numerical Optimization in Engineering and Sciences*, Advances in Intelligent Systems and Computing 979,
https://doi.org/10.1007/978-981-15-3215-3_35

find initial basic feasible solutions which can later be improved to become optimum solutions.

Some methods like north-west corner (NWC) and least cost (LC) are computationally simple to obtain initial solutions; however, it will take longer time to improve them to optimum solutions. Other methods like Vogel's approximation method (VAM) are computationally more laborious initially, but the initial solutions obtained by these methods will be optimal in many cases or may take lesser number of iterations to reach optimum in the other cases. Methods like stepping stone and modified distribution (MODI) have been developed for this purpose.

There are various transportation models, and the simplest of them was first presented by Hitchcock [6] which was later developed by Koopmans [10] and Dantzig [3]. A transportation problem is called balanced if the total supply is equal to the total demand, otherwise it is unbalanced. Several extensions of the transportation models and the methods have been developed by several authors subsequently. Vogel's approximation method (VAM) [14] is widely used to solve transportation problems to find optimal or near optimal allocation of commodities to various routes which minimizes the total transportation cost. Charnes and Cooper [2] developed stepping stone method to find optimal solution of a transportation problem. Srinivasan and Thomson [17] proposed area cost operator method or an improved version of it named cell cost operator method (CCOM) by solving the dual problems of the transportation problem and proved the convergence of the method. However, in this approach, it is assumed that the transportation problem is balanced which may not be the case always. Some researchers proposed algorithms based on the dual problems of the transportation problem, and an interested reader can refer to the works of Sharma and Sharma [15] and Ji and Chu [7]. The present paper is aimed to solve only the primal problem and therefore not discussed.

Shimshak et al [16] pointed out that when the transportation problem is unbalanced, VAM fails to provide efficient solutions and several alternatives can be developed. By using a modified heuristic method (SVAM), Shimshak et al [16] illustrated with a numerical example and presented an improved solution over VAM. Goyal [4] remarked that if the transportation problem is unbalanced, VAM should be modified in calculating penalties of the allocating cells. With the help of some numerical examples, Goyal [4] illustrated his variant GVAM which is improved in the sense of obtaining optimum solution. Ramakrishnan [13] improved the method GVAM for unbalanced transportation problem. Balakrishnan [1] further modified the methods SVAM and GVAM and demonstrated with numerical examples. Kirca and Satir [9] describe a heuristic method called total opportunity cost method (TOM) wherein the allotment of quantity is made to the cell based on the total opportunity cost of that cell. Here, the sum of row opportunity cost and column opportunity cost gives the total opportunity cost. It is illustrated with a numerical study that TOM outperforms VAM in case of unbalanced transportation problems. On the other hand, VAM was found superior to TOM to solve balanced transportation problems.

The comparison of TOM with VAM is further carried out by Goyal [5] who suggested that the largest opportunity cost should be assigned to the dummy cells. The VAM is considered to be very efficient to find an optimal or near optimal solution to the TP. However, in case of unbalanced TP, VAM fails to give efficient solutions and many alternate methods can be proposed which are available in the works of Shimshak [16], Goyal [5], etc. Mathirajan and Meenakshi [12] have also proposed some variants to VAM, and the solutions have been analyzed with experimental analysis. Kulkarni and Datar [11] suggested that if a TP is unbalanced and if there is an excess in supply than in demand, then the excess supply will be added to the demands of one of the destinations. Similarly, if there is an excess in demand, it will be added to the supply of one of the sources. However, this method gives the costs higher than allotting the excess quantity to some dummy source or destination and hence is not preferred. Juman and Hoque [8] used a method that the sum of allocations in a row is greater than the corresponding supply and then transfers the maximum excess quantity supply from least unit cost cell to the next least-cost cell in a column.

In all the above methods, the nonbasic variable is selected based on the cell cost but not on the quantity, and hence, a method can be developed in which the entering variable is selected not only on the cost but also on the quantity of shipment.

The main objective of this paper is to propose a method which selects the nonbasic cell based on both lowest cost and also quantity of shipment. Since the quantity of shipment will be the minimum of supply and demand corresponding to the cell, the method should take supplies and demands into consideration. An algorithm is developed for the proposed method and implemented in R software version 3.3.1 on Windows O.S. The method is illustrated with several numerical examples discussed in the literature, and the results are compared with its competitors in terms of number of computations. The proposed method is introduced in Sect. 2 to solve both balanced and unbalanced transportation problems. Numerical illustration is provided in Sect. 3, and a comparative study is done in Sect. 4.

2 Proposed Method

Let there be m sources $S_i, i = 1, \ldots, m$, each can supply a_i units of a certain commodity to n destinations $D_j, j = 1, \ldots, n$ which has the demand b_j each.

Let $c_{ij}, i = 1, \ldots, m, \ j = 1, \ldots, n$ be the transportation cost to be incurred in shipping one unit of the commodity from ith source to jth destination.

The decision variable is the quantity $x_{ij}, i = 1, \ldots, m, j = 1, \ldots, n$ to be shipped from ith source to jth destination, so that the total transportation cost is the minimum. Pictorially the transportation problem is shown as follows (Table 1).

The transportation problem can be modeled as a linear programming problem (LPP) as below

$$\text{Min} \sum_{i=1}^{m} \sum_{j=1}^{n} c_{ij} x_{ij}$$

Subject to

Table 1 A typical transportation problem

	D_1	D_2	D_j	D_n	$\sum$
S_1	c_{11}	c_{12}	c_{1j}	c_{1n}	a_1
S_2	c_{21}	c_{22}	c_{2j}	c_{2n}	a_2
S_i	c_{i1}	c_{i2}	c_{ij}	c_{in}	a_i
S_m	c_{m1}	c_{m2}	c_{mj}	c_{mn}	a_m
$\sum$	b_1	b_2	b_j	b_n	

$$\sum_{j=1}^{n} x_{ij} = a_i, \quad i = 1, \ldots, m$$
$$\sum_{i=1}^{m} x_{ij} = b_j, \quad j = 1, \ldots, n$$
$$a_i, b_j, x_{ij} \geq 0, \quad \forall i \text{ and } j$$

If the total supply is equal to the total demand

$$\sum_{i=1}^{m} a_i = \sum_{j=1}^{n} b_j$$

then the TP is called balanced otherwise it is unbalanced.

Algorithm

Step 1. Check whether the given problem is balanced. If not make it balanced.
Step 2. Reducing cost matrix to a matrix having at least one zero in each row and column. This can be done by subtracting minimum cost of each row from the elements of that row followed by subtracting minimum cost of each column from the elements of that column.
Step 3. **Selecting row or column**:
Find the minimum nonzero value among the supplies or demands and select the row or column corresponding to that supply or demand. This may lead to the following cases

1. If minimum nonzero value appears uniquely among the supplies (demands), then select the corresponding row (column)
2. If minimum nonzero value appears at more than one place among the supplies (demands), then select the row (column) for which the sum of row (column) costs is the maximum.
3. If minimum nonzero value appears at more than one place among both supplies and demands, then select the row if its sum of row costs is the maximum or the column if its sum of column costs is the maximum.
4. If tie occurs further, then break it arbitrarily.

Step 4. Selecting the cell for allotment:
Once a row or column is selected in Step 3, then select the cell as follows

1. If a cell with zero cost occurs uniquely, then select that cell.
2. If a cell with zero cost occurs at more than one place, then select the cell for which the sum of costs in that row or column is larger.
3. If a cell with zero cost is not available (or already allotted some quantity), then choose the cell with next least cost. If there is further tie in the least costs, then break it arbitrarily.

Step 5. Allotment to cell:
Allot some quantity of supply or demand whichever is the minimum to that cell. Block the row whose supply is exhausted or column whose demand is satisfied from future allotments.
Step 6. Stopping condition:
If there are any nonzero supplies or demands which remain un-allotted then go to step 3. Otherwise, stop.
The above algorithm can be applied on both balanced and unbalanced TPs. In the next section, we illustrate the usage of the algorithm on some numerical examples discussed in the literature.

3 Numerical Illustration

Example-1 JHM (2015)

Destinations / Sources	D1	D2	D3	Supply
S1	6	8	10	150
S2	7	11	11	175
S3	4	5	12	275
Demand	200	100	300	600 / 600

Step 1:

Find the minimum cost among the row costs and subtract it from the cost elements of that row.

Destinations / Sources	D1	D2	D3	Supply
S1	0	2	4	150
S2	0	4	4	175
S3	0	1	8	275
Demand	200	100	300	600 / 600

Step 2:

Choose the minimum cost among the column costs and subtract it from the cost elements of that column.

Destinations / Sources	D1	D2	D3	Supply
S1	0	1	0	150
S2	0	3	0	175
S3	0	0	4	275
Demand	200	100	300	600 / 600

Step 3:

The smallest value among the supplies and demands is 100 corresponding to the destination D2. Among the costs of D2 column, the smallest cost is 0 corresponding to the cell (3, 2) against which allotment is to be done.

Sources \ Destinations	D1	D2	D3	Supply
S1	0	1	0	150
S2	0	3	0	175
S3	0	0 **100**	4	175
Demand	200	0	300	500 \ 500

Step 4:

The smallest value among the supplies and demands is 150 corresponding to the source S1, and therefore, S1 row is chosen for allotment. Among the costs of S1 row, the smallest cost is 0 corresponding to the cells (1, 1) and (1, 3). Out of these, (1, 3) cell is chosen as the sum of costs in that column which is maximum (0 + 0 + 4 = 4) against the allotment which is to be done.

Sources \ Destinations	D1	D2	D3	Supply
S1	0	1	0 **150**	0
S2	0	3	0	175
S3	0	0 **100**	4	175
Demand	200	0	150	350 \ 350

Step 5:

The smallest value among the supplies and demands is 150 corresponding to the destination D3, and therefore, D3 column is chosen for allotment. Among the costs of D3 column, the smallest cost is 0 corresponding to the cells (1, 3) (against which allotment has already been done in Step-4) and (2, 3). Therefore, the cell (2, 3) is considered.

Sources \ Destinations	D1	D2	D3	Supply
S1	0	1	0 (**150**)	0
S2	0	3	0 (**150**)	25
S3	0	0 (**100**)	4	175
Demand	200	0	0	200 / 200

Step 6:

The smallest value among the supplies and demands is 25 corresponding to the source S2, and therefore, S2 row is chosen for allotment. Among the costs of S2 row, the smallest cost is 0 corresponding to the cells (2, 1) and (2, 3) (against which allotment has already been done in Step-5). So, choose the cell (2, 1) against which allotment is to be done.

Sources \ Destinations	D1	D2	D3	Supply
S1	0	1	0 (**150**)	0
S2	0 (**25**)	3	0 (**150**)	0
S3	0	0 (**100**)	4	175
Demand	175	0	0	175 / 175

Step 7:

The smallest value among the supplies and demands is 175 corresponding to the source S3, and therefore, S3 row is chosen for allotment. Among the costs of S3 row, the smallest cost is 0 corresponding to the cell (3, 1) against which allotment is to be done

Destinations / Sources	D1		D2		D3		Supply
S1	0		1		0	**150**	0
S2	0	**25**	3		0	**150**	0
S3	0	**175**	0	**100**	4		0
Demand	0		0		0		0 / 0

Since all the supplies exhausted and demands fulfilled, the algorithm is stopped, and the solution is $x_{13} = 150$, $x_{21} = 25$, $x_{23} = 150$, $x_{31} = 175$ and $x_{32} = 100$. Multiplying each variable with the corresponding cell cost and adding give the cost of transportation as 4525.

By using the JHM method of unmet rows and met rows, it is verified that the above solution is optimum. It may be remarked here that the above solution coincided with solution obtained by Juman and Hoque [8] by applying JHM method.

4 Conclusion

In this paper, a new method is proposed which uses both costs and as well as quantities while choosing the entering nonbasic variable. The solution obtained by the proposed method has been shown to be improved over the existing methods VAM, etc., in terms of obtaining minimum cost or better starting solution to reach optimality in terms of taking lesser number of iterations. Finding the initial feasible solution by JHM method is lengthy and time-consuming. The proposed method is very simple, efficient, easily understood and reliable, when compared with the existing methods. Table 2 presents a comparison of various solutions.

Table 2 Comparison of solutions by various methods

S. no.	Problem taken from	Minimum cost	Minimum cost by the proposed method	Optimal cost	No. of iterations to reach optimality
1	Balakrishnan [1]	1650	1650	1650	0
2	Goyal [4]	1665	1650	1650	0
3	Goyal [5]	1565	1590	1565	1
4	Kirca et al. [9]	1480	1480	1480	0
5	Ramakrishnan [13]	1650	1650	1650	0
6	Shimshak et al. [16]	1695	1650	1650	0
7	Srinivasan et al. [17]	880	880	880	0

References

1. Balakrishnan, N.: Modified Vogel's approximation method for the unbalanced transportation problem. Appl. Math. Lett. **3**(2), 9–11 (1990)
2. Charnes, A., Cooper, W.W., Henderson, A.: An Introduction to Linear Programming. Wiley, New York (1953)
3. Dantzig, G.B.: Linear Programming and Extensions. Princeton University Press, Princeton, N. J. (1963)
4. Goyal, S.K.: Improving VAM for unbalanced transportation problems. J. Oper. Res. Soc. **35**(12), 1113–1114 (1984)
5. Goyal, S.K.: A note on a heuristic for obtaining an initial solution for the transportation problem. J. Oper. Res. Soc. **42**(9), 819–821 (1991)
6. Hitchcock, F.L.: The distribution of a product from several sources to numerous localities. J. Math. Phys. 20, 224–230 (1941)
7. Ji, P., Chu, K.F.: A dual-matrix approach to the transportation problem. Asia-Pac. J. Oper. Res. **19**(1), 35–45 (2002)
8. Juman, Z.A.M.S., Hoque, M.A.: An efficient heuristic to obtain a better initial feasible solution to the transportation problem. Appl. Soft Comput. **34**, 813–826 (2015)
9. Kirca, O., Satir, A.: A heuristic for obtaining an initial solution for the transportation problem. J. Oper. Res. Soc. **41**(9), 865–871 (1990)
10. Koopmans, T.C.: Optimum utilization of transportation system. Econometrica. 17, (1949)
11. Kulkarni, S.S., Datar, H.G.: On solution to modified unbalanced transportation problem. Bull. Marathwada Math. Soc. **11**(2), 20–26 (2010)
12. Mathirajan, M., Meenakshi, B.: Experimental analysis of some variants of Vogel's approximation method. Asia-Pac. J. Oper. Res. **21**(4), 447–462 (2004)
13. Ramakrishnan, C.S.: An improvement to Goyal's modified VAM for the unbalanced transportation problem. J. Oper. Res. Soc. **39**(6), 609–610 (1988)
14. Reinfeld, N.V., Vogel, W.R.: Mathematical Programming, pp. 59–70. Prentice Hall, Englewood Cliffs, N.J. (1958)

15. Sharma, R.R.K., Sharma, K.D.: A new dual based procedure for the transportation problem. Eur. J. Oper. Res. **122**(3), 611–624 (2000)
16. Shimshak, D.G., Kaslik, J.A., Barclay, T.D.: A modification of Vogel's approximation method through the use of heuristics. INFOR **19**, 259–263 (1981)
17. Srinivasan, V., Thompson, G.L.: Cost operator algorithms for the transportation problem. Math. Program. **12**, 372–391 (1977)

Heat Transfer to Peristaltic Transport in a Vertical Porous Tube

V. Radhakrishna Murthy and P. Sudam Sekhar

Abstract Peristaltic transport of a Newtonian fluid, with heat transfer, in a vertical porous axisymmetric tube under long-wavelength approximation is considered. Closed-form solution is obtained as an asymptotic expansion in terms of porosity and free convection parameters. Expressions, for velocity, temperature, coefficient of heat transfer and pressure–flow relationship at the boundary wall of the tube, are derived. It is observed that pressure drop increases as amplitude ratio increases. Further, it has been observed that for some specific values of different parameters under consideration, the mean flux significantly increases by about 8–10% as Grashof number increases from 1 to 2. This relates to optimization of heat transfer in certain processes.

Keywords Peristalsis · Heat transfer · Porosity · Pressure drop · Mean flux

1 Introduction

Peristalsis is a mechanism for fluid transport which is achieved by the passage of area contraction and expansion waves along the length of the distensible tube. It is known that peristalsis emerges from optimization principles. It is one of the main mechanisms for fluid transport in physiological systems. Based on this principle, a blood pump in dialysis is designed to prevent the transported fluid from being contaminated. Peristaltic transport of toxic liquid is used in nuclear industries so as not to contaminate outside environment.

In view of its importance, studies of peristaltic transport have been carried out theoretically and experimentally by various authors [1–12]. In particular, Radhakrishnamacharya [1] investigated peristaltic pumping in an axisymmetric tube under long-wavelength approximation using power-law fluid model. Takabatake et al. [2]

V. Radhakrishna Murthy (✉) · P. Sudam Sekhar
Division of Mathematics, Vignan's Foundation for Science, Technology and Research (Deemed to be University), Guntur, India
e-mail: vadapalli.radhakrishna@gmail.com

D. Dutta and B. Mahanty (eds.), *Numerical Optimization in Engineering and Sciences*, Advances in Intelligent Systems and Computing 979,
https://doi.org/10.1007/978-981-15-3215-3_36

have developed complete numerical solutions for peristaltic pumping and its efficiency in Cartesian and axisymmetric geometries. Recently, the thermodynamical aspects of peristalsis have received attention [3–7] as it might be relevant in certain processes like oxygenation and hemodialysis. A drug pump, with optimization pump motor, is configured to treat a variety of medical conditions and works on peristalsis principle.

Translocation of water in tall trees is a phenomenon which is not well understood by scientists over centuries. It is speculated that peristalsis might be involved in this process since diameters of trunks of the trees are found to vary with time. Hence, some authors [10, 11] have investigated the peristalsis phenomena particularly in the case of transport of water in trees. Further, it is observed that flow of water takes place through the porous matrix of the tree.

Keeping in view of the above observation, in this paper, an attempt has been made to study heat transfer for the Newtonian fluid flow under the action of peristalsis in a vertical axisymmetric porous tube. Assuming long-wavelength approximation, analytic expressions for velocity, temperature, pressure drop and heat transfer coefficient have been obtained in terms of porosity (σ^2) and free convection (G_{m}) parameters. From the analysis, it has been observed that for fixed values of other parameters, a small change in free convection mean flux increases significantly.

2 Mathematical Formulation

The flow of a Newtonian incompressible fluid through an axisymmetric vertical tube filled with porous material is considered. Peristaltic waves of very large wavelength are assumed to travel down the wall of the tube. In this problem, the cylindrical polar coordinate (X, R) is chosen where X represents axial coordinate and R is radial coordinate.

The simplified zeroth-order equations under long-wave approximation governing the flow [8–10] are

$$0 = -\frac{\partial p}{\partial x} + \frac{\mu}{R}\frac{\partial}{\partial R}\left(R\frac{\partial W}{\partial R}\right) - \frac{\mu}{k_0}W + \rho g\beta\,(T - T_0) \tag{1}$$

$$0 = \frac{\partial W}{\partial R} + \frac{U}{R} + \frac{\partial U}{\partial R} \tag{2}$$

$$0 = \frac{K}{R}\frac{\partial}{\partial R}\left(R\frac{\partial T}{\partial R}\right) + \mu\left(\frac{\partial W}{\partial R}\right)^2 + \frac{\mu}{k_0}W^2 \tag{3}$$

The equation of the tube wall is given by

$$H(X,t) = a + b\,\mathrm{Sin}\left(\frac{2\pi}{\lambda}\right)(X - \mathrm{ct}) \tag{4}$$

where the fluid velocity components are W and U in the direction X and R, respectively, T is the fluid temperature, T_0 is the boundary temperature, 'p' is the pressure, ρ is the density, the coefficient of expansion is β, k stands for thermal conductivity of the fluid, μ is viscosity coefficient, g is the acceleration due to gravity, c is the phase speed of the wave, k_0 is the permeability of the medium, a is the mean radius of the tube, b is the amplitude and the wavelength is λ. The assumed boundary conditions for the problem under considerations are

$$T = T_0 \text{ and } W = 0 \text{ at } R = \pm H \tag{5}$$

Relative to the laboratory frame, the wave frame of reference moves with a constant speed c. In the wave frame, the measurement of the variables x and r are defined by

$$x = X - \mathrm{ct},\ r = R \tag{6}$$

The corresponding velocity components of the fluid are

$$w = W - c,\ u = U \tag{7}$$

The governing system of equations of the fluid flow in the wave frame of reference is written as

$$0 = -\frac{\partial p}{\partial x} + \frac{\mu}{r}\frac{\partial}{\partial r}\left(r\frac{\partial w}{\partial r}\right) - \frac{\mu}{k_0}(w + c) + \rho g\beta(T - T_0) \tag{8}$$

$$0 = \frac{\partial w}{\partial R} + \frac{u}{r} + \frac{\partial u}{\partial r} \tag{9}$$

$$0 = \frac{k}{r}\frac{\partial}{\partial r}\left(r\frac{\partial T}{\partial r}\right) + \mu\left(\frac{\partial w}{\partial r}\right)^2 + \frac{\mu}{k_0}(w + c)^2 \tag{10}$$

The boundary conditions are

$$w = -c \text{ and } T = T_0 \text{ at } r = a + b\sin\left(\frac{2\pi}{\lambda}\right)x \tag{11}$$

The non-dimensional quantities can be introduced as

$$x' = \frac{x}{\lambda},\ r' = \frac{r}{a},\ w' = \frac{w}{c},\ u' = \frac{\lambda}{ac}u,\ \theta = \frac{T - T_0}{T_0},\ p' = \frac{p}{\mu c\lambda/a^2} \tag{12}$$

Into Eqs. (8)–(11), we get (after dropping the primes)

$$0 = -\frac{\partial p}{\partial x} + \frac{1}{r}\frac{\partial}{\partial r}\left(r\frac{\partial w}{\partial r}\right) - \sigma^2(w + 1)G_m\theta \tag{13}$$

$$0 = \frac{\partial w}{\partial x} + \frac{u}{r} + \frac{\partial u}{\partial r} \tag{14}$$

$$0 = \frac{1}{r}\frac{\partial}{\partial r}\left(r\frac{\partial \theta}{\partial r}\right) + E_m\left(\frac{\partial w}{\partial r}\right)^2 + \sigma^2 E_m (w+1)^2 \tag{15}$$

The boundary conditions for the dimensionless quantities are

$$w = -1 \text{ and } \theta = 0 \text{ at } r = \pm\eta(x) \tag{16}$$

where

$$\eta(x) = 1 + \varepsilon \sin 2\pi x \tag{17}$$

$$\sigma^2 = \frac{a^2}{k_n}(\text{Porosity parameter}),\ G_m = \frac{g\beta T_0 a^3}{\upsilon^2}(\text{Grashof number})$$
$$E_m = \frac{\mu c^2}{K T_0}(\text{Eckert number}) \text{ and } \varepsilon = \frac{b}{a}(\text{amplitude ratio})$$

3 Analysis

Equations (13) and (15) are simultaneous nonlinear equations; hence, for arbitrary values of all the parameters, it is quite impossible to get an exact solution. So, we opt for perturbation method in the form of a series as

$$F = (F_{00} + G_m F_{01} + \cdots) + \sigma^2(F_{10} + \cdots + \cdots \tag{18}$$

where the flow variable is denoted as F.

By solving the resultant equation which is obtained by using (18) in Eqs. (13), (15) and (16), the solutions for velocity component w and temperature θ can be obtained under the suitable boundary conditions as follows

$$w = (w_{00} + G_m w_{01} + \cdots) + \sigma^2(w_{10} + \cdots) + \cdots \tag{19}$$

$$\theta = (\theta_{00} + G_m \theta_{01} + \cdots) + \sigma^2(\theta_{10} + \cdots) + \cdots \tag{20}$$

where

$$w_{00} = -1 + \frac{1}{4}\alpha\left(r^2 - \eta^2\right),$$

$$w_{01} = \frac{1}{4}\tau\left(r^2 - \eta^2\right) + \frac{1}{64}E_m\alpha^2\left[\frac{r^6 - \eta^6}{36} - \frac{\eta^4\left(r^2 - \eta^2\right)}{4}\right]$$

$$w_{10} = \frac{1}{4}\gamma\left(r^2 - \eta^2\right) + \frac{1}{4}\alpha\left[\frac{r^4 - \eta^4}{16} - \frac{\eta^2\left(r^2 - \eta^2\right)}{4}\right]$$

$$\theta_{00} = \frac{-1}{64}E_m\alpha^2\left(r^4 - \eta^4\right),$$

$$\theta_{01} = \frac{-1}{32}E_m\left[\alpha\tau\left(r^4 - \eta^4\right) + \frac{E_m}{768}\left(r^8 - 12r^4\eta^4 + 11\eta^8\right)\right]$$

$$\theta_{10} = \frac{-1}{32}E_m\left[\alpha\gamma\left(r^4 - \eta^4\right) + \frac{\alpha^2}{36}\left(2r^6 - 9r^4\eta^2 + 7\eta^6\right)\right]$$
$$- \frac{\alpha^2}{72}\left(2r^6 - 9r^4\eta^2 + 18r^2\eta^4 - 20\eta^6\right)$$

$$\alpha = \frac{8\left(\overline{Q_{00} + \pi}\right)}{\pi\left(1 - 2\eta^2\right)},$$

$$\tau = \frac{1}{\pi\left(1 - 2\eta^2\right)}\left[32\overline{Q}_{01} - \frac{1}{9}E_m\alpha^2\left(140\eta^6 - 18\eta^4\right) + 1\right]$$

$$\gamma = \frac{1}{\left(1 - 2\eta^2\right)}\left[\frac{8}{\pi}\overline{Q}_{10} - \frac{\alpha}{24}\left(1 - 21\eta^4 - 6\eta^2\right)\right]$$

The pressure drop over one wavelength is defined by

$$\Delta p_\lambda = \int_0^\lambda \frac{\partial p}{\partial x}\mathrm{d}x \tag{21}$$

By substituting the relation for $\frac{\partial p}{\partial x}$ from Eqs. (13) in (21) and using expressions for velocity and temperature from Eqs. (19) and (20), the non-dimensional pressure drop can finally be obtained as

$$\Delta p = \frac{\Delta p_\lambda}{\frac{\mu c\lambda}{a^2}} = \left(\Delta p_{00} + G_m\Delta p_{01} + \cdots\right) + \sigma^2\left(\Delta p_{10} + \cdots\right) + \cdots \tag{22}$$

where

$$\Delta p_{00} = \frac{32\varepsilon^2\left(1 - \frac{\varepsilon^2}{16}\right) - 8\bar{Q}_{00}\left(1 + \frac{3\varepsilon^2}{2}\right)}{\left(1 - \varepsilon^2\right)^{7/2}}$$

$$\Delta p_{01} = \frac{5}{6}E_m\left[1 + \left(\bar{Q}_{00} - \left(1 + \frac{\varepsilon^2}{2}\right)\right)^2\right)\left(\frac{\left(1 + \frac{3\varepsilon^2}{2}\right)}{\left(1 - \varepsilon^2\right)^{7/2}} + \frac{\bar{Q}_{00} - \left(1 + \frac{\varepsilon^2}{2}\right)}{\left(1 - \varepsilon^2\right)^{9/2}}\right)$$

$$- 8\bar{Q}_{01}\frac{\left(1+\frac{3\varepsilon^2}{2}\right)}{(1-\varepsilon^2)^{7/2}}$$

$$\Delta p_{10} = -6\bar{Q}_{10}\left[\frac{2+\varepsilon^2}{2(1-\varepsilon^2)^{5/2}}\right] - \frac{7}{20}\left[1+\frac{\bar{Q}_{00}-1}{(1-\varepsilon^2)^{9/2}}\right]$$

and $\overline{Q}$ is the dimensionless mean flux.

The non-dimensional form of heat transfer coefficient Z on the boundary of the tube is given by

$$Z = \left(\frac{\partial\theta}{\partial x} + \frac{\partial\theta}{\partial r}\frac{\partial\eta}{\partial x}\right) \text{at}\, r = \eta \tag{23}$$

which in view of (18) and can be expressed as

$$Z = (Z_{00} + G_m Z_{01} + \cdots) + \sigma^2(Z_{10} + \cdots) + \cdots \tag{24}$$

where

$$Z_{00} = \frac{\partial\theta_{00}}{\partial r}\frac{\partial\eta}{\partial x},\ Z_{01} = \frac{\partial\theta_{01}}{\partial r}\frac{\partial\eta}{\partial x},\ Z_{10} = \frac{\partial\theta_{10}}{\partial r}\frac{\partial\eta}{\partial x}.$$

4 Results and Discussion

Analytic expressions for velocity, temperature, pressure drop and the coefficient of heat transfer are given by the Eqs. (19), (20), (22) and (24), respectively. To study explicitly, temperature, pressure drop and coefficient of heat transfer have been numerically evaluated, and the results are shown in Figs. 1, 2, 3 and 4 and presented in Tables 1, 2, 3 and 4. Here, the effects of various parameters on these flow variables are depicted clearly.

Figures 1, 2, 3 and 4 show the variation of temperature versus X with respect to various parameters. It is observed from Figs. 1, 2, 3 and 4 that for a fixed value of certain parameters, temperature first increases down the tube and then decreases. This may be due to the effects of peristalsis. From Figs. 1 and 2, we can see that the temperature increases as Eckert number (E_m) or Grashof number (G_m) increases when other parameters are made fixed. Further, the temperature increases as σ^2 or ε increases, i.e., the temperature increases as the tube becomes more porous or peristaltic wave amplitude increases (Figs. 3 and 4).

It has been observed that the mean flux, $\overline{Q}$, increases by about 8–10% as G_m increases from 1 to 2 when some other parameters are given a fixed value during the computational process.

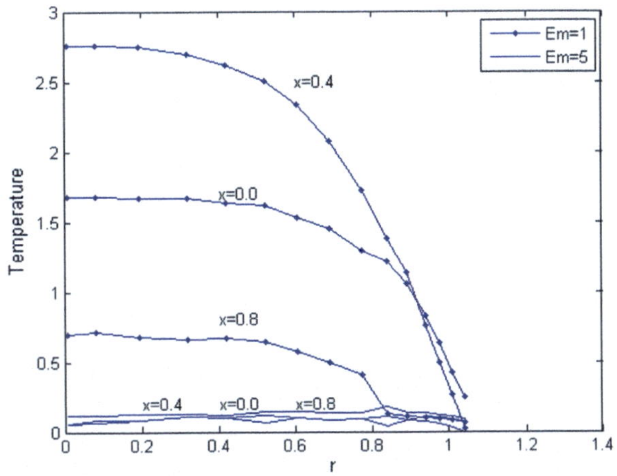

Fig. 1 Temperature variation with E_m for the fixed values of $G_m = 3, \sigma^2 = 2$ and $\varepsilon = 0.1$

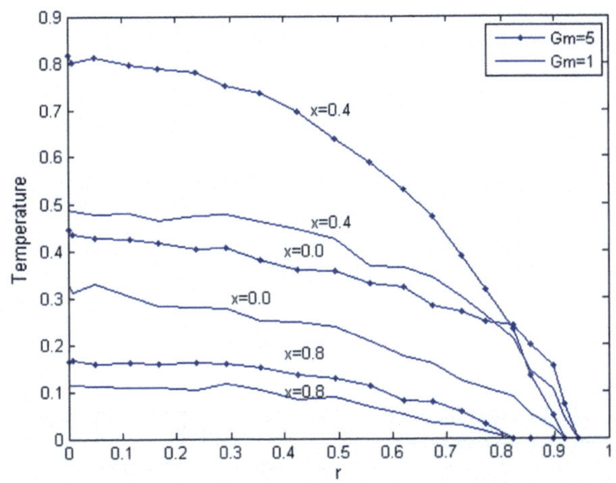

Fig. 2 Temperature variation with G_m for the fixed values of $\varepsilon = 0.1, \sigma^2 = 2$ and $E_m = 3$

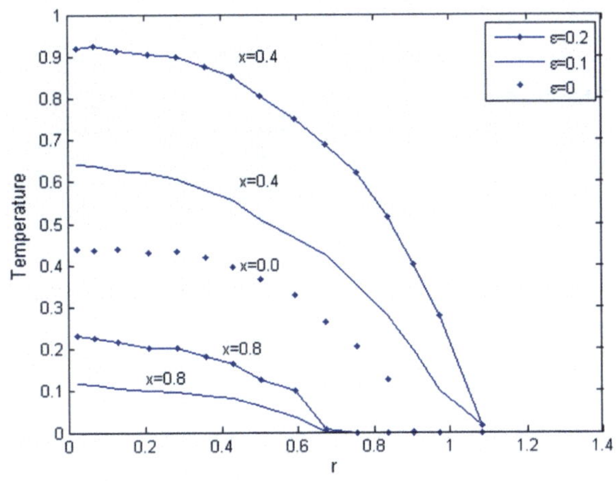

Fig. 3 Temperature variation with ε for the fixed values of $G_m = 3, \sigma^2 = 2$ and $E_m = 3$

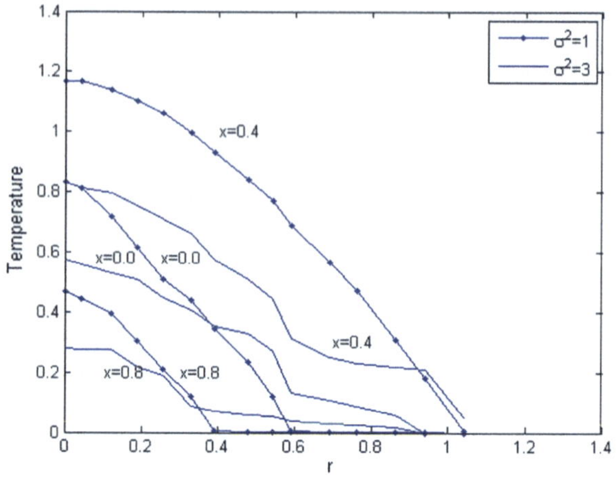

Fig. 4 Temperature variation with σ^2 for the fixed values of $G_m = 3$, $\varepsilon = 0.1$ and $E_m = 3$

Table 1 Heat transfer variation with E_m ($G_m = 3$, $\sigma^2 = 2$, $\varepsilon = 0.1$)

x	$E_m = 1$	$E_m = 3$	$E_m = 5$
0.0	2.42945	22.0037	61.17046
0.4	2.84512	25.72656	71.150905
0.8	0.39168	3.56543	9.91642

Table 2 Variation of heat transfer with respect to G_m ($\sigma^2 = 2$, $\varepsilon = 0.1$, $E_m = 3$)

x	$G_m = 1$	$G_m = 3$	$G_m = 5$
0.0	7.30205	22.0037	36.70534
0.4	8.54303	25.72656	42.910105
0.8	1.18121	3.56543	5.94964

Table 3 Variation of heat transfer with respect to σ^2 ($E_m = 3$, $G_m = 3$, $\varepsilon = 0.1$)

x	$\sigma^2 = 1$	$\sigma^2 = 2$	$\sigma^2 = 3$
0.0	22.42945	22.0037	21.99219
0.4	25.94013	25.72656	25.451301
0.8	3.76532	3.56543	3.47135

Table 4 Heat transfer variation with ε ($E_m = 3$, $G_m = 3$, $\sigma^2 = 2$)

x	$\varepsilon = 0$	$\varepsilon = 0.1$	$\varepsilon = 0.2$
0.0	0.0	22.0037	44.0073
0.4	0.0	25.72656	72.89057
0.8	0.0	3.56543	3.763

Heat transfer coefficient, Z, on the boundary of the tube is evaluated numerically, and the results are presented in Tables [1, 2, 3 and 4]. Z increases down the tube and then decreases, as in the case of temperature, which may be due to the peristalsis.

From Tables 1 and 2, we can see that by making some parameters fixed during the computation, there is an increase in the heat transfer coefficient as Eckert number or Grashof number increases. Tables 3 and 4 show that Z increases with ε, while it decreases with porosity [11, 12].

5 Conclusion

In the above results, it has been observed that temperature increases for some specific values of different parameters under consideration which is shown in Figs. 1, 2, 3 and 4. Further, it has been observed that the absolute value of the temperature significantly increases by about 8–10% as increase in 1–2% in Grashof number (G_m) which is given in Tables 1, 2, 3 and 4.

References

1. Radhakrishnamacharya, G.: Long wave length approximation to peristaltic motion of power-law fluid. Rheol. Acta **21**, 30–35 (1982)
2. Takabatake, S., Ayukawa, K., Mori, A.: Peristaltic pumping in circular cylindrical tubes: a numerical study of fluid transport and its efficiency. J. Fluid Mech. **193**, 267–283 (1988)
3. Radhakrishnamacharya, G., Radhakrishna Murthy, V.: Heat transfer to peristaltic transport in a non-uniform channel. Def. Sci. J. **43**(3), 275–280 (1993)
4. Shukla, J.B., Parihar, R.S., Rao, B.H.P., Gupta, S.P.: Effect of peripheral layer viscosity on peristaltic transport of bio-fluid. J. Fluid Mech. **97**, 225–237 (1980)
5. Sivaiah, R., Hemadri Reddy, R.: Magnetohydrodynamic peristaltic motion of a newtonian fluid through porous walls through suction and injection. IOP Conf. Ser. Mater. Sci. Eng. **263**, 062007 (2017)
6. Singh, U.P., Medhavi, A., Gupta, R.S., Bhatt, S.S.: Analysis of peristaltic transport of non-newtonian fluids through nonuniform tubes: Rabinowitsch fluid model. Zeitschrift Für Naturforschung A **72**(7), 601–608 (2017)
7. Nadeem, S., Akbar, N.S.: Influence of heat transfer on peristaltic transport of a Johnson-Segalman fluid in an inclined asymmetric channel. Commun. Nonlinear Sci. Numer. Simul. **15**(10), 2860–2877 (2010)
8. Hayat, T., Saleem, N., Ali, N.: Effect of induced magnetic field on peristaltic transport of a carreau fluid. Commun. Nonlinear Sci. Numer. Simul. **15**(9), 2407–2423 (2010)
9. Vajravelu, K., et al.: Peristaltic transport of a williamson fluid in asymmetric channels with permeable walls. Nonlinear Anal. Real World Appl. **13**(6), 2804–2822 (2012)
10. Radhakrishnamacharya, G., Srinivasulu, C.: Influence of wall properties on peristaltic transport with heat transfer. Comptes Rendus Mecanique **335**(7), 369–373 (2007)
11. Ellahi, R., Mubashir Bhatti, M., Vafai, K.: Effects of heat and mass transfer on peristaltic flow in a non-uniform rectangular duct. Int. J. Heat Mass Transfer **71**, 706–719 (2014)
12. Bhatti, M.M., et al.: Mathematical modeling of heat and mass transfer effects on MHD peristaltic propulsion of two-phase flow through a Darcy-Brinkman-Forchheimer porous medium. Adv. Powder Technol. **29**(5), 1189–1197 (2018)

Geometrical Effects on Natural Convection in 2D Cavity

H. P. Rani, V. Narayana and K. V. Jayakumar

Abstract In the present work, free convective airflow in the rectangular cavity with three different aspect ratios (AR <1, =1, >1) is investigated using direct numerical simulation. The bottom wall of the cavity has higher temperature than the top wall and the other two vertical walls are assumed to be thermally insulated. Finite volume method is employed to solve non-dimensional governing equations. An attempt has been made to analyze the flow behavior inside the cavity using streamlines, isotherms and energy streamlines. When Rayleigh number (Ra) is 10^3, vertical energy streamlines are observed in the cavity. As Ra is further increased, free energy streamlines at the boundary and trapped energy streamlines at the center are observed due to the dominant heat transfer mechanism, which switches from conduction to convection states, inside the cavity.

Keywords Energy streamlines · Rayleigh number · Aspect ratio

1 Introduction

Free convection in rectangular enclosures has been paid attention for decades both numerically and experimentally due to their extensive applications in nuclear reactors, solar collectors, power collectors and electronic device cooling (de Vahl Davis and Jones [1]). For a rectangular cavity, the heat transfer and fluid flow are profound to the experimental conditions along with boundary conditions, whereas, for small absolute velocity values, accurate experimental work is limited. In several cases of flow analysis in cavity, while computing heat transfer coefficient, the aspect ratio (AR) plays a key role and depends on the geometry and its topology. Differentially heated cubical cavity was studied by Lee et al. [2] to investigate the effects of AR with 1 ≤AR ≤2 for Ra varying between 10^6 and 10^8. Anil et al. [3] studied the flow

H. P. Rani (✉) · V. Narayana
Department of Mathematics, National Institute of Technology, Warangal 506004, India
e-mail: hprani@nitw.ac.in

K. V. Jayakumar
Department of Civil Engineering, National Institute of Technology,
Warangal, Telangana 506004, India

D. Dutta and B. Mahanty (eds.), *Numerical Optimization in Engineering and Sciences*,
Advances in Intelligent Systems and Computing 979,
https://doi.org/10.1007/978-981-15-3215-3_37

inside the cavities that are heated from above and below along with the radiation effect and showed the presence of unicell flow pattern for AR $\geq$ 1. From the literature survey, it can be observed that plotting of velocity vectors, streamlines and isotherms are the general visualization tools for the fluid flow problems. The other visualization techniques are heat function and heat lines and are proposed by Kimura and Bejan [4]. Later Costa [5, 6] presented a similar approach to visualize the physical aspects of the flow with the aid of the heat function and heat lines. Similar to heat lines, Mahmud and Fraser [7] presented the energy streamlines related to the convective heat transfer. Despite the past investigations, in this paper, an attempt is made to analyze the AR-dependence of flow features such as the vortex formation in the convection-dominated regime and the corresponding rate of heat transfer with the help of energy streamlines. Emphasis is made to understand the influence of AR and Ra on convection between the horizontal walls when the vertical walls are adiabatic. Thus, in this work, the behavior of energy streamlines for different Ra has been explained and calculated for different AR.

2 Governing Equations

The considered rectangular cavity is of length L and height H with the AR = L/H. The gravitational force, g, is assumed to act along the negative vertical direction on the cavity. The temperature of the bottom wall, T_H, of the cavity is assumed to be more than the top wall temperature, T_C, and then the non-dimensional temperature is given by $T^* = 1$ at $Y = 0$ and $T^* = 0$ at $Y = 1$. The vertical walls are assumed to be adiabatic and no-slip conditions are assumed on all the walls. The present problem is governed by the following equations (Basak and Roy [8]):

$$\nabla.\boldsymbol{V} = 0 \tag{1}$$

$$\boldsymbol{V}.\nabla\boldsymbol{V} = -\nabla P + \mathrm{Pr}\nabla^2\boldsymbol{V} + \mathrm{RaPr}T^* \tag{2}$$

$$\boldsymbol{V}.\nabla T^* = \nabla^2 T^* \tag{3}$$

where $\boldsymbol{V}$, P and T^* denote velocity, pressure and temperature variables, respectively. The non-dimensional control parameters are given by Prandtl number, Pr, and Rayleigh number, Ra. The heat transfer rate at the hot wall is calculated by the Nusselt number.

3 Results and Discussion

The finite volume method is adapted to discretize Eqs. (1–3). Established open-source CFD package OpenFOAM has been utilized. The geometry, volume, boundary and initial conditions are set in the buoyantBoussinesqSimpleFoam. The spatial derivatives are discretized by employing the second-order upwind linearization technique. The Laplacian and divergent terms are discretized by the Gauss linear and QUICK schemes, respectively. The Conjugate Gradient Scheme was employed to accelerate the convergence. With the aid of streamlines, isotherms and energy streamlines, the flow behavior is visualized and analyzed in the subsequent paragraphs with respect to Ra and AR. The simulated results are validated with those available in the literature and are shown in Table 1. A good agreement between the two is noted.

In Fig. 1, the simulated characteristics of streamlines for different of Ra and AR are depicted. As anticipated due to the presence of isothermal horizontal walls, the Bénard cells or vortices are formed in the vicinity of hot bottom and cold top walls. Due to the buoyancy effect, the symmetric rolls are formed in the clockwise and anticlockwise directions. For Ra $= 10^3$, the magnitude of stream function is very small and heat transfer rate occurs purely by conduction. Thus, conduction mechanism is dominant for small values of Ra.

When AR $= 0.5$, Fig. 1a, d shows the streamlines for Ra: 10^3–10^4. It can be observed from Fig. 1a, when Ra is 10^3 (subcritical flow), four Bénard cells showed their presence because of the dominant conduction mode of heat transfer. As Ra further increases to 10^4, there is a formation of secondary flow. When AR $= 1$ (Fig. 1b, e), a large primary cell is formed almost in the middle of the cavity because of the convection mechanism. For AR $= 2$, the above similar pattern was observed in vertical direction (Fig. 1c, f). As Ra increases, the buoyancy force controls the viscous force; hence, the mechanism of convection dominates. Thus, the circulation close to the central regimes becomes stronger. Thus, it can be observed that the formation of Bénard cells varies along with the characteristics of developing flow. With increasing Ra, from the enormous output data, following three different phases were detected in the formation of secondary flow: formation, appearance and maturity. These results show that the AR plays a key role in the convection-dominated flows.

Figure 2 depicts the simulated isotherms for different values of Ra and AR. When Ra $= 10^3$, the conduction mode of heat transfer and convection mode of heat transfer are comparable with each other. As Ra further increased to 10^4, the mechanism of convection in heat transfer becomes more prominent, and therefore, the thermal boundary layer in the vicinity of the walls turns to be thinner. Furthermore, a plume

Table 1 Validation of 2D results in terms of $\overline{\text{Nu}}$ for AR $= 1$

Authors	Ra $= 10^3$	Ra $= 10^4$
Present work	1.068	2.039
de Vahl Davis [1]	1.116	2.234
Ouertatani et al. [9]	1.0004	2.158

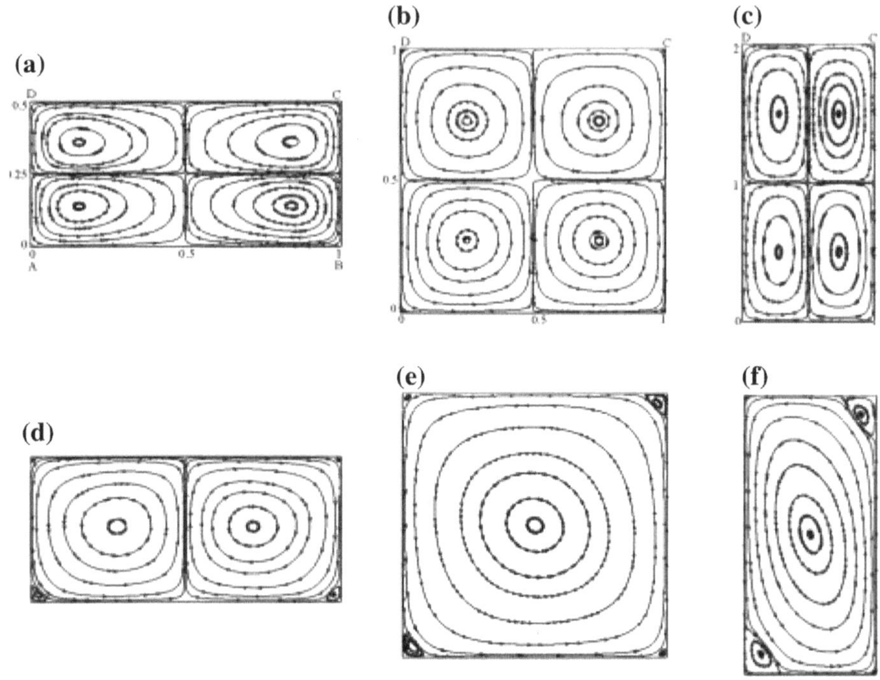

Fig. 1 Streamlines for Ra = 10^3 (**a**, **b**, **c**) and 10^4 (**d**, **e**, **f**) for different AR = 0.5 (**a**, **d**), 1 (**b**, **e**) and 2 (**c**, **f**)

starts to appear and feeble secondary recirculating zone is formed over the top of the rectangular cavity. The primary recirculation cell occupies the entire space and grows in size. Also, the secondary recirculating cell formed at the top of the cavity. In this case, the prominent point is that there exists a bigger isothermal section on the upper half of enclosure. For small fixed Ra and increasing AR, isotherms appear parallel to the horizontal wall. This is due to conduction mode of heat transfer as displayed in Figs. 2a, b, c.

With fixed AR = 0.5 (Figs. 2a, d) and when Ra is increased from 10^3 to 10^4, it is noticed that the shape of isotherms become curved and also two of the Bénard cells start to appear in the horizontal direction. The flow is dominated by recirculating motion in the core region (Rincon-Casado et al. [10]). When AR is increased and fixed at 1, it is noticed that the flow is symmetrical as shown in Fig. 2b and the contours are more distorted (Fig. 2e). This result is due to the strengthening of convective heat transport in the enclosure with increasing AR. When AR = 2 and as Ra increases (Fig. 2c, f), the shape of isotherms become curved, and there is a formation of three Bénard cells in the vertical direction due to the fact that temperature influence of opposing walls is high (Rincon-Casado et al. [10]).

By solving the Poisson equation of the type, $\nabla^2\Phi = \left(\nabla \times \overline{E}\right).\bar{k}$, the energy streamlines are obtained. Here, Φ represents the energy stream function, $\bar{k}$ is the unit

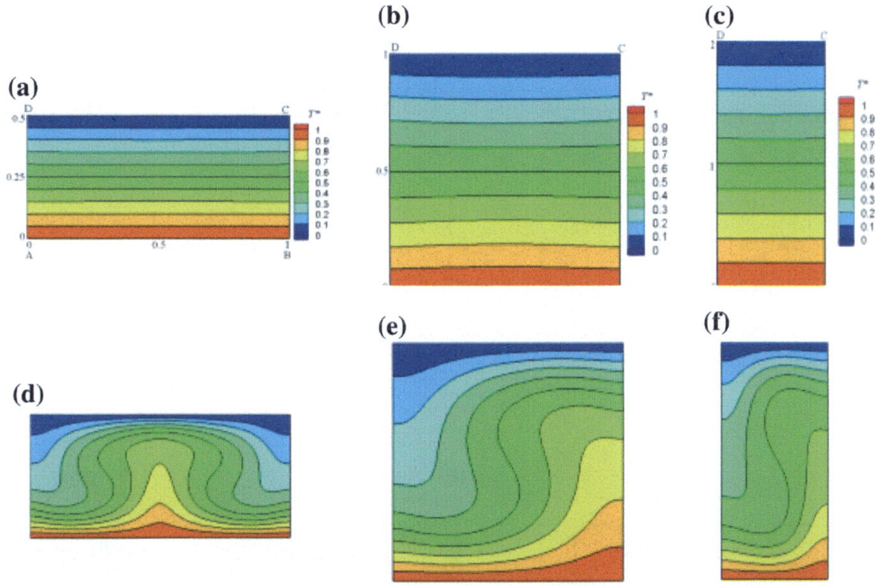

Fig. 2 Isotherms for Ra = 10^3 (**a**, **b**, **c**), 10^4 (**d**, **e**, **f**) for different AR = 0.5 (**a**, **d**), 1 (**b**, **e**) and 2 (**c**, **f**)

vector and $\overline{E}$ is the energy flux density vector, which is given by:

$$\overline{E} = \rho \boldsymbol{V}\left(\frac{1}{2}\boldsymbol{V}^2 + h\right) - \boldsymbol{V}.\overline{\sigma} - K\nabla T \tag{4}$$

where the ρ, h, $\overline{\sigma}$ and K denote the density, enthalpy, stress tensor and thermal conductivity. Energy streamlines shown in Eq. (4) include the contribution of energy due to the surface forces and energy fluxes. They can provide a complete view for configurations where these impacts are significant. Therefore, the simultaneous utilization of energy streamlines and heat lines is helpful to examine the quantitative details regarding the participation of the extra energy fluxes from the energy stream function.

Figure 3 shows the energy streamlines for Ra: 10^3–10^4 with AR = 0.5–2. For Ra = 10^3, the strength of recirculation is very small. The intensity of the primary recirculation cell increases with the increasing Ra, which contributes moderately high energy as compared to the small values of Ra. The adiabatic walls are not active for energy in terms of heat transfer. The energy streamlines on the left vertical wall extend downward depending on increasing Ra. This result can be attributed to the increase in the temperature gradient on the left vertical wall due to the high strength of the fluid recirculation in the enclosure. Energy obtained from the fluid friction also increases on the top wall with the increasing values of Ra. It can be observed that the formation of trapped energy streamlines varies deeply with the characteristics

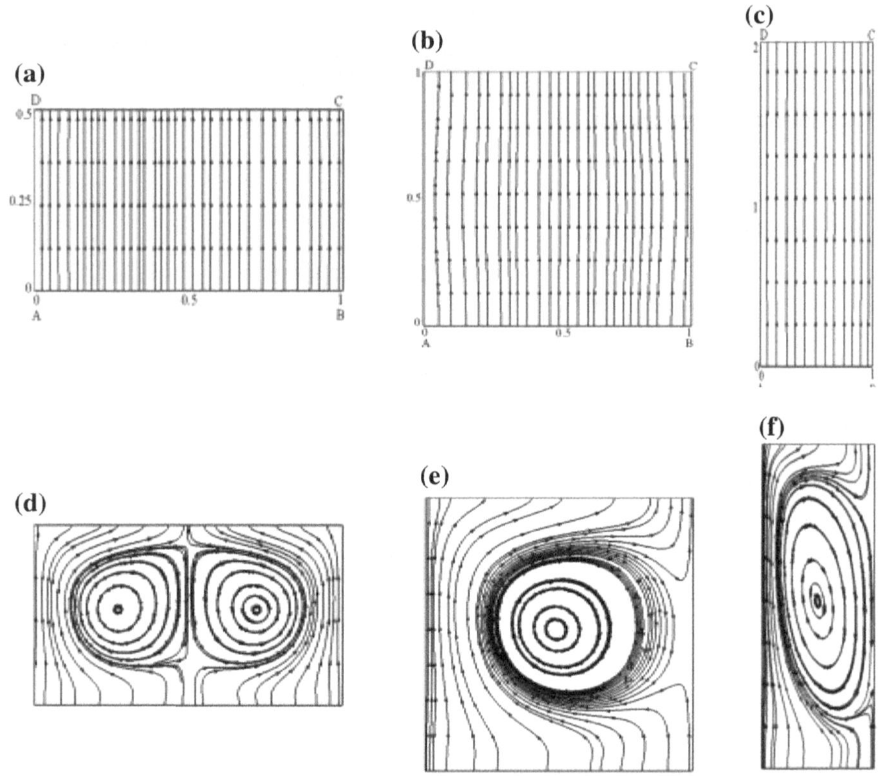

Fig. 3 The energy streamlines for Ra = 10^3 (**a**, **b**, **c**), 10^4 (**d**, **e**, **f**) for different AR = 0.5 (**a**, **d**), 1 (**b**, **e**), and 2 (**c**, **f**)

of developing flow. For all values of AR, i.e., the flow started at the hot wall moves through the fluid and then terminated at the cold wall. Figure 3b, e depicts the energy flow behavior in the enclosures with fixed AR = 1. It can be observed that when Ra increases, there is a formation of secondary flow (trapped energy streamlines) and also two emerging eddies at the top left and top right corners of the cavity. When AR = 2, the vertical energy streamlines with vortices of equal size are observed in vertical directions.

4 Conclusion

The present paper deals with the energy streamlines which are used to visualize the flow and thermal characteristics. The laminar flow regimes in rectangular enclosure are simulated for AR = 0.5, 1.0 and 2.0 with Ra = 10^3 and 10^4. When Ra = 10^3, the vertical energy streamlines are observed irrespective of AR. The trapped energy streamlines occupy more space in the cavity when Ra increases. Also when AR < 1, horizontal energy streamlines appeared, whereas vertical energy streamlines appeared only for AR > 1.

References

1. de Vahl Davis, G., Jones, I.P.: Natural convection in a square cavity: a comparison exercise. Int. J. Numer. Meth. Fluids **3**, 227–248 (1983)
2. Lee, T.S., Son, G.H., Lee, J.S.: Numerical predictions of three-dimensional natural convection in a box. Proc. KSME-JSME Therm. Fluids Eng. Conf. **2**, 278–283 (1988)
3. Anil, K.S., Velusamy, K., Balaji, C., Venkateshan, S.P.: Conjugate turbulent natural convection with surface radiation in air filled rectangular enclosures. Int. J. Heat Mass Transf. **50**, 625–639 (2007)
4. Kimura, S., Bejan, A.: The Heatline visualization of convective heat transfer. J. Heat Transfer **105**, 916–919 (1983)
5. Costa, V.: Unification of the streamline, heatline and massline methods for the visualization of two-dimensional transport phenomena. Int. J. Heat Mass Transfer **42**, 27–33 (1999)
6. Costa, V.: Unified streamline, heatline and massline methods for the visualization of two-dimensional heat and mass transfer in anisotropic media. Int. J. Heat Mass Transfer **46**, 1309–1320 (2003)
7. Mahmud, S., Fraser, R.: Visualizing energy flows through energy streamlines and pathlines. Int. J. Heat Mass Transf. **50**, 3990–4002 (2007)
8. Basak, T., Roy, S.: Role of Bejans Heatlines in heat flow visualization and optimal thermal mixing for differetially heated square enclosures. Int. J. Heat Mass Transfer **51**, 3486–3503 (2008)
9. Ouertatani, N., Cheikh, N.B., Beya, B.B., Lili, T., Campo, A.: Mixed convection in a double lid-driven cubic cavity. Int. J. Therm. Sci. **48**, 1265–1272 (2009)
10. Rincon-Casado, A., Sanchez de la Flor, F.J., Chacon Vera, E., Sanchez Ramos, J.: New natural convection heat transfer correlations in enclosures for building performance simulation. Eng. Appl. Comput. Fluid Mech. **11**(1), 340–356 (2017)

Convection Dynamics of SiO_2 Nanofluid

Rashmi Bhardwaj and Meenu Chawla

Abstract Investigation in this research discusses the nonlinear stability and convection dynamics for the magnetic and temperature-based disparity on electrical conductivity of nanofluid. The system in Cartesian coordinates comprises of a fluid layer exposed towards the external-magnetic field, gravity along with the heat in a cavity. From the equation of conservation of momentum and energy the partial differential equations have been obtained, which are converted to three-dimensional differential equation sets of nonlinear system similar to Lorenz equations. Applying stability, phase portrait and time series analysis, the effect of temperature and magnetic field variation via Rayleigh and Hartmann number simulating chaos have been discussed for SiO_2 (Silicon Di-Oxide) nanofluid. Some kind of magnetic cooling has been observed which is indicated by the stabilization of chaos in nanofluid convection with the increase in the applied magnetic field or the Hartmann number. As the value of Rayleigh number increases then system transit from stable to chaotic stage and once the chaotic phase begin system stability cannot be restored by controlling Rayleigh number. It is concluded that variations in temperature and magnetic field causes the transition of system from stable to chaotic and back to stable state along with the electrical conductivity for different nanofluids decreases and increases respectively and this phenomenon has a wider application in pharmacy, biosciences, health sciences, environment and in all fields of engineering.

Keywords Phase portrait · Chaotic phase · Rayleigh number · Hartman number · Nanofluid

---

R. Bhardwaj (✉)
University School of Basic & Applied Sciences (USBAS), Non-Linear Dynamics Research Lab, Guru Gobind Singh Indraprastha University, Dwarka, Delhi 110078, India
e-mail: rashmib22@gmail.com

M. Chawla
Echelon Institute of Technology, Faridabad, India

D. Dutta and B. Mahanty (eds.), *Numerical Optimization in Engineering and Sciences*, Advances in Intelligent Systems and Computing 979,
https://doi.org/10.1007/978-981-15-3215-3_38

1 Introduction

Chaos in convection dynamics plays a significant role and has a vast comprehensive application into appreciating the evolution of dynamical systems in electrical, mechanical, magneto mechanical, biological, chemical reaction or fluid flows. Bhardwaj and Bangia [1, 2] developed and explained various real-world models. They include dynamical study of meditating body and spread of human immunodeficiency virus (HIV). Bhardwaj and Meenu [3] gave a detailed investigation on blood flow containing Fe_3O_4 nanoparticles in the vessels. Bhardwaj and Das [4] extended the study for the CuO nanoparticles in the convection flow at chaos. The extensive study on method required for chaotic scenario in the fluid layer is discussed by Lorenz [5]. Studies began on different problems adopting the ideas derived by Lorenz. The critical value related to harm of linear-stability for Lorenz equation was modelled and discussed by Sparrow [6]. Kimura et al. [7] applied pseudo-spectral numerical-scheme plus discussed convection fluids into saturated porous layer. Pecora and Carroll [8] explained case-studies applying synchronisation of fractional-chaotic setups.

The significance of the study is applying nanofluids as coolants in reactors having higher-temperatures exposure could result in chaotic convection of fluids plus harms the cooling-setup of the device. Prevention is feasible by attaining stability, i.e., steady state of the convection is essential to monitor it in the case where fluid flow becomes chaotic. None of the author has studied the chaotic behaviour of SiO_2.

2 Mathematical Modeling

Considering the hollow opening that is infinitesimal to be analysed on the coordinate system where electrical conduct in nanofluids is passing. The horizontal fluid layer in its flow is imperiled towards heat and magnetic contact along with the gravity. Two elongated walls are sustained at temperature t_h and t_c, whereas short end walls are thermally insulated. Vertical axis z *is* taken collinearly along with gravity which means $\hat{e}_g = -\hat{e}_z$. The unvarying magnetic field, (taken as $\boldsymbol{B}$) is streamlined routinely on the heated side of the hollow opening can be referred as cavity as shown in Fig. 1.

The change in fluid density is caused by variation in temperature which is owing to heat transfer plus interaction of the magnetic field with convective motion. Magnetic Reynold's number considered to be small, so as to induce negligible magnetic field and virtually do not have any effect onto magnetic field. Time derivative cannot be neglected in Darcy's equation lower value of Prandtl number as for a matter of fact, its law deliberates towards regulating fluidic flow and Boussinesq approximation that is implied for density disparities into gravity part of momentum equation. Thus, equations monitoring laminar flow are:

$$\nabla . \upsilon = 0 \tag{1}$$

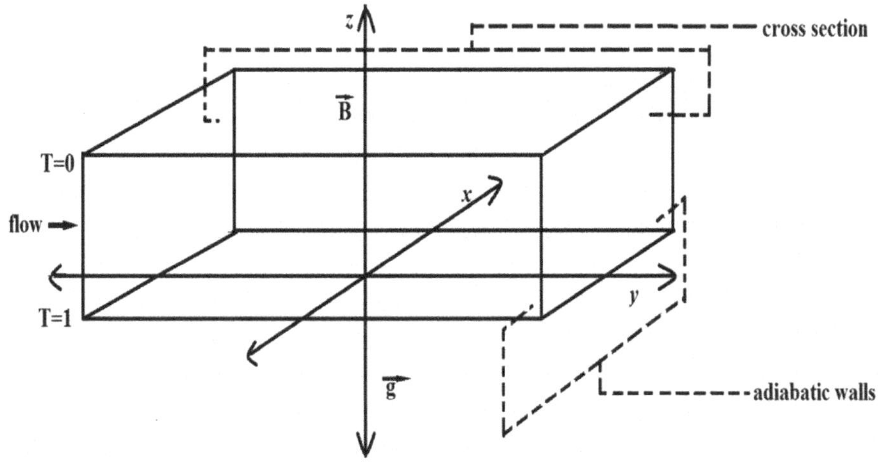

Fig. 1 Graphic representation of the cavity along with nanofluids

$$\left\{\frac{\partial \upsilon}{\partial t_*} + \upsilon.\nabla\upsilon\right\} = -\frac{1}{\rho_{nf}}\nabla P^* + \nu_{nf}\nabla^2\upsilon + \vec{\psi} \times \vec{B}_* - \frac{(\rho\gamma)_{nf}}{\rho_{nf}}\vec{g}(t_e - t_c) \quad (2)$$

$$\vec{\nabla}.\vec{\psi} = 0\,\&\,\vec{\psi} = \theta\left(-\nabla\omega + \upsilon \times \vec{B}_*\right) \quad (3)$$

$$\frac{\partial t_e}{\partial t_*} + \upsilon.\nabla t_e = \alpha_{nf}\nabla^2 t_e \quad (4)$$

υ	velocity
P^*	pressure
ω	electric potential
ψ	electric current density
B_*	magnetic field
α_{nf}	thermal diffusivity
$(\rho\gamma)_{nf}$	thermal expansion coefficient
t_e	temperature
γ	thermal expansion coefficient
ν	fluid viscosity
θ	electric conductivity of base fluid
ρ_{nf}	effective density of nanofluid
g	gravity
$(t_e - t_c)$	temperature difference

In Eq. (3), electric potential for steady state vanishes to zero $\nabla^2\omega = 0$ and then Lorenz force reduces to a systematic damping factor.

Simplifying Eq. (2) and to eliminate pressure, taking curl on both sides:

$$\left[\frac{\partial}{\partial t}(\nabla \times v) + \nabla \times (v.\nabla v) - \frac{\nu_{\mathrm{nf}}}{\alpha_{\mathrm{f}}}\nabla \times \left(\nabla^2 v\right) + \theta B^2 \frac{h_*^4}{\alpha_{\mathrm{f}}}(\nabla \times v)\right]$$
$$= -\frac{(\rho\gamma)_{\mathrm{nf}}}{\rho_{\mathrm{nf}}}\vec{g}\Delta T_{\mathrm{c}}\frac{h_*^3}{\alpha_{\mathrm{f}}^2}(\nabla \times T)$$

Using boundary conditions: $T = 1$ at $z = 0$ and $T = 0$ at $z = 1$,
Stress-free condition: $\frac{\partial u}{\partial Z} = \frac{\partial v}{\partial Z} = \frac{\partial^2 w}{\partial Z^2} = 0$,
Impermeability condition: $v.\hat{e}_{\mathrm{n}} = 0$
Stream function: $u = \frac{\partial \psi}{\partial Z}$ & $w = -\frac{\partial \psi}{\partial X}$
The following equations are formed:

$$\left[\frac{1}{\mathrm{Pr}}\left(\frac{\partial}{\partial t} + \frac{\partial \psi}{\partial Z}\frac{\partial}{\partial X} - \frac{\partial \psi}{\partial X}\frac{\partial}{\partial Z}\right) - \upsilon\nabla^2 + \eta\right]\nabla^2\psi = -\gamma\,\mathrm{Ra}\frac{\partial T}{\partial X} \tag{5}$$

$$\&\frac{\partial T}{\partial t} + \frac{\partial \psi}{\partial Z}\frac{\partial T}{\partial X} - \frac{\partial \psi}{\partial X}\frac{\partial T}{\partial Z} = \alpha\left(\frac{\partial^2 T}{\partial X^2} + \frac{\partial^2 T}{\partial Z^2}\right) \tag{6}$$

where
$\mathrm{Pr} = \upsilon_f/\alpha_f$ (Prandtl number),

$$\upsilon = \upsilon_{nf}/\upsilon_f$$

$\mathrm{Ra} = \left[\gamma_f \vec{g}\Delta T h_*^3\right]/\left[\alpha_f \upsilon_f\right]$ (Rayleigh number),
$\gamma = (\rho\gamma)_{nf}/\rho_{nf}\gamma_f$
$\mathrm{Ha} = B(\theta k/\upsilon_f)^{1/2}$ (Hartmann number),
$\eta = \theta B^2 h_*^4/\upsilon_f = Ha^2 h_*^4/k$

Now, taking stream function and temperature in the form

$\psi = A_1 \sin(kx)\sin(\pi z)$ and
$T = 1 - z + B_1\cos(kx)\sin(\pi z) + B_2\sin(2\pi z)$.

This representation is equivalent to Galerkin expansion of the solution in the x- and z-directions.

Solving Eqs. (5) and (6), the following set of equations is obtained:

$$\frac{\mathrm{d}A_1}{\mathrm{d}\tau} = -\mathrm{Pr}\upsilon\left[\frac{\gamma}{\upsilon}B_1 + A_1\left(1 + \frac{\eta}{\upsilon(\pi^2 + k^2)}\right)\right] \tag{7}$$

$$\frac{dB_1}{d\tau} = -RA_1 + 2\pi A_1 B_2 - \alpha B_1 \tag{8}$$

$$\frac{dB_2}{d\tau} = \frac{\pi}{2}A_1 B_1 - \alpha\lambda B_2 \tag{9}$$

On rescaling time and amplitudes w.r.t their convective fixed points and following points are obtained:

$$X_{11} = \frac{A_1}{\sqrt{\frac{\alpha\gamma\lambda}{\pi^2\upsilon}\left(\frac{\alpha\upsilon}{\gamma} - \frac{R}{L}\right)}} \tag{10}$$

$$X_{12} = \frac{-B_1}{L\sqrt{\frac{\alpha\lambda\upsilon}{\gamma}}}\sqrt{\left(\frac{\alpha\upsilon}{\gamma} - \frac{R}{L}\right)} \tag{11}$$

$$X_{13} = -\frac{B_2}{\frac{L}{2}\left(\frac{\alpha\upsilon}{\gamma} - \frac{R}{L}\right)} \tag{12}$$

where

$$L = \left(1 + \frac{\eta}{\upsilon(\pi^2 + k^2)}\right)$$

Solving, we get

$$\dot{x}_{11} = c(Nx_{12} - x_{11}) \tag{13}$$

$$\dot{x}_{12} = Mx_{11} - ax_{12} + Tx_{11}x_{13} \tag{14}$$

$$\dot{x}_{13} = s(Gx_{11}x_{12} - x_{13}) \tag{15}$$

where $m = \left(1 + \frac{\eta}{\upsilon(\pi^2+k^2)}\right)$, $s = a\lambda$, $a = \alpha$, $\Pr = p$ and $c = p\upsilon m = \Pr \upsilon m$

$$T = \frac{\gamma}{\upsilon}\left(\frac{\alpha\upsilon}{\gamma} - \frac{R}{L}\right)^2 = \frac{b}{v}U^2, N = \frac{\pi}{\left(\frac{\alpha\upsilon}{\gamma} - \frac{R}{L}\right)} = \frac{d}{U},$$

$$G = \left(\frac{\alpha\upsilon}{\gamma} - \frac{R}{L}\right)^{-1} = \frac{1}{U}, M = \frac{R\gamma}{\pi\upsilon L}\left(\frac{\alpha\upsilon}{\gamma} - \frac{R}{L}\right) = \frac{Rb}{dvm}U$$

2.1 Stability Analysis

The fixed points obtained are: (0, 0, 0),

$$\left(\sqrt{\frac{a - \text{MN}}{\text{TG}}}, \frac{1}{N}\sqrt{\frac{a - \text{MN}}{\text{TG}}}, \frac{a - \text{MN}}{\text{TG}}\right) \& \left(-\sqrt{\frac{a - \text{MN}}{\text{TG}}}, -\frac{1}{N}\sqrt{\frac{a - \text{MN}}{\text{TG}}}, \frac{a - \text{MN}}{\text{TG}}\right).$$

1. **For the point (0, 0, 0)**, it is observed that (0, 0, 0): stable asymptotic if MN < a; unstable if MN > a; for MN = a, it is a critical case.
2. **For the point** $\left(\sqrt{\frac{a-\mathrm{MN}}{\mathrm{TG}}}, \frac{1}{N}\sqrt{\frac{a-\mathrm{MN}}{\mathrm{TG}}}, \frac{a-\mathrm{MN}}{\mathrm{TG}}\right)$:

$$\text{Jacobian } J = \begin{bmatrix} -c-\lambda & cN & 0 \\ M+Tx_{13} & -a-\lambda & Tx_{11} \\ sGx_{12} & sGx_{11} & -s-\lambda \end{bmatrix}$$

Characteristic polynomial is given as:

$$\lambda^3 + \lambda^2(s+a+c) + \lambda(cs + \mathrm{NM}s) + cs(\mathrm{NM} - 2a) = 0$$
$$\Rightarrow \mathrm{NM} = -c(3a+c+s)/(a+s) = (\mathrm{NM})_c$$

Observe that $\left(\sqrt{\frac{a-\mathrm{MN}}{\mathrm{TG}}}, \frac{1}{N}\sqrt{\frac{a-\mathrm{MN}}{\mathrm{TG}}}, \frac{a-\mathrm{MN}}{\mathrm{TG}}\right)$ is stable if $\mathrm{NM} < (\mathrm{NM})_c$; critical when $\mathrm{NM} = (\mathrm{NM})_c$ and chaotic when $\mathrm{NM} > (\mathrm{NM})_c$.

3. **For point** $\left(\sqrt{\frac{a-\mathrm{MN}}{\mathrm{TG}}}, \frac{1}{N}\sqrt{\frac{a-\mathrm{MN}}{\mathrm{TG}}}, \frac{a-\mathrm{MN}}{\mathrm{TG}}\right)$

$$J = \begin{bmatrix} -c-\lambda & cN & 0 \\ M+Tx_{13} & -a-\lambda & -Tx_{11} \\ -sGx_{12} & -sGx_{11} & -s-\lambda \end{bmatrix}$$

and the characteristic polynomial is given as:

$$\lambda^3 + \lambda^2(s+a+c) + \lambda(cs + \mathrm{NT}s) + cs(\mathrm{NT} - 2a) = 0$$
$$\Rightarrow \mathrm{NT} = -c(3a+c+s)/(a+s) = (\mathrm{NT})_c$$

It is observed that the point $\left(-\sqrt{\frac{a-\mathrm{MN}}{\mathrm{TG}}}, -\frac{1}{N}\sqrt{\frac{a-\mathrm{MN}}{\mathrm{TG}}}, \frac{a-\mathrm{MN}}{\mathrm{TG}}\right)$ is stable if $\mathrm{NM} < (\mathrm{NM})_c$; critical if $\mathrm{NM} = (\mathrm{NM})_c$; chaotic if $\mathrm{NM} > (\mathrm{NM})_c$.

3 Results and Discussions

To investigate the dynamical behaviour of electrical conductivity with the effect of magnetic field and temperature, nonlinear equations have been numerically solved. Thermophysical properties of nanofluid are given in Table (1).

The values pr = 10, $k = 2.2$, $H = 0.25$, $\lambda = 8/3$ are used for the computation. The different values of Ra and Ha are tabulated in Table 2, at which the different stages of nanofluid convection for SiO_2 nanofluid have been observed. When the Rayleigh number is increased then phase transition from stable to chaotic stage is observed and on further increasing the temperature or the value of *Ra* the chaotic stage continues

Table 1 Thermophysical properties of water and SiO_2 nanoparticle

Substance	ρ (kgm^{-3})	k ($Wm^{-1}K^{-1}$)	C_p ($Jkg^{-1}K^{-1}$)	γ	α	ν
H_2O	997.1	0.613	4179	–	–	–
SiO_2	2200	1.4	745	0.898	1.078	1.0724

Table 2 Phases for different nanofluids with variation in parameter

Stable to chaotic phase with variation in parameter Ra with Ha = 0.7			
Substance	Stable spiral phase	Limit cycle phase	Chaotic phase
SiO_2	≤ 25.7	25.8–31.5	≥ 31.6
Chaotic to stable phase with variation in Ha with Ra = 35			
Substance	Chaotic phase	Limit cycle phase	Stable spiral phase
SiO_2	≤ 0.52	0.53–0.86	≥ 0.87

which are shown in Fig. 2. However, on increasing Hartmann number during chaotic stage stability is restored for the nanofluids which are shown in Fig. 3.

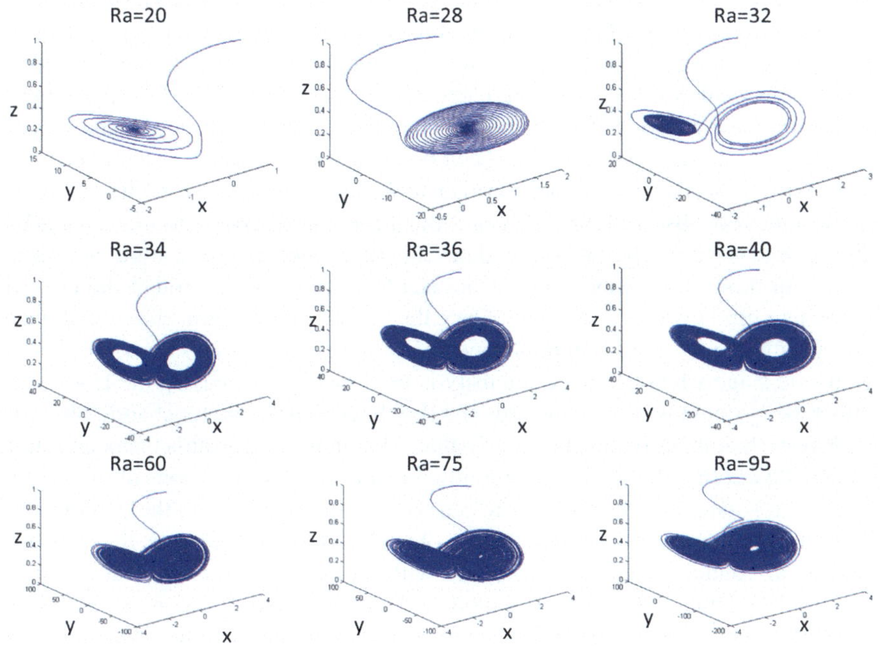

Fig. 2 Transitions in phase portraits of SiO_2 for Ha = 0.7 with variation in Ra

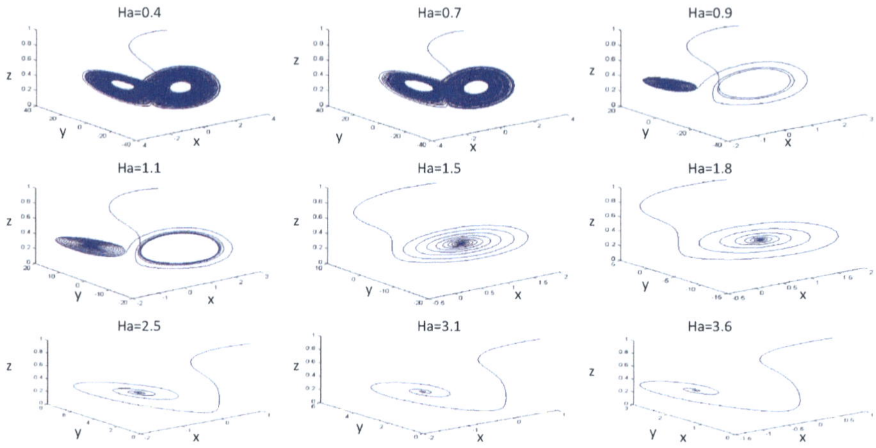

Fig. 3 Transitions in phase portraits of SiO_2 for Ra = 35 with variation in Ha

4 Conclusion

In this work, the nanofluidic flow of SiO_2 nanofluid in a rectangular hollow opening (cavity) together with exposure to temperature and magnetic field variation has been studied. From equations of conservation of mass, momentum and energy the partial differential equations of fluid convection are obtained which are then transformed to three-dimensional ordinary differential equation system using Galerkin approximation for nonlinear analysis of the nanofluid convection. From the stability analysis of the system, the critical condition has been obtained beyond which system becomes chaotic. Different stages, i.e., stable, critical and chaotic phases of nanofluid convection dynamics are observed for SiO_2 nanofluid through phase portraits plots. It can be deduced that as Rayleigh number escalates, the difference in temperature increases, and system transits to chaotic stage from stable stage passing through the critical stage where limit cycle are observed. Once the chaotic phase begin chaos continues to grow with an increase in value of Rayleigh parameter. Also, it is observed that in the chaotic stage when magnetic intensity is increased by increasing the Hartmann number the system transforms towards stability but it is possible for chaotic phase at high Rayleigh number below threshold value. This indicates towards the occurrence of magnetic cooling that facilitates modified machinery towards control of chaotic phase of SiO_2 nanofluid convection which has its significance into the deployment of nanofluids as coolants in different areas of application. Furthermore, the decrease in electrical conductivity of the nanofluid with an increase in temperature as suggested by Wiedemann–Franz Law is modelled and observed. It is concluded that as parameters increase the value of Ra for which system retains stability also increases while increase in value of R for which the system is stable and it also impacts the value of Ha beyond which stability of the system is again restored from the chaotic phase at high Ra values.

References

1. Bhardwaj, R., Bangia, A.: Complex dynamics of meditating body. Indian J. Ind. Appl. Math. **7**(2), 106–116 (2016)
2. Bhardwaj, R., Bangia, A.: Statistical time series analysis of dynamics of HIV. Jnanabha **48**, 22–27 (2018)
3. Bhardwaj, R., Chawla, M.: Convection dynamics of Fe3O4 nanoparticles in blood fluid flow. Indian J. Ind. Appl. Math. **9**(1), 23–36 (2018)
4. Bhardwaj, R., Das, S.: Chaos in nanofluidic convection of CuO nanofluid. In: Manchanda P. et al. (eds.) Industrial Mathematics and Complex Systems, pp. 283–293
5. Lorenz, E.N.: Deterministic non-periodic flow. J. Atmos. Sci. **20**, 130–141 (1963)
6. Sparrow, C.: The Lorenz Equations: Bifurcations. Chaos and Strange Attractors, Springer, New York (1982)
7. Kimura, S., Schubert, G., Straus, J.M.: Route to chaos in porous-medium thermal convection. J. Fluid Mech. **166**, 305–324 (1986)
8. Pecora, L.M., Carroll, T.L.: Synchronization in chaotic systems. Phys. Rev. Lett. **64**, 821–824 (1990)

Development of a Simple Gasifier for Utilization of Biomass in Rural Areas for Transportation and Electricity Generation

Mainak Bhaumik, M. Laxmi Deepak Bhatlu and S. M. D. Rao

Abstract The development of a simple low-cost gasifier is described with a focus to extract wood gas also called syngas by burning waste biomass in rural areas. In many rural areas of India, wood (biomass) is used for cooking and other purposes. If this could be efficiently burnt by pyrolysis in a gasifier, the gases could be separated and stored in cylinders for further utilization such as cooking, running automobiles and electricity generation. The experimental setup consists of a typical gasifier utilizing the process of pyrolysis for the burning of biomass in the absence of air. The pyrolyzing gases are pumped out of the system through appropriate processes such as drying, pyrolyzing, reduction and combustion to obtain a tar-free syngas. The obtained syngas mostly consists of CO and H_2 in addition to small amounts of methane and CO_2. In the initial experiments, ejector pump locally developed has been used to pump out the gases for burning outside the system. A four-stroke petrol engine could be run by directly connecting the gas outlet to the carburetor with proper modification. The gas outlet is connected to a compressor to fill into cylinders for further application in cooking and petrol engines. While the gas is pumped out, about 40% of the biomass is left behind as biochar. This could be used for many applications such as farming, medicine, smoke-free fuel and others.

Keywords Synthesis gas (syngas) · Hydrocarbon · Hydrogen (H_2) · Carbon monooxide (CO) · Wooden chips · Combustion chamber

M. Bhaumik (✉)
Mechanical Engineering Department, MGM's College of Engineering and Technology, Kamothe, Navi Mumbai 41020, India
e-mail: mainak.bhaumik04@gmail.com

M. L. D. Bhatlu
Chemical Engineering Department, Karpagam Academy of Higher Education, Coimbatore 641021, India

S. M. D. Rao
Visiting Specialist, IOP, Academia Sinita, Taipei, Taiwan

D. Dutta and B. Mahanty (eds.), *Numerical Optimization in Engineering and Sciences*, Advances in Intelligent Systems and Computing 979,
https://doi.org/10.1007/978-981-15-3215-3_39

1 Introduction

Till date in most of rural India, biomass consisting of wood, dry leaves, bushes, animal dung, etc., is being used for cooking, heat and energy generation purpose [1, 2]. Smoke evolving from the biomass which consists of CO_2, CO, methane, H_2 and other minor components is responsible for the flames we see. This is called wood gas, and a refined form of this consisting of CO and H_2 is called synthetic gas or syngas [3–7]. During the Second World War, Georges Imbert developed a gasifier to capture this gas and use it for running motor vehicles. Following this, several investigators have investigated the composition of the wood gas and refining it [8–10]. Furthermore, several investigators have made gasifiers of the same principle but different designs. One of the common types is the downdraft gasifier developed by Imbert [11–13]. We describe here a simple gasifier based on the designs of Luke and Mason.

Processes taking place in the machine:

Different steps in gasification

As the biomass is heated, it gets dried by giving out moisture of H_2O. This stage is called drying. As the temperature increases, pyrolysis takes place in the absence of oxygen to give out tar and charcoal. On further heating, combustion takes place in the presence of air resulting in tarry gas and charcoal. On further increasing the temperature, in the reduction reaction, these volatile products pass over the burning charcoal at high temperature to give out CO and H_2. The ensuing reactions occurring at different stages are summarized below:

1. Biomass (HCO) → Heat in Air 80–200 °C → H_2O
2. HCO → Heat in the absence of air 300–400 °C → Char Coal + Tar
3. Char Coal + Tar → In the presence of air 600 °C → $H_2O + CO_2$
4. When the mixture of $CO_2 + H_2O$ passes over hot coal above 800 °C, the following reactions take place:

$$CO_2 + C \rightarrow 2CO \quad \text{and}$$
$$H_2O + C \rightarrow H_2 + CO$$

Thus leaving a mixture of H_2 and CO which is a clean fuel.

Thus, with proper temperature management, it is possible to burn biomass and obtain a very efficient gas mixture for efficient heating and other applications. The present gasifier is designed to achieve this objective.

2 Experimental Setup

The design features are given in Fig. 1. It consists of two chambers. The inner chamber is used to burn the biomass and the outer one to collect the gas and the left out coal from the reactions in the combustion.

Biomass is introduced from the top in small amounts. Initially, charcoal is placed at the bottom in the combustion zone and ignited, while the suction pump is switched on. The coals burn with flare being sucked inside. Then, small amount of biomass is added which slowly burns by pyrolysis evolving gases which increase the combustion. As more and more biomass is added, additional zones develop in varying thickness with time. This may be represented by the different regions represented in Fig. 1 above, and the reactions taking place in each zone in the equations above contribute to the formation of the gas in the end.

The photograph of the system is shown in Fig. 2.

The experimental setup used for the experiments is shown in Fig. 2. It consists of a gasification chamber, a cyclone, an automatic feeding system, a filter and ejector system for pumping the gases out.

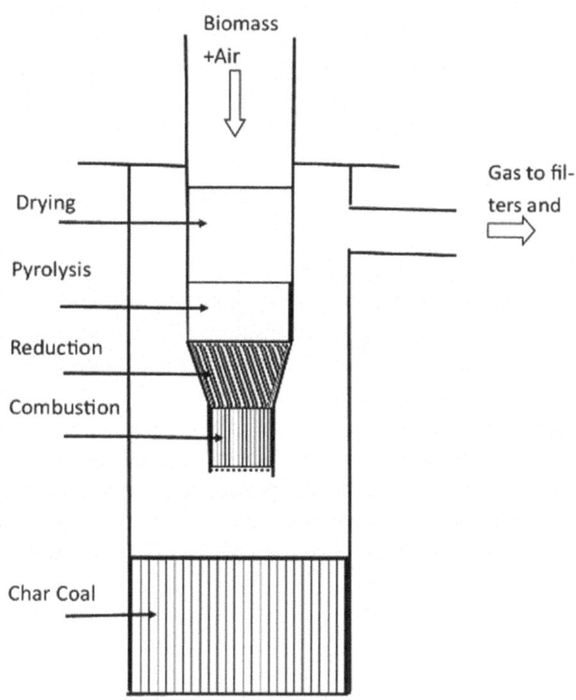

Fig. 1 A schematic of the working of the gasifier employed in the present development

Fig. 2 Fabricated syngas generation setup

3 Results and Discussions

Before starting the system, the system is initially loaded with some charcoal and ignited after starting the ejector pump. Bright burning coals are seen. Then, biomass chips are slowly added to the burning coals. Then, smoke coming out of the ejector can be seen. Initially, this smoke does not catch fire if ignited due to large amount of moisture. In a short while, it will ignite with a reddish yellow flame. The automatic feeder is started to feed the biomass. In a short time, the flame slowly changes to yellow–blue flame which ultimately becomes blueish like a home cooking stove. At this time, we close the valve leading to the ejector and connect the gas line to the automobile engine. The automobile engine could be operated for a short time demonstrating that the gas is of automobile quality. Further modifications are needed to the automobile engine to work efficiently with the wood gas or syngas.

While the continuous operation of the system has been achieved with the present gasification system, more improvements are being made to automate the starting and ignition and shutdown. The changeover from the ejector to the engine is presently manual with a three-way valve. This will also be automated (Fig. 3).

Since this system is simple and can be operated totally on biomass with little initial electricity needs, it is proposed to develop this to be operated in remote rural areas with suitable changes such as using as battery start and add a bottling system to fill gas cylinders so that the gas could be used for cooking, transport vehicles and electricity generation.

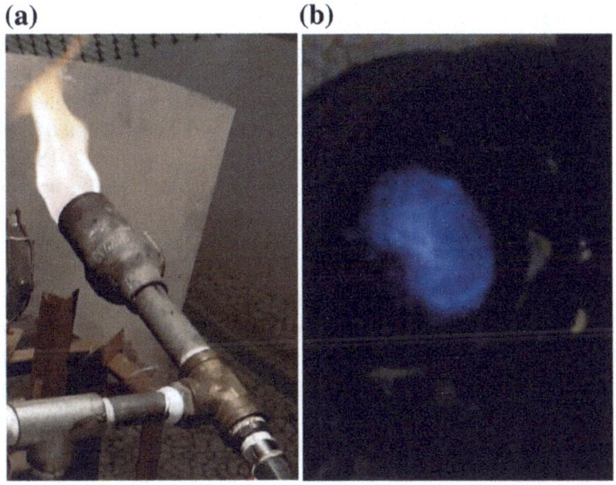

Fig. 3 Flame of the gas at different stages of the gasification. **a** Initial stages, **b** after 20 min of operation

4 Conclusion

We have designed and operated a simple biomass gasification system. The gas produced is of good quality judging from the color of the flame. The gas could run a motor vehicle engine confirming the quality of the gas. However, laboratory tests will be carried out to assess the gas composition.

References

1. Dubey, A., Kolekar, S.K., Gopinath, C.S.: Back cover: C-H activation of methane to syngas on MnxCe1–x–yZryO2: a molecular beam study. ChemCatChem **8**(13), 2307 (2016). https://doi.org/10.1002/cctc.201600778
2. Banerjee, P., Hazra, A., Ghosh, P., Ganguly, A., Murmu, N.C., Chatterjee, P.K.: Solid waste management in India: a brief review. In: Waste Management and Resource Efficiency, pp. 1027–1049. Springer, Singapore (2019). https://doi.org/10.1007/978-981-10-7290-1_86
3. Huang, C., Xu, C., Wang, B., Hu, X., Li, J., Liu, J., Li, C.: High production of syngas from catalytic steam reforming of biomass glycerol in the presence of methane. Biomass Bioenerg. **119**, 173–178 (2008). https://doi.org/10.1016/j.biombioe.2018.05.006
4. Xie, Y., Wang, X., Bi, H., Yuan, Y., Wang, J., Huang, Z., Lei, B.: A comprehensive review on laminar spherically premixed flame propagation of syngas. Fuel Process. Technol. **181**, 97–114 (2018). https://doi.org/10.1016/j.fuproc.2018.09.016
5. Maisano, S., Urbani, F., Cipitì, F., Freni, F., Chiodo, V.: Syngas production by BFB gasification: experimental comparison of different biomasses. Int. J. Hydrogen Energy **44**(9), 4414–4422 (2019). https://doi.org/10.1016/j.ijhydene.2018.11.148
6. Shahirah, M.N.N., Gimbun, J., Lam, S.S., Ng, Y.H., Cheng, C.K.: Synthesis and characterization of a LaNi/α-Al_2O_3 catalyst and its use in pyrolysis of glycerol to syngas. Renew. Energy **132**, 1389–1401 (2019). https://doi.org/10.1016/j.renene.2018.09.033
7. Luo, G., Jing, Y., Lin, Y., Zhang, S., An, D.: A novel concept for syngas biomethanation by two-stage process: Focusing on the selective conversion of syngas to acetate. Sci. Total Environ. **645**, 1194–1200 (2018). https://doi.org/10.1016/j.scitotenv.2018.07.263

8. Hagos, F.Y., Aziz, A.R.A., Sulaiman, S.A., Mamat, R.: Engine speed and air-fuel ratio effect on the combustion of methane augmented hydrogen rich syngas in DI SI engine. Int. J. Hydrogen Energy **44**(1), 477–486 (2019). https://doi.org/10.1016/j.ijhydene.2018.02.093
9. Li, K., Liu, J.L., Li, X.S., Lian, H.Y., Zhu, X., Bogaerts, A., Zhu, A.M.: Novel power-to-syngas concept for plasma catalytic reforming coupled with water electrolysis. Chem. Eng. J. **353**, 297–304 (2018). https://doi.org/10.1016/j.cej.2018.07.111
10. Nadaleti, W.C.: Utilization of residues from rice parboiling industries in southern Brazil for biogas and hydrogen-syngas generation: heat, electricity and energy planning. Renew. Energy **131**, 55–72 (2019). https://doi.org/10.1016/j.renene.2018.07.014
11. Neto, A.F., Marques, F.C., Amador, A.T., Ferreira, A.D., Neto, A.M.: DFT and canonical ensemble investigations on the thermodynamic properties of syngas and natural gas/syngas mixtures. Renew. Energy **130**, 495–509 (2019). https://doi.org/10.1016/j.renene.2018.06.091
12. Sun, X., Atiyeh, H.K., Kumar, A., Zhang, H., Tanner, R.S.: Biochar enhanced ethanol and butanol production by clostridium carboxidivorans from syngas. Biores. Technol. **265**, 128–138 (2018). https://doi.org/10.1016/j.biortech.2018.05.106
13. Xu, Z., Jia, M., Li, Y., Chang, Y., Xu, G., Xu, L., Lu, X.: Computational optimization of fuel supply, syngas composition, and intake conditions for a syngas/diesel RCCI engine. Fuel **234**, 120–134 (2018). https://doi.org/10.1016/j.fuel.2018.07.003

Identification of Parameters in Moving Load Dynamics Problem Using Statistical Process Recognition Approach

Shakti P. Jena, Dayal R. Parhi and B. Subbaratnam

Abstract The present work is focused to develop an indirect approach for structural health monitoring analogy for moving load dynamics problem using the concepts of statistical process recognition (SPR) approach. The objectives of the proposed analogy are to identify and locate the subsistence of cracks on the structure in supervised manner. The statistical process reorganization approach is based on the concepts of the time domain auto-regressive (AR) method. Numerical studies are carried out for a damaged simply supported beam under a moving load for the exactness of the proposed method. The numerical study evidenced that the proposed method is perceptive to structural damage identification parameters for moving load dynamics problem.

Keywords AR · Moving load · Crack

1 Introduction

Structural health monitoring problem has received major attention in the present scenario of civil and mechanical industries. The early detection of fault in structures becomes an interesting topic for engineers and researchers for structural integrity and healthy environment. Jena and Parhi [1–4] have carried out several studies like numerical, finite element analysis (FEA) and experimental to determine the moving load-induced dynamic responses on the structures. Apart from various studies, many researchers also focused their work on statics-based approach for structural health monitoring problem. Xue et al. [5] have applied the Monte Carlo and likelihood estimation methods to identify the damage parameters on structures. Mao et al. [6] have developed a statistical crack detection approach based on the concept of sensitivity of dynamic response. Farahani and Penumadu [7] investigated a noble

S. P. Jena (✉) · B. Subbaratnam
Depertment of Mechanical Engineering, Vardhaman College of Engineering, Hyderabad, India
e-mail: shaktipjena@gmail.com

D. R. Parhi
Depertment of Mechanical Engineering, National Institute of Technology, Rourkela, India

D. Dutta and B. Mahanty (eds.), *Numerical Optimization in Engineering and Sciences*,
Advances in Intelligent Systems and Computing 979,
https://doi.org/10.1007/978-981-15-3215-3_40

damage detection procedure for girder bridge structure using vibration data and time series analysis. Mechbal et al. [8] have developed an algorithm based upon multi-class pattern reorganization approach for structural health monitoring problem. Yang et al. [9] explored a static algorithm for localization of damage in structure by traversing mass. Ma et al. [10] have considered the measurement of the consequence of noise and performance of sensors for detection of damage in structures.

As per the authors' knowledge is concerned, the literatures in the moving load induced dynamic response in the domain of statistical process reorganization approach for structural health monitoring era is a little. The novelty of the present approach is to develop a structural health monitoring analogy in the domain of AR method by using the measured responses of the structure. The present analogy has done in supervised manner.

2 The Problem Definition

For the inverse approach, a problem has been formulated. As per Fig. 1, a simply supported beam with two cracks subjected to a transit mass has been analyzed in this work. The governing equation of motion of the system has been articulated as per Euler–Bernoulli's equation, i.e.,

$$\mathrm{EI}\frac{\partial^4 y}{\partial x^4} + m\frac{\partial^2 y}{\partial t^2} = F(x,t), \text{ where } F(x,t) = P(t)\delta(x-\beta) + r(x,t) \quad (1)$$

The solution of the governing Eq. (1) has been already obtained by Jena and Parhi [1] earlier which is represented in Eq. (2), i.e.,

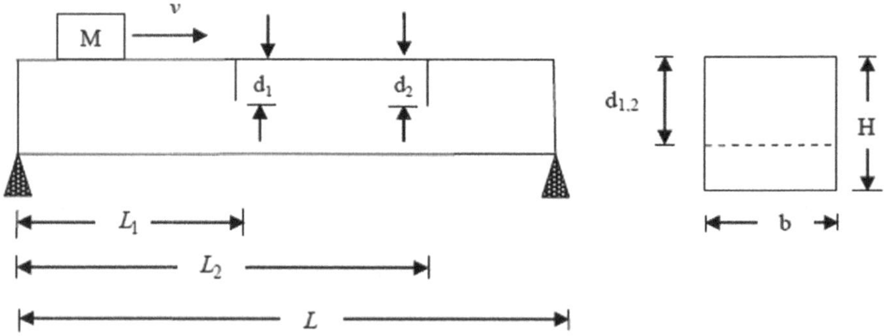

Fig. 1 Cracked simply supported structure under moving mass

$$\mathrm{EI}\lambda_n^4 T_n(t) + mT_{n,tt}(t) - \left(\frac{M}{V_n}\right)\left[g - \sum_{q=1}^{\infty}\left(\frac{\partial}{\partial t} + v\frac{\partial}{\partial \beta}\right)^2 w_q(\beta)T_q(t)\right]w_n(\beta) = 0 \quad (2)$$

where $F(x, t)$ is the applied force, $P(t)$ is the driving force, δ is the Dirac delta function, $\beta = vt$, is the position of the transit mass, w_q is the shape function of the beam, $T_n(t)$ is the amplitude function, v is the speed and M is the transit mass. The detailed procedures are already discussed [1]. For the said analysis, a numerical problem has been formulated.

Beam dimensions = (150 × 5 × 0.5) cm, $\alpha_{1,2} = d_{1,2}/H$ = Relative crack depth, $M = 2.5$ kg, $v = 6.3$ m/s, $\eta_{1,2} = L_{1,2}/L$ = Relative crack position. $L_{1,2}$ = Position of cracks. L = Length of the structure. $\alpha_{1,2} = 0.35, 0.32$ and $0.42, 0.25$, $\eta_{1,2} = 0.3, 0.5$ and $0.4, 0.67$.

For the response analysis of the system, one numerical example has been described in Fig. 2. From the observed responses, it has been proven that there are sudden rises in amplitudes at certain positions of the structure. The sudden rise in amplitude and the corresponding positions indicate the existence and positions of crack in that particular structure, respectively. The said problem has been considered in statistical process control approach in a supervised manner.

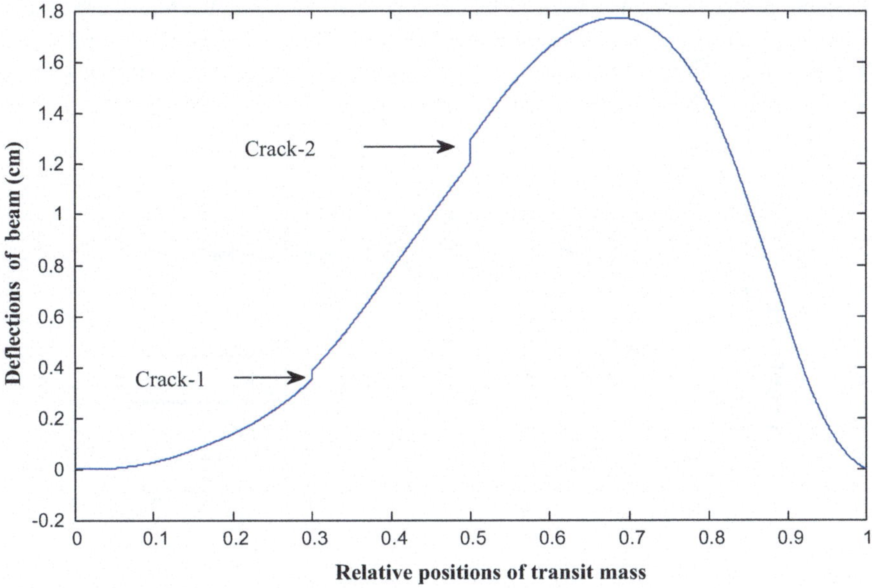

Fig. 2 For deflections versus relative positions $\eta_{1,2} = 0.3, 0.5$, $\alpha_{1,2} = 0.35, 32$, $M = 2.5$ kg, $v = 6.3$ m/s

3 Statistical Process Recognition (SPR) Approach

Statistics-based methodology is the compilation of methods for taking the pronouncement about an approach that depends on the exploration of data controlled in a specified sample. The SPR approach presents the primary features that a product can be sampled, trained and monitored in a specified manner. The proposed fault detection approach is developed with the implementation of time series analysis in the SPR domain. The mechanism of control chart analysis has been applied to detect the existence of cracks on the structure.

$$\text{Lower control Limit (LCL)} = \mu_s - z\frac{\sigma_s}{\sqrt{n}} \tag{3a}$$

$$\text{Upper control Limit (UCL)} = \mu_s + z\frac{\sigma_s}{\sqrt{n}} \tag{3b}$$

where μ_s and σ_s are the mean and standard deviation of the sample size (n), respectively. z is the sample statistics. The value of z is normally obtained from different sample sizes [11]. The mean and standard deviations of the samples are determined from beam responses at various damage parametric conditions.

The values of the center line (CL), LCL and UCL in the control chart analysis (Fig. 3) are evaluated by considering the response analysis of undamaged beam under transit mass only. The points below LCL and above UPL exhibit the feasible presence of cracks on the structure. The implementation of AR process is here to locate the crack positions. The general equation of AR approach at a given time by considering the time history data (Beam deflection vs. time) of the given sample size can be articulated as:

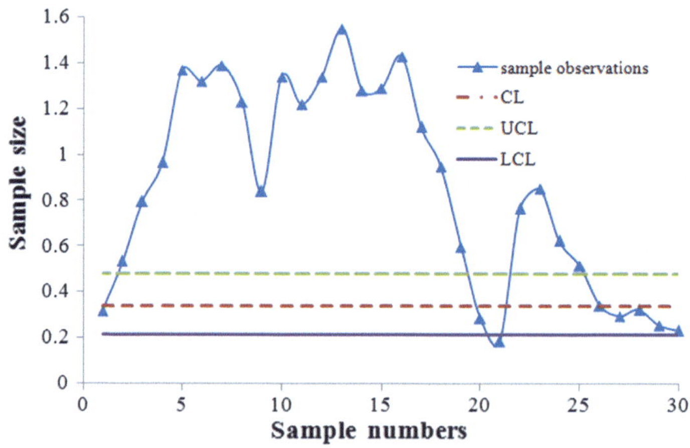

Fig. 3 Control chart analysis of the beam under transit mass

$$\psi_i(t_j) = [\psi_1(t_j) \quad \psi_2(t_j) \ldots y_n(t_j)]^{\mathrm{T}} \tag{4}$$

where $\psi_i(t_j)$ is the data from the time history at 'n' number of observations and sampled at 'r' number of intervals. The time history data are correlated with the previous data along with each other's data also. The above equation can be rearticulated as:

$$\psi_t = [\psi_{1t} \quad \psi_{2t} \ldots y_{nt}]^{\mathrm{T}} \tag{5}$$

ψ_t is the response data from the time history at time 't.'

Now, the covariance (ϕ) which is $n \times n$ matrix among all the observed points over the whole time interval can be represented as:

$$\phi = \sum_{j=1}^{n} \psi(t_j)\psi(t_j)^{\mathrm{T}} \tag{6}$$

Using the concepts of feature extraction approach with the covariance, the damage-sensitive parameters can be obtained from the measured time history data which can differentiate between damaged and undamaged states. In the AR (q) approach, the current data in the time series study is characterized as the linear combination of the precedent 'q' data. The AR model with order 'q,' AR (q), can be expressed as:

$$\psi_t = \sum_{j=1}^{q} \varphi_j \psi_{t-q} + a_t \tag{7}$$

where a_t is shock value with zero mean and constant variance.

ψ_t is the coefficient of AR approach which acts like damage-sensitive parameters.

$$\psi_t = \varphi_1 y_{t-1} + \varphi_2 y_{t-2} + \cdots \varphi_q y_{t-q} + a_t \tag{8}$$

The values of 'φ_j' are obtained by feeding the AR (q) form to the time vs. deflection data by using the Yule–Walker approach [11], where, the coefficients of the AR process are calculated with the liner linear least squares regression analysis by corroborating the Gaussianity and randomness of the predicted errors by trial and error methods.

The displacement-time data of the undamaged and damaged states are compared with each other to obtain the damage sensitive parameter. The entire undamaged data set is divided into two groups; reference data sample (DS^R) and healthy data sample (DS^H), while those of damaged sets are named as DS^D. The DS^R has been applied to get the damage-sensitive parameters for all sets of data for further comparison. The damage-sensitive parameters for DS^H and DS^D are also found out. The damage-sensitive parameters or coefficients of AR (q) model from both the DS^H and DS^D states are corroborated with DS^R. By observing deviation of damage-sensitive parameters that take place in the anticipated coefficients of AR model indicates the continuation of cracks in that system. The Fisher criterion approach has been implemented to obtain the authentic disparity of coefficients of AR model for both the DS^H and DS^D. The Fisher criterion approach is as follows:

$$F_{\text{criterion}} = \frac{(\mu_D - \mu_H)^2}{\phi_D + \phi_H} \tag{9}$$

where 'μ_D and μ_H' are the values of mean, 'ϕ_D and ϕ_H' are the values of variances of damage-sensitive parameters of damaged and undamaged states, respectively. The Fisher criterion values are greater at the potential crack locations where the sudden rise in the structure's response occurs which indicates the probable locations of cracks.

4 Results and Discussions

In the present work, a noble damage detection procedure has been elaborated in the concepts SPR approach for moving load dynamics problems. The proposed approach includes two parts that is training and monitoring. The control chart approach is established to recognize the probable existence of cracks. Later, the AR process and Fisher criterion methodologies are applied to identify the potentiality of cracks at any locations. For the present analogy, 300 numbers of patterns are generated, out of which 250 patterns are from healthy and 50 patterns are damaged states, respectively. From the control chart analysis, it can be predicted that system is stable or not, i.e., the presence of cracks. The AR approach along with the implementation of Fisher criterion has predicted the probable locations of cracks. To corroborate the authentication of the said approach, a numerical example has been formulated. The same example which has been taken in the problem definition part is also considered for authentication of SPR approach. The predicted results from SPR approach has been found out and compared with those of the numerical example. It has been traced that the variation of results on crack location from SPR approach converges (Table 1) with those of the numerical with an average discrepancy about 4.64% which seemed to be well.

Table 1 Comparison of SPR results with numerical data

Numerical Data		SPR Results	
η_1	η_2	η_1	η_2
0.3	0.5	0.284	0.479
0.4	0.67	0.3806	0.6413
Average percentage of error		5.07	4.22
Total percentage of error		4.64	

5 Conclusion

In the present investigation, a structural health monitoring problem in the domain of SPR approach has been developed for damage detection in moving load-induced structural dynamics problem. The entire study has been carried out in a supervised manner. The control chart mechanism along with AR process with the implementation of Fisher criterion has been applied for the proposed problem to spot the potential existence of damage in structure. Numerical examples has been devised and compared with the predicted results from SPR approach. It has been proven that the SPR approach converges well with the supervised problem. The SPR method can also be applied for structural health monitoring problem in unsupervised manner and also for online condition monitoring of structure.

References

1. Jena, S.P., Parhi, D.R.: Parametric study on the response of cracked structure subjected to moving mass. J. Vib. Eng. Technol. **5**(1), 11–19 (2017)
2. Jena, S.P., Parhi, D.R., Mishra, D.: Comparative study on cracked beam with different types of cracks carrying moving mass. Struct. Eng. Mech. Int. J. **56**(5), 797–811 (2015)
3. Jena, S.P., Parhi, D.R.: Dynamic and experimental analysis on response of multi-cracked structures carrying transit mass. J. Risk Reliab. **231**(1), 25–35 (2017)
4. Jena, S.P., Parhi, D.R.: Dynamic response and analysis of cracked beam subjected to transit mass. Int. J. Dyn. Control **6**(3), 961–972 (2018)
5. Xue, S., Wen, B., Huang, R., Huang, L., Sato, T., Xie, L., Tang, H., Wan, C.: Parameter identification for structural health monitoring based on Monte Carlo method and likelihood estimate. Int. J. Distrib. Sens. Netw. **14**(7), 1–9 (2018)
6. Mao, L., Weng, S., Li, S., Zhu, H., Sun, Y.: Statistical damage identification method based on dynamic response sensitivity. J. Low Freq. Noise Vib. Control (2018). https://doi.org/10.1177/1461348418784828
7. Farahani, R.V., Penumadu, P.: Damage identification of a full-scale five-girder bridge using time-series analysis of vibration data. Eng. Struct. **115**, 129–139 (2016)
8. Mechbal, N., Uribe, J.S., Rbillat, M.: A probabilistic multi-class classifier for structural health monitoring. Mech. Syst. Signal Process. **60**, 106–123 (2015)
9. Yang, Q., Liu, J.K., Sun, B.X., Liang, C.F.: Damage localization for beam structure by moving load. Adv. Mech. Eng. **9**(3), 1–6 (2017)
10. Maa, S., Jiang, S.-F., Li, J.: Structural damage detection considering sensor performance degradation and measurement noise effect. Measurement **131**, 431–442 (2019)
11. Mantgomery, D.C.: Introduction to Statistical Quality Control, 5th edn. Wiley Publications, Hoboken

TIG Welding Process Parameter Optimization for Aluminium Alloy 6061 Using Grey Relational Analysis and Regression Equations

A. Arul Marcel Moshi, D. Ravindran, S. R. Sundara Bharathi, F. Michael Thomas Rex and P. Ramesh Kumar

Abstract The present research enumerates the process parameter optimization of TIG welding process with aluminium alloy 6061 (AA6061). The input parameters such as weld current, shielding gas flow rate and double 'V' groove angle have been considered in view of optimizing the output parameters such as tensile strength and hardness of the TIG-welded specimens. The experiments are conducted based on L_9 orthogonal array. Grey relational analysis (GRA) has been employed to get the common optimal combination of considered input factors which yields better results for both the output responses. Optimal regression equations have been generated with the help of Design-Expert software. The influence of the input parameters on the output results has been analysed using ANOVA using Minitab software. 3D surface plots have been plotted to show the relationship between combinations of input factors with the output responses.

Keywords TIG welding · Aluminium alloy 6061 · Regression analysis · Grey relational analysis · Double 'V' groove angle

1 Introduction

Different types of welding processes are carried out in joining of metals at various manufacturing environments; each one has its own process parameters. Various parameters of welding determine the quality of weldments in turn the quality of the components. Tungsten inert gas (TIG) welding is one of the commonly used metal joining processes owing to its advantage of quality weldments as it is carried out in inert gas environment. Aluminium alloys are difficult weld materials, and TIG welding process is commonly employed for such alloys as the concerned parameters are controllable in this process. Various parameters such as weld current (WC),

A. A. M. Moshi (✉) · D. Ravindran · S. R. Sundara Bharathi · F. M. T. Rex
Department of Mechanical Engineering, National Engineering College, Kovilpatti, Tuticorin District, Tamil Nadu 628503, India
e-mail: moshibeo2010@gmail.com

P. R. Kumar
L&T Construction, Panchsheel Park, Pune, Maharashtra 411007, India

D. Dutta and B. Mahanty (eds.), *Numerical Optimization in Engineering and Sciences*, Advances in Intelligent Systems and Computing 979,
https://doi.org/10.1007/978-981-15-3215-3_41

shielding gas flow rate (SGFR) and double 'V' groove angle (DVGA) are some of the important process parameters that influence the quality of the weld greatly.

Aluminium alloy 6061 (AA6061) sets the standard for a medium-to-high strength, lightweight and economical material. The properties of AA6061 include its structural strength and toughness, its machinability and its ability to be easily welded and joined. AA6061 is used extensively as a construction material and most commonly in manufacturing of automotive components. The commonly used aluminium alloy of grade AA6061 has been chosen in this work for joining using TIG welding process in view of optimizing various process parameters mentioned above. In order to study and analyse the process parameters and to optimize the same, the AA6061 material specimens have been welded using TIG welding process. The trial of experiments has been planned as per Taguchi's L_9 orthogonal array by considering the input parameters of WC, SGFR and DVGA at three different levels. The output parameters such as tensile strength and hardness at the weld zones have been considered. The results have been analysed using grey relational analysis (GRA) and formulating regression equations. The best optimized input parameter combination of TIG welding process for AA6061 material in terms of the considered output responses has been suggested. Further, the effect of influence of each input parameter has been predicted with respect to the output responses by carrying out Analysis of Variance (ANOVA) with Minitab software of version 18.0.

In view of understanding the progress made in this area, the published literatures have been collected and thoroughly analysed. Some of them have been presented with their highlights. G. Magudeeswaran et al. had performed on the activated TIG (ATIG) welding process by focusing the depth of penetration of weldment [1]. K. S. Pujari et al. optimized the gas tungsten arc welding process parameters of aluminium alloy 7075-T6 welded joints. Further, ANOVA was employed by the authors to reveal the influence of individual parameters in obtaining the optimal results [2]. M. Balasubramanian et al. used lexicographic method to optimize the gas tungsten arc welding process parameters for titanium alloy material, and the authors suggested the optimal factor combination in order to obtain the strong weld pool geometry [3]. Arun Kumar Srirangan et al. had an attempt on optimizing the TIG welding process parameters with the Inconel alloy 800HT by considering current, welding speed and voltage as the influencing parameters [4]. A. Kumar et al. had an attempt to analyse the tungsten inert gas welding process parameters on Al–Mg–Si alloy welded joints in order to improve their mechanical parameters, and their results revealed that the impact strength and notch tensile strength were inversely proportional to each other [5]. G. Rambabu et al. had developed mathematical model for the corrosion resistance of aluminium alloy 2219 friction stir-welded joints by setting profile of tool pin, axial force, rotational and welding speeds as the input parameters for the work [6]. S. C. Juang, et al. had analysed the optimal selection of input parameters to set the geometry of the weld pool on the TIG-welded stainless steel specimens [7]. A. Kumar et al. had published another work with 5456 aluminium alloy weldments and improved mechanical properties through pulsed TIG welding process. Taguchi method was employed by them for optimizing the influencing factors of pulsed TIG welded on AA5456 joints [8]. Joby Joseph et al. focused in their published work on optimizing the activated TIG welding controllable factors for achieving better strength with AISI 4135 PM steel TIG-welded specimens by means of genetic algorithm (GA). Further, they reported that voltage and current had maximum influence on obtaining maximum tensile strength in A-TIG-welded joint [9]. RaviShanker Vidyarthy et al. investigated the TIG welding process parameters on 9–12% chromium ferritic stainless steel using central composite design and reported that welding speed, welding current and flux coating density had the direct influence on the weld strength [10]. A. Balaram Naik et al. had an optimization work on TIG welding process parameters with 2205 stainless steel material. The authors used neural network technique and ANOVA for setting out the optimal input combinations [11].

It is inferred from the literatures that the optimization of process parameters with respect to TIG welding process with various materials has been considered by various authors. The AA6061 material is one of the materials used for noncorrosive environment and has got the scope for further study with TIG welding process. The input parameters such as WC, SGFR and DVG have a scope for further analysis with respect to output responses of tensile strength and hardness. The oxidizing nature of aluminium alloys limits the weldability of AA6061, and TIG welding is one of the convenient processes to overcome the same. The study of optimizing the process parameters would help the manufacturers involved in use of such alloys.

2 Experimental Work

The proposed material AA6061 of size 150 X 75 mm with the thickness of 6 mm has been chosen as masterpiece, and V groove of various angles was formed at the edges after cleaning the surface. The welding was performed using TIG welding process along the length of 150 mm. The test specimens were prepared using wire-cut electrical discharge (WEDM) machining process from the masterpiece as per ASTM standard E-8 M for performing tensile test. The Vickers hardness test was performed over the welded zone. The chemical composition and the mechanical properties of Al6061 alloy are presented in Tables 1 and 2.

2.1 Design of Experiments (DOE)

In view of optimizing the strength and hardness of the weldment, the input parameters of TIG welding such as weld current (WC), shielding gas flow rate (SGFR) and double 'V' groove angle (DVGA) were considered for the analysis at three different levels which are presented in Table 3. The experiments to be performed were decided based on Taguchi's L_9 orthogonal array table. The combinations of the selected parameters for performing experimentation on samples with the input parameters and their levels are presented in Table 4.

Table 1 Chemical composition of Al 6061

Element	Cr	Cu	Fe	Si	Ti	Zn	Mg	Mn	Al
wt%	0.15	0.21	0.60	0.38	0.03	0.01	1.03	0.06	Bal.

Table 2 Base material properties

Properties	Values
Ultimate tensile strength (MPa)	310
Yield strength (MPa)	270
Modulus of elasticity (GPa)	70–80

Table 3 Selection of levels and parameters

S. No.	Parameters	Units	Level 1	Level 2	Level 3
1	WC	Amps	150	170	190
2	SGFR	Lpm	6	8	10
3	DVGA	Degrees	30	40	45

Table 4 Taguchi's L_9 orthogonal array

S. No.	Sample No.	Input parameters		
		WC	SGFR	DVGA
1	SA01	150	6	45
2	SA02	150	8	40
3	SA03	150	10	30
4	SA04	170	6	40
5	SA05	170	8	30
6	SA06	170	10	45
7	SA07	190	6	30
8	SA08	190	8	45
9	SA09	190	10	40

2.2 *Specimen Preparation*

In each category of samples, three specimens were prepared and tested. Hence, for nine set of parameters, as presented in Table 4, totally 27 samples were prepared using TIG welding process. The test samples were prepared from the master sample for tensile and hardness tests using wire-cut electrical discharge machine (WEDM) following ASTM Standards. The welded zone was grinded so as to make smooth surface by removing the burrs.

2.3 *Tensile Test*

Tensile tests on the samples were performed with computer interfaced universal testing machine of 60 tones capacity. It had the provision to extract the test data to

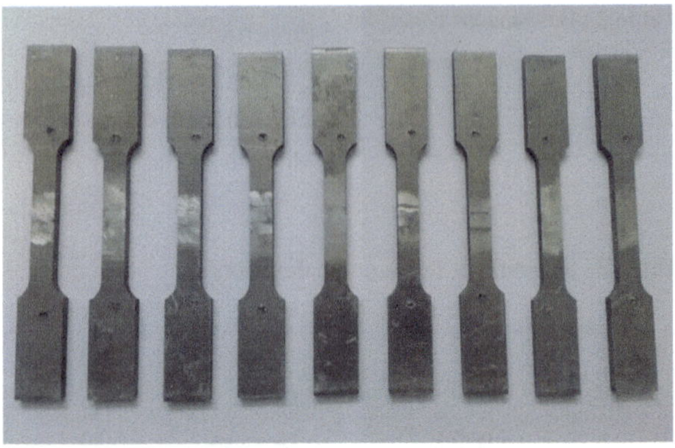

Fig. 1 Specimens for the tensile testing

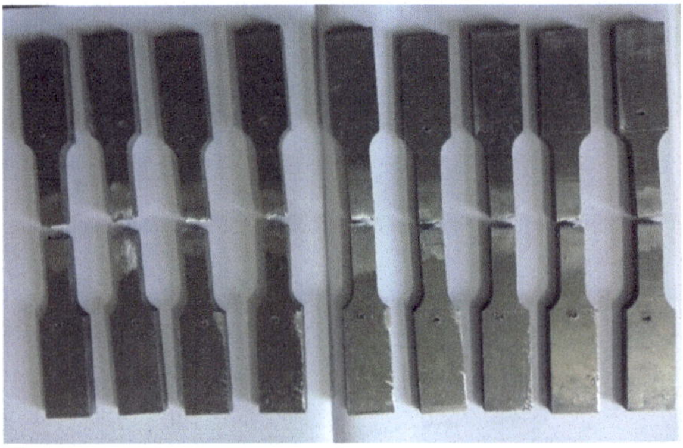

Fig. 2 Tensile tested specimens

a computer interfaced with the digital data logger. The tensile test specimens before and after the testing are shown in Figs. 1 and 2.

2.4 Hardness Test

Hardness test was conducted on Vickers hardness testing machine. The hardness test was made on all the test specimens at the welded portion.

2.5 Optimization Techniques

Grey relational analysis (GRA) is one of the multi-objective optimization techniques since it provides the optimized combination of input parameters for obtaining desirable results for all the output parameters. With the grey relational grade values obtained from the analysis, Taguchi's approach was used to confirm the results.

Regression analysis provides the relationship between the input parameters with each output parameter in the form of statistical equations. By conducting more number of experiments, we can frame the regression equation with better accuracy.

3 Results and Discussion

The results of tensile strength test and hardness test are presented in Table 5.

3.1 Grey Relational Analysis (GRA)

With the available test results, the technique of GRA has been employed for predicting the best set of parameters that yields best tensile and hardness value.

The output parameters of tensile and hardness properties will have different units. In order to consider them together, the normalization of the parameters is inevitable. The normalization of the above parameters was carried out as the first step of GRA. Followed by, the deviation sequence values for the normalized values of considered output responses were predicted. Table 6 represents the normalized values and deviation sequences of each output response.

The grey relational coefficient values and grey relational grade values were computed with the help of regular formulae, which are presented in Table 7.

Taguchi analysis was then made to finalize the optimal combination of input responses from the grey relational grade values with the help of Minitab software 18.0. The main effect plot for the S/N ratio values of the grey relational grades obtained from the Taguchi analysis is shown in Fig. 3.

As observed from Fig. 3, the optimal parameter combination for the TIG welding process on Alloy 61 s regarding tensile strength and hardness was noted as: current (150 A), gas flow rate (10 lpm) and double 'V' groove angle (45°). The confirmation test was conducted with the arrived optimal combination of input parameters, and the results were arrived for the tensile strength value as 136.4 MPa and hardness value as 87 VHN. Analysis of Variance was performed for the grey relational grades corresponding to the levels of input parameters. The ANOVA table obtained for the proposed case is as shown in Table 8. From the results of ANOVA table, it was palpable that weld current influenced more in getting optimal output results, and the least influencing factor was identified as double 'V' groove angle.

Table 5 Tensile test results

S. No.	Sample No.	Ultimate strength (MPa)				Vickers hardness number (VHN)			
		T_1	T_2	T_3	T_{avg}	T_1	T_2	T_3	T_{avg}
1	SA01	123.50	122.70	123.10	123.10	87.60	86.60	87.10	87.10
2	SA02	99.80	98.80	99.30	99.30	86.20	87.00	86.60	86.60
3	SA03	132.60	133.40	133.00	133.00	86.10	85.30	85.70	85.70
4	SA04	103.90	105.50	104.70	104.70	84.50	80.40	82.45	82.45
5	SA05	110.80	111.20	111.00	111.00	82.90	84.30	83.60	83.60
6	SA06	121.00	120.80	120.90	120.90	87.40	82.20	84.80	84.80
7	SA07	106.20	105.00	105.60	105.60	86.50	82.20	84.35	84.35
8	SA08	127.50	125.10	126.30	126.30	84.20	81.80	83.00	83.00
9	SA09	138.40	136.50	137.45	137.45	84.20	83.10	83.65	83.65

Table 6 Normalized values and deviation sequences of responses

S. No.	Sample No.	Normalized values		Deviation sequences	
		σ_u	VHN	σ_u	VHN
1	SA01	0.624	1.000	0.376	0.000
2	SA02	0.000	0.892	1.000	0.108
3	SA03	0.883	0.699	0.117	0.301
4	SA04	0.142	0.000	0.858	1.000
5	SA05	0.307	0.247	0.693	0.753
6	SA06	0.566	0.505	0.434	0.495
7	SA07	0.165	0.409	0.835	0.591
8	SA08	0.708	0.118	0.292	0.882
9	SA09	1.000	0.258	0.000	0.742

Table 7 Grey relational coefficient and grey relational grade values of output responses

Sample No.	Grey relational coefficient		Grey relational grade	Rank
	σ_u	VHN		
SA01	0.571	1.000	0.785	1
SA02	0.333	0.823	0.578	4
SA03	0.811	0.624	0.718	2
SA04	0.368	0.333	0.351	9
SA05	0.419	0.399	0.409	8
SA06	0.535	0.503	0.519	5
SA07	0.375	0.458	0.416	7
SA08	0.631	0.362	0.496	6
SA09	1.000	0.403	0.701	3

3.2 Regression Analysis

The tensile and hardness test results were fed into the Design-Expert software with their corresponding combination of input parameters. The optimized statistical relationship between the input parameters and tensile strength as well as with hardness values was arrived as represented in Eqs. 1 and 2 with the *R*-squared values of 0.9694 and 0.8768.

$$\begin{aligned}\textbf{Tensile strength} = & -2331.10018 + (9.45267 * \text{WC}) + (148.96481 * \text{SGFR}) \\ & + (46.43422 * \text{DVGA}) - (0.29051 * \text{WC} * \text{SGFR}) \\ & -(0.15793 * \text{WC} * \text{DVGA}) - (2.30006 * \text{SGFR} * \text{DVGA}) \quad (1)\end{aligned}$$

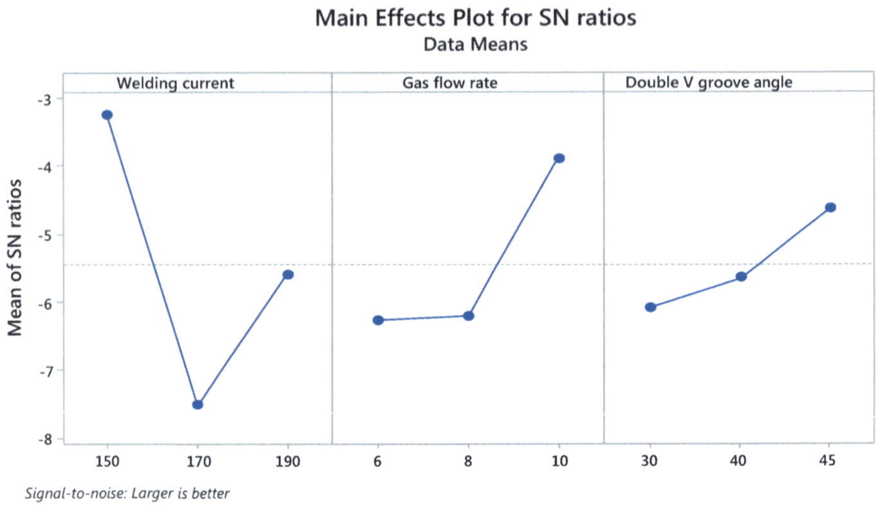

Fig. 3 Main effects plot for S/N ratio values of grey relational grade values

Table 8 ANOVA results for the considered process parameters

Main control factors	Degree of freedom (DF)	Sum of squares (SS)	Mean of squares (MS)	*P*-value	Contribution, C (%)
WC (A)	2	0.10820	0.054099	0.209	57.46
SGFR (lpm)	2	0.04009	0.020043	0.417	21.29
DVGA (degree)	2	0.01139	0.005695	0.715	6.05
Error	2	0.02861	0.014307	–	15.20
Total	8	0.18829	–	–	100.00

$$\mathbf{Hardness} - 160.73099 + (1.11580 * \mathrm{WC}) + (12.77027 * \mathrm{SGFR}) + (4.91578 * \mathrm{DVGA}) - (0.037804 * \mathrm{WC} * \mathrm{DVGA}) - (0.021506 * \mathrm{WC} * \mathrm{DVGA}) - (0.14157 * \mathrm{SGFR} * \mathrm{DVGA}) \tag{2}$$

The 3D response plots were drawn and analysed the effects of varying the levels input factors on the responses, which are shown in Figs. 4 and 5. The 3D plots revealed that by increasing the value of weld current, it can be achieved the TIG-welded specimens with more tensile strength. Tensile strength also increased with the increase in gas flow rate. In case of achieving better hardness results, double 'V' groove angle had less significant effect, and the increase in the gas flow rate and increase in VHN were of directly proportional.

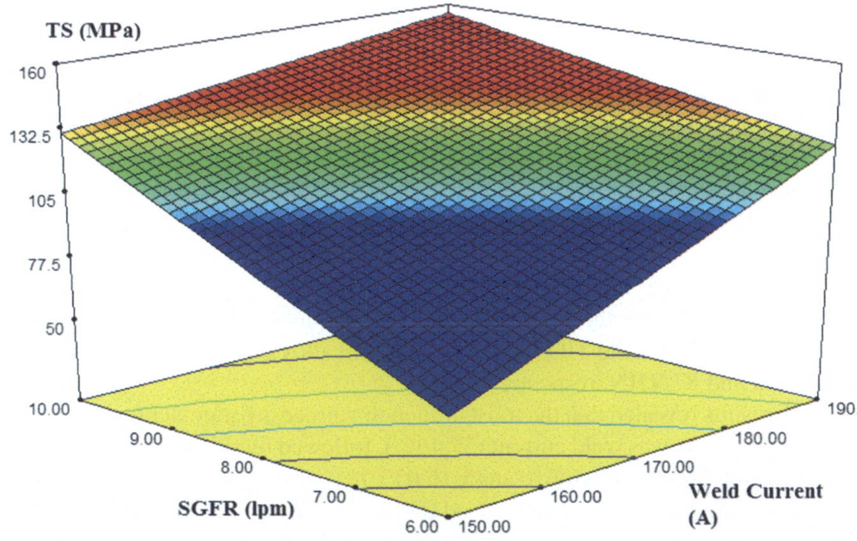

Fig. 4 3D surface plot of TS versus WC and SGFR

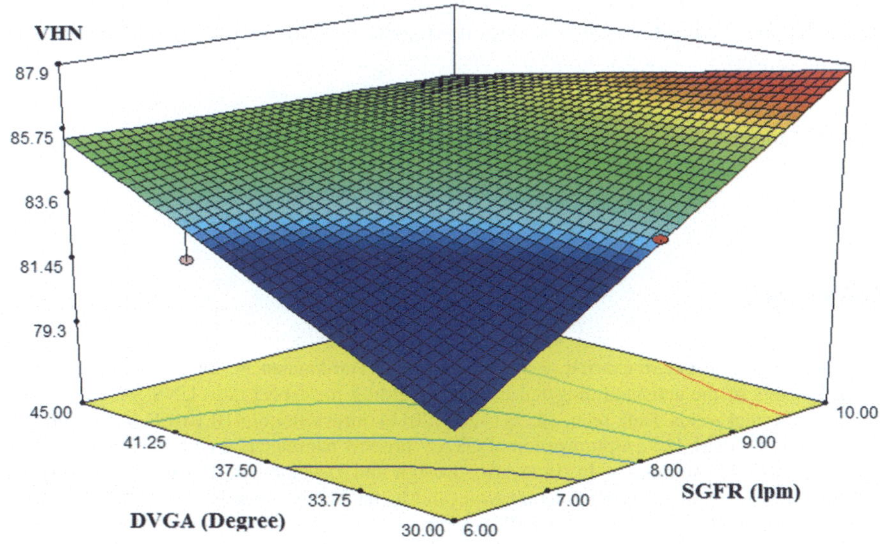

Fig. 5 3D surface plot of VHN versus SGFR and DVGA

4 Conclusion

- In this work, optimization of TIG welding process parameters was performed using Taguchi-based grey relational analysis and regression analysis. Three parameters with three levels were considered for the analysis. Experiments have been conducted as per the parameter combination of L_9 orthogonal array.
- From the grey relational analysis, the optimal parameter combination for the TIG welding process on AA6061 is: current (150 A), gas flow rate (10 lpm) and double 'V' groove angle (45°).
- The optimal results for the tensile strength and hardness values were noted, and the optimal results for the TIG-welded aluminium alloy 6061 joints were noted as 136.4 MPa and 87 VHN, respectively.
- ANOVA results revealed that the most influencing process parameter for achieving the optimal results was weld current (57.46%), followed by shielding gas flow rate (21.29%) and double 'V' groove angle (6.05%).
- The statistical relationships were arrived for the input parameters with each output response with the help of regression analysis with the R-squared values of 0.9694 and 0.8768.
- The 3D relationship plots were drawn and analysed. The results showed that tensile strength will be higher while increasing the values of weld current and gas flow rate. By increasing the gas flow rate, the hardness value of the welded specimens got increased.

Conflict of Interest The authors declare that there is no conflict of interest in publishing this research work.

References

1. Magudeeswaran, G., Sreehari, R., Nair, R., Sundar, L., Harikannan, N.: Optimization of process parameters of the activated tungsten inert gas welding for aspect ratio UNS S32205 duplex stainless steel welds. Def. Tech. **10**, 251–260 (2014). https://doi.org/10.1016/j.dt.2014.06.006
2. Pujari, K.S., Patil, D.V.: Optimization of GTAW process parameters on mechanical properties of AA7075-T6 weldments. In: Innovative Design and Development Practices in Aerospace and Automotive Engineering, Lecture Notes in Mechanical Engineering pp. 187–195 (2017). https://doi.org/10.1007/978-981-10-1771-1_23
3. Balasubramanian, M., Jayabalan, V., Balasubramanian, V.: Prediction and optimization of pulsed current gas tungsten arc welding process parameters to obtain sound weld pool geometry in titanium alloy using lexicographic method. J. Mater. Eng. Perform. **18**(7), 871–877 (2009). https://doi.org/10.1007/s11665-008-9305-6
4. Srirangan, A.K., Paulraj, S.: Multi-response optimization of process parameters for TIG welding of Incoloy 800HT by Taguchi grey relational analysis. Eng. Sci. Tech., Int. J. **19**(2), 811–877 (2016). https://doi.org/10.1016/j.jestch.2015.10.003
5. Kumar, A., Sundarrajan, S.: Effect of welding parameters on mechanical properties and optimization of pulsed TIG welding of Al–Mg–Si alloy. Int. J. Adv. Manuf. Tech. **42**(1–2), 118–125 (2009). https://doi.org/10.1007/s00170-008-1572-8

6. Rambabu, G., Balaji Naik, D., Venkada Rao, C.H., Srinivasa Rao, K., Madhusudan Reddy, G.: Optimization of friction stir welding parameters for improved corrosion resistance of AA2219 aluminum alloy joints. Def. Tech. **11**, 330–337 (2015). https://doi.org/10.1016/j.dt.2015.05.003
7. Juang, S.C., Tarng, Y.S.: Process parameter selection for optimizing the weld pool geometry in the tungsten inert gas welding of stainless steel. J. Mater. Process. Tech. 33–37 (2002). https://doi.org/10.1016/s0924-0136(02)00021-3
8. Kumar, A., Sundarrajan, S.: Optimization of pulsed TIG welding process parameters on mechanical properties of AA 5456 Aluminum alloy weldments. Mater. Des. 1288–1297 (2009). https://doi.org/10.1016/j.matdes.2008.06.055
9. Joseph J., Muthukumaran, S.: Optimization of activated TIG welding parameters for improving weld joint strength of AISI 4135 PM steel by genetic algorithm and simulated annealing. Int. J. Adv. Manuf. Tech. **93**, 23–34 (2017). https://doi.org/10.1007/s00170-015-7599-8
10. Vidyarthy, R.S., Dwivedi, D.K., Muthukumaran, V.: Optimization of A-TIG process parameters using response surface methodology. Mater. Manuf. Process. **33**, 709–717 (2017). https://doi.org/10.1080/10426914.2017.1303154
11. Balaram Naik, A., Chennakeshava Reddy, A.: Optimization of tensile strength in TIG welding using the Taguchi method and analysis of variance (ANOVA). Therm. Sci. Eng. Prog. **8**, 327–339 (2018). https://doi.org/10.1016/j.tsep.2018.08.005

Mathematical Modeling in MATLAB for Convection Through Porous Medium and Optimization Using Artificial Bee Colony (ABC) Algorithm

A. Siva Murali Mohan Reddy and Venkatesh M. Kulkarni

Abstract The innovative mathematical model is employed to ascertain the mathematical equation by means of the recognized input and output experimental values taken from Reddy et al. (Int Org Sci Res J Mech Civ Eng 13(4):131–140, [1]). The recognized inputs comprise the length of the porous medium (L), diameter (D), porosity (ε), heat flux (q) and mass flow rate (m). They are employed by the mathematical modeling to assess the ideal outputs of the pressure drop (Δp) and Nusselt number (Nu). The innovative mathematical modeling with artificial bee colony technique is amply applied to obtain ideal arrangements of the input limits; in the preparation of the procedure, 80% of the dataset is used for training action, and testing of the scientific model is done with the help of remaining 20% of the dataset. In the mathematical modeling, the optimization procedure is used to reach the best values of α and β of the problem for scaling down the value of error of the network. Several optimization approaches like the artificial bee colony algorithm (ABC), ant colony optimization (ACO), genetic algorithm (GA) and particle swarm optimization (PSO) are efficaciously utilized to learn the best weight of the system. The mathematical modeling ABC emerges with least error by bringing in the best weight α and β.

Keywords Heat transfer · Porous medium · Pressure drop · Temperature · Weights · Artificial bee colony algorithm · Length · Heat flux · Porosity

1 Introduction

Popularity of meta-experience-based thinking sets of computer instructions has incremented in over the last few years. A notable number of new meta-experience-based thinking algorithms [1–3] are being used for elucidating optimizing problems

A. S. M. M. Reddy (✉)
Department of Mechanical Engineering, Sapthagiri College of Engineering, Bangalore, Karnataka, India
e-mail: mohan047@gmail.com

V. M. Kulkarni
Department of Thermal Power Engineering, VTU Centre for Postgraduate Studies, Kalaburagi, Karnataka, India

D. Dutta and B. Mahanty (eds.), *Numerical Optimization in Engineering and Sciences*, Advances in Intelligent Systems and Computing 979,
https://doi.org/10.1007/978-981-15-3215-3_42

of different nature like population predicated, repetitive predicated, random-based predicated approaches. Karaboga [4] developed an incipient optimization technique called artificial bee colony (ABC) algorithm in the year 2015. Over the last few years, it became a well-liked technique, because of its potential for easy approach to solve practical engineering quandaries. ABC is an expeditious convergent and facile-handling technique, which is directly applied in different fields like electronics circuit design, production, civil and electrical engineering, heat transfer and fluid flow. For instance, an incipient shell and tube heat exchanger design optimization using artificial bee colony (ABC) technique was developed by Şahin et al. [5]. This reduced the total cost of the equipment. Hetmaniok et al. [6] have made an attempt to solve two-phase axisymmetric, unidimensional inverse Stefan problem employing third-kind boundary condition using ABC algorithm technique. The procedure includes the reformation of the function which illustrates the heat transfer coefficient appearing in boundary condition of the third kind in such a way that the modified values of temperature would be as contiguous as possible to the quantifications of temperature given in culled points of the solid. Hetmaniok et al. [7] made use of culled swarm perspicacity algorithms like the artificial bee colony algorithm and the ant colony optimization algorithm for solving the inverse heat conduction quandary The analyzed quandary consists of reformed temperature distribution in the given domain and the form of heat flow coefficient appearing in the third-kind boundary condition. Karaboga, Bahriye Akay [8] have utilized ABC for improving an astronomically immense set of numerical test functions, and the results engendered by ABC algorithm are compared with the results got by particle swarm optimization algorithm, genetic algorithm, differential evolution algorithm and evolution strategies. Results show that the performance of the ABC is homogeneous to those of other population predicated techniques with the utilization of less control parameters. Porous media are utilized in a wide variety of natural and artificial systems including biological systems, cooling in electronic circuits, air filters, heat exchangers, thermal insulation and thermal storage devices [9], gives a brief idea for the efficacious utilization of porous media for fluid flow and heat transfer problems. Zhang-Jing [10] developed an incipient optimization technique known as genetic algorithm (GA) to obtain heat transfer enhancement in a tube flow by optimizing the configurations of porous insert. The results conclude that the thermal-hydraulic performance of the enhanced tube can be ameliorated efficaciously by utilizing the optimized porous insert, concretely utilizing the optimized multiple layers of porous insert. However, there is a congruous layer number of porous inserts to ascertain the optimal performance of the enhanced tube for a given set of parameters.

2 Mathematical Modeling

The perceived input parameters like the length of the porous medium (L), diameter (D), porosity (ε), heat flux (q) and mass flow rate (m). They are employed by the mathematical modeling for determining the excellent outputs of the pressure drop

(Δp) and Nusselt number (Nu). At the beginning, the arbitrary weights are sanctioned in the system within the inhibitions. The datasets are formulated by the system for achieving the base slip by using the weights α and β, which are changed for the purport of figuring the velocity of the input restrictions. Then the velocity by changing the weights α and β duly accomplishes the pressure drop. After preparation, the prepared system is utilized for analyzing the dataset with the 80:20 proportions. Optimization technique such as the genetic algorithm (GA), particle swarm optimization (PSO) and artificial bee colony (ABC) optimization and ant colony optimization (ACO). Systems are efficaciously used to arrive at the ideal weights of the desired function, which is given by the distinction between the tested and forecast values. In mathematical modeling, the kenned inputs with the optimal weights are made up and taken as equation

$$F = \sum_{j=1}^{h} \alpha_j * \left(\frac{1}{1 + \exp\left(- \sum_{i=1}^{N} X_i \beta_{ij} \right)} \right) \tag{1}$$

where X is the input parameters, α and β indicates weights, i denotes number of inputs, j indicates number of weights, whereas N is the number of the input data and h is the number of obnubilated neurons. Arithmetical demonstration of mathematical modeling is conventionally predicated on several optimizations of the weights. Among these, the best weights are achieved by artificial bee colony (ABC) algorithm.

2.1 ABC Algorithm

The ABC technique, in turn, makes skillful utilization of the scout bee, onlooker bee and employed bee. This description revolves around the examination of the comportment of genuine bees toward ascertaining the nectar amount and bringing the data of sustenance sources to the parallel bees in the hive. Our innovative technique has prosperously evaluated the dimension of pressure drop, which paves the way for the guessing of excellent parameters such as the length, diameter, porosity, heat flux and mass flow rate predicated pressure drop by skillful employment of the ABC algorithm.

3 Result and Discussion

Sundry inputs such as the diameter, heat flux, length, porosity and fluid flow rate, and output values represented by the pressure drop and Nusselt Number are taken from the document with their corresponding input and output values. In the first stages, the mathematical model is trained in such a way that 80% of the dataset is utilized for training the network model whereas the remnant representing 20% of the dataset is used for authentication. The best solutions of α and β can be obtained by cumulating the mathematical model with artificial bee colony (ABC) optimization technique, which performs the fascinating function of finding, and the best solution with input restrictions are arrived at with the help of the astounding artificial bee colony (ABC) algorithm. The output is changed for least error value by the mathematical model. The major goal of the model is to achieve the differential error between real time output and the procured output from mathematical model, which is found to be nearly tantamount to zero. With the result, the cognate output evaluated by using the velocity is used to obtain the temperature.

3.1 Mathematical Modeling with Optimization Techniques

In the present work, among the mathematical modeling with the optimization techniques like the ABC, GA, PSO and ACO, artificial bee colony (ABC) technique gives the least error value for the best equation with best weights α and β. The least error is achieved in artificial bee colony (ABC) technique, facing parallel optimization approaches such as GA, PSO and ACO. Figure 2 elegantly exhibits the least error value of the pressure drop for the mathematical modeling with the artificial bee colony (ABC) technique. Particle swarm optimization (PSO), artificial bee colony (ABC) and ant colony optimization (ACO) approaches are a lot employed to figure out the error value of the mathematical modeling.

Figure 1 makes it completely and totally clear that the mathematical model along with ABC turns out the minimum error values of the pressure drop ultimately ushering in the overall least error value. ABC algorithm gives the least error value of 0.0121, which is carefully studied and contrasted with the GA the error minimized at 96.63% and after that contrasted with the PSO method error, minimized at 96.98% and

Fig. 1 Error graph for different algorithms

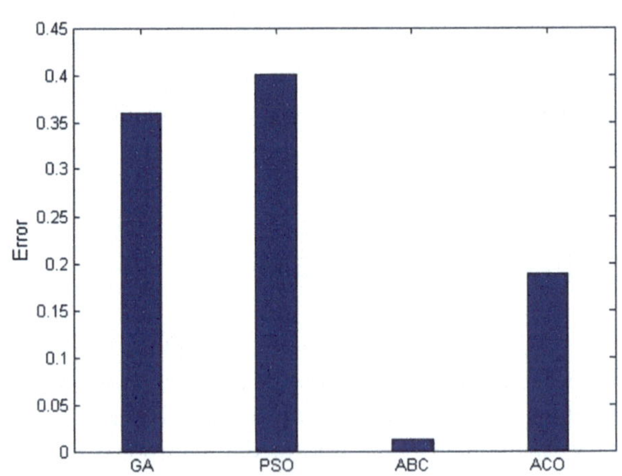

evaluated with the ACO at 93.59% whose error will be scaled down. In the overall performance, the lowest possible error value of ABC is contrasted with parallel techniques at 95.733% the pressure drop error will be reduced to the least in trail and forecast values.

3.2 Iteration-Based Graph for Different Algorithms

The graphs appearing in Fig. 2 efficaciously demonstrate the pressure drop graph for each and every iteration of the GA, PSO, ABC and ACO by weight altering in the range of negative 500 to positive 500, and appropriately the values of error are discovered The error chart is plotted with the iteration along *x*-axis and error values in *y*-axis. Figure 3 illustrates the error value of the pressure drop for several techniques such as the ABC, GA, PSO and ACO oriented by duly altering the weight. In each algorithm, the minimum error value is realized in the iteration number 100. In the case of the ABC algorithm, the minimum error value is 0.0121 which when compared with the pristine value the error difference is 0.045%, and the initial error value is minimized as 99.95%. Thereafter, in GA, the error is minimized as 2.280%, which when compared with the pristine values the difference in ABC and GA the error value is 98.02% minimized. In the case of the PSO technique the initial error value is 13.34, minimized error value is 0.40088 with the error minimized as 3.005%, and when the value is compared with the ABC the error difference is 98.50%, in the ACO method, initial iteration the error value is 5.3243 and the minimum error value

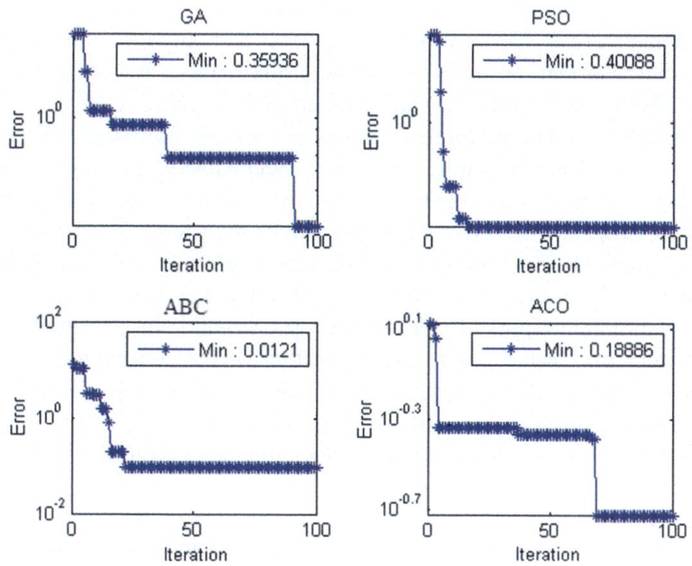

Fig. 2 Iteration versus error graph for different algorithms

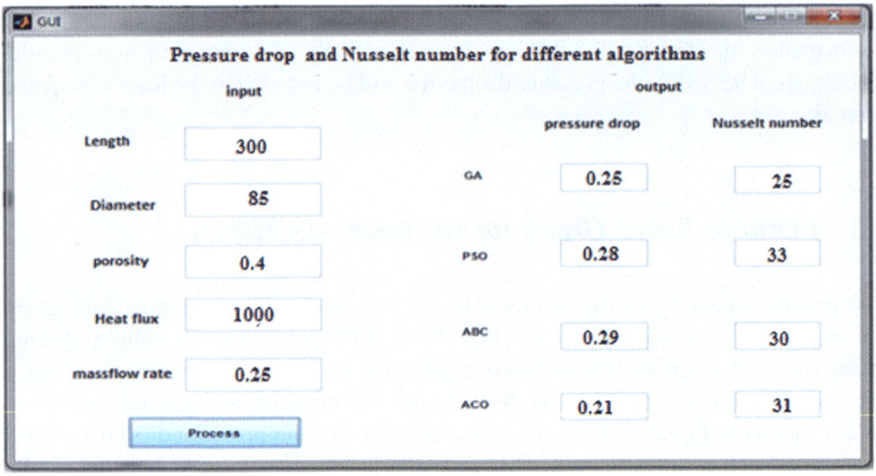

Fig. 3 MATLAB output

is 0.18886 procured in 100th iteration. The error value is minimized at 3.547% of initial value, and when compared with the pristine value, the difference is 96.45% in the ACO in comparison with the ABC the error is minimized as 98.73%.

In the mathematical modeling, 80% of data available is aptly utilized by duly altering the weights and the remnant 20% data is efficaciously used for testing. In this process, the pressure drop is ascertained by modifying the weight in the range of −500 to 500, and likewise, the Nusselt number is evaluated by transmuting the weight in the range of −5 to 5 for different techniques like the ABC, PSO, GA and ACO. The practical dataset values of the input and output parameters grasp the length (L), diameter (D), porosity (ε) and heat flux (q) and mass flow rate (m) and the outputs like pressure drop (Δp) and Nusselt number (Nu).

This elegantly exhibits the forecast values of pressure drop and Nusselt number by means of the 20% test data utilized. In the cognate testing procedure, six diverse types of data are taken into account to ascertain the pressure drop and Nusselt number for several techniques. Further, the forecast values are tested and harmonized with the pristine values to come up with the least error of each approach in reverence of the inputs like length (L), diameter (D), porosity (ε) and heat flux (q) and mass flow rate (m) and the outputs like pressure drop (Δp) and Nusselt number (Nu). In the introductory data values, in reverence of the forecast pressure drop in all the techniques, the average difference in pristine value obtained is 0.05234. In the case of Nusselt number, the corresponding value is observed to be 3.57 in all techniques, where as for the pressure drop and Nusselt number (Nu), the least error value is realized in the ABC technique.

The average error value of pressure drop in several methods against the pristine value in the testing process is found to be 97.91%. In the case of the ABC algorithm pressure drop values compared with other approaches like the GA, PSO and

ACO the pressure drop error value differences are found to be 58.55%, 82.62%, and 76.31%, respectively. The average error value of Nusselt number in ABC algorithm in cognation to the pristine value is estimated to be 90.19%. In the case of the ABC algorithm, Nusselt number values vis-a-vis the parallel techniques like the PSO, GA, ACO, the Pressure drop error value difference is observed to be 69.83, 74.41 and 69.19%, correspondingly.

Figure 3 shows a set of input values of the operation during the performance technique which is carried out in the MATLAB programming designated on the relative chart. At this juncture, the required constraints encompass the length, diameter, porosity, heat flux and mass flow rate. With the input data based on several techniques, the pressure drop and Nusselt number values are estimated. For different testing data values, the output values are obtained in several methods. Table 1 contains first data input values, whereas the matching output values are indicated in Fig. 3 in this process, the input data-based pressure drop and Nusselt number in ABC algorithm is estimated to be 79.12% and the value vis-à-vis the parallel approaches like the GA, PSO ACO the difference values are found to be 41.32%, 28.47% and 57.21%, respectively. In this graphical user interface-based procedure, the values of input parameters are changed and the corresponding output pressure drop and Nusselt number are evaluated.

4 Conclusion

This paper sophisticatedly gives an idea about the mathematical modeling technique inaugurate with the robust artificial bee colony (ABC) optimization method which surprisingly reaches the accurate ideal values of the weights in mathematical model. Multiracial development issues in the global best solution and illustrates the ability to change to choose the appropriate design parameters based on the weights. During the transaction of the system, the pressure drop and Nusselt number values are evaluated with the datasets. The authentic output results are approached to the dataset lowest possible error value of pressure drop obtained in the ABC method of optimization. In the imminent era, investigators will look toward further incredible improvement ways of doing things for the harvests of reduced errors with their sterling procedures for skilled chasing after their probing parades.

Table 1 Predicted values of testing data for different algorithms

Inputs					Outputs (original)		ABC		GA		PSO		ACO	
L m	D m	q W/m^2	m Kg/s	ε	Δp bar	Nu	Δp bar	Nu	Δp bar	Nu	Δp bar	Nu	Δp bar	Nu
0.3	0.85	1000	0.25	0.4	0.29	29.5	0.29	30	0.25	25	0.28	33	0.21	31
0.3	0.85	1000	0.30	0.4	0.35	31	0.34	31	0.30	29	0.31	38	0.29	31
0.3	0.85	1000	0.35	0.4	0.41	33	0.40	33	0.38	30	0.36	31	0.31	32
0.3	0.85	1000	0.40	0.4	0.46	34	0.45	34	0.39	32	0.40	32	0.32	32
0.3	0.85	1000	0.45	0.4	0.54	36	0.56	36	0.43	33	0.44	33	0.45	33
0.3	0.85	1000	0.5	0.4	0.62	38	0.63	37	0.59	35	0.58	36	0.49	34

References

1. Reddy, S.M.M., Kulkarni, V.M.: An experimental investigation of heat transfer performance for forced convection of water in a horizontal pipe partially filled with a porous medium. Int. Org. Sci. Res. J. Mech. Civ. Eng. **13**(4), 131–140 (2016). Ver. III. ISSN: 2278-1684
2. Kennedy, J., Eberhart, R.: Particle swarm optimization. In: IEEE International Conference on Neural Networks, Perth, Australia, IEEE Service Center, Piscataway, NJ (1995)
3. Dorigo, M.: Positive feedback as a search strategy. Technical Report 91-016, Politecnico di Milano, Italy (1991)
4. Karaboga, D.: An Idea Based On Honey Bee Swarm for Numerical Optimization. From Technical Report (TR06), University of Erciyes (2005)
5. Şahin, A.Ş., et al.: Design and economic optimization of shell and tube heat exchangers using ABC algorithm. Energy Convers. Manag. **52**, 3356–3362 (2011)
6. Hetmaniok, E., Słota, D., Zielonka, A.: Identification of the heat transfer coefficient in the inverse Stefan problem by using the ABC algorithm. Arch. Foundry Eng. **12**, 27–32 (2012). ISSN (1897-3310)
7. Hetmaniok, E., et al.: Application of swarm intelligence algorithms in solving the inverse heat conduction problem. Comput. Assisted Methods Eng. Sci. 361–367 (2012)
8. Karaboga, D and Bahriye Akay.: A comparative study of ABC algorithm. Appl. Math. Comput. **214**(1), 108–132 (2009)
9. Mahmoudi, Y., Karimi, N.: Numerical investigation of heat transfer enhancement in a pipe partially filled with a porous material under local thermal non-equilibrium condition. Int. J. Heat Mass Transf. 161–173 (2014)
10. Zheng, Z.-J., Li, M.-J., He, Y.-L.: Optimization of porous insert configurations for heat transfer enhancement in tubes based on genetic algorithm and CFD. Int. J. Heat Mass Transf. **87**, 376–379 (2015)

Utility Theory Embedded Taguchi Optimization Method in Machining of Graphite-Reinforced Polymer Composites (GRPC)

Vikas Kumar and Rajesh Kumar Verma

Abstract The present work shows the hybrid optimization approach by the use of utility embedded Taguchi philosophy to optimize machining constraints with multiple characteristics in machining (milling) operation of graphite-reinforced polymer composite (GRPC). Taguchi-based L_{16} orthogonal array has been used to perform milling operation. Vertical milling center machine has been used for the milling operation with input factor, speed (S), feed rate, depth of cut (DoC) and weight percentage (wt.) of graphite. The output response considered is metal removal rate, thrust, torque and surface roughness (Ra). The purpose of using this technique is to convert the multi-response to single response function (U) finally optimized by Taguchi method. In utility theory, the initial step is to find the preference number and then overall utility. From this study, it has been proven that the utility theory based on Taguchi method has accomplished the effective milling environments in order to minimize thrust (T), torque (TR) and surface roughness (Ra) during machining. This technique has been suited for off-line quality control of the product development as well as mass production shapes. Utility theory-based Taguchi method has considered optimizing the multiple responses simultaneously, which results in the predicted S/N ratio value of **17.4111**, the mean value is **6.65345**, and most favorable machining value is speed 500 mm, feed rate 25 mm/rev., depth of cut −1.5 mm and weight percentage 40%, respectively. This approach can be forwarded for quality monitoring (offline/online) of manufacturing industries.

Keywords Graphite-reinforced polymer composite · Taguchi method · Utility theory · Surface roughness

V. Kumar (✉) · R. K. Verma
Department of Mechanical Engineering, MMM University of Technology, Gorakhpur 273010, India
e-mail: vikas.sahni95@gmail.com

R. K. Verma
e-mail: rkvme@mmmut.ac.in

D. Dutta and B. Mahanty (eds.), *Numerical Optimization in Engineering and Sciences*, Advances in Intelligent Systems and Computing 979,
https://doi.org/10.1007/978-981-15-3215-3_43

1 Introduction

Recently, optimization plays a very important role in the interdisciplinary area of research work, and the variety of investigation has done in the area of machining by pioneer researchers. The main application of graphite-reinforced polymer composites in the area of aerospace, space vehicle, automobile sector, electronic insulator and fuselage component of the airplane and so on. The machining (milling) of graphite-reinforced polymer performed with four input parameter like—speed (S), feed rate, depth of cut (DOC) and weight percentage and output response like—MRR, thrust, torque and Ra. The diameter of the milling cutter is 5 mm, and the tool material is HSS. Taguchi L_{16} orthogonal array design has been used for vertical milling machine center to perform the milling operation. The term utility means the usefulness of a product/procedure in reference to the expectations levels of the customers while performing the machining operation of the number of output characteristics. So, the overall quality has been essential to scale its complete performance. Utility theory gives the procedural framework for the assessment of other features complete through manufacturing phases. According to the utility theory, the decision-maker provides the best possible decision. Taguchi method has been used for finding the S/N ratio and optimal setting of the experimental data.

2 Literature Review

Nowadays, the yarn strengthening polymer compound has been used in many areas like aerospace application, automobile, sports goods, marine, the as well as in electronic industries. Mostly, the fiber-reinforced polymer composites have been categorized through the high strength, low specific weight as well as a tough chemical confrontation in contradiction to corrosion [1–3]. Epoxy is known as polymer resin which has currently helped to make the matrix constituents for the fiber-reinforced compounds. It has been extensively applied in numerous uses like adhesives, coatings as well as laminates [4–6]. Graphite has one of the most adaptable constituents which is extensively applied in manufacturing engineering due to the aforementioned surprising physical as well as chemical possessions. The electrical conductivity of graphite tends to a metal which has been widely applied in electrodes for batteries as well as electrolysis reaction. There has been a tremendously great current constancy of graphite (700 K in the air as well as more than 3000 K in an inert atmosphere) which shows a great use for manufacturing of receptacles for holding melted metals. Higher conductivity has been subjected in heat sinks (graphite) intended for PC computers which retain calm during convertible load. It has cast-off as a filler of electrically conductive material aimed at the development of conductive polymer compounds [7, 8]. Optimization of machining parameters has one of noticeable concerns in the area of manufacturing, wherever the low cost of machining operation shows an important part of effectiveness in the market place. Owing to great investment as well as machining prices of NC apparatus. Numerous investigator has given out the optimization of machining factor for turning operation as well as the graphical techniques have been used to define the best speed as well as feed rate [9–13]. Limited investigators have focused on the machining operations of multi-point tool as well as resolved through forced mathematical programming approaches [14]. The various procedures have stated in the works to improve the machining constraints of face milling processes.

3 Experimental Details

In this study for finding the machining behavior and machinability aspects of composite, an extensive investigation has been done through statical tools and optimization modules. Here, L_{16} orthogonal array has been applied to perform the experiments of machining (milling). Process parameter considered is speed (S), feed rate (F), depth of cut (DOC) and weight percentage (wt.) varied at four different levels as depicted in Table 1. DOE includes the set of experiments (Table 2) which has varied into a consecutive manner aimed at assessing the experimental measurement of responses. In Table 2, the input data is considered according to the Taguchi L_{16} orthogonal array-based DOE, and (−) negative sign in depth of cut shows that the milling operation of GRPC sample has done in a downward direction in CNC **vertical machining center** successively.

Milling operation has been used for machining of the graphite-based composite. In machining, the work part has fed a rotating cylinder-shaped tool by use of numerous rotating cutter edge. The tool arrangement is perpendicular to the feed mode, and the name of the tool is known as a milling cutter. Usually, the surface of the plane has shaped through the machining process. The operation of machining (milling) has used to machining the plane surface. Ultimately, machining operation has considered as upright or rough flanks finished beneficial to the workpiece stand to a flat revolving plate. Milling operation is categorized like knee-type as well as cutter cover a figure of cutting parameters. The accurateness of the milling machining has much superior in comparison with the parent operation of machining.

$$\textbf{Size of the mould} = (\textbf{8.6 cm} \times \textbf{7 cm} \times \textbf{1 cm}).$$

Table 1 Input factors and levels

Factors	Unit	Level 1	Level 2	Level 3	Level 4
Speed	[RPM]	500	1000	1500	2000
Feed rate	[mm/rev]	10	15	20	25
Depth of cut	[mm]	0.5	1	1.5	3
Weight %	%	10	20	30	40

Table 2 Taguchi-based L_{16} orthogonal array (DOE)

S. No.	Speed (rpm)	Feed rate (mm/rev.)	Depth of cut (mm)	Weight (wt.) %
1	500	10	−0.5	10
2	500	15	−1	20
3	500	20	−1.5	30
4	500	25	−3	40
5	1000	10	−1	30
6	1000	15	−0.5	40
7	1000	20	−3	10
8	1000	25	−1.5	20
9	1500	10	−1.5	40
10	1500	15	−3	30
11	1500	20	−0.5	20
12	1500	25	−1	10
13	2000	10	−3	20
14	2000	15	−1.5	10
15	2000	20	−1	40
16	2000	25	−0.5	30

3.1 Materials Used for Fabrication Work

A. **Collection of Matrix material**: Matrix material select, Epoxy resin LAPOX (L-12) and HARDER K-6 as a binder for the resin.
B. Reinforcement of natural graphite.
C. Requirements for the fabrication of composites—Epoxy resin, hardener and mold for making the size of the specimen (Figs 1 and 2).

Fig. 1 Steel mold

Fig. 2 Sample of graphite-reinforced polymer composite

3.2 *Specification of CNC Vertical Machining Center*

The CNC vertical machining center manufactured by Bharat Fritz Werner Ltd. Figure 3 shows the setup of CNC vertical machining for milling operation (Figs. 4 and 5).

Tool material—HSS MILLING CUTTER (Diameter of milling cutter = 5 mm).

Fig. 3 CNC vertical machining center

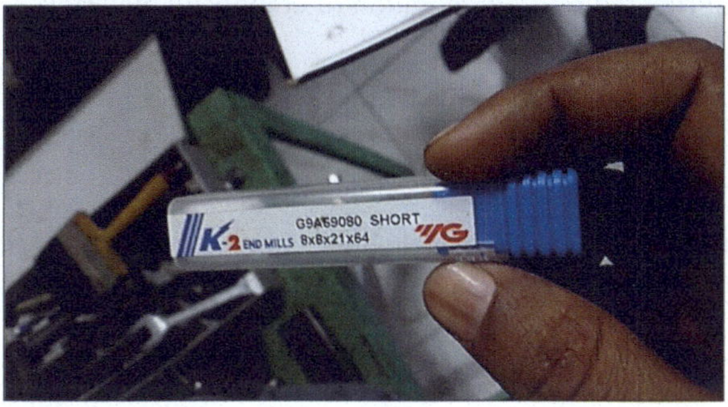

Fig. 4 HSS milling cutter

Fig. 5 Milling operation of graphite-reinforced polymer composite (GRPC)

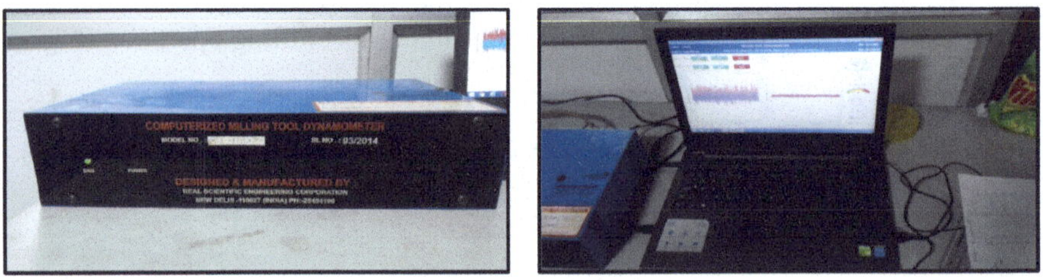

Fig. 6 Setup of computerized milling tool dynamometer

3.3 *Equipment Used for Measuring Responses (Thrust and Torque) During Machining*

Computerized tool dynamometer has been used to measure force and thrust during the milling operation of graphite-based composite polymer. The dynamometer has been used for obtaining the graph, and also, it gives the average value of thrust and torque. Figure 6 which we have been clicked when storing the responses and graph during the milling operation of composite material.

3.4 *Metal Removal Rate (MRR)*

Generally, the concept behind material removal is stated that considered material volume has eliminated to the machining time and density.

$$\text{Metal Removal Rate} = \frac{\text{initial weight} - \text{final weight}}{\text{time}(t) \times \text{density}(\rho)} \left(\text{mm}^3/\text{s}\right)$$

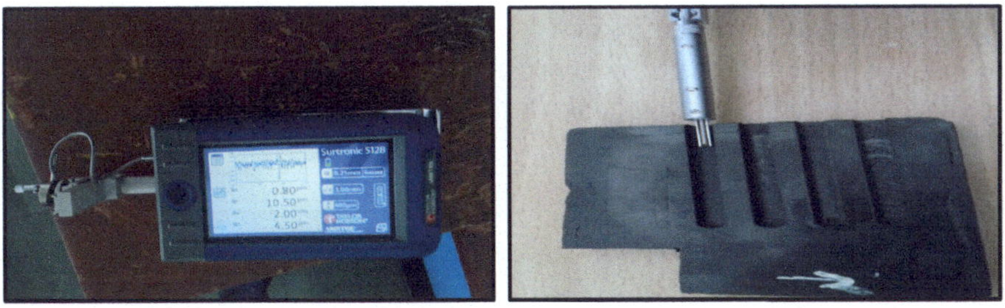

Fig. 7 Surface measurement of composite

3.5 Surface Roughness (Ra)

The surface roughness of the machined area of the GRPC sample has been measured with the use of **TAYLOR HOBSON** device. If the large deviation occurs means surface of the machined area is rough, and when the deviation is small means, surface of the machined area is smooth. In concerned about the **surface metrology**, unevenness has been measured as the high-frequency, short-wavelength constituent toward dignified exteriors (Fig. 7).

Milling operation has been performed in CNC vertical machining center and HSS milling cutter used for machining, and the obtained experimental parameters such as MRR, thrust (T), torque (TR) and then surface roughness (Ra) have specified in Table 3.

4 Parametric Optimization: Utility Theory

The term utility means the usefulness of a procedure in orientation to the expectations levels of the customers. When doing the machining operation, the number of output characteristics has been used to evaluate the performance. So, the joint amount has been essential to scale its complete performance which is used to the account of the comparative involvement of totally to the excellence features. The overall utility of the process is indicated by the index of the composite. Utility theory gives the procedural framework for the assessment of other features complete through individuals, organizations as well as groups. Basically, the utility theory mentions the amount of satisfaction in which each element gives to the choice maker. Therefore, utility maximization principle has been used for any type of decision. According to the utility theory, the decision-maker provides the best possible decision. Based on the utility concept, X_I is the helpfulness portion of an attribute or excellence features, there are N points to be used for evaluating consequence interplanetary. Now, combined function of utility has defined as follows: Now, $U(X_1 \ldots X_2 \ldots X_3 \ldots X_N) = f(U_1(X_1), U_2(X_2) \ldots U_N(X_N))$. Here, $U_I(X_I)$ shows Ith utility point. The function of total utility was a summation of single utility, if points were self-determining as well

Table 3 Experimental data (output response)

S. No.	MRR (mm^3/s)	Thrust (N)	Torque (N-mm)	Ra (μm)
1	0.6287	0.28	0.1	1.40
2	1.29385	0.31	0.18	0.8
3	2.60233	0.5	0.08	0.9
4	5.5404	0.36	0.13	0.9
5	0.826183	0.42	0.05	1.1
6	0.597195	0.39	0.11	0.8
7	5.7360	0.44	0.11	1.20
8	3.19013	0.39	0.25	1.1
9	1.17894	0.31	0.06	0.7
10	4.01136	0.42	0.16	0.9
11	0.958750	0.44	0.13	1.6
12	2.1732	0.28	0.16	1.50
13	2.8615	0.5	0.19	2.4
14	2.30162	0.33	0.13	1.0
15	1.8593	0.36	0.09	0.7
16	1.15410	0.25	0.13	1.4

as was given in the below: $U(X_1 \ldots X_2 \ldots X_3 \ldots X_N) = \sum U_I X_I$ Where, $I = 1 \ldots 2 \ldots 3 \ldots N$, The total function of utility next hiring the weights to the points may express such as:

$$U(X_1, X_2, X_3 \ldots X_N) = \sum W_I \cdot U_I X_I \quad \text{Where, } I = 1, 2, 3 \ldots N$$

Utility value determination: Utility value has been evaluated based on the preference scale of quality characteristic. Basically, there are two random mathematical values which have referred to the **preference number** which lies on the **0** as well as **9**. The value has been assigned to the acceptable range as well as the finest worth of the quality characteristic correspondingly. Gupta and Murthy [15–18] have been planned that the first choice (**Preference number "$\mathbf{P_i}$"**) could state in the logarithmic gauge has given below.

$$\mathbf{P_i = A \times log\left(X_i/X_i^*\right)}$$

Now, the value of X_i of several excellence features or characteristic i, X_i was impartial satisfactory data of excellence features i as well as A has shown as a constant [19]. The data of A may search through the state as given below. If the $X_i = X^*$ (where X^* has been the best value), now $\mathbf{P_i = 9}$. So,

$\mathbf{A = 9/log\ (X_I/X_I^*)}$, total function of utility has been shown as follows: $\mathbf{U = \sum W_i P_i}$.

Where $i = 1, 2, 3 \ldots n$, as for the concern to the state: $\sum \mathbf{W_i} = \mathbf{1}$, where $i = 1, 2, 3 \ldots n$. The total index of utility has designed for the optimization of a single objective function. The W_i is the weight of response which is assigned equally in this study. Among the different type of quality characteristics like lower-the-better, higher-the-better, as well as nominal-the-best (NB), mentioned through, when using utility method, the obtained data of single output response has been collected for computing the total index of utility. The overall index of utility has been severed for the optimization of the unique function of objective.

5 Results and Discussions

The observed data for thrust, torque, MRR and Ra depicted in Table 3 and preference number (P_I) which has been used in the utility theory. Taguchi method has applied to find optimal setting value than predicted the main data, and the predicted value and overall utility are shown in Table 4. The analysis of the signal to noise ratio is shown in Fig. 8. The value of PSNRA1 AND PMEAN1 is calculated for the predicted setting.

Figure 8 which shows the signal to noise ratio obtained from Taguchi method which gives the desired optimal setting values such as speed 500 rpm, feed rate 25 mm/rev., depth of cut −1.5 (negative sign shows in the downward direction)

Table 4 Preference number and overall utility

S. No.	Pi thrust	Pi torque	Pi MRR	Pi Ra	*U* overall	SNRA1	MEAN1
1	7.528511	5.123911	0.204526	3.937015	4.198491	12.46186	4.198491
2	6.206939	1.836999	3.075757	8.024641	4.786084	13.59961	4.786084
3	0	6.371733	5.855732	7.164313	4.847945	13.71115	4.847945
4	4.265381	3.656766	8.861971	7.164313	5.987108	15.54434	5.987108
5	2.263849	9	1.291247	5.698546	4.56341	13.18579	4.56341
6	3.226086	4.590935	0	8.024641	3.960415	11.95481	3.960415
7	1.659821	4.590935	9	5.062985	5.078435	14.1146	5.078435
8	3.226086	0	6.665931	5.698546	3.89764	11.81604	3.89764
9	6.206939	7.980455	2.705749	9	6.473286	16.2225	6.473286
10	2.263849	2.495644	7.577234	7.164313	4.87526	13.75996	4.87526
11	1.659821	3.656766	1.883273	2.961656	2.540379	8.097971	2.540379
12	7.528511	2.495644	5.138817	3.433068	4.64901	13.34721	4.64901
13	0	1.534655	6.233427	0	1.94202	5.765076	1.94202
14	5.395159	3.656766	5.367221	6.394724	5.203467	14.32586	5.203467
15	4.265381	5.713088	4.5182	9	5.874167	15.37893	5.874167
16	9	3.656766	2.621032	3.937015	4.803703	13.63152	4.803703

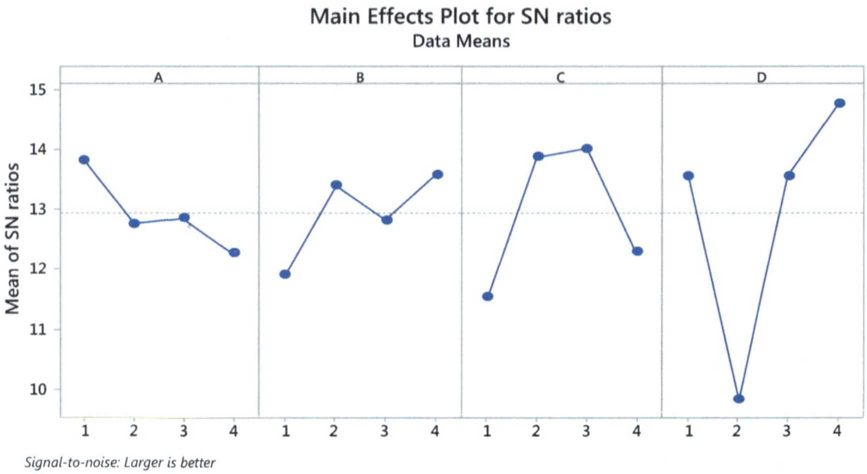

Fig. 8 Graph of signal to noise ratio

Table 5 Predicted value and optimal setting

Optimal setting		Predicted value
Speed	500 rpm	S/N ratio **17.4111**
Feed rate	25 mm/rev.	
Depth of cut	−1.5 mm	
Weight %	40%	

and wt. percentage 40%. Table 5 shows the optimal setting parameters and predicted value which is generated the graph of S/N ratio from Minitab 18a. The optimal results have been verified through a confirmatory test which shows the satisfactory results.

6 Conclusion

The present research study mainly focused on the machining performance optimization of quality and productivity characteristics.

- This study aims to maximize the metal removal rate and minimizes the thrust (T), torque (TR) and surface roughness (Ra) for quality and productivity improvement.
- Aggregation of multi-performance into a single function called as overall utility (U) been done fruitfully which is not possible by traditional Taguchi method.
- The optimal results and predicted value show the effectiveness of hybrid approach, i.e., utility embedded Taguchi approach.
- This robust approach can be forwarded to check the quality (offline/online) of product and process development in industries and manufacturing sectors.

References

1. Borrego, L.P., Costa, J.D.M., Ferreira, J.A.M., Silva, H.: Fatigue behavior of glass fibre reinforced epoxy composites enhanced with nanoparticles. Compos. B Eng. **62**, 65–72 (2014)
2. Guan, F.L., Gui, C.X., Zhang, H.B., Jiang, Z.G., Jiang, Y., Yu, Z.Z.: Enhanced thermal conductivity and satisfactory flame retardancy of epoxy/alumina composites by combination with graphene nano platelets and magnesium hydroxide. Compos. B Eng. **98**, 134–140 (2016)
3. Xu, Y., Van Hoa, S.: Mechanical properties of carbon fibre reinforced epoxy/clay nanocomposites. Compos Sci Technol. **68**, 854–861 (2008)
4. Park, S.J., Seok, S.J., Min, B.G.: Thermal and mechanical properties of epoxy/polyurethane blend system initiated by cationic latent thermal catalyst. Solid State Phenom. **119**, 215–218 (2007)
5. Nagai, Y., Lai, G.C.: Thermal conductivity of epoxy resin filled with particulate aluminium nitride powder. J. Ceram. Soc. Jpn. **105**, 197–200 (1997)
6. May, C.: Epoxy resins: chemistry and technology (1987)
7. Chen, G., Wu, D., Weng, W., Wu, C.: Carbon **41**, 619 (2003)
8. Chen, L., Lu, D., Wu, D.J., Chen, G.H.: Polym. Compos. **28**, 493 (2007)
9. Brewer, R.C., Reuda, R.A.A.: A simplified approach to the optimum selection of machining parameters. Eng. Dig. **24**(9), 131–151 (1963)
10. Colding, B.N.: Machining economics and industrial data manuals. Ann. CIRP **17**, 279–288 (1969)
11. Ermer, D.S.: Optimization of the constrained machining economics problem by geometric programming. Trans ASME J Eng Ind **93**, 1067–1072 (1971)
12. Lwata, K., Murotsa, Y., Jwotsubo, T., Fuji, S.: A probabilistic approach to the determination of the optimum cutting conditions. Trans. ASME J. Eng. Ind. **94**, 1099–1107 (1972)
13. Gopalakrishnan, B., Faiz, A.K.: Machining parameter selection for turning with constraints: an analytical approach based on geometric programming. Int. J. Prod. Res. **29**, 1897–1908 (1972)
14. Rao, S.S., Hati, S.K.: Computerized selection of optimum machining conditions for a job requiring multiple operations. Trans. ASME J. Eng. Ind. **100**, 356–362 (1978)
15. Gupta, V., Murthy, P.N.: An introduction to engineering design methods. Tata McGraw Hill, New Delhi (1980)
16. Ishikawa, K., Seki, M., Tobe, S.: Application of thermal spray coatings to prevent corrosion of construction in Japan. In: Proceedings of the 5th National Thermal Spraying Conference, Anaheim, CA, USA, pp. 679–684 (1993)
17. Knotek, O.: Thermal spraying and detonation spray gun processes (Chap. 3). In: Bun Shah, R.F. (ed.): Handbook of Hard Coatings: Deposition Technologies, Properties and Applications, pp. 77–107. Noyes Pub. Park Ridge, New Jersey, USA; William Andrew Publishing, LLC, Norwich, New York, USA (2001)
18. Kumar, P., Barua, P.B., Gaindhar, J.L.: Quality optimization (multi-characteristics) through Taguchi technique and utility concept. J. Qual. Reliabil. Eng. Int. **16**, 475–485 (2000)
19. Verma, R.K., Pal, P.K., Kandpal, B.C.: Machining performance optimization in drilling of GFRP composites: a utility theory (UT) based approach. In: International Conference on Control Computing Communication and Materials (ICCCCM-2016) 21st and 22nd October (2016)

Optimization of Micro-electro Discharge Drilling Parameters of Ti6Al4V Using Response Surface Methodology and Genetic Algorithm

Pankaj Kumar and Manowar Hussain

Abstract In the present investigation, an organized study with optimization of the process parameters for the fabrication of micro-holes and their surface integrity is carried out using the response surface methodology (RSM). The influence of variable parameters, such as machining voltage and machining on time, on the recast material layer and micro-hardness of the machined sample were investigated. The RSM is used to establishe a regression equation to predict output parameters such as micro-hardness and thickness of recast materials of the fabricated holes. From the developed model, the effects of the input variable parameters on the micro-hardness, thickness of recast materials and change in the chemical composition are accomplished with the optimized results. In order to get minimum values of recast layer thickness and micro-hardness of the fabricated micro-holes, a mathematical model was established using response surface methodology (RSM), and subsequently, genetic algorithm (GA) was utilized to reach a set of input machining parameters. Machining input parameters such as gap voltage (V) and machining on time (T_{on}) were selected. The analysis of variance (ANOVA) result indicates that developed models are adequate. The genetic algorithm method in conjugation with RSM is able to identify a particular set of machining parameters which gives minimum values of recast layer thickness and micro-hardness. Confirmation test is also carried out and found that the difference between predicted and measured value is insignificant.

Keywords Mirco-EDM · Recast layer thickness · Micro-hardness · RSM · GA

P. Kumar (✉)
Department of Mechanical Engineering, Centre for Materials and Manufacturing,
S R Engineering College, Warangal, India
e-mail: pikupankaj82@gmail.com

M. Hussain
Department of Mechanical Engineering, CBIT,
Gandipet, Hyderabad 500075, Telangana, India

D. Dutta and B. Mahanty (eds.), *Numerical Optimization in Engineering and Sciences*,
Advances in Intelligent Systems and Computing 979,
https://doi.org/10.1007/978-981-15-3215-3_44

1 Introduction

In the fabrication of micro-devices using titanium-based alloy (Ti6Al4V) by conventional machining processes like turning, milling and drilling, all of these process have problems of poor machinability [1, 2]. However, with the help of advanced micromachining process like micro-electro discharge machining (micro-EDM), micron-size feature can be achieved successfully on these alloys as there is no direct contact between the tool and work piece [3]. Nowadays, it becomes one of best machining process for difficult to cut material such as Ti6Al4V with high dimensional accuracy [4, 5].

The surface integrity of EDMed surface can be estimated in terms of surface roughness, recast layer thickness, heat affected zone, micro-cracks, residual stress, micro-hardness, microstructures and metallurgical transformation under different machining conditions [6, 7]. Ndaliman et al. [8] conducted experimentations with distilled water as dielectric medium. Titanium and its alloys are used in various engineering fields, such as aviation, automobile and biomedical, because of their high strength-to-weight ratio. In the present study, the effect of the machining parameters, such as pulse on time and gap voltage, on the surface integrity parameters of titanium alloys (Ti6Al4V) has been analyzed and reported for machining of micro-holes. Such surface integrity parameters include recast layer thickness, micro-hardness, microstructure and chemical composition of the machined micro-holes [9, 10]. A mathematical model was established between the input machining parameters. The model was further tested for quality, and optimization was performed using genetic algorithm (GA) method. Confirmation test was also done at the optimized parameters.

2 Experimentation Details

In this study, all the designated experiments were conducted on a laboratory scale in house developed micro-EDM setup which is shown in Fig. 1. In this research work, transistor type of pulse generator has been used for conducting experiments.

2.1 Work Piece, Tool Material and Dielectric Materials

In the fabrication of micro-devices using titanium-based alloy (Ti6Al4V) by conventional machining processes like turning, milling and drilling, all of these process have problems of poor machinability [11]. However, with the help of advanced micromachining process like micro-electro discharge machining (micro-EDM), micron-size feature can be achieved successfully on titanium-based alloy as there is no direct contact between the tool and workpiece [12]. Titanium alloys (Ti6Al4V) have been used as work piece material with dimensions of 60 mm × 50 mm × 1 mm. Pure tungsten rod of diameter of 0.5 mm has been used as tool material and 'EDM oil' as the dielectric liquid. The dielectric used in this experiment was 'EDM oil' due to

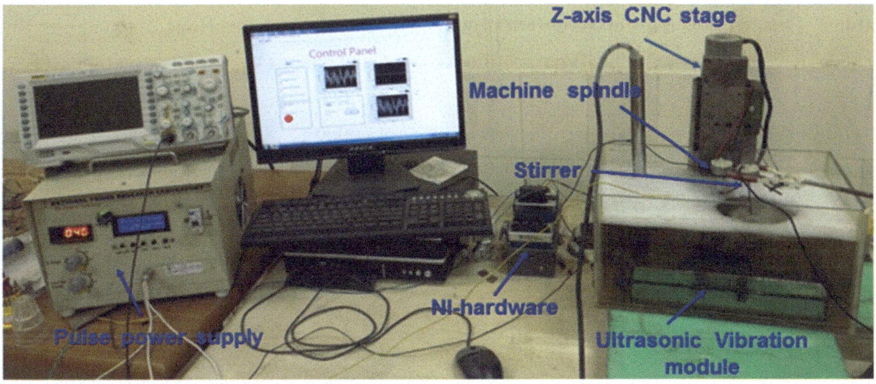

Fig. 1 Photograph of integrated micro-EDM setup

its high auto-ignition temperature, relatively high dielectric strength and high flash point. The gap voltage, pulse on time and frequency were selected after performing trial experiments, and it was observed that these parameters give optimal results (Table 1).

In response surface methodology, central composite design (CCD) with three factors, eight cube points, six center points in cube, six axial points and in total 20 experimental run is used. Table 2 represents 20 CCD parameters settings for estimation of second-order response surface using statically experimental design approach.

This design of experiments (DOE) reduces total number of experimentation and gives optimal results, thereby giving a correlation between the input and output machining parameters in an optimal way. A second-order regression equation which represents the process is presented as follows:

$$\begin{aligned} Y = \beta_0 + \beta_1 X_1 + \beta_2 X_2 + \beta_3 X_3 + \beta_1 X_1 + \beta_{12} X_1 X_2 + \beta_{13} X_1 X_3 \\ + \beta_{23} X_2 X_3 + \beta_{11} X_1 X_1 + \beta_{22} X_2 X_2 + \beta_{33} X_3 X_3 \end{aligned} \quad (1)$$

where Y represents output variable, X_1, X_2, X_3 are the input machining variables, β_0 is a constant, β_1, β_2 and β_3 are the linear coefficients whereas β_{12}, β_{13} and β_{23} are the interaction coefficients.

Table 1 Parameter settings for the fabrication of micro-holes

Work piece material	Ti6Al4V sheet
Tool material	W: ϕ 500 μm
Dielectric liquid	EDM oil
Gap voltage (V)	30, 40, 50, 60
Pulse on time (μs)	30, 40, 50, 60
Pulse frequency	10 kHz

Table 2 Design of experiments (DOE) of central composite design using RSM

StdOrder	RunOrder	PtType	Blocks	Gap voltage (V)	Pulse on time (μs)
15	1	0	1	45.0000	45.0000
14	2	−1	1	45.0000	45.0000
18	3	0	1	45.0000	45.0000
1	4	1	1	30.0000	30.0000
12	5	−1	1	45.0000	70.2269
4	6	1	1	60.0000	60.0000
2	7	1	1	60.0000	30.0000
16	8	0	1	45.0000	45.0000
19	9	0	1	45.0000	45.0000
6	10	1	1	60.0000	30.0000
17	11	0	1	45.0000	45.0000
11	12	−1	1	45.0000	19.7731
5	13	1	1	30.0000	30.0000
13	14	−1	1	45.0000	45.0000
7	15	1	1	30.0000	60.0000
10	16	−1	1	70.2269	45.0000
9	17	−1	1	19.7731	45.0000
8	18	1	1	60.0000	60.0000
20	19	0	1	45.0000	45.0000
3	20	1	1	30.0000	60.0000

3 Analysis of Variance

3.1 Estimation of Recast Layer Thickness, Change in Micro-Hardness Using ANOVA

A second-order model response surface model was established using trial version of Minitab 17 software considering all responses and based on all experimental measured data. An empirical second-order regression equation between output parameters and input parameters was established and represented by Eqs. (2) and (3).

Regression equation in uncoded units

$$Y_1 = 2.5 + 0.286X_1 + 0.274X_2 - 1.54X_3 + 0.00044X_1 * X_1 - 0.00168X_2 * X_2 + 0.0633X_3 * X_3 + 0.00011X_1 * X_2 - 0.0050X_1 * X_3 + 0.0027X_2 * X_3 \quad (2)$$

Table 3 p-value of the factor and interaction for both recast layer thickness and micro-hardness

Term	p-value of the factor and interaction	
	Recast layer thickness	Micro-hardness
Gap voltage (V)	0.000	0.000
Pulse on time (μs)	0.002	0.001
Frequency (μs)	0.043	0.013
Gap voltage (V) * gap voltage (V)	0.086	0.007
Pulse on time (μs) * pulse on time (μs)	0.921	0.262
Frequency (μs) * frequency (μs)	0.706	0.837
Gap voltage (V) * pulse on time (μs)	0.135	0.262
Gap voltage (V) * frequency (μs)	0.985	0.794
Pulse on time (μs) * frequency (μs)	0.780	0.896

$$
\begin{aligned}
Y_2 = {} & 169.4 - 0.342X_1 + 0.469X_2 - 1.69X_3 + 0.00734X_1 * X_1 \\
& - 0.00130X_2 * X_2 + 0.0660X_3 * X_3 + 0.00222X_1 * X_2 \\
& + 0.0033X_1 * X_3 - 0.0167X_2 * X_3
\end{aligned} \tag{3}
$$

where Y_1 represents recast layer thickness, Y_2 represents micro-hardness, X_1 is gap voltage (V), X_2 is pulse on time(μs), and X_3 is frequency (μs). p-value of the factors and interaction for both recast layer thickness and micro-hardness is presented in Table 3.

If $p < 0.05$, then significant else insignificant.

In this investigation, the recast layer thickness was measured from the SEM images of the machined micro-holes using image J software which is presented in Fig. 2.

3.2 *Optimization Using Genetic Algorithm*

GA is commonly used optimization technique which utilizes the evolutionary natural principles for working. Due to its fast response and accuracy, it is most widely used by the researchers. The GA code was developed in MATLAB for optimizing

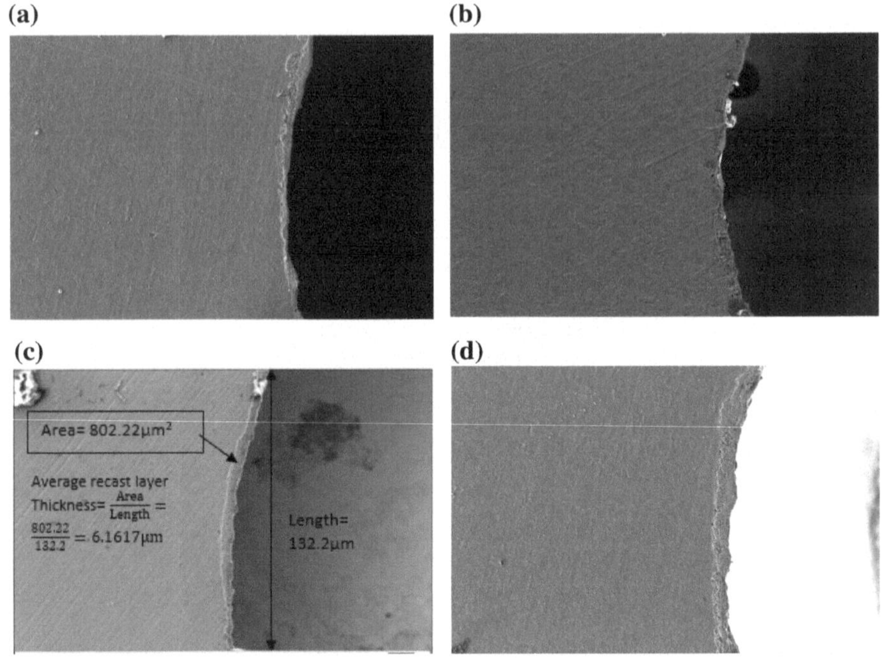

Fig. 2 FESEM images of a few of the samples (**a**) and (**b**) at 30-μs pulse on time, 40-V gap voltage (**c**) and (**d**) at gap voltage of 30 V, 40-μs pulse on time

the micro-EDM input parameters to achieve minimum recast layer thickness and micro-hardness.

The flowchart describing the various steps involved in execution of the GA is given in Fig. 3. The range of the input parameters such as gap voltage (V) and pulse on time (T_{on}) lies within which GA searches for optimal solution. This range is defined as bounds in MATLAB program as follows:

Bounds on Voltage: $30 \leq V \leq 60$.
Bounds on T_{on}: $30 \leq T_{on} \leq 60$.

The motive of this investigation is to minimize the recast layer thickness (RLT) and micro-hardness (MH) of the fabricated hole. The intended size of hole is 6 mm. The objective function was defined as followed

A = Minimize (RLT Targeted − MH Experimental)
A = Objective function for GA, RLT Targeted = Desire surface roughness (6 μm), MH Experimental = Actual micro-hardness.

The genetic algorithm code was executed to optimize the objective function. The speediness of convergence of the GA is subject to on population size, number of generations, crossover type, crossover rate and mutation rate. Number of generations was taken as 90 to limit the GA. One hundred and ninety iterations were accomplished

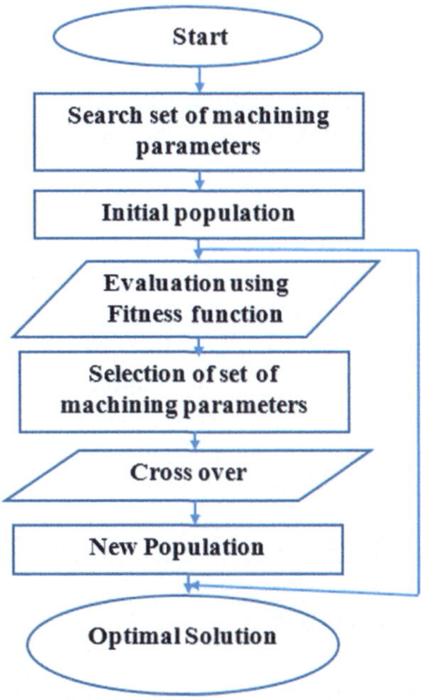

Fig. 3 Flowchart describing the various steps involved in execution of the GA

to attain local minimum value. Confirmation examination was done by setting the optimized input machining parameters, and the recast layer thickness (RLT) and micro-hardness were once more measured and matched.

4 Results and Discussions

The optimal solution of objective function, i.e., minimum difference in recast layer thickness (RLT) and micro-hardness achieved was 6.8 μm and 175 HV, respectively. The corresponding input machining parameters such as voltage (V) and pulse on time (T_{on}) were 30 and 40, respectively. With these input parameters, confirmation test was accomplished, and the measured values were found to be 7.9 μm and 169 HV, respectively. The measured values were in good agreement with projected value. From the obtained solution, only local minimum value of fitness function is observed. Therefore, this set of value gives the better solution than any other set of parameters. Consequently, one can take optimal value of machining input parameters according to the requirements. Hence, from the present investigation, the developed genetic algorithm model results in optimization the objective function of minimum recast layer thickness (RLT) and micro-hardness accurately.

5 Conclusion

In this investigation, the micro-holes were machined on Ti6Al4V sheet through micro-EDM process, the surface integrity of the machined micro-holes studied. This study has mainly focused on the analysis of the recast layer thickness and micro-hardness of the fabricated micro-holes. The following conclusions can be drawn from this study:

The optimal values of recast layer thickness (RLT) and micro-hardness achieved were 6.8 μm and 175 HV, respectively. The corresponding input machining parameters such as voltage (V) and pulse on time (T_{on}) were 30 and 40, respectively.

With these input parameters, confirmation test was accomplished, and the measured values were found to be 7.9 μm and 169 HV, respectively.

The developed genetic algorithm model results in optimization the objective function of minimum recast layer thickness (RLT) and micro-hardness accurately.

References

1. De Lacalle, L.N.L., et al.: Using high pressure coolant in the drilling and turning of low machinability alloys. Int. J. Adv. Manuf. Technol. **16**(2), 85–91 (2000)
2. Fatima, A., Mativenga, P.T.: A review of tool–chip contact length models in machining and future direction for improvement. Proc. Inst. Mech. Eng., Part B: J. Eng. Manuf. **227**(3), 345–356 (2013)
3. Liu, Q., Zhang, Q., Zhu, G., Wang, K., Zhang, J., Dong, C.: Effect of electrode size on the performances of micro-EDM. Mater. Manuf. Process. **31**(4), 391–396 (2016)
4. Kumar, A., Kumar, V., Kumar, J.: Surface integrity and material transfer investigation of pure titanium for rough cut surface after wire electro discharge machining. Proc. Inst. Mech. Eng., Part B: J. Eng. Manuf. **228**(8), 880–901 (2014)
5. Amorim, F.L., Stedile, L.J., Torres, R.D., Soares, P.C., Laurindo, C.A.H.: Performance and surface integrity of Ti6Al4V after sinking EDM with special graphite electrodes. J. Mater. Eng. Perform. **23**(4), 1480–1488 (2014)
6. Ekmekci, B.: Residual stresses and white layer in electric discharge machining (EDM). Appl. Surf. Sci. **253**(23), 9234–9240 (2007)
7. Prakash, V., Shubham Kumar, P., Singh, P.K., Das, A.K., Chattopadhyaya, S., Dixit, A.R.: Surface alloying of miniature components by micro-electrical discharge process. Mater. Manuf. Process. **33**(10), 1051–1061 (2018)
8. Ndaliman, M.B., Khan, A.A., Ali, M.Y.: Influence of electrical discharge machining process parameters on surface micro-hardness of titanium alloy. Proc. Inst. Mech. Eng., Part B: J. Eng. Manuf. 0954405412470443 (2013)
9. Das, A.K., Kumar, P., Sethi, A., Singh, P.K., Hussain, M.: Influence of process parameters on the surface integrity of micro-holes of SS304 obtained by micro-EDM. J. Braz. Soc. Mech. Sci. Eng. **38**(7), 2029–2037 (2016)
10. Prakash, V., Kumar, P., Singh, P.K., Hussain, M., Das, A.K., Chattopadhyaya, S.: Micro-electrical discharge machining of difficult-to-machine materials: a review. Proc. Inst. Mech. Eng., Part B: J. Eng. Manuf. **233**(2), 339–370 (2019)
11. Cardoso, P., Davim, J.P.: A brief review on micromachining of materials. Rev. Adv. Mater. Sci **30**, 98–102 (2012)
12. Câmara, M.A., Rubio, J.C., Abrão, A.M., Davim, J.P.: State of the art on micromilling of materials, a review. J. Mater. Sci. Technol. **28**(8), 673–685 (2012)

Multi-response Optimization of 304L Pulse GMA Weld Characteristics with Application of Desirability Function

Rati Saluja and K. M. Moeed

Abstract Effective optimization of multiple responses is equally essential in the development of predictive mathematical models. A higher value of geometrical characteristics increases few of the mechanical properties but apprehends susceptibility to unwanted metallurgical discrepancies. To conquer this problem, multi-response optimization technique must be implemented to assure equilibrium among geometrical, metallurgical and mechanical properties that help in the formation of quality and sound welds. Therefore, the present research addresses the application of desirability function approach to attain optimum characteristics for Grade 304L pulse gas metal arc welds. The predictive mathematical models for weld reinforcement height, weld width, weld penetration depth (geometrical characteristics), ferrite number (metallurgical characteristic), micro-hardness and ultimate tensile strength (mechanical characteristics), are crafted for concerning gas metal process variables (welding current, arc voltage and shielding gas flow rate) by implementation of central composite rotatable design employing Design-Expert software. Statistical analysis of variance and confirmatory tests assured the robustness of conducted investigation.

Keywords Austenitic stainless steel · Multi-response optimization · Pulse gas metal arc welding

1 Introduction

Grade 304L steels are an iron–chromium–nickel-based alloy that excessively favoured in industries due to confinement of their outstanding features at variable temperatures [1]. Gas metal arc welding (GMAW) is a widely adapted multi-input multi-response technique for joining of Grade 304L austenitic stainless steels (ASSs). GMAW is preferred over several other welding techniques for fabrication of Grade 304L steel of thicker cross sections in a single-/multi-pass weld joints because of low to moderate heat application, higher deposition rates and superior quality welds. GMAW was first discovered, in 1940, with an interminable supply of aluminium

R. Saluja (✉) · K. M. Moeed
Department of Mechanical Engineering, Integral University, Lucknow, UP 226028, India
e-mail: ratisaluja@gmail.com

D. Dutta and B. Mahanty (eds.), *Numerical Optimization in Engineering and Sciences*, Advances in Intelligent Systems and Computing 979,
https://doi.org/10.1007/978-981-15-3215-3_45

filler wire under the absolute shielding of argon gas in the Battelle Memorial Institute, USA, Soon after, welding experts commenced their focus on development of metal transfer modes of GMAW in order to diminish weld irregularities and defects in addition to limit the filler metal spattering. Consequently, in 1950, pulsed GMAW was introduced that further took another ten years to attain its commercialization. In 1990, the Lincoln Company made several welding equipments that enhanced and extended the use of automatic/semi-automatic pulsed GMA equipment in several manufacturing industries. That further expanded the use of both polarities (DCEP and DCEN) in one pulse welding equipment. Sooner, pulse GMAW machines were globally preferred for welding due to their capability to provide sound welds with easy application and low production and maintenance cost [2].

Maintaining the trend, welding experts focused towards new opportunities by maximizing utilization of welding sources along with waste minimization. However, the literature review reveals that precedent researchers focused on optimizing GMAW process variables for no more than one output variable. This further advances the immediate focus of Harrington, who recommended the application of desirability objective function for multi-response optimization of output variables over single response optimization techniques. In this methodology, the entire objectives were hooked for one desirability objective function for numerous responses and input variables. This mathematical optimization approaches a focal point termed as 'Desirability'. The desirability (D) is an objective function that translates each response from zero to one ($0 \rightarrow 1$) using their geometric mean. Thus, a simultaneous objective function is a geometric mean of all transformed responses. Importance (weight) from '+' to '+++++' could be assigned to each process variables/response according to the significance of factor. Thus, desirability could be 'minimized', 'maximized' or kept 'within range' for each response/variable separately [3]. Later, various optimization techniques were applied by succeeding investigators to concentrate on the same topic using several other optimization methods [4].

Various researchers had attempted optimization of multiple welding parameters and optimized output variables using genetic algorithm, fuzzy logic, artificial neural network, finite element methods, particle swarm optimization and grey relational analysis for various welding techniques [1, 5–10]. Dutta et al. executed Taguchi multi-response optimization approach for submerged arc process variables for determination of optimal bead geometry [5]. Later, Sreeraj et al. optimized GMA process parameters to attain optimal bead geometry of carbon steel using particle swarm optimization methodology [6]. Sudhakaran et al. had also attempted optimization of gas tungsten process parameters for Grade 202 steel bead geometry, using genetic algorithm [1]. Later, Kundu et al. implemented grey relational analysis for optimization of friction stir process variables to attain optimal mechanical properties [7]. Soon, Satheesh et al. also conducted multi-response optimization of submerged arc process variables implementing fuzzy logic [8]. Al-Jader et al. simulated results of spot welding process variables for weld nugget growth curve of galvanized steel used for application in automobile industries [9]. Similarly, Lee et al. had predicted optimal process variables for GMA welds with the implementation of multiple regressions technique. They validated the outcome of maximal and minimal grey relational results for mechanical properties with the metallurgical characteristics [10].

Most of the researchers had spotted for optimal solutions of geometrical/mechanical/metallurgical features, but the rarest literature was available for optimization of overall weld characteristics using pulsed GMAW with implementation of response surface methodology (RSM). Pulsed mode of transfer could be achieved using welding current beyond transition current (220A) for ASSs, within a voltage range (18–34 V) using 98% argon + 2% oxygen shielding [11]. Response surface methodology (RSM)-based desirability technique is recommended for optimization due to its easy application, accessibility and flexible approach to attain optimum solutions. RSM not only provides a relaxed approach for crafting predictive mathematical models via experimental investigations or computer-based simulations, but RSM also allows the user to implement optimization criteria during investigational procedures for each response in an organized manner [12]. Complied moves even conquer the shortcomings of conventional optimization techniques. Hence, current research presents robustness and litheness in explaining multiple criterion optimal solution for pulse GMA process variables (welding current, voltage and shielding gas flow rate) using Design-Expert software [13]. Weld reinforcement height, weld width, weld penetration (near to base metal thickness), micro-hardness, ferrite number (FN) are optimized to attain their minimum values whether ultimate tensile strength (UTS) of Grade 304L steel pulse gas metal arc welds is maximized.

2 Experimental Methodology

ESAB Aristo MIG advanced synergic welding machine is used for joining of Grade 304L austenite plates of size (300 × 150 × 6) mm. U5000I is USA make three-phase, 50 Hz, 500 A, 100% duty cycle welding equipment with weld cloud 2.0 software that permits variation in welding current (I), voltage (U) during welding. As 'I' and 'U' are preferred as an independent variable, welding speed was restricted to a fixed value. Shielding gas flow rate (F) was varied by regulating the pressure regulator of a cylinder containing 98% argon + 2% oxygen. Variation in process parameters alters the mechanical, metallurgical and geometrical features of AISI 304L welds.

2.1 *Fixing the Range of Independent Process Variables*

The experimental range of the pulsed GMA variables was quantified via trial runs by varying one of the parameters keeping others at fixed values. Five levels of process variables are enumerated as displayed in Table 1 for development of full factorial central composite rotatable design. Equation 1 is implemented for determination of levels of current (I), voltage (U) and shielding gas flow rate (F). Design matrix

Table 1 Process parameters with their units, notation and levels

Process variables	Units	Notation	Levels				
			−1.682	−1	0	1	1.682
Voltage	Volt	*U*	25	26	27	28	29
Shielding gas flow rate	Litres/min	*F*	10	12	14	16	18
Welding current	Amperage	*I*	220	226	235	244	250

consists of eight factorial (±1, ±1, ±1), six centre (0, 0, 0) and six axial {(±1.682, 0, 0), (0, ±1.682, 0), (0, 0, ±1.682)} runs; thus, total 20 investigational runs are conducted for measurement of responses.

$$X_i = 1.682\left[\frac{2X - (X_{\max} + X_{\min})}{(X_{\max} - X_{\min})}\right] \quad (1)$$

Here, X_i = entailed coded value of preferred parameter, X = common value of the preferred parameter within maximal and minimal range ($X_{\min} < X < X_{\max}$) [1].

2.2 Measurement of Weld Geometrical, Metallurgical and Mechanical Characteristics

Weld width was measured with digital vernier calliper (Make: Mitutoyo, model no: 500-196-30), and weld penetration and reinforcement height were quantified using Vernier height gauge (Make: Kristeel Shinwa industries ltd. Model no: 2918) as mentioned in Table 2. The ferrite number (FN) was measured with ferrite scope (Make: Fischer Technology Inc., model no. FMP30). Micro-hardness was measured through Vicker's hardness tester (Make: BIE). Ultimate tensile strength was calculated with universal testing machine (Make: MCS and FIE, model: UTE 40 as detailed in Table 2.

2.3 Development of Predictive Model

The response surface function representing weld characteristics could be stated as *f(I, U, F)*. If b_0 = free term of the regression equation, b_i = linear term, b_{ii} = full quadratic terms, b_{ij} = interaction terms used for process variable X_i and ε = inaccuracy of developed mathematical model [1]. The metallurgical, mechanical and geometrical properties of Grade 304 welds could be represented as a second-order polynomial response surface as mentioned in Eq. 2. The regression coefficients were computed with the implementation of Design-Expert 11.0 software [13]. '*P*' values more than 0.05 point towards insignificancy of regression terms. After removing the

Table 2 Coded design matrix with observed responses for 304L weld

STD	Run order	Coded parameters			Ultimate tensile strength	Micro-hardness	Ferrite number	Weld reinforcement	Weld width	Weld penetration
		I	*U*	*F*						
1	2	−1	−1	−1	606.359	161	7.9	0.830	9.000	6.00
2	10	1	−1	−1	580.193	159	6.2	1.060	9.764	6.52
3	13	−1	1	−1	575.551	160	6.8	0.965	9.290	6.27
4	9	1	1	−1	554.326	159	4.8	1.230	9.740	6.91
5	15	−1	−1	1	612.006	162	7.3	0.810	8.810	6.27
6	3	1	−1	1	589.960	160	5.6	1.140	9.340	6.84
7	16	−1	1	1	579.104	161	7.5	0.960	10.070	6.32
8	18	1	1	1	566.052	160	5.4	1.340	10.410	6.89
9	7	−1.682	0	0	598.796	162	8.0	0.886	8.700	6.20
10	19	1.682	0	0	562.794	159	4.9	1.420	9.670	7.21
11	12	0	−1.682	0	613.940	160	6.7	0.890	8.890	6.29
12	11	0	1.682	0	562.153	160	5.7	1.240	10.100	6.66
13	17	0	0	−1.682	571.471	160	6.8	0.890	9.700	6.14
14	4	0	0	1.682	586.398	160	6.9	0.990	10.04	6.43
15	20	0	0	0	574.236	162	6.8	1.130	9.128	6.46
16	6	0	0	0	574.058	159	6.8	1.130	9.120	6.44
17	14	0	0	0	574.325	160	6.8	1.120	9.140	6.53
18	1	0	0	0	574.058	160	6.8	1.120	9.152	6.46
19	8	0	0	0	574.013	159	6.7	1.160	9.160	6.46
20	5	0	0	0	573.995	160	6.8	1.090	9.300	6.46

Table 3 Significance testing (ANOVA) of developed model

Parameter	First-order term		Second-order term		Third-order term		Error term		F test	p test
	SS	DOF	SS	DOF	SS	DOF	SS	DOF		
Ultimate tensile strength	4672.02	3	43.1623	3	463.09	3	0.09	5	647.81	<0.0001
Micro-hardness	1022.85	3	229.50	3	35.650	3	0.8333	5	338.29	<0.0001
Ferrite number	13.123	3	0.8437	3	0.8914	3	0.0083	5	1031.6	<0.0001
Weld reinforcement	0.9646	3	0.0071	3	0.0721	3	0.0026	5	115.03	<0.0001
Weld width	7.0851	3	0.5791	3	1.1204	3	0.0224	5	152.23	<0.0001
Weld penetration	3.0197	3	0.0411	3	0.1651	3	0.0019	5	311.90	<0.0001

insignificant regression terms, the final predictive mathematical models concerning selected input variables are hereby presented in Eqs. (3)–(8).

$$Y = b_o + \sum_{i=1}^{k} b_i X_i + \sum_{i=1}^{k} b_{ii} X_i^2 + \sum_{i=1}^{k} b_{ij} X_i X_j + \varepsilon \quad (2)$$

$$\begin{aligned}\text{Ultimate tensile strength} = {} & 574.12 - 10.47(I) - 14.69(U) + 4.09(F) + 2.34\left(I^2\right) \\ & + 4.89\left(U^2\right) + 1.67\left(F^2\right) + 1.74(IU) + 1.53(IF) \quad (3)\end{aligned}$$

$$\begin{aligned}\text{Micro-hardness} = {} & 186.81 - 8.01(I) - 2.74(U) + 1.79(F) + 1.07\left(I^2\right) - 0.9 \\ & \left(U^2\right) + 0.72\left(F^2\right) + 3.25(IU) - 2.75(IF) - 3.25(UF) \quad (4)\end{aligned}$$

$$\begin{aligned}\text{Ferrite-number} = {} & 6.7846 - 0.931(I) - 0.306(U) - 0.0126\left(I^2\right) - 0.214\left(U^2\right) \\ & - 0.088(IU) + 0.314(UF) \quad (5)\end{aligned}$$

$$\begin{aligned}\text{Weld reinforcement} = {} & 1.13 + 0.154(I) + 0.09(U) + 0.024(F) - 0.022\left(U^2\right) \\ & + 0.067\left(F^2\right) \quad (6)\end{aligned}$$

$$\begin{aligned}\text{Weld-width} = {} & 9.17 + 0.272(I) + 0.339(U) + 0.103(F) + 0.119\left(U^2\right) \\ & + 0.252\left(F^2\right) - 0.063(IU) + 0.258(UF) \quad (7)\end{aligned}$$

$$\begin{aligned}\text{Weld-penetration} = {} & 6.466 - 0.293(I) - 0.101(U) + 0.0811(F) + 0.088\left(I^2\right) \\ & - 0.0604\left(F^2\right) - 0.07(UF) \quad (8)\end{aligned}$$

2.4 Analysis of Variance for Developed Predictive Models

F values of predictive mathematical models signify towards accuracy of developed model as revealed in Table 3. The coefficient of determination (R^2) of adjusted predictive model is evaluated against R^2 of full predictive model. The standard error of predictive model is approximately 2% for FN, 6.2% for micro-hardness, 3.3% for UTS and found below 2% for weld geometrical variables experimental work that is lesser than the standard error of adjusted model and predicted model. Computed sound-to-noise ratios also ensure robustness of predictive models as the value of S/N ratio above four is attained as mentioned in Table 4.

Table 4 Fit statistics of mathematical model for prediction of responses

Response	Coefficient of determination (R^2)			Standard error ($1 - R^2$)			Adequate precision (S/N ratio)
	Full model	Adjusted model	Predicted model	Full model	Adjusted model	Predicted model	
Ultimate tensile strength	0.9983	0.9967	0.9865	0.0017	0.0033	0.0105	88.5022
Micro-hardness	0.9967	0.9938	0.9792	0.0033	0.0062	0.0208	76.0905
Ferrite number	0.9989	0.9980	0.9952	0.0011	0.0020	0.0048	112.6356
Weld reinforcement	0.9904	0.9818	0.9572	0.0096	0.0182	0.0428	38.2701
Weld width	0.9928	0.9862	0.9757	0.0072	0.0138	0.0243	42.0613
Weld penetration	0.9965	0.9933	0.9786	0.0035	0.0067	0.0214	72.9074

3 Optimization of Process Parameters/Responses with RSM-Based Desirability Approach

RSM-based multi-objective optimization was carried out implementing desirable goal set for process variables for 304L GMA welds as mentioned in Table 5. To attain multi-response optimization solutions, the independent variables and their associated responses are further resolved to define particular criteria, based on the best desirability solutions as mentioned in Eq. 9. If 'G_i' is the goal value of 'ith' response desirability, then overall desirability 'D' could be represented as.

$$D = (d_1\, X d_2\, X d_3 \ldots X d_n)^{1/N} = \left(\prod_{i=1}^{n} d_i\right)^{1/N} \tag{9}$$

Table 5 Goals set and constraints for multi-objective optimization

Factors/response	Goal	Lower limit	Upper limit	Importance
Welding current	Minimize	220 A	250 A	5
Arc voltage	Minimize	25 V	29 V	4
Shielding gas flow rate	Is in range	10 L/min	18 L/min	3
Ultimate tensile strength	Maximize	554.326 MPa	613.94 MPa	4
Ferrite number	Minimize	4.8	8	5
Micro-hardness	Minimize	176 HV	200 HV	3
Weld reinforcement	Minimize	0.81 mm	1.42 mm	3
Weld width	Minimize	8.7 mm	9.1 mm	5
Weld penetration	Minimize	6.0 mm	7.2 mm	4
Ferrite number	Minimize	4.8 FN	8 FN	3

Here, d = desirability of 'nth' response in the experiment having weight (W_i) for 'n' number of responses and '$N = \sum W_i$'. Whichever of the response drop outside the desirability range, the overall desirability 'D' becomes zero. When a response lies within ideal range or approaches towards its ideal value, the desirability is equal to one. In the meantime, if a response drops inside the tolerance range but not the ideal value, the desirability between zero and one ($0 < d < 1$) could be achieved. Thus, several performance features of optimization are rejoined and revolutionize as one optimization problem as displayed in Fig. 2 for conducted investigations [12]. In the desirability approach, the solution with the highest desirability is preferred. The optimum solutions for responses are specified in Table 6 for welding current at 234.86 A, voltage value at 25 V and shielding gas flow rate at 14.57 L/min. Combined desirability value of '$D = 0.591$' was achieved as displayed in Fig. 1.

Table 6 Optimum solution of responses with confirmatory results

Ultimate tensile strength (MPa)			Micro-hardness (HV)			Ferrite content (FN)		
Predicted	Observed	Error %	Predicted	Observed	Error %	Predicted	Observed	Error %
613.939	605.0	−1.46	195.96	200.5	2.32	6.58	6.51	−1.06
Weld reinforcement (mm)			Weld width (mm)			Weld penetration (mm)		
Predicted	Observed	Error %	Predicted	Observed	Error %	Predicted	Observed	Error %
0.906	0.918	1.33	8.863	8.746	−1.32	6.352	6.410	0.91

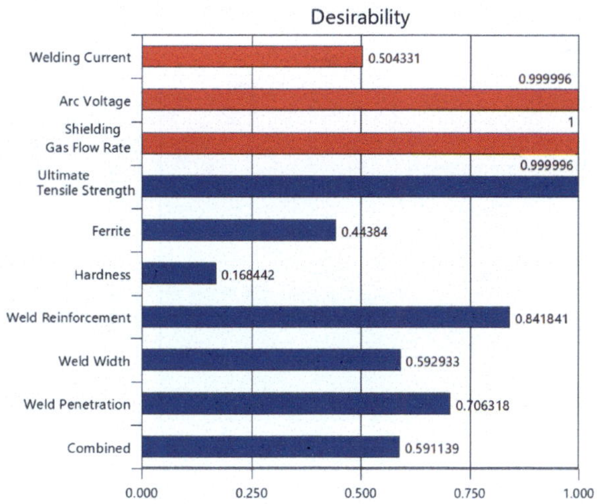

Fig. 1 Optimum process variables and responses with their achieved desirability solutions

4 Effect of Preferred Process Variables on Desirability

A perturbation plot presents the direct effect of preferred process variables for response in a single plot. Figure 2a. displays that desirability decreases with variation in welding current. Desirability also reduces, if arc voltage was increased beyond its optimal value. The outcomes of the graphical optimization are displayed through overlay plots in Fig. 2b–d. Overlay plots are exceptionally useful for immediate assessment of the welding process variables values which attain desired optimal responses. The yellow-coloured zone of the overlay plots is the areas that achieve optimum criteria.

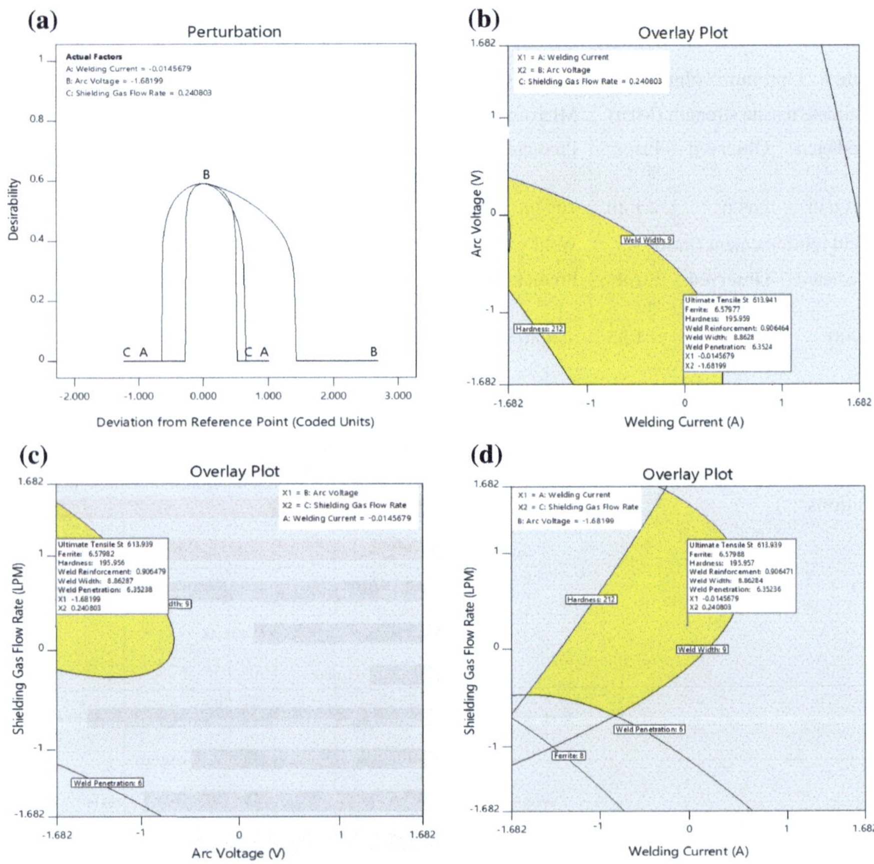

Fig. 2 **a** Perturbation plot displaying deviation of optimal process variables (coded units) from reference points. **b** Overlay plot of '*IU*' for optimal responses at fixed '*Q*' = 14.17 L/min. **c** Overlay plot '*QU*' for optimal responses at fixed '*I*' = 235.01 A. **d** Overlay plot '*QI*' for optimal responses at fixed '*U*' = 25.02 V

5 Conclusion

The second-order quadratic mathematical model has been created as a function of welding current, arc voltage and shielding gas flow rate for prediction of Grade 304L weld characteristics. The experimental investigation is accomplished for optimization of mechanical, geometrical and metallurgical characteristics for 304L GMA welds using desirability function with application of RSM. An optimum combination of preferred process variables is obtained for '*I*' = 234.86 A, '*V*' = 25 V and '*Q*' = 14.57 L/min. for desirability value (*D*) of '0.591'. The optimum UTS = 613.939 MPa (*d* = 0.999996), micro-hardness = 195.96 HV (*d* = 0.168442), ferrite number = 6.58 FN (*d* = 0.44384), weld reinforcement height = 0.906 mm (*d* = 0.841841), weld width = 8.863 mm (*d* = 0.592933) and depth of penetration = 6.352 mm (*d* = 0.706318). The predicted results have been compared with the investigated results, and the deviation falls within the limit of 99.5% level of significance. Concluding elucidation summarized that RSM-based multi-response optimization could be employed for selected range of process parameters Grade 304L austenitic steel pulsed welds.

References

1. Sudhakaran, R., VeL Murugan, V., Senthilkumar, K.M.: Effect of welding process parameters on weld bead geometry and optimization of process parameters to maximize depth to width ratio for stainless steel gas tungsten arc welded plates using genetic algorithm. Eur. J. of Sci. Res. **62**, 76–94 (2011)
2. Kyriakongonas A.P.: M.Sc. thesis, 3D Numerical modeling of austenitic stainless steel 316 l multi-pass butt welding and comparison with experimental results. National technical university of Athens school of Naval Architecture and Marine Engineering
3. Harrington, E.C.: The desirability function. Ind. Qual. Contr. **21**, 494–498 (1965)
4. Derringer, G., Suich, R.: Simultaneous optimization of several response variables. J. Qual. Tech. **12**, 214–219 (1980)
5. Datta, S., Bandyopadhyay, A., Pal, P.K.: Desirability function approach for solving multi-objective optimization problem in submerged arc welding. Int. J. Manuf. Sci. Prod. **7**, 127–135 (2006)
6. Sreeraj, P., Kannan, T., Maji, S.: Optimization of GMAW process parameters using particle swarm optimization. Hindawi Publishing Corporation ISRN Metallurgy (2013)
7. Kundu, J., Singh, H.: Friction stir welding: multi-response optimisation using Taguchi-based GRA. Prod. Manuf. Res. **4**, 228–241 (2016)
8. Satheesh, M., Dhas, J.E.R.: Multi objective optimization of weld parameters of boiler steel using fuzzy based desirability function. J. Eng. Sci. Tech. Rev. **7**, 29–36 (2014)
9. Al-Jader, M.A., Cullen, J.D., Athi, N., Al-Shamma, A.I.: Experimental and computer simulation results of the spot welding process using SORPAS software. In: Journal of Physics: Conference Series, pp. 1–9 (2009)
10. Lee, J.L., Rhee, S.: Prediction of process parameters for gas metal arc welding by multiple regression analysis. Proc. Inst. Mech. Eng. B. **14**, 443–449 (2000)
11. Lienert, T., Siewert, T., Babu, S., Acoff, V.: Welding fundamentals and processes, ASM handbook, 6A. ASM International, Materials Park, Ohio (1996)
12. Cornell, J.A., Khuri, A.I.: Response surfaces: designs and analysis. Marcel Dekker, New York (1987)

13. Design Expert Version 11.0. Stat-ease Inc. (2018)
14. Montgomery, D.C., Runger, G.C.: Applied statistics and probability for engineers. Wiley, New York (1994)

Simulation Study on the Influence of Blank Offset in Deep Drawing of Circular Cups

Araveeti C. Sekhara Reddy and S Rajesham

Abstract The alignment of blank in sheet metal forming is very much essential for quality production. In most of the cases, the blank alignment is a complex and time-consuming process. An attempt has been made using the FE simulations to predict the quality of the deep drawing process to enhance the process efficiency by lowering the development time and tool cost associated with it. The tests are conducted through numerical simulation on the effects of blank offset for the quality of the cup produced through the equivalent stresses and strains induced in the cup formed in deep drawing process. The deep draw tests are conducted using FE code PAM-STAMP using aluminum AA6111 alloy for different offset values of 0, 2.5, 5, 7.5 and 10 mm. It has observed from the results that the induced stresses increase as the blank offset increases. It has found from the results that the induced stresses drastically increases for the blank offset value of more than 5 mm with the use of 80 mm diameter punch and 120 mm diameter blank.

Keywords Deep drawing · FE simulation · PAM-STAMP · Blank offset

1 Introduction

The deep drawing is an extensively used sheet metal forming process having a variety of industrial applications such as beverages, automobiles and household appliances. The experience gained in the past can be used for designing new and complex parts. The optimization of the process using Finite Element methods and a good understanding of different parameters involved in it leads to a considerable saving of time and money during the initial stages of the product design [1, 2]. The magnitude of stresses and strains induced at different zones of the blank during the forming process

A. C. Sekhara Reddy (✉)
Osmania University, Hyderabad, Telangana, India
e-mail: acsreddy64@gmail.com

Sreyas Institute of Engineering and Technology, Hyderabad, Telangana, India

S. Rajesham
Department of Mechanical Engineering, RGUKT, Basar, Telangana, India

D. Dutta and B. Mahanty (eds.), *Numerical Optimization in Engineering and Sciences*, Advances in Intelligent Systems and Computing 979,
https://doi.org/10.1007/978-981-15-3215-3_46

is the main reason for the failure of the formed component. The stress development and progression in the deep drawing process is a complex phenomenon, involving compression in the circumferential direction and tension in the longitudinal direction. The inferior design leads to the production of defective parts and damages in the tool setup. Vrolijk et al. [3] discussed a detailed procedure on the prediction of deformations in tool setup using a standard forming simulation tool, PAM-STAMP.

2 Tool Setup Design

In this work, the modeling of deep drawing setup has performed using the computer-aided design tool, CATIA V5. The functional surfaces of the tool are created using CATIA and saved in *.igs format. The tool setup designed using CATIA V5 has imported into PAM-STAMP 2G software for numerical simulation.

The material considered was aluminum AA6111 of 0.9 mm of thickness. The mechanical properties of the material are determined through tensile testing in the UTM machine available in the laboratory. The tool setup modeled is shown in Fig. 1, and the physical and yield properties are shown in Table 1. The mesh has created with a 2 mm element, as the mesh size highly influences on the computation time and quality of the product produced [4]. The size of the blank considered is 120 mm, and the tests are conducted with blank offset values 0, 2.5, 5, 7.5 and 10 mm. The effect of offset on the physical and mechanical properties of the component generated are simulated using PAM-STAMP.

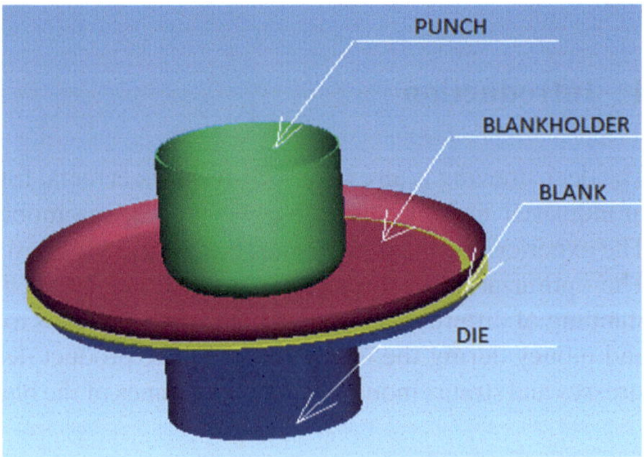

Fig. 1 Modeled tool setup for deep drawing

Table 1 Tool parameters used for modeling and simulation

Tool	Parameter	Qty
Punch	Diameter	80.0 mm
	Nose radius	5.50 mm
Die	Diameter	81.50 mm
	Shoulder radius	5.50 mm
Blank holder	Inner diameter	85.0 mm
	Outer diameter	250.0 mm
Blank	Diameter	120 mm
	Thickness	0.9 mm
	Young's modulus	70 Gpa
	Poison's ratio	0.3
	Plasticity law	Hill
	Material type	Isotropic
	Hardening law	Hollomon

3 Simulation Tests

The simulation process consists of preprocessing, simulation and post-processing. The preprocessing consists of defining the model of the physical problem. The simulation, which normally runs on background process, is the stage in which PAM-STAMP solves the numerical problem, and the results are shown for different offset values in Figs. 2, 3, 4, 5 and 6.

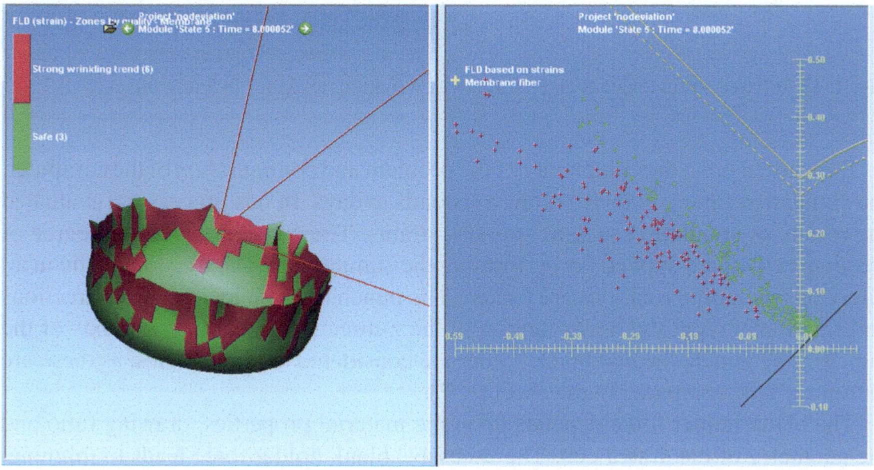

Fig. 2 FLD-based strain diagram and the zones of wrinkling tendency with zero offset

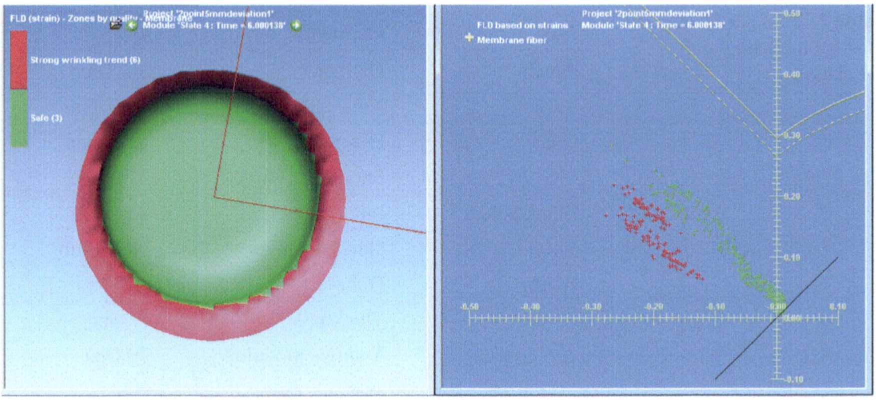

Fig. 3 FLD-based strain diagram and the zones of wrinkling tendency in 2.5 mm offset

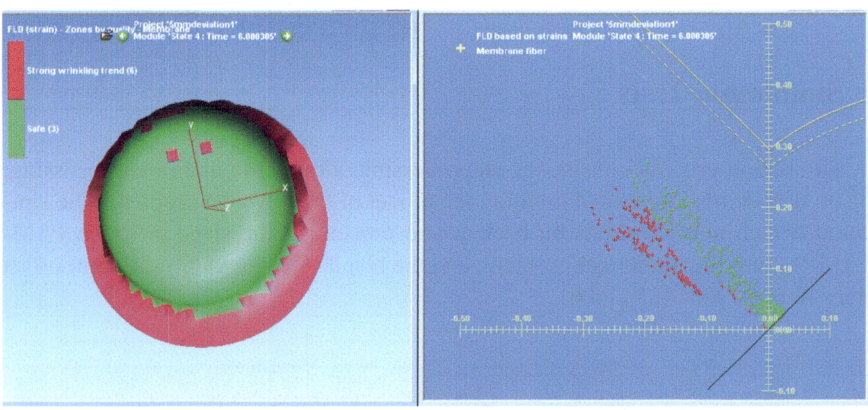

Fig. 4 FLD-based strain diagram and the zones of wrinkling tendency in 5 mm offset

Depending upon the complexity of the problem and configuration of the computer, the simulation time will be from few a seconds to days. In postprocessing, evaluated the results of simulation such as stresses, strains. The constraints for the degree of freedom (DOF) are applied for performing the simulations. The freedom to die in all directions including rotations are locked. The punch has also locked in all directions and rotations except the linear motion in the z-direction (i.e., along the axis of the punch). The punch, die and blank holder are considered as rigid bodies, as these are extremely stiff compared to the sheet [5–7].

The blank holder force depends upon the material properties, drawing ratio and the geometry of the drawn cup. The excessive blank holder force leads to thinning, and low value leads to wrinkling. A constant blank holder force of 50 N and a punch force of 3000 N are applied during the simulation. It has observed from simulation results that the thickness is not uniformly distributed in the drawn part produced in

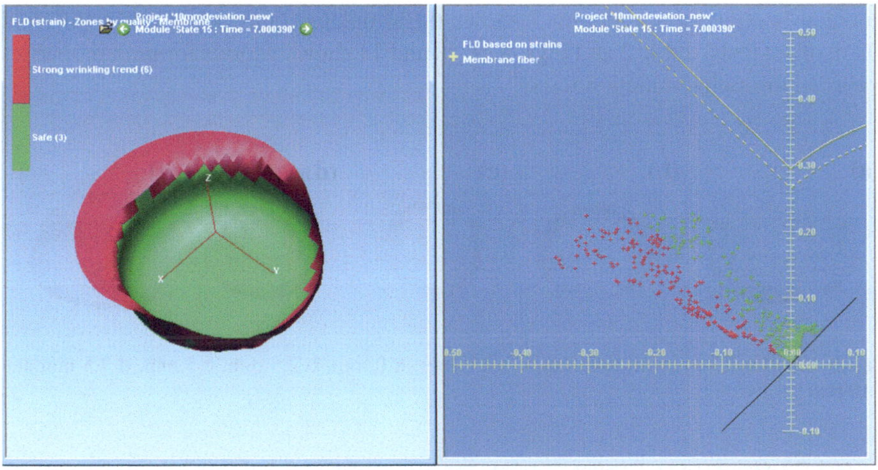

Fig. 5 FLD-based strain diagram and the zones of wrinkling tendency in 7.5 mm offset

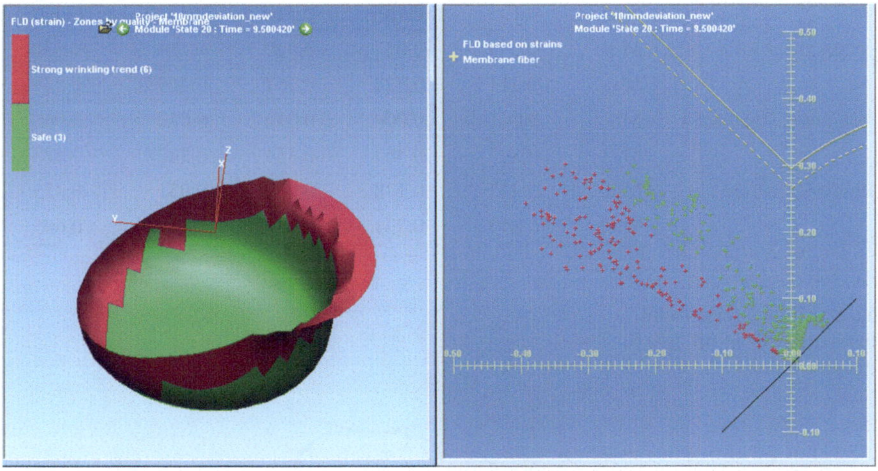

Fig. 6 FLD-based strain diagram and the zones of wrinkling tendency in 10 mm offset

deep drawing. It was uniform at the bottom, lower at the wall and higher at the flange area. But, due to the incorporation of alignment error in the process, the thickness was unevenly distributed throughout the cup [8–10].

It has observed that the cup formed is not uniform, and the strain diagram shows the strain values at bottom of the cup which is less at the top of the cup in comparison with the punch nose region. The cup has the tendency to fail in further deformation at the punch nose region. The FLD diagram shows that the top region of the cup is subjected to more negative strain, i.e., compressive strain. At the bottom region, the minor and major strains are almost equal as the blank at the center is subjected to equal biaxial strains. Figure 7 shows the simulated cups produced in different blank offset tests. Table 2 and Figs. 8, 9, 10 and 11 shows various forming properties recorded through simulation tests.

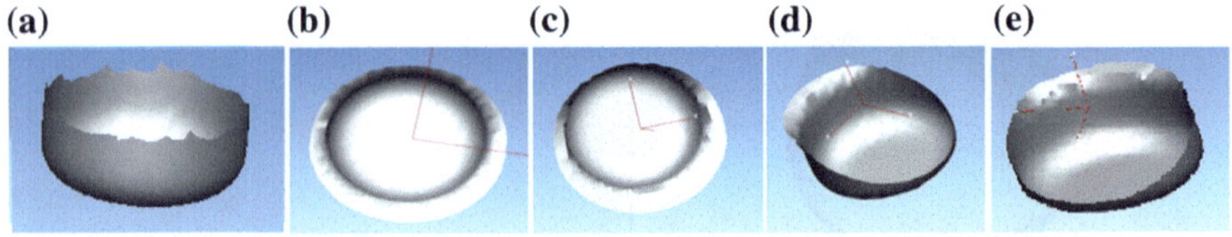

Fig. 7 Simulated cups with different offset values: **a** 0 mm, **b** 2.5 mm, **c** 5 mm, **d** 7.5 mm and **e** 10 mm

Table 2 Maximum and minimum stress and strains recorded in simulation tests

Blank deviation		0 mm	2.5 mm	5.0 mm	7.5 mm	10.0 mm
Minor stress GPa	Min.	0.240	0.143	0.085	−0.201	−0.478
	Max.	0.195	0.188	0.210	0.3520	0.450
Major stress GPa	Min.	−0.039	−0.02	−0.020	−0.184	−0.332
	Max.	0.311	0.314	0.367	0.4698	0.724
Von mises stress GPa	Min.	0.001	0.004	0.036	0.032	0.060
	Max.	0.377	0.364	0.354	0.3827	0.423
Equivalent strain	Min.	0.443	0.138	0.028	−0.051	−0.425
	Max.	0.849	0.721	0.677	0.657	0.605

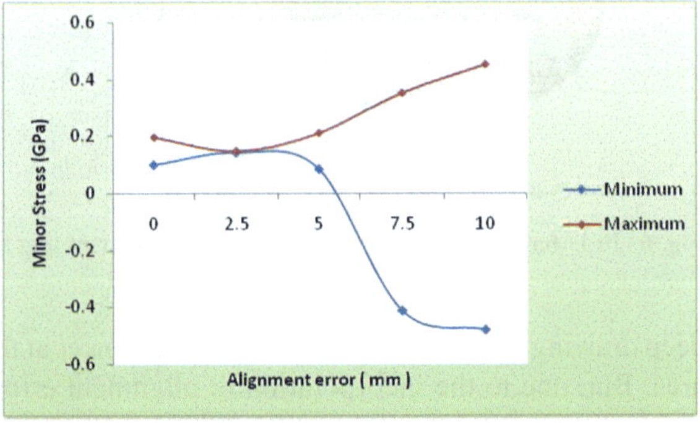

Fig. 8 Minor stresses with alignment error of nil, 2.5, 5, 7.5 and 10 mm

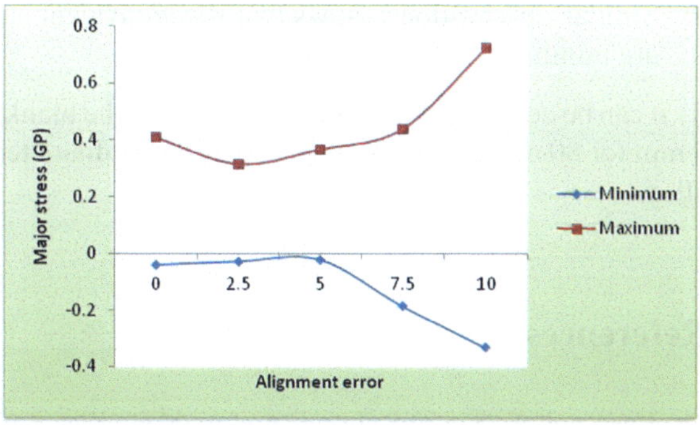

Fig. 9 Major stresses with alignment error of nil, 2.5, 5, 7.5 and 10 mm

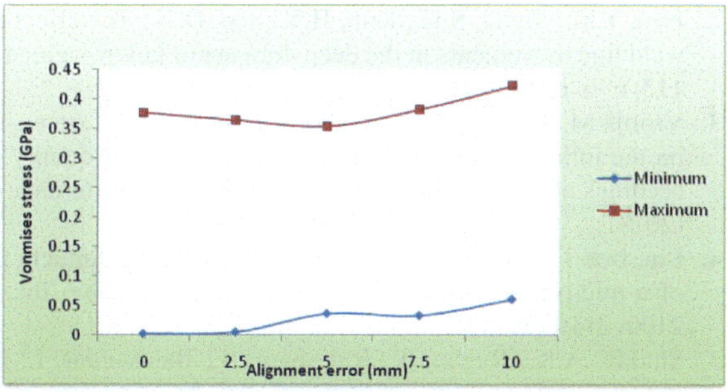

Fig. 10 Von misses stresses with alignment error of nil, 2.5, 5, 7.5 and 10 mm

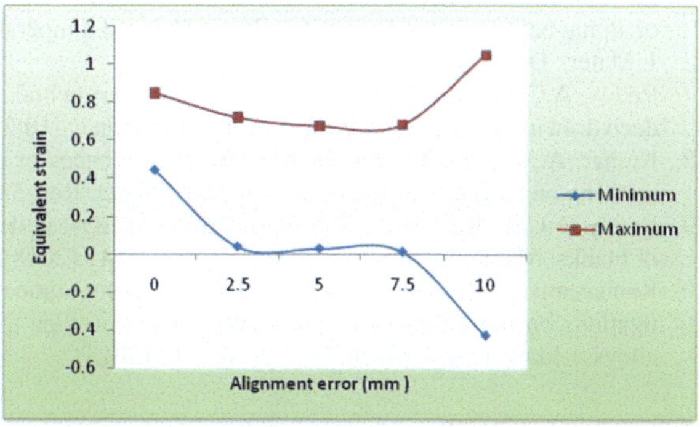

Fig. 11 Equivalent strains for alignment error of nil, 2.5, 5, 7.5 and 10 mm

4 Conclusion

Using the numerical simulation, the following conclusions are derived.

1. The lower values of minor stresses vary considerably for offset values of 5 mm and above.
2. The major stresses also vary considerably at 5 mm and above offset values.

3. Similar observations shows that the equivalent stresses, and von mises stresses are minimum at 5 mm offset.

It can be concluded from these results that the blank offset should not be more than 5 mm for 80 mm diameter cup using 120 mm diameter blank of AA6111 aluminium alloy.

References

1. Nguyen, N.T., Seo, O.S., Lee, C.A., Lee, M.G., Kim, J., Kim, H.Y.: Mechanical behavior of AZ31B mg alloy sheets under monotonic and cyclic loadings at room and moderately elevated temperatures. Materials **7**(2), 1271–1295 (2014)
2. Heo, Y.M., Wang, S.H., Kim, H.Y., Seo, D.G.: The effect of the drawbead dimensions on the weld line movements in the deep drawing of tailor–welded blanks. J. Mater. Process. Technol. **113**, 686–691 (2001)
3. Vrolijk M., Ogawa, T., Camanho, A., Biasutti, M., Lorenz, D.: A study with ESI PAM-STAMP on the influence of tool deformation on final part quality during a forming process. In: Proceedings of the 21st International ESAFORM Conference on Material Forming, pp. 291–296 (2018)
4. Lin, Bor-Tsuen, Yang, Cheng-Yu.: Applying the Taguchi method to determine the influences of a microridge punch design on the deep drawing. Int. J. Adv. Manuf. Technol. **88**(5–8), 2109–2119 (2017)
5. Shukla, A.K., Raghu, T., Rajesham, S., Balasundar, I.: Analysis of significant parameters influencing formability of titanium alloy by using over all evaluation criteria and new matrix model based on Taguchi method. Trans. Indian Inst. Metals **67**(5), 721–730 (2014)
6. Shukla, A.K., Raghu, T., Rajesham, S., Balasundar, I.: Optimization of formability parameters of alpha–beta-processed titanium alloy at elevated temperature. Proc. Inst. Mech. Eng. Part L: J. Mater.: Des. Appl. **228**(3), 231–240 (2014)
7. Reddy, A.C.S., Rajesham, S., Reddy, P.R.: Experimental and simulation study on the warm deep drawing of AZ31 alloy. Adv. Prod. Eng. Manag. **10**(3), 153–161 (2015)
8. Kumar, A., Raghu, T.: Rajesham S (2012) Influences of temperature of thermo mechanical working on hardness of titanium alloy. Adv. Mater. Res. **585**, 381–386 (2012)
9. Reddy, A.C.S., Rajesham, S.: Determination of LDR in deep drawing using reduced number of blanks. Mater. Today: Proc. **5**(13), 27136–27141 (2018)
10. Kandasamy, J., Hussain, M.M., Rajesham, S.: Computational analysis and experimental investigations on the influence of the FSW tool pin profiles towards superior IMC in aluminum alloys. i-Manager's J. Mech. Eng. **2**(2), 7–12 (2018)

PCA-GRA Coupled Multi-criteria Optimisation Approach in Machining of Polymer Composites

Vikas Kumar and Rajesh Kumar Verma

Abstract The present study deals with the machining of graphite-reinforced polymer composite (GRPC) using grey relation analysis (GRA) embedded principal component analysis (PCA). L_{16} orthogonal array established for creating the experimental procedure. The evaluation of grey relation grade based on the process of GRA and PCA technique is used to find the correlation array and to find the Eigen value as well as Eigen vector finally for calculating the overall value of GRG. The combination of GRA-PCA is used to optimise the machining parameter and get the best optimal setting parameter and to identify the significant factor. ANOVA method is used to find the residual plot for GRG. In this study, it has proven that the GRA embedded PCA is an efficient method for giving the effective machining environment in order to minimise thrust (T), torque (TR), surface roughness (Ra) and maximise the metal removal rate. The optimal setting value through the Taguchi method is speed 500 rpm, feed rate 25 mm/rev. depth of cut −1 mm and weight percentage 40% simultaneously. The ANOVA is used towards identify the significant factors influence the respective response, i.e. in this case weight percentage (49.27%) is most significantly affects GRG followed by speed (8.41%), feed rate (8.30%) and depth of cut (1.64%).

Keywords Grey relation analysis (GRA) · Principal component analysis (PCA) · Analysis of variance (ANOVA) · Machining · Optimisation

1 Introduction

Machining (milling) operation is widely applied in the area of manufacturing industry such as aerospace, automobile sector, space vehicle, fuselage component, which is the main body of the airplane, electronic insulator. The importance of doing machining

V. Kumar (✉) · R. K. Verma
Department of Mechanical Engineering, MMM University of Technology, Gorakhpur 273010, India
e-mail: vikas.sahni95@gmail.com

R. K. Verma
e-mail: rkvme@mmmut.ac.in

D. Dutta and B. Mahanty (eds.), *Numerical Optimization in Engineering and Sciences*, Advances in Intelligent Systems and Computing 979,
https://doi.org/10.1007/978-981-15-3215-3_47

(milling) operation is because it has accomplished making the various types of goods with intricate geometry. Therefore, for manufacturing, it is very important to choose the machining parameter in a way that the need for quality must be maintained without losing any cost or profit. When using the machining (milling) of graphite-reinforced polymer composite, there is some important factor which has been considered like speed (*S*), feed rate, depth of cut and weight percentage (wt.). The output responses have been considered such as metal removal rate (MRR), thrust (*T*), torque (TR) and surface roughness (Ra). When doing machining (milling) process, the diameter of the milling cutter is to be taken 5 mm. Before doing the machining operation, Taguchi-based L_{16} orthogonal array must be considered since it gives a precise and effective number of the process for the experimental work. Taguchi method is used for generating the single response optimisation and gives the S/N ratio. GRA-PCA method has been used to transform the multi-response to single response optimisation, GRA is applied to find the correlation among multiple responses, and then GRA has been proposed by Deng Julong in 1988. Initially, PCA has proposed through Pearson and hotelling. ANOVA is applied to get individual optimisation after GRA-based Taguchi method is used for obtaining the S/N ratio and effective optimisation. PCA has been used to find the correlation as well as calculate the Eigen value plus Eigen vector, where accountability proportion (AP) has been used to calculate the grey relation grade (GRG), Eigen value and Eigen vector have been obtained through the use of grey relation coefficient (GRC), and after that Taguchi method is used to get the best possible solution.

1.1 Literature Review

Though the power of fascination among the separate coatings in graphite was single weak van der Waals forces and exfoliation of graphite which is arbitrarily distributed few layers has an exciting task. Therefore, numerous chemical intercalation processes were grown to support the dispersion of graphitic layers [1]. There is subsequent fast heating of the graphite intercalation composites can reason of individual carbon coatings to change a part as well as growth of graphite crystals through the hundred times in a manner normal to the coatings [2, 3]. The subsequent extended graphite along using small thickness as well as great temperature confrontation was a significant manufacturing fresh material. Simply treated through lamination, it has been applied to create great areas as well as shaped parts, like gaskets as well as applications of seals for high temperature [4]. The common forms of improved graphitic were graphite oxide, graphite-intercalated compounds as well as extended graphite (EG) [5]. Graphene oxide has been called as graphite oxide has typically ready through the action of graphite chips along with oxidising agents therefore the polar collections has presented on the surface of graphite that's why spreading the interlayer arrangement of the planes of graphene [6]. GICs have been usually normally shaped through the addition of molecular coatings of various chemical types among coatings of graphite matrix [7, 8]. When inserted graphite has flamed previous serious high heat, then a huge development of chips happens with the c-axis and then in in-plane technique, and this product has been called as extended graphite (EG) [9, 10]. Recently, the intumescent flame retardant (IFR) flavourings were found since they have not individual additional naturally approachable and then the halogenated holding glow retardants then also have been advanced glow retardant competence than the inorganic paddings. Therefore, the form of graphite (natural graphite, extended graphite as well as graphene oxide) especially expanded graphite has been extensively used to grow carbonise produce as well as encourage updraft stability owing plentiful foundation for modest groundwork as well as little price for numerous types of polymer compounds plus EVA, polyethylene, polyolefin, polyurethane, acrylic resin, unsaturated polyester as well as polylactide [11, 12]. There are dissimilar types of paddings like carbon nanotubes, metal, carbon fibre as well as graphite that are used to make thermally conductive compounds constituents. Mostly, it has been used due to its small price, great updraft conductivity (110–130 W/mk) as well as less in heaviness. Pioneer has applied L_{16} orthogonal array as well as use of ANOVA method to learning the effect the constraints like feed rate (*f*), speed (*S*), thrust (*T*), torque (TR) as well as damage factor to minimise these responses [13].

2 Experimental Detail

Taguchi-based L_{16} orthogonal array (DOE) has been used to perform the experiment. In this experiment, we have considered four factors like speed (*S*), feed rate (*f*), depth of cut (DOC) as well as weight percentage (wt.) which are displayed in Table 1.

The software Minitab 18 has been used for graphical analysis and optimisation of machining parameter. The size of the mould is (8.6 cm × 7 cm × 1 cm), and the sample is based on the graphite-reinforced polymer composite (GRPC), and Table 2 shows the output parameter based on DOE. The operation of milling along with numerous cutters is committed on the arbor at a similar time, therefore rising MRR for permitting several sides of machining at the corresponding period and establishing the precision of the machining. Now, Table 2 has shown the whole step of the experiment, i.e. L_{16} orthogonal array which is based on the Taguchi-based design of the experiment. We have considered four input parameters for the milling operation such as speed (*S*), feed rate (*f*), depth of cut (DOC) as well as weight percentage (wt.).

Table 1 Process parameter and levels

Factors	Unit	Level 1	Level 2	Level 3	Level 4
Drill speed	[RPM]	500	1000	1500	2000
Feed rate	[mm/rev]	150	200	250	300
Depth of cut	[mm]	0.5	1	1.5	3
Weight %		10	20	30	40

Table 2 DOE based on L_{16} orthogonal array

S. No.	Speed (rpm)	Feed rate (mm/rev.)	Depth of cut (mm)	Weight %
1	500	10	−0.5	10
2	500	15	−1	20
3	500	20	−1.5	30
4	500	25	−3	40
5	1000	10	−1	30
6	1000	15	−0.5	40
7	1000	20	−3	10
8	1000	25	−1.5	20
9	1500	10	−1.5	40
10	1500	15	−3	30
11	1500	20	−0.5	20
12	1500	25	−1	10
13	2000	10	−3	20
14	2000	15	−1.5	10
15	2000	20	−1	40
16	2000	25	−0.5	30

The purpose of taking these four-machining parameters is because our machining sample is based on the graphite-based composite polymer (GRPC), and in various researches it has been found that using these parameters will give the better output responses. The purpose of taking the depth of cut because the milling cutter works through the downward direction which has shown through negative sign.

Tool material – HSS milling cutter (milling cutter diameter = 5 mm).

Machine specification = CNC vertical machining centre manufactured by BHARAT FRITZ WERNER LTD.

CNC vertical machine is one of the best modern machines for various important manufacturing operations such as milling operation and drilling operation. It has produced various difficult shapes with ease and also at minimum time in case of milling and drilling operation, and it has been shown in Figs. 1 and 2 during the milling operation of the sample of graphite-reinforced polymer composite (GRPC).

Machining (milling) operation has been performed in CNC vertical machining centre and HSS milling cutter used for machining, and the obtained experimental parameters such as MRR, thrust (T) torque (TR) and then surface roughness (Ra) have been given in Table 3. Usually, Table 3 has shown the experimental or output response that is obtained through the taken input parameters for the milling operation, and the obtained output data is based on the Taguchi-based design of experiments successively.

Fig. 1 CNC vertical machining centre

Fig. 2 Milling process of GRPC

3 Concept of GRA and PCA

Deng Julong in 1988 has proposed the concept of grey relation analysis which shows the interrelationships among the multi-response of changing parameters. In this method, the grey relation grade (GRG) has been obtained through examining the degree of relation of the multiple output responses. There are following steps which have been considered for getting the value of grey relation grade (GRG).

1. Experiment Parameters
2. Normalised Data (N)
3. Deviation Sequence (D)
4. Grey Relation Coefficient (GRC)
5. Grey Relation Grade (GRG)

Table 3 Calculation of experimental data

S. No.	MRR (mm^3/s)	Thrust (N)	Torque (N-mm)	Ra (μm)
1	0.6287	0.28	0.1	1.40
2	1.29385	0.31	0.18	0.8
3	2.60233	0.5	0.08	0.9
4	5.5404	0.36	0.13	0.9
5	0.826183	0.42	0.05	1.1
6	0.597195	0.39	0.11	0.8
7	5.7360	0.44	0.11	1.20
8	3.19013	0.39	0.25	1.1
9	1.17894	0.31	0.06	0.7
10	4.01136	0.42	0.16	0.9
11	0.958750	0.44	0.13	1.6
12	2.1732	0.28	0.16	1.50
13	2.8615	0.5	0.19	2.4
14	2.30162	0.33	0.13	1.0
15	1.8593	0.36	0.09	0.7
16	1.15410	0.25	0.13	1.4

Normalised data:

First, the output responses based on the L_{16} orthogonal array are normalised between the values 0–1. The preferable design features such as MRR, thrust (T), torque (TR) and surface roughness (Ra) have three aspects such as nominal is best, higher the better, lower the better, and Eq. 1 as well as 2 is used to normalise the machining parameter.

Higher the better (HB)

$$x_i^*(j) = \frac{x_i(j) - \min x_i(j)}{\max x_i(j) - \min x_i(j)} \tag{1}$$

Lower the better (LB)

$$x_i^*(j) = \frac{\min x_i(j) - x_i(j)}{\max x_i(j) - \min x_i(j)} \tag{2}$$

where $x_i(j)$ is comparability sequence and $x_i{*}(j)$ sequence after data pre-processing.

Deviation Sequences

$$\Delta_{oi}(k) = \left|\chi_i^*(j) - x_i^*(j)\right| \tag{3}$$

where $\Delta_{oi}(j)$ is the deviation sequence of the reference sequence $x_o{}^*(j)$ as well as the comparability sequence $x_i{}^*(j)$.

Grey Relational Coefficient

The equation of GRC

$$\xi_i(k) = \frac{\Delta_{\min} + \xi\,\Delta_{\max}}{\Delta oi + \xi\,\Delta_{\max}} \tag{4}$$

where ξ is defined as an identification coefficient.

Methodology of PCA:

PCA (Hotelling 1993) has been a multivariate numerical technique mainly used to identify the correlation among various responses.

Collection of multi-response

$$x = \begin{bmatrix} x_1(1) & x_1(2) & \ldots & x_1(n) \\ x_2(1) & x_2(2) & \ldots & x_2(n) \\ x_m(1) & x_m(2) & \ldots & x_m(n) \end{bmatrix}$$

$X_i(j)$, where $i = 1, 2, 3, \ldots m$ and $j = 1, 2, 3, \ldots n$, where "m" has shown the numeral assessment trial also the "n" has shown the number of the output response.

Considering the correlation coefficient array: The array of correlation coefficient of the standardised parameter collection has been assessed as given below: R_{ji} = Covar. $X_i{}^*(j)$, $X^*(l)/\sigma_{xi}{}^*(j)^* \times \sigma_{xi}{}^*(1)$, wherever, Covari. $X_i^*(j)$, $X^*(l)$ is the covariance of sequences $x_i^*(j)$ and $x_i^*(l)$; $\sigma_{xi}{}^*(1)$: The standard deviation of sequence is $x_i{}^*(l)$.

Considering the Eigen values as well as the Eigen vectors: The Eigen values as well as the Eigen vectors have been considered from the array of association factor.

$$(R - \lambda_{k}I_{m})V_{ik} = 0 \qquad \sum \lambda_k = n \qquad \text{where, } k = 1\ 2\ 3 \ldots n$$

where, λ_k is shows the Eigen values and the $V_{ik} = [ak1\ ak2 \ldots \text{akn}]^T$: Eigen vectors conforming to the Eigen value of λ_k.

Assessment of principal modules

The uncorrelated principal components are given below:

$$Y_{mk} = \sum X_m^*(i)V_{ik}$$

Y_{mk} is called the principal component, where $i = 1, 2 \ldots n$.

4 Results and Discussions

Initially, normalised experimental data such as MRR, thrust (T), torque (TR) as well as Ra has been presented in Table 3 by applying the GRA method and then by calculating the deviation sequence. After finding, the deviation data is used for calculating the GRC value and that GRC value which helps to find the Eigen value and Eigen vector for calculating grey relation grade (GRG) and whole calculation of experimental data is based on the combination of data obtained from the Taguchi-based DOE technique, and the obtained data has been shown in Table 4 which shows the calculated value of GRC and GRG. PCA analysis of experimental data is shown in Table 5, the best optimal setting parameters are shown in Table 6, and Fig. 2 shows the main effect plot of S/N ratio. ANOVA is applied to calculate the involvement of

Table 4 Calculations of GRC and GRG

Grey relation coefficient				Grey relation grade
MRR	Thrust	Torque	Ra	GRG
0.334701	0.806452	0.666667	0.548387	0.144434
0.366453	0.675676	0.434783	0.894737	0.137578
0.45053	0.333333	0.769231	0.809524	0.129797
0.929259	0.531915	0.555556	0.809524	0.178829
0.343539	0.423729	1	0.68	0.144589
0.333333	0.471698	0.625	0.894737	0.126799
1	0.396825	0.625	0.62963	0.171335
0.5023	0.471698	0.333333	0.68	0.121269
0.360544	0.675676	0.909091	1	0.162395
0.598364	0.423729	0.47619	0.809524	0.136856
0.349738	0.396825	0.555556	0.485714	0.105283
0.419002	0.806452	0.47619	0.515152	0.142082
0.471978	0.333333	0.416667	0.333333	0.070063
0.427964	0.609756	0.555556	0.73913	0.138055
0.398598	0.531915	0.714286	1	0.14515
0.359292	1	0.555556	0.548387	0.157243

Table 5 PCA analysis

Eigenvalue	1.4274	1.3006	0.7363	0.5357
Accountability proportion (AP)	0.357	0.325	0.184	0.134
Cumulative accountability proportion (CAP)	0.357	0.682	0.866	1

Table 6 Evaluation of optimal setting

Level	Machining parameter			
	Speed (*S*)	Feed (*f*)	DOC	wt%
1	0.053547	0.13037	0.13344	0.148976
2	0.089695	0.134822	**0.14235**	0.108548
3	0.023537	0.137891	0.137879	0.142121
4	**0.127628**	**0.149856**	0.139271	**0.153293**

Table 7 ANOVA table for GRG

Source	DF	Seq SS	Contribution	Adj SS	Adj MS	*F*-value
Speed	3	0.000846	8.41%	0.000846	0.000282	0.26
Feed	3	0.000835	8.30%	0.000835	0.000278	0.26
Doc	3	0.000164	1.64%	0.000164	0.000055	0.05
Wt	3	0.004954	49.27%	0.004954	0.001651	1.52
Error	3	0.003256	32.38%	0.003256	0.001085	
Total	15	0.010055	100.00%			

the considered constraint, and obtained data is shown in Table 7. The residual plot for GRG as well represents the variation between residual and their expected value as shown in Fig. 3. The value of GRG is calculated by GRA procedure, and the higher value of GRG decides the rank which is closer to the optimal setting value.

As we know, the Taguchi method is applied for single response optimisation, therefore Fig. 3 has shown the optimal setting value of GRG for single response optimisation on high S/N ratio such as speed 500 rpm, feed 25 mm/rev., depth of cut is −1 mm (downward) and wt. percentage 40%, and Fig. 4 has been obtained through the ANOVA technique which gives the residual plot for GRG, i.e. expandable case of GRG, and both techniques are applied in the Minitab18a software for getting the desired value as well as graph to show the correct variation of certain parameters. Basically, Fig. 4 shows the residual plots for GRG value that has been obtained by ANOVA technique and the purpose of using this technique to know how exact the machining value near to the reference line of the subsequent plot for GRG. ANOVA also gives the information about how much part is affected during the machining (MRR, *T*, TR, Ra). ANOVA has been used to know about the contribution of percentage of the considered parameters.

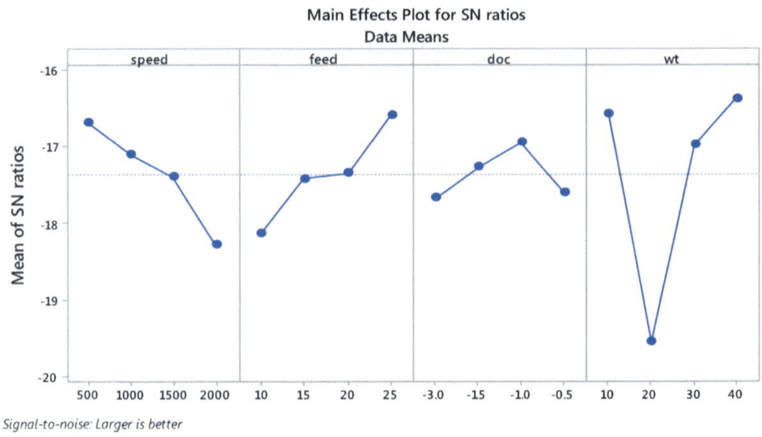

Fig. 3 S/N ratio plot for GRG

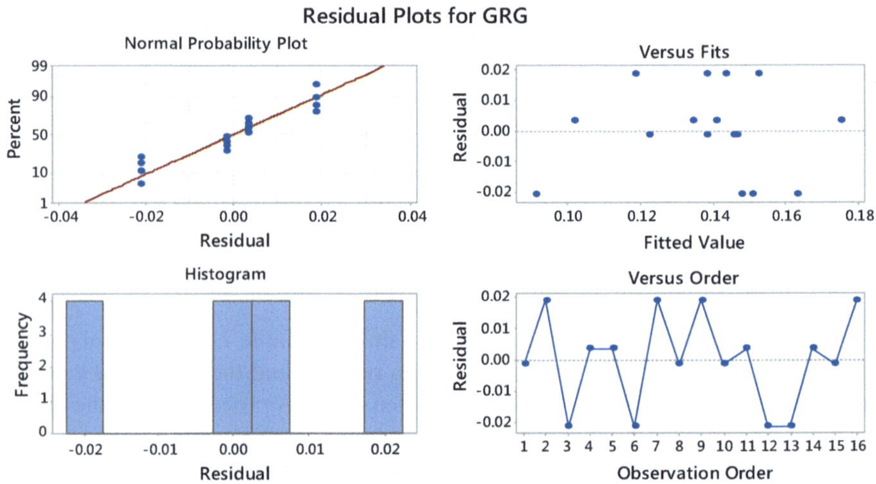

Fig. 4 Residual graph for GRG

5 Conclusion

Generally, when using optimisation of machining (milling), data with multi-performance characteristics (high MRR, low thrust, torque and Ra) for the milling operation of GRPC and to get the higher optimal setting value for GRG is obtained through Taguchi-based main effect of S/N ratio plot which shows the optimal setting constraints like speed (S) 500 rpm, feed rate 25 mm/rev., depth of cut −1 mm as well as weight percentage (wt.) 40%. After that ANOVA method coupled GRG for getting the residual plot for GRG and also obtained the significant factor for surface roughness (Ra), thrust (*T*), torque (TR) and MRR, and it has been found that the most effective factor for MRR is the depth of cut. ANOVA for Ra has accepted the goodness of fitting and their percentage contribution in the considered parameter such as weight percentage 49.27%, speed 8.41%, feed rate 8.30% and depth of cut 1.64%. Feed rate effect Ra of composite material during machining. It has been found that the GRA-based Taguchi technique coupled with PCA is the most effective and desirable for solving the problem of quality during the milling machining of GRPC material.

References

1. Dresselhaus, M.S., Dresselhaus, G.: Adv. Phys. **51**, 1 (2002)
2. Li, X., Wang, X., Zhang, L., Lee, S., Dai, H.: Science **319**, 1229 (2008)
3. Malik, S., Vijayaraghavan, A., Erni, R., Ariga, K., Khalakhan, J., Hill, J.P.: Nanoscale **2**, 2139 (2010)
4. Chung, D.D.L.: J. Mater. Sci. **22**, 4190 (1987)
5. Sengupta, R., Bhattacharya, M., Bandyopadhyay, S., Bhowmick, A.K.: Prog. Polym. Sci. **36**, 638 (2011)
6. Gao, J., Shen, K., Bao, F., Yin, J., Wang, D., Ma, R., Yan, C., Chen, T., Wang, G., Liu, X., Zhang, X., Zheng, D.: Polym. Plast. Technol. Eng. **52**, 553 (2013)
7. Kim, S.R., Poostforush, M., Kim, J.H., Lee, S.G.: Xpress Polym. Lett. **6**, 476 (2012)
8. Zhang, H., Lerner, M.M.: Nanoscale **8**, 4608 (2016)
9. Zhang, F., Zhao, Q., Yan, X., Li, H., Zhang, P., Wang, L., Zhou, T., Li, Y., Ding, L.: Food Chem. **197**, 943 (2016)
10. Chen, G.H., Wu, D.J., Weng, W.G., He, B., Yan, W.: Polym. Int. **50**, 980 (2001)
11. Wu, X., Wang, L., Wu, C., Lu, J., Xie, L., Wang, G., Jiang, P.: Polym. Degrad. Stab. **97**, 54 (2012)
12. Zhang, P., Hu, Y., Song, L., Ni, J., Xing, W., Wang, J.: Sol. Energ. Mat. Sol. Cells **94**, 360 (2010)
13. Verma, R.K., Abhishek, K., Datta, S., Pal, P.K., Mahapatra, S.S.: Multi-response optimization in machining of GFRP (Epoxy) composites: an integrated approach. J. Manuf. Sci. Prod. **15**(3), 267–292 (2015). ISSN: 2191-0375

FEA-Based Electrothermal Modeling of a Die-Sinker Electro Discharge Machining (EDM) of an Aluminum Alloy AA6061

Suresh Gudipudi, Vipul Kumar Patel, N. Selvaraj, S. Kanmani Subbu and C. S. P. Rao

Abstract Electric discharge machining (EDM) is the most promising one to machine conductive materials irrespective of their hardness with good accuracy and precision. The plasma discharge phenomenon during EDM involves complex physics which needs to be understood. Several attempts are being tried by researchers in this direction but still, it required further efforts. The present study made to develop an FEA-based electrothermal model to simulate the temperature distribution and material removal rate in EDM of the aluminum alloy AA6061. The cylindrical disk-type heat source was assumed for the current model. To have more realistic machining conditions, the model has introduced a discharge current term in addition to discharge duration in defining the plasma radius and used the Gaussian heat flux. Also, both the conductive (due to plasma) and convective heat transfer (due to dielectric) modes were taken into consideration. The theoretical material removal rate (MRR) was estimated by assuming the cavity produced in the model as a hemispherical type. The EDM experimentation was performed on the AA6061 work material at similar electrical conditions used in the model. The experimental MRR was calculated from the weight difference (before and after machining) method. The results showed that the difference in theoretical (understated assumptions) and experimental MRR as 12%. The reasons for the MRR difference were also discussed in detail.

Keywords Finite element analysis (FEA) · Plasma · Discharge · Gaussian heat flux · Electric discharge machining (EDM) · Material removal rate (MRR)

S. Gudipudi (✉) · V. K. Patel · N. Selvaraj
Department of Mechanical Engineering, National Institute of Technology Warangal, Warangal, Telangana 506004, India
e-mail: gudipudi.suresh@gmail.com

N. Selvaraj
e-mail: selva@nitw.ac.in

S. Kanmani Subbu
Discipline of Mechanical Engineering, Indian Institute of Technology Palakkad, Palakkad, Kerala 678557, India
e-mail: sksubbu@iitpkd.ac.in

C. S. P. Rao
National Institute of Technology Andhra Pradesh, Tadepalligudem, Andhra Pradesh 534102, India

D. Dutta and B. Mahanty (eds.), *Numerical Optimization in Engineering and Sciences*, Advances in Intelligent Systems and Computing 979,
https://doi.org/10.1007/978-981-15-3215-3_48

1 Introduction

Electric discharge machining (EDM) is one of the most important non-conventional machining processes mainly used for machining hard materials with accuracy and precision. This process is based on the electrothermal process in which the potential difference is applied across the two electrically conductive electrodes (work and tool materials) which are kept at a small gap (in micrometers) in a dielectric medium. When the applied voltage across these two electrodes reaches more than the dielectric strength of the dielectric medium, it gets ionize and the spark is generated for few microseconds. The heat energy of this spark causes melting and vaporization of the material from the work and the tool. The analytical and numerical EDM models [1–12] under various assumptions which differ from one to another were reported in the published literature. The assumptions majorly based in the consideration of heat source shape, mode of heat transfer, an approximation of the heat flux radius, boundary conditions which control the heat flux, variation in thermophysical properties with temperature rise, and the fraction of heat energy available at the work material. These models were compared by experimentally by Yeo et al. [13]. Authors observed that the Di Bitonto's model [6, 7] which considers a point heat source with an average thermophysical properties of the material, discharge power as boundary condition, and fraction of heat energy available at the work as 0.18 could predict the dimensional geometry and erosion rate of the process with a good accuracy. Authors also suggest disk heat source models have the potential to be further developed. Salonitis et al. [14] used a uniformly distributed heat source to model the EDM for determination of material removal and surface roughness. The authors also validated the numerical results with the experimental results by assuming the volume of the crater as circular paraboloid geometry. Tao et al. [15] modeled the anode crater formation by considering the two consecutive plasma heating and bubble collapsing phases. Authors also found that the experiments were good agreement with the developed model. Sarikavak et al. [16] developed a 3-D transient thermal model for a single discharge to calculate the molten metal radius and material removal rate (MRR). Authors validated this model with published literature and found good approximation. A single spark FEA EDM model for a composite material was developed by Vishwakarma et al. [17] to calculate the temperature for a single discharge and was extended to a second discharge. Authors estimated the material removal for multi-discharge by calculating the number of pulses. Guo et al. [18] developed an electrothermal model to predict the effect of pulse-on time and pulse current on the crater volume. Hinduja e al. [19] extensively reviewed the thermal models that have been developed to simulate for the plasma arc in EDM to determine the discharge location. Bai Shao et al. [20] developed an FEA-based electrothermal model to predict crater formation in micro-electric discharge machining. Somashekhar et al. [21] simulated the miro-EDM model to determine the effect of pulse-on time and pulse-off time on temperature distribution in work material. Authors verified the simulation results with experimental results. The numerical model based on the finite difference method which takes into account the growth of the plasma channel and the instantaneous removal of material throughout the electric discharge was reported by Adnene Tlili et al. [22] and validated experimentally.

From the reported literature, it was observed that the successful EDM electrothermal modeling of any work material dependent on the assumed conditions and introduced realistic machining conditions. Also, the validation with experimental is dependent on the EDM setup and machining environment. Hence, the present work was performed to model the EDM of AA6061 for MRR and experimentally compared.

2 Numerical Modeling of EDM Process

The present model was developed under the following assumptions.

1. A two-dimensional axisymmetric domain of the workpiece is considered. 2. Material is removed because of the single spark only. 3. Thermal properties of the work material are temperature-dependent. 4. Transfer of heat from the tool to the work material is by both conduction and convection. 5. Tool wear effect is neglected. 6. Formation of the recast layer is neglected. 7. Heat flux is supposed to be the Gaussian distributed.

In the present model as shown in Fig. 1, a two-dimensional axisymmetrical domain of a cylindrical workpiece is considered to which heat flux is applied through a single spark. The dimensions of the model were considered as 0.2 mm (PR) and 0.1 mm (PT) in *r*- and *z*-coordinate directions, respectively.

The more realistic condition of the heat transfer from tool and work is considered to be the function of critical electrical parameters pulse-on time, T_{on} and the current, I. Hence, the radius of discharge plasma is expressed according to the Eq. 1. [9].

$$R_p = 2.04 \times I^{0.43} \times t_{\text{on}}^{0.44} \tag{1}$$

where R_p is the plasma radius (μm), I is the current (A), t_{on} is the discharge time (μs).

The intensity of heat source is considered to be Gaussian distributed and is given by Eq. 2. [23]

$$Q_{(r,t)} = \frac{4.45VIF}{\pi R_p^2} \mathrm{e}^{-4.5\left(\frac{r}{R_p}\right)^2} \tag{2}$$

where F is the fraction of energy transferred from the heat source to the work material. Researchers have suggested that this value may vary from 8 to 50% depending upon the type of electrode material and the dielectric medium. The present work is considered this value as 26%.

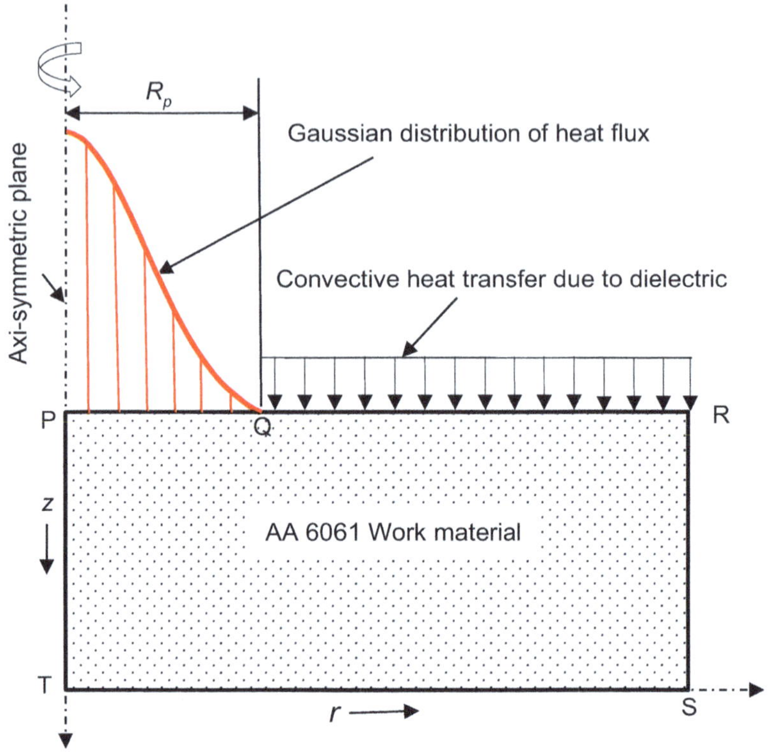

Fig. 1 Representation of thermoelectric model for EDM

The following boundary conditions are used for simulating the model to study the distribution of the temperature, T in the work material with a variation of time (t) and coordinate (r and z) distances.

At the initial time, $t = 0$

$$T(r, z, 0) = T_0 \tag{3}$$

where T_0 is the ambient temperature.

At $t > 0$

At the boundary PQR

$$T = -K\frac{\partial T(r, 0, t)}{\partial z} \begin{cases} Q(r, t), r \leq R_p \\ h(T - T_0), r > R_p \end{cases} \tag{4}$$

As shown in Fig. 1, the conductive heat transfer mode takes place up to the PQ due to direct contact of plasma with the work material. The convective heat transfer mode was considered at QR because of the work material was surrounded by a dielectric medium. Generally, the very type of the dielectric fluid will decide the convective heat transfer coefficient, h.

The boundaries RS, TS, and PT are insulated. Therefore, the change in temperature with time is set to zero as shown in Eq. 5. Axis symmetry is about the boundary PS.

$$\frac{\partial T}{\partial t} = 0 \tag{5}$$

The governing heat transfer differential equation without internal heat generation written in a cylindrical coordinate of an axisymmetric thermal model is given by Eq. 6. [24]

$$\frac{1}{r}\left[\frac{\mathrm{d}}{\mathrm{d}r}\left(r\frac{\mathrm{d}T}{\mathrm{d}r}\right)\right] + \frac{\partial}{\partial z}\left(\frac{\partial T}{\partial z}\right) = \frac{\rho C_p}{K}\left(\frac{\partial T}{\partial t}\right) \tag{6}$$

where T is the temperature, ρ is the density of the work material, C_p is the heat capacity of the work material, K is the thermal conductivity of the work material, and r, z are coordinating axes.

The multi-physics finite element solver COMSOL was used to simulate the developed model. Two physics "*Heat Transfer in solids*" and "*Deformed geometry*" were coupled to get the desired simulation. The Gaussian distributed heat flux up to the plasma radius, R_p, and the conduction heat transfer to the work material was obtained by the former one. The latter one is used on the surface PQR to get the crater volume by removing the elements having the temperature more than the melting temperature of the work material. The simulation of the model is performed by using a free triangular mesh of the fine size. The work material properties, simulation, and experimental conditions were represented in Table 1.

3 Results and Discussions

3.1 *Calculation of the Theoretical Material Removal Rate, MRR_{th} (Mm^3/Min)*

The radius of plasma discharge, R_p and the Gaussian heat flux distribution, Q for a single discharge expressed by Eqs. (1) and (2) in the model were shown in Figs. 2 and 3, respectively. The crater produced for a single discharge with the temperature distribution obtained from the simulation was shown in Figs. 4 and 5. The crater is assumed to be spherical dome shape to estimate its volume was represented in Fig. 6 [17]. Hence, the theoretical volume of the crater, V_c for single discharge was calculated by using Eq. 7. The radius, r_c and the depth, h_c values of the crater were observed as 0.004045 mm and 0.013264 mm, respectively, from the simulation plot.

Table 1 Representation of properties of work material, simulation, and experimental conditions

The properties of work material AA6061	
Density, ρ (kg/m^3)	2700
Thermal conductivity, K (W/m K)	180
Hear capacity, C_p (J/kg K)	897
Simulation conditions of the developed model	
Discharge current, I (A)	8
Pulse duration on time, t_{on} (μs)	25
Voltage, V (V)	55
Fraction of energy considered at work, (%)	26
Melting/evaporation temperature of work, T_{ev} (°C)	750
Response variable-1 (simulation output)	Temperature distribution
Response variable-2 (simulation output)	Volume of the crater
EDM experimental conditions	
Tool electrode material	Copper
Work material	AA6061 (as cast)
Thickness of the tool, mm	1.00 ± 0.04
Discharge current, I (A)	8
Pulse duration on time, t_{on} (μs)	45
Pulse duration off time, t_{off} (μs)	24
Voltage, V (V)	55
Machining time, min	4
Dielectric medium	EDM spark pureol
EDM setup model	PSR 20 Die-sinker EDM
Experimental response variable, (mm^3/min)	Material removal rate, MRR

$$V_c = \frac{\pi}{6} h_c \left(3r_c^2 + h_c^2\right) \tag{7}$$

The crater volume, Vc is about 1.15524E-06 mm3 per single discharge is used to estimate the theoretical material removal rate (MRR_{th}) (mm^3/min) which consists of multiple discharges. The number of multiple discharges/cycles/pulses, NOD is obtained for 1 min from Eq. 8 [14].

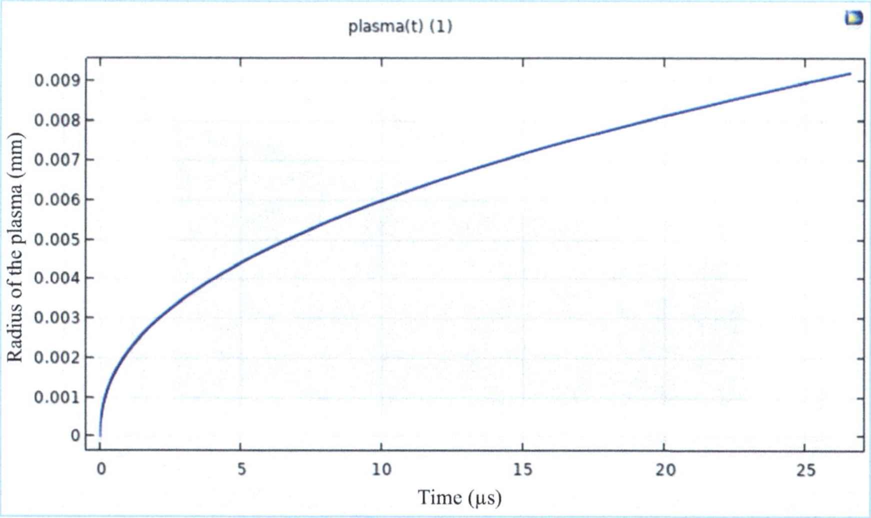

Fig. 2 Plot of radius of plasma discharge R_p

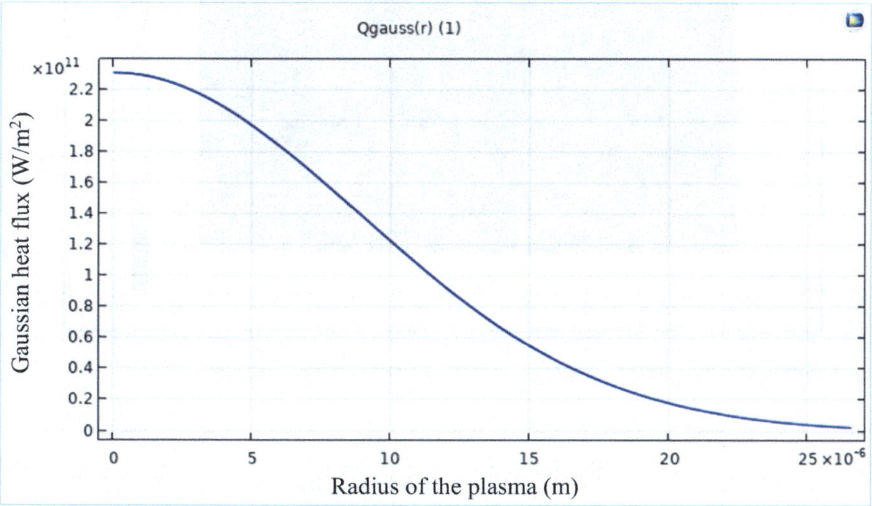

Fig. 3 Plot of Gaussian heat flux distribution, Q

$$\text{NOD} = \frac{\text{Total machining time}}{t_{\text{on}} + t_{\text{off}}} \tag{8}$$

The discharge pulse-off time, t_{off} was assumed as 24 μs. Hence, *NOD* was approximately yielded to 869565. The MRR_{th} was calculated from Eq. 9 and is 1.004 mm^3/min.

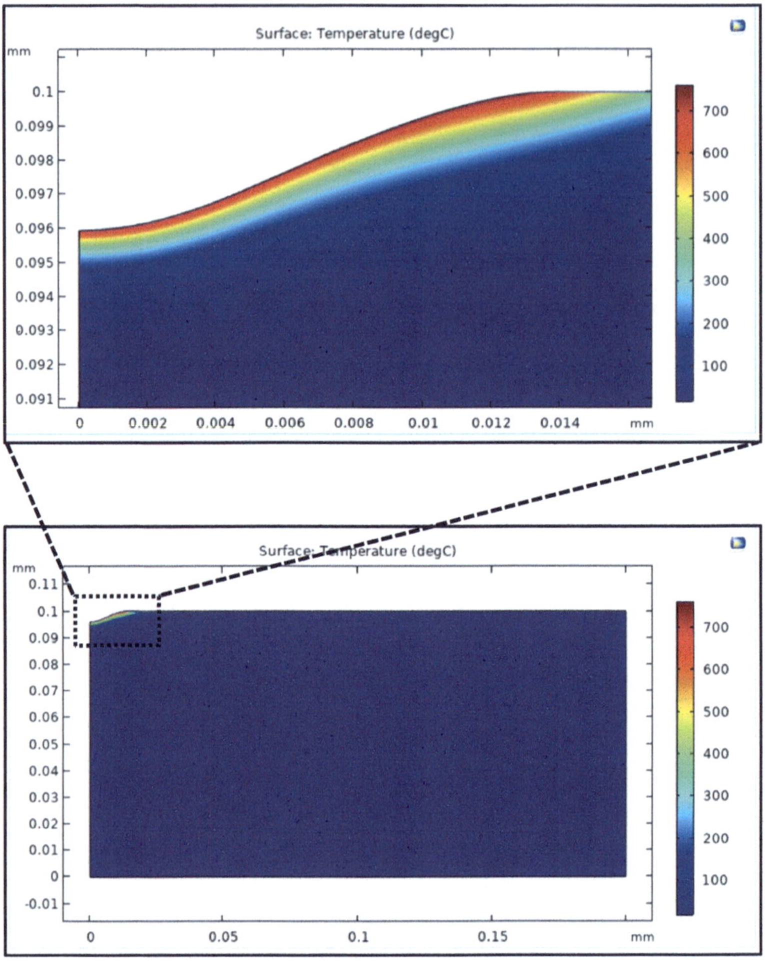

Fig. 4 Simulated crater due to single discharge with temperature variation

$$\mathrm{MRR_{th}} = \frac{V_c \times \mathrm{NOD}}{\text{Total machining time}} \tag{9}$$

3.2 *Calculation of the Experimental Material Removal Rate, MRR_{exp} (Mm^3/Min)*

The die-sinking EDM was performed on the AA6061 with a copper electrode and the conditions were represented in Table 1. The photographs of the EDM experimental

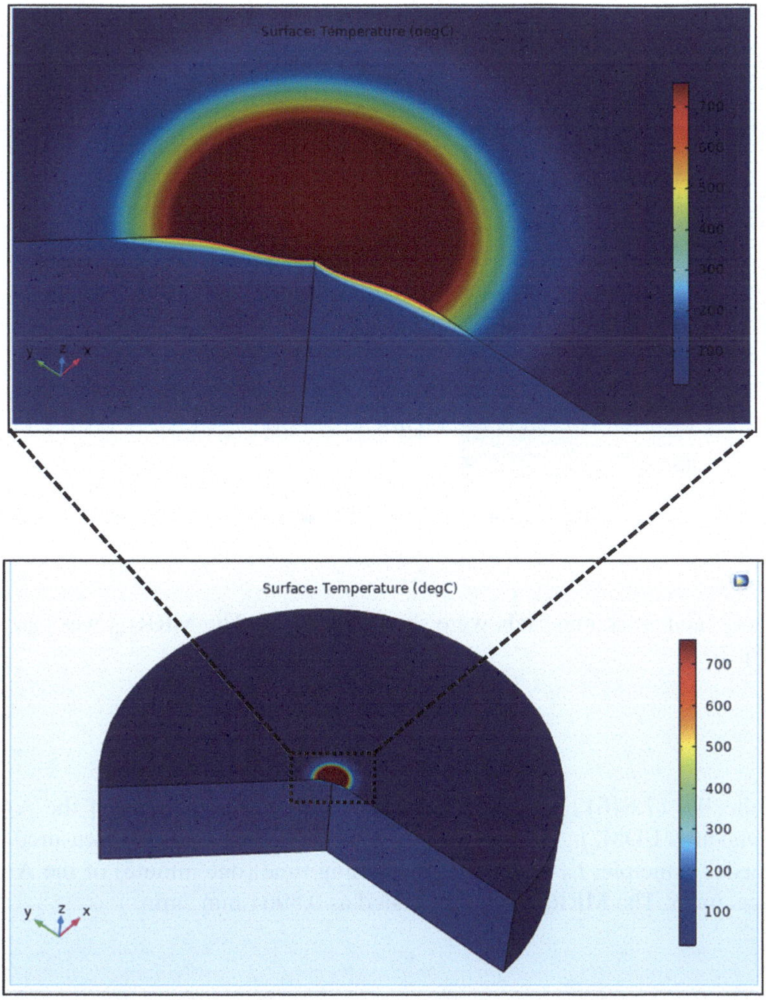

Fig. 5 Three-dimensional view of simulated crater due to single discharge with temperature variation

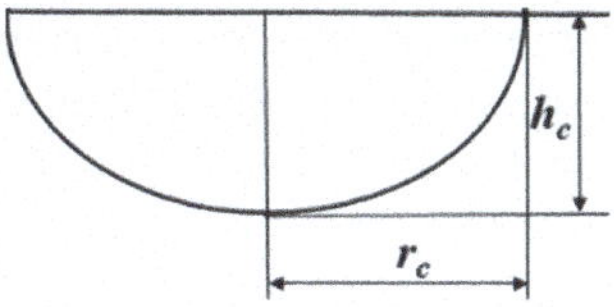

Fig. 6 Schematic of assumed spherical dome model for calculating the crater volume per single discharge

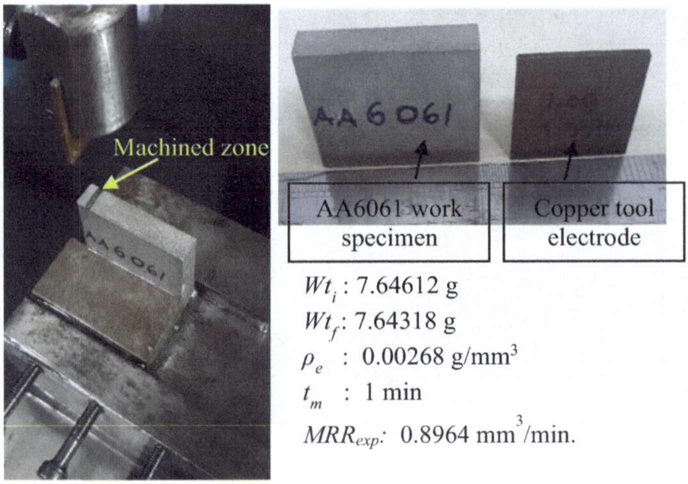

Fig. 7 Photographs of EDM experimental setup, tool and work materials with representation of MRR value

setup, tool, and work materials were shown in Fig. 7. The MRR_{exp} was calculated from Eq. 10.

$$\text{MRR}_{\text{exp}} = \frac{Wt_i - Wt_f}{\rho_e \cdot t_m} \tag{10}$$

where the Wt_i (7.64612 g) and Wt_f (7.64318 g) are the weights of the AA6061 before and after EDM. ρ_e is the measured density (0.00268 g/mm^3) measured by the Archimedes principle. t_m is the total machining time (one minute) of the AA6061 work specimen. The MRR_{exp} was calculated as 0.8964 mm^3/min.

4 Conclusion

The electrothermal model for sinking EDM was developed for simulating the temperature distribution and volume of the material removal. The experimentation was also performed to validate the model. It was observed that MRR values for simulation and experimentation were 1.004 mm^3/min and 0.8964 mm^3/min, respectively. The difference in these values is attributed to the lack of introducing the more realistic conditions such as dielectric fluid flow characteristics, analysis of debris, exact geometry for evaluating the crater volume, appropriate constant values, etc. Validation of the present model with published literature and improving the model by considering this above-said feature may help in reducing the MRR difference is a future scope.

Acknowledgements The authors would like to thank the Department of Science and Technology-Science and Engineering Research Board (DST-SERB), New Delhi, India, for supporting this work under the Early Career Research Scheme (File No. ECR/2017/001320). Authors also would like to thank Dr. I. A. Palani, Associate Professor, IIT Indore, India, for helping in the simulation.

References

1. Van Dijck, F.S., Dutré, W.L.: Heat conduction model for the calculation of the volume of molten metal in electric discharges. J. Phys. D Appl. Phys. **7**(6), 899–910 (1974). https://doi.org/10.1088/0022-3727/7/6/316
2. Jilani, S.T., Pandey, P.C.: Analysis and modelling of edm parameters. Precis. Eng. **4**(4), 215–221 (1982). https://doi.org/10.1016/0141-6359(82)90011-3
3. Das, S., Klotz, M., Klocke, F.: EDM simulation: finite element-based calculation of deformation, microstructure and residual stresses. J. Mater. Process. Technol. **142**(2), 434–451 (2003). https://doi.org/10.1016/S0924-0136(03)00624-1
4. Katz, Z., Tibbles, C.J.: Analysis of micro-scale EDM process. Int. J. Adv. Manuf. Technol. **25**, 923–928 (2005). https://doi.org/10.1007/s00170-003-2007-1
5. Jilani, S.T., Pandey, P.C.: An analysis of surface erosion in electrical discharge Machining. Wear **84**, 275–284 (1983). https://doi.org/10.1016/0043-1648(83)90269-7
6. Pandit, S.M., Rajurkar, K.P.: A stochastic approach to thermal modeling applied to electro-discharge machining. J. Heat Transf. **105**, 555–562 (2016). https://doi.org/10.1115/1.3245621
7. Pandey, P.C., Jilani, S.T.: Plasma channel growth and the resolidified layer in EDM. Precis. Eng. **8**(2), 104–110 (1986). https://doi.org/10.1016/0141-6359(86)90093-0
8. Patel, M.R., Barrufet, M.A., Eubank, P.T., DiBitonto, D.D.: Theoretical models of the electrical discharge machining process. II. The anode erosion model. J. Appl. Phys. **66**(9), 4104–4111 (1989). https://doi.org/10.1063/1.343995
9. DiBitonto, D.D., Eubank, P.T., Patel, M.R., Barrufet, M.A.: Theoretical models of the electrical discharge machining process. I. A simple cathode erosion model. J. Appl. Phys. **66**(9), 4095–4103 (1989). https://doi.org/10.1063/1.343994
10. Eubank, P.T., Patel, M.R., Barrufet, M.A., Bozkurt, B.: Theoretical models of the electrical discharge machining process. III. The variable mass, cylindrical plasma model. J. Appl. Phys. **73**(11), 7900–7909 (1993). https://doi.org/10.1063/1.353942
11. Hocheng, H., Lei, W.T., Hsu, A.S.: preliminary study of material removal in electrical-discharge machining of SiC/Al. J. Mater. Process. Technol. **63**, 813–818 (1997). https://doi.org/10.1016/S0924-0136(96)02730-6
12. Singh, A., Ghosh, A.: A thermo-electric model of material removal during electric discharge machining. Int. J. Mach. Tools Manuf **39**, 669–682 (1999). https://doi.org/10.1016/S0890-6955(98)00047-9
13. Yeo, S.H., Kurnia, W., Tan, P.C.: Critical assessment and numerical comparison of electro-thermal models in EDM. J Journal of Materials Processing Technology **203**, 241–251 (2008). https://doi.org/10.1016/j.jmatprotec.2007.10.026
14. Salonitis, K., Stournaras, A., Stavropoulos, P., Chryssolouris, G.: Thermal modeling of the material removal rate and surface roughness for die-sinking EDM. Int. J. Adv. Manuf. Technol. **40**, 316–323 (2009). https://doi.org/10.1007/s00170-007-1327-y
15. Tao, J., Ni, J., Shih, A.J.: Modeling of the anode crater formation in electrical discharge machining. J. Manuf. Sci. Eng. **134**(011002), 1–11 (2012). https://doi.org/10.1115/1.4005303
16. Sarikavak, Y., Cogun, C.: Single discharge thermo-electrical modeling of micromachining mechanism in electric discharge machining. J. Mech. Sci. Technol. **26**(5), 1591–1597 (2012). https://doi.org/10.1007/s12206-012-0305-y

17. Vishwakarma, U.K., Dvivedi, A., Kumar, P.: FEA modeling of material removal rate in electrical discharge machining of Al6063/SiC composites. Int. J. Mech. Aerosp. Ind. Mech. Manuf. Eng. **6**(3), 586–91 (2012). ISNI: 0000000091950263
18. Guo, Y.B., Klink, A., Klocke, F.: Multiscale modeling of sinking-EDM with Gaussian heat flux via user subroutine. Proced. CIRP **6**, 438–443 (2013). https://doi.org/10.1016/j.procir.2013.03.047
19. Hinduja, S., Kunieda, M.: Modelling of ECM and EDM processes. CIRP Ann. Manuf. Technol. **62**(2), 775–797 (2013). https://doi.org/10.1016/j.cirp.2013.05.011
20. Shao, B., Rajurkar, K.P.: Modelling of the crater formation in micro-EDM. Proced. CIRP **33**, 376–381 (2015). https://doi.org/10.1016/j.procir.2015.06.085
21. Somashekhar, K.P., Panda, S., Mathew, J., Ramachandran, N.: Numerical simulation of micro-EDM model with multi-spark. Int. J. Adv. Manuf. Technol. **76**(1–4), 83–90 (2013). https://doi.org/10.1007/s00170-013-5319-9
22. Tlili, A., Ghanem, F.: Salah N Ben (2015) A contribution in EDM simulation field. Int. J. Adv. Manuf. Technol. **79**(5–8), 921–935 (2015). https://doi.org/10.1007/s00170-015-6880-1
23. Yadav, V., Jain, V.K., Dixit, P.M.: Thermal stresses due to electrical discharge machining. Int. J. Mach. Tools Manuf **42**, 877–888 (2002). https://doi.org/10.1016/S0890-6955(02)00029-9
24. Rajput, R.K.: Heat and mass transfer. S. Chand and Company Ltd., 7361, New Delhi, India. ISBN: 81-219-2617-3

Modeling of Material Removal Rate and Hole Circularity on Soda–Lime Glass for Ultrasonic Drilling

Abhilash Kumar, Sanjay Mishra and Sanjeev Kumar Singh Yadav

Abstract Soda–lime glass possesses high hardness and brittleness. Analysis of machining of soda–lime glass becomes essential as it is now being extensively used material for different engineering and medical purposes. The intrinsic characteristics of soda–lime glass make it difficult to machine materials. In this study, experimental modeling for material removal rate (MRR) and hole circularity has been performed using the response surface methodology (RSM) for ultrasonic drilling (USD) of 2 mm diameter hole in 3 mm thick soda–lime glass. The effect of input process parameters like ultrasonic power, frequency, and abrasive grit number has been analyzed on MRR and hole circularity. It was experiential that the effect of ultrasonic power on both MRR and hole circularity for different grit size is not very important. Abrasive with high grit number rises the MRR but declines the MRR after initial rise up to grit number of 40. At a specific ultrasonic power, the increase of frequency continuously increases the hole circularity, but in case of MRR, it begins to decrease after 23 kHz. The result of ultrasonic power on both MRR and hole circularity is nonlinear gradual decline up to 80 W further increase later.

Keywords Response surface methodology · Soda–lime glass · Silicon carbide · Circularity · Ultrasonic drilling

1 Introduction

The rapid development in the field of medical, electrical, and optical devices leads to the development of microdevices where glasses are widely used as an engineering material due to its chemical and electrical resistivity, temperature stability, optical transparency, and biocompatibility. But the amorphous structure of these materials which provide high hardness and brittleness makes it difficult to machine material.

A. Kumar (✉) · S. Mishra
Department of Mechanical Engineering, Madan Mohan Malaviya University of Technology, Gorakhpur, Uttar Pradesh 273010, India

S. K. S. Yadav
Department of Mechanical Engineering, Harcourt Butler Technical University, Uttar Pradesh, Kanpur, India

D. Dutta and B. Mahanty (eds.), *Numerical Optimization in Engineering and Sciences*, Advances in Intelligent Systems and Computing 979,
https://doi.org/10.1007/978-981-15-3215-3_49

The soda–lime glass which is prepared by melting silica with soda, dolomite, lime, and alumina at the temperature of approx. 1700 °C is used in production of MEMS, microfluidic devices, opt mechatronic systems, data storage disks, touch screens, photomasks, etc. The wide application of soda–lime glass required creation of small diameter holes in the glass substrates. The conventional machining of glass is very difficult due to its high brittleness, low ductility, and fracture toughness. Therefore, the feasibility of machining of soda–lime glass with good geometrical accuracy and productivity is of utmost importance to manufacturing engineers.

Ultrasonic machining (USM) is a non-traditional machining method which uses the mechanical means in the form of ultrasonic vibration to remove the material from both conductive and non-conductive materials. Ultrasonic drilling (USD) removes the material by brittle fracture produced by the repeated hammering action of the abrasive grain. The materials are continuously removed from the machining zone by the abrasive slurry. Machining of the glasses by conventional machining methods and alternative non-traditional machining methods like abrasive water jet machining has the shortcomings like higher cutting forces, stray cutting, high tool wear rate with severe surface deterioration, etc (Kataria et al. [1]). The drilled glass parts are to be assembled to main product. Since the quality of final assembled parts directly depends on the geometrical accuracy of the hole along with its surface morphology, selecting a suitable machining method is of utmost importance in fabrication of components using glass.

Cong et al. [3] performed an analysis in order to assess the result of several processing parameters on quality features in rotary ultrasonic machining (RUM) of CFRP/Ti stacks. The effect of ultrasonic power and tool rotation on MRR at constant feed rate was investigated. It was found to be higher in comparison with conventional drilling. Zhang et al. [2] performed an analysis to examine the effect of the process parameters of RUM in optical K9 glass. Coolant used in the experiment was compressed air. Resultant surface roughness (SR) was found to increase with increasing feed rate and ultrasonic power, and decrease with increase in spindle speed. Zhang et al. [4] used the compressed air as coolant and K9 as glass to study the effect of various process inputs in contrast to power consumption by RUM components and the total RUM system. On varying ultrasonic power (0–50%), a nonlinear increase of ultrasonic power consumption was observed. Also, with increasing spindle speed (2000–5000 rpm), nearly a linear decrease in power consumption was noted. Electrical voltage, current, and feed rate were the key elements influencing power consumption of ultrasonic power supply. Churi et al. [5] studied cutting force during RUM of Ti alloy. They described that the effect of spindle speed on cutting force is also dependent on workpiece. Zhang et al. [6] reported the effect of spindle speed and feed rate RUM while machining K9 glass. They observed that low feed rate and high spindle rotation may cause low value of chipping size. Also, the effect of low feed rate on the chipping size was stronger as compared to high spindle speed. Lv. et al. [7] observed that subsurface damages of lower order can be attained at a spindle speed at 12000 rpm and feed rate of 200 mm/min in RUM of BK7 glass. Cong et al. [8] performed experiment on RUM of CFRP composites. They assessed cutting force under the effect of two separate coolants. They used cutting fluid and cold air as coolants. The difference in cutting force for both coolants was unaffected with ultrasonic power (from 0 to 80%). It was also observed that the cutting force decreased with an increase in spindle speed (1000–5000 rpm). Cong et al. [9] investigated twist drilling and RUM of CFRP composites. They reported spindle speed as insignificant for material removal rate (MRR) during the machining of CFRP composites with RUM. Linear increase in MRR was noted with increasing feed rate. Enhancement and increase in MRR were also observed at higher feed rates. Spindle speed and ultrasonic power had insignificant outcome on MRR. Ya et al. [10] experimentally analyzed the influence of machining variables on MRR while performing RUM of glass and found that MRR was higher with RUM than that produced with stationary USM process. Zou et al. [11] investigated cutting temperature under the deviation of ultrasonic power. They applied novel fiber optic sensors during RUM of titanium, spindle speed, and feed rate remaining constant. Ultrasonic power was found to have a substantial result on cutting temperature [12].

Kumar and Singh investigated the effect of spindle speed, ultrasonic power, and feed rate on the chipping width and taper during rotary ultrasonic drilling process of ceramic glasses "BK-7" using response surface method (RSM). The authors found that federate plays the most dominant role in controlling both these parameters [13]. Sindhu et al. applied the central composite design technique of RSM to conduct experiments for analyzing the effect of tool feed rate, tool rotational speed, and ultrasonic power on material removal rate and surface roughness during rotary ultrasonic drilling of quartz material [14]. Murti and Philip investigated the influence of ultrasonic irradiation of spark gap on EDM characteristics for both planing and drilling operations. Experiments were conducted using response surfaces to determine the effect of various input variable on machining rate, tool wear rate, corresponding wear ratio, and surface roughness [15].

The above literature survey indicates that even though experimental work has been carried out to analyze the performance of USD for different set of input parameters, specific study to investigate the impact of ultrasonic power, frequency, and grit size on the MRR has not been reported. Therefore, in this study, experimental model has been developed using RSM approach to study the above aspects. Furthermore, the combined effect of various input parameters on MRR has been elaborated using RSM surface plots.

2 Experimental Details

The experiments were performed on 3 mm thick soda–lime glass using mild steel tool of diameter 2 mm using the USM set up as shown in Fig. 1. The various properties of soda–lime glass are shown in Tables 1, 2, 3, and 4.

For determining the viable range of input parameters at which a quality hole drilling could be done, pilot experiments were performed. Black silicon carbide of three different sizes was used to prepare slurry with water. The details of the input process parameter have been shown in Table 5.

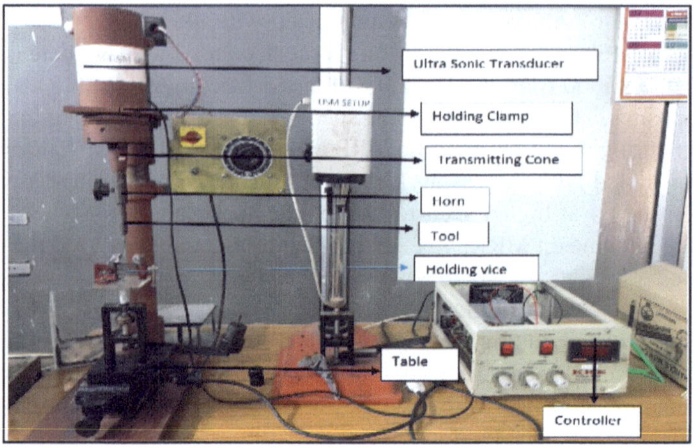

Fig. 1 Ultrasonic drilling setup

Table 1 Selected properties of soda–lime glass [16]

Property	Value
Refractive index, *n*	1.46
% Transmittance (at 1 μm)	70–80
Density, ρ (g/cm^3)	2.5
Thermal conductivity, *k* (W/m °C)	1.06
Specific heat, CP (J/g °C)	0.87
Softening point (°C)	700
Melting temperature, TM (°C)	1000
Price	0.8–1.7 USD/kg

Table 2 Mechanical properties of soda–lime glass [16]

Young's modulus	68–72 GPa
Yield strength (elastic limit)	30–35 MPa
Tensile strength	31–35 MPa
Elongation	0%
Hardness—Vickers	439–484 HV
Fatigue strength at 107 cycles	29.4–32.5 MPa
Fracture toughness	0.55–0.7 MPa m$^{1/2}$

Table 3 Thermal properties of soda–lime glass [16]

Maximum service temperature	443–673 K
Thermal conductor or insulator?	Poor insulator
Thermal conductivity	0.7–1.3 W/m K
Specific heat capacity	850–950 J/kg K
Thermal expansion coefficient	9.1–9.5 μstrain/°C

Table 4 Electrical properties of soda–lime glass [16]

Electrical conductor or insulator?	Good insulator
Electrical resistivity	$7.94 \times 1017 - 7.94 \times 1018$ μΩ cm
Dielectric constant	7–7.6
Dissipation factor	0.007–0.01
Dielectric strength	$12\text{–}14 \times 10^6$ V

Table 5 Input process variables

Parameters	Symbol	Level 1	Level 2	Level 3
Ultrasonic power (W)	*A*	70	80	90
Frequency (KHz)	*B*	21.5	22.5	25.5
Grit size	*C*	30	46	60

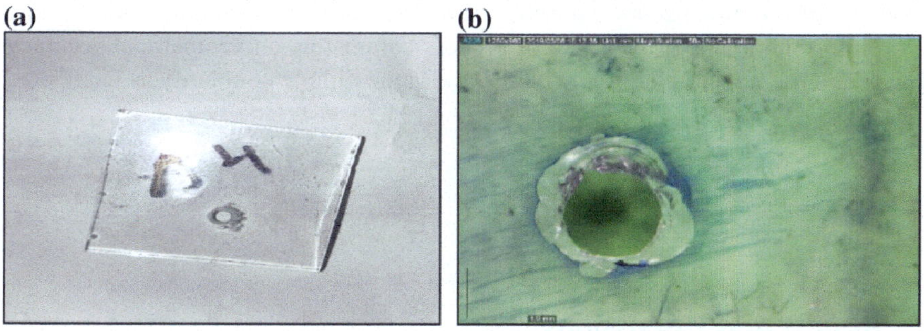

Fig. 2 **a** Actual drilled hole and **b** microscopic image of drilled hole

The workpiece of dimension 25 mm × 25 mm was used to drill hole as shown in Fig. 2a. To determine the MRR, the weight of the workpiece before and after performing the drilling operation were measured using the electronic weighing balance (ACZET make) of maximum weighing capacity of 220 g and minimum readability of 0.0001 g. The hole dimension was measured using digital microscope (Dino-Lite make). The image of the drilled hole is shown in Fig. 2b.

3 Methodology

Response surface methodology (RSM) is an experimental plan which helps us to devise an economical procedure for collecting data. The collected data is further used to determine a multiple regression equation. Thus, it helps to describe the functional relationship between a predicted response and the experimental variables. It also assists us in developing experimental designs to estimate response surface polynomials in the most efficient, economical, and flexible manner. Here, RSM is used to develop the experimental model of MRR using Box–Behnken approach. It is a second-order regression model to establish the relation between the input and output using design of experiment (DOE) approach. Box–Behnken approach requires only three levels to run an experiment. It does not contain any points at the vertices of the experiment region. This approach is particularly advantageous when the points on the corners of the cube represent level combinations that are prohibitively expensive or impossible to test because of physical process constraints. Using the Box–Behnken approach, 15 experiments were performed for three input parameters at three levels. The weight of the workpiece before and after the machining was measured using scientific weighing balance, and MRR was calculated using Eq. (1).

$$\text{MRR} = \frac{\text{initial weight} - \text{final weight}}{\text{time}} \text{g/min} \tag{1}$$

Hole circularity at entry (C_{entry}) was calculated as the ratio of minimum to maximum diameter of hole at entry side, using digital microscope. Since most of the geometrical variation occurs at the entry side, circularity at the entrance will provide good information about the effect of process parameters on geometrical accuracy. The geometrically circular hole will give the value of hole circularity as one.

$$C_{\text{entry}} = \frac{d_{min}}{d_{max}} \tag{2}$$

4 Results and Discussions

The experimental test result obtained from experimental planning using Box–Behnken approach of RSM for MRR and hole circularity is shown in Table 6.

The mathematical model as shown in Eqs. (3) and (4) has been obtained using RSM approach to correlate the various input parameters, i.e., ultrasonic power, ultrasonic frequency, and grit size with MRR and hole circularity through Minitab software.

$$\begin{aligned}\textbf{MRR} = {} & 0.223 - 0.00119\,A + 0.0233\,B - 0.000020\,C + 0.000009\,A * A \\ & - 0.000476\,B * B + 0.000004\,C * C - 0.000007\,A * B \\ & - 0.000001\,A * C\,0.000009\,B * C\end{aligned} \tag{3}$$

$$\begin{aligned}\textbf{HOLE CIRCULATORY} = {} & -1.26 - 0.02251\,\text{A} + 0.2428\,\text{B} \\ & + 0.00761\,\text{C} + 0.000117\,\text{A} * \text{A} \\ & - 0.00503\,\text{B} * \text{B} - 0.000071\,\text{C} * \text{C} \\ & + 0.000119\,\text{A} * \text{B} + 0.000044\,\text{A} * \text{C} \\ & - 0.000229\,\text{B} * \text{C}\end{aligned} \tag{4}$$

Table 6 Experimental plan showing the input values and observe responses

S. No.	Ultrasonic power	Frequency	Abrasive grit no.	Initial weight	Final weight	MRR	Hole circularity at entry
1	70	21.5	46	3.8831	3.8497	0.00209	0.9338
2	90	21.5	46	3.593	3.5579	0.00289	0.9365
3	70	25.5	46	3.7571	3.721	0.00188	0.9450
4	90	25.5	46	3.7043	3.6717	0.00161	0.9670
5	70	22.5	30	3.5367	3.4935	0.00248	0.9385
6	90	22.5	30	3.2972	3.2579	0.00213	0.9533
7	70	22.5	60	3.3904	3.3529	0.00708	0.9123
8	90	22.5	60	3.3071	3.2806	0.00624	0.9548
9	80	21.5	30	3.2894	3.2505	0.00138	0.9100
10	80	25.5	30	3.831	3.7891	0.00144	0.9470
11	80	21.5	60	3.7647	3.7373	0.00269	0.9036
12	80	25.5	60	3.5193	3.4794	0.00300	0.9136
13	80	22.5	46	3.7313	3.7004	0.00203	0.9425
14	80	22.5	46	3.5053	3.4691	0.00232	0.9340
15	80	22.5	46	3.3928	3.3519	0.00393	0.9540

where *A*, *B*, and *C* are the ultrasonic power, frequency, and grit size, respectively. The statistical probability of regression model was displayed through *F*-value and *p*-value as shown in Tables 7 and 8. The analysis of variance (ANOVA) of the response surface of quadratic model is also shown in Tables 7 and 8 for MRR and hole circularity, respectively.

Analysis of variance (ANOVA), *p*-test, and *F*-test have been performed to evaluate the adequacy of the developed model for MRR (Table and hole circularity at entrance. Since the experiment has been performed at 95% confidence level, therefore, *p*-value of less than 0.05 indicates the "significance" of the considered parameter and vice versa. ANOVA table of MRR shows the *p*-values of the model and liner effects as 0.032 and 0.023 which are quite smaller than 0.05. Therefore, the developed second-order regression model is significant and the linear effect of the process parameters is also significant. Similarly, the *p*-values of the model, linear effect, and square effect of model for circularity are quite smaller than 0.05 which again corroborate the significance of the model with linear and square effect.

The adequacy of the developed model can also be checked by analyzing the *p*-values of the "Lack-of-Fit" in the ANOVA tables for both parameters. *P*-values of higher than 0.05 will indicate that the respective terms is insignificant." Therefore, the obtained higher *p*-values of MRR and hole circularity indicate the adequacy of the developed model. R^2 (coefficient of determination) at 82.23 and 91.59% with R^2 (adj) value of 50.25% and 76.47%, respectively, for MRR and hole circularity indicates that the capability of independent variable to predict the variance of output.

Table 7 ANOVA result for MRR

Source	Contribution (%)	*F*-value	*P*-value
Model	91.50	5.98	0.032
Linear	51.88	8.03	0.023
UP	4.75	1.10	0.343
Frequency	5.88	0.72	0.436
Grit no.	41.26	22.37	0.005
Square	26.02	5.11	0.056
UP * UP	5.67	3.24	0.132
Frequency * frequency	12.44	6.46	0.052
Grit no. * grit no.	7.91	4.66	0.083
2-way interaction	13.60	2.67	0.159
UP * frequency	2.90	1.81	0.236
UP * grit no.	10.67	6.28	0.054
Frequency * grit no.	0.02	0.01	0.915
Error	8.50		
Lack-of-fit	5.70	1.36	0.451
Pure error	2.80		
Total	100.00		

Table 8 ANOVA result for hole circularity at entry

Source	Contribution (%)	*F*-value	*P*-value
Model	91.59	6.05	0.031
Linear	37.53	9.94	0.015
Up	16.86	9.68	0.027
Frequency	11.44	12.20	0.017
Grit no.	9.22	8.22	0.035
Square	45.75	9.12	0.018
UP * up	14.13	6.07	0.057
Frequency * frequency	12.85	9.00	0.030
Grit no. * grit no.	18.78	11.17	0.021
2-way interaction	8.32	1.65	0.291
UP * frequency	0.55	0.30	0.605
UP * grit no.	3.50	2.08	0.209
Frequency * grit no.	4.26	2.54	0.172
Error	8.41		
Lack-of-fit	4.36	0.72	0.626
Pure error	4.04		
Total	100.00		

The lower S-value for both the output variables as shown in Tables 9 and 10 indicates that the data is well fitted in the developed model.

The combined effect of various input parameters on MRR and hole circularity has been depicted though response surface plots. Figure 3a depicts that with the increase of grit no., the MRR increases very rapidly for all values of ultrasonic power. The comparative effect of ultrasonic power on MRR is not significant and it shows a nonlinear variation with a reduction in MRR at initial value but with the increase of power, it again shows slight increment in MRR. Similarly, the effect of ultrasonic power on hole circularity is not very dominant. As the grit no. increases, the effect of power on circularity shows slight decreasing and then increasing trend (Fig. 3b). As the grain size decreases (grit no. increases), the circularity initially rises but very small grain size decreases the circularity. The magnitude of variation in circularity due to increase of grit number decreases at high ultrasonic power.

It can be observed from Fig. 4a, b that the comparative effect of frequency on both MRR and circularity is more compared to grit no. With the decrease of grain size, the circularity decreases but the MRR increases at a particular frequency. Both the figures depict that up to 23 kHz both the MRR and circularity increase after which it begins to decrease. Similarly, up to grit no. 40, the MRR decreases, whereas circularity increases. The combined effect of ultrasonic power and frequency on MRR and hole circularity has been elaborated in Fig. 5a, b, respectively. The MRR increases till 23 kHz after which it shows a decline trend. For all range of frequency, the MRR shows a decline trend up to 80 and later on it gradually increases. The frequency

Table 9 Model summary for MRR

S	R-sq (%)	R-sq(adj) (%)
0.0011821	82.23	50.25

Table 10 Model summary for hole circularity at entry

S	R-sq (%)	R-sq(adj) (%)
0.0091534	91.59	76.47

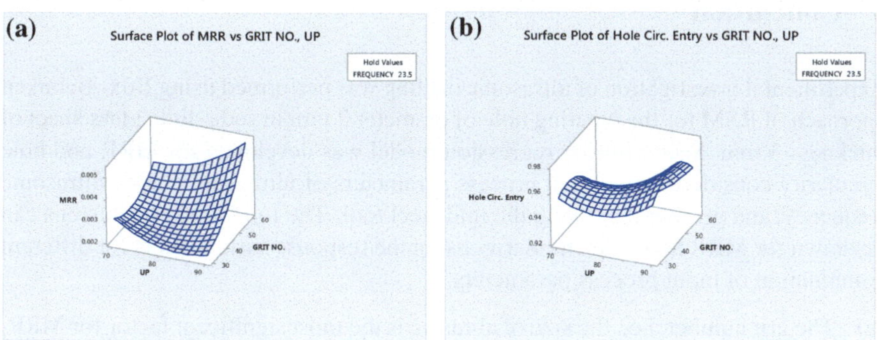

Fig. 3 Effect of ultrasonic number on **a** MRR and **b** circularity

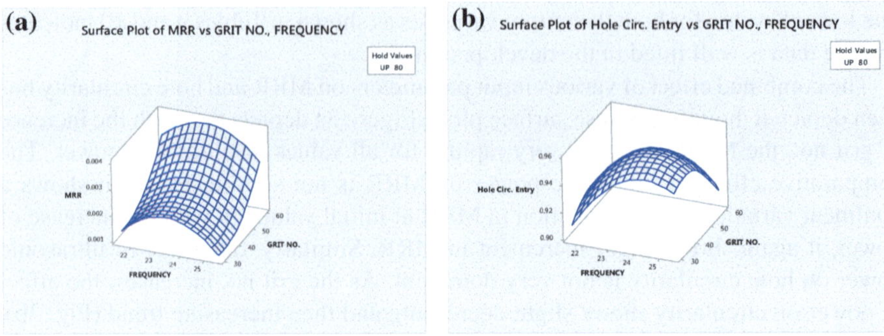

Fig. 4 Effect of frequency and abrasive grit number on **a** MRR and **b** circularity

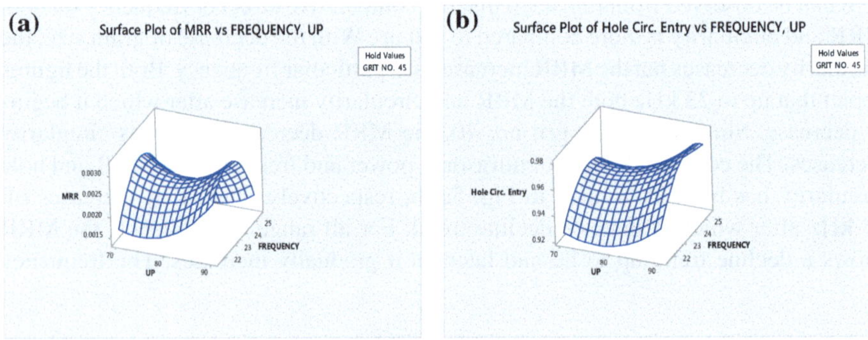

Fig. 5 Effect of ultrasonic power and frequency on **a** MRR and **b** circularity

has more dominant impact on the circularity with an increasing trend. The rate of variation of frequency on circularity increases with the increase of ultrasonic power.

5 Conclusion

Experimental investigation of ultrasonic drilling was performed using Box–Behnken approach of RSM for the creating hole of diameter 2 mm in soda–lime glass sheet of thickness 3 mm. Second-order regression model was developed for MRR and hole circularity considering the input process parameters of ultrasonic power, ultrasonic frequency, and grit number using the mid steel tool. The following conclusions can be drawn for MRR and hole circularity using the response surface plots for different combination of input process parameters.

(a) The grit number, i.e., the size of abrasive is the most significant factor for MRR, whereas hole circularity is mostly effected by ultrasonic frequency followed by ultrasonic power and grit size.

(b) Abrasive with high grit number increases the MRR but abrasives beyond grit no. 40 reduces the MRR. The interactive effect of ultrasonic power and grit number also play vital role in MRR.
(c) The combined analysis of grit number and frequency shows that both MRR and circularity are highly affected by frequency, and up to 23 kHz, both the parameters increase after which it declines. Both these parameters show contrasting effect after grit size number 40.
(d) At a particular ultrasonic power, the increase of frequency continuously increases the hole circularity but in case of MRR it begins to decrease after 23 kHz. The effect of ultrasonic power on both MRR and hole circularity is nonlinear gradual decrease up to 80 W further increase afterward.

References

1. Kataria, R., Kumar, J., Pabla, B.S.: Experimental investigation and optimization of machining characteristics in ultrasonic machining of WC-Co composite using GRA method. Mater. Manuf. Process. 1–9 (2015)
2. Zhang, C., Feng, P., Zhang, J., Wu, Z., Yu, D.: Theoretical and experimental research on the features of cutting force in rotary ultrasonic face milling of K9 glass. Appl. Mech. Mater. Vols. Trans. Tech. Publ. **158**, 1674–167 (2012)
3. Cong, W.L., Pei, Z.J., Treadwell, C.: Preliminary study on rotary ultrasonic machining of CFRP/Ti stacks. Ultrason. **54**(6), 1594–1602 (2014)
4. Zhang, C., Cong, W., Feng, P., Pei, Z.: Rotary ultrasonic machining of optical K9 glass using compressed air as coolant: a feasibility study. Proc. Inst. Mech. Eng. Part B: J. Eng. Manuf. **228**(4), 504–514 (2014)
5. Churi, N.J., Pei, Z.J., Treadwell, C.: Rotary ultrasonic machining of titanium alloy: effects of machining variables. Mach. Sci. Technol. **10**(3), 301–321 (2006)
6. Zhang, C.L., Feng, P.F., Wu, Z.J., Yu, D.W.: An experimental study on processing performance of rotary ultrasonic drilling of K9 glass. Adv. Mater. Res. **230**, 221–225 (2011)
7. Lv, D., Huang, Y., Tang, Y., Wang, H.: Relationship between subsurface damage and surface roughness of glass BK7 in rotary ultrasonic machining and conventional grinding processes. Int. J. Adv. Manuf. Technol. **67**(1–4), 613–622 (2013)
8. Cong, W.L., Feng, Q., Pei, Z.J., Deines, T.W., Treadwell, C.: Rotary ultrasonic machining of carbon fiber-reinforced plastic composites: using cutting fluid versus cold air as coolant. J. Compos. Mater. **46**(14), 1745–1753 (2011)
9. Cong, W.L., Pei, Z.J., Feng, Q., Deines, T.W., Treadwell, C.: Rotary ultrasonic machining of CFRP: a comparison with twist drilling. J. Reinf. Plast. Compos. **31**(5), 313–321 (2012)
10. Ya, G., Qin, H.W., Xu, Y.W., Zhang, Y.S.: An experimental investigation on rotary ultrasonic machining. Key Eng. Mater. **202**, 277–280 (2001)
11. Zou, X., Cong, W., Wu, N., Pei, Z.J., Wang, X.: Novel fiber optic sensors and their application on cutting temperature measurement in rotary ultrasonic machining of titanium. In: Proceedings of the AME/ISCIE International Symposium on Flexible Automation, Missouri, USA (2012)
12. Hu, P., Zhang, J.M., Pei, Z.J., Treadwell, C.: Modeling of material removal rate in ultrasonic machining: designed experiments. J. Mater. Process. Technol. **129**, 339–344 (2002)
13. Kumar, V., Singh, H.: Investigation of hole quality in rotary ultrasonic drilling of borosilicate glass using RSM. J. Braz. Soc. Mech. Sci. Eng. **41**, 36 (2019)
14. Sindhu, D., Thakur, L., Chandna, P.: Parameter optimization of rotary ultrasonic machining on quartz glass using response surface methodology (RSM), Silicon (2019). https://doi.org/10.1007/s12633-019-00160-2

15. Murti, V.S.R., Philip, P.K.: A comparative analysis of machining characteristics in ultrasonic assisted EDM by the response surface methodology. Int. J. Prod. Res. (2007). http://dx.doi.org/10.1080/00207548708919838
16. Hanrahan, G., Lu, K.: Crit. Rev. Anal. Chem. **36**, 141 (2006)

Experimental Investigation on Chemical-Assisted AISI 52100 Alloy Steel Using MAF

Ankita Singh, Swati Gangwar and Rajneesh Kumar Singh

Abstract Surface finish is a crucial aspect to decide the quality of a manufactured product. With the technological advancement, alloy steel having high strength and hardness is preferred over conventional materials. In some cases, a very fine surface of alloy steel is needed which cannot be done by conventional method. In the current investigation, surface finishing of AISI 52100 steel has been done using magnetic abrasive finishing assisted by chemical oxidiser. Hydrogen peroxide is used as an oxidising agent. The effect of input parameters such as voltage, tool rotation, finishing time has been investigated on surface roughness output parameter. Experiments have been performed and analysed using response surface methodology (RSM), and to obtain the significant factors, ANOVA was used, while empirical model was developed to examine the surface roughness (R_a) in terms of significant process variables.

Keywords Response surface methodology (RSM) · Chemically treated AISI 52100 alloy steel · ANOVA · Magnetic abrasive finishing (MAF)

1 Introduction

Due to the advancement of manufacturing technology, the feasibility of fine surface finish with improved surface finish can be an attribute for a large range of industrial applications [1]. It is very difficult to finish the advanced materials like steels, ceramics, and non-ferrous metal using conventional methods like honing, lapping, and super finishing to a higher extent. This causes the advance finishing technology to come into existence. Magnetic abrasive finishing (MAF) is one of the dominant super finishing methods to produce a surface with minimum surface roughness [2].

In MAF, the work metal surface is finished by removing the materials in microchips forms with the help of magnetic field produced in finishing zone. The working gap between tool and workpiece is filled with a mixture of abrasive powder

A. Singh (✉) · S. Gangwar · R. K. Singh
Department of Mechanical Engineering, Madan Mohan Malaviya University of Technology, Gorakhpur, Uttar Pradesh 273010, India
e-mail: Singhankita53@gmail.com

D. Dutta and B. Mahanty (eds.), *Numerical Optimization in Engineering and Sciences*, Advances in Intelligent Systems and Computing 979,
https://doi.org/10.1007/978-981-15-3215-3_50

and ferromagnetic particles [3]. Chemical mechanical polishing (CMP) integrates the mechanical and chemical forces to finish the surface in which combination of chemical etches over workpiece and abrasive polishing. During CMP process, an abrasive and chemical slurry is utilised along with free abrasive polishing. This abrasive slurry was imposed on work material surface so that a soft oxide layer forms on the surface and can be easily removed in comparison to parent material due to softness [4].

El-Taweel used the combination of magnetic abrasive finishing along with electrochemical turning that leads to reduction in surface roughness and increase in material removal rate. 6061 Al/Al_2O_3 was used as workpiece. After adding both processes results in increment of machining efficiency by 147% and 33% increment in surface roughness (R_a) in MAF [5].

Sihag et al. analysed finishing characteristics for Tungsten work piece experimentally through chemical-assisted magnetic abrasive finishing. The experimentation was planned by Taguchi L9 orthogonal array [6].

Mulik and PM Pandey introduced a new hybrid process called ultrasonic-assisted magnetic abrasive finishing (UAMAF). This technique combines the ultrasonic vibrations and magnetic abrasive finishing (MAF) method for finishing of the surfaces to nano-range within a limited time period. The surface roughness observed was as small as 22 nm in 80 s on a (AISI 52100) hardened steel by UAMAF [7].

Thus, the present article aimed to investigate the effect of process variables such as voltage, finishing time, and tool rotation on surface roughness for chemical-assisted magnetic abrasive finishing of AISI 52100 alloy steel.

2 Experimental Detail

A series of experiments have been done according to response surface methodology (RSM). The particulars about the workpiece, experimental set-up, selection of input process parameters, and their working range, design of experiment have been discussed in the following sections.

2.1 Work Material

There is a growing demand for alloy steels which give an excellent performance within severe conditions. Being a high carbon-chromium steel, AISI 52100 alloy steel has broad areas of applications. Thus, in the present work, AISI 52100 alloy steel has been selected as workpiece material in all the experiments; 60 mm × 60 mm × 3 mm dimensions of workpiece have been considered. The properties of AISI 52100 are (Table 1):

Table 1 Workpiec composition of AISI 52100 [8]

Elements	Chemical composition
C	0.95–1.05
Cr	1.35–1.60
Mn	0.25–0.45
Si	0.15–0.35
Ni	0.25
Fe	Rest

2.2 *Experimental Set-up*

In the current study, all the experimental trials have been performed on the MAF set-up. The set-up comprises an electromagnet mounted over drill machine to provide rotary motion to electromagnet. The electromagnet consists of copper windings wounded on a circular rod of mild steel having a diameter of 4 cm. The electromagnet generates a magnetic field intensity ranging from 0.2 to 0.9 T between the tool and workpiece surface when it was connected to a DC power supply (1–5 V). For the generation of oxide layer over the workpiece surface, H_2O_2 was used as chemical oxidiser. Before the experimentation, the workpiece was dipped in oxidiser for half an hour. Finishing gap taken was 1.5 mm. A blend of ferromagnetic (iron) particles (300 in mesh number) and SiC abrasive powder (400 mesh number) was filled in the working gap between magnet and workpiece surface which leads to the formation of magnetic abrasive brush by which finishing action was achieved. The experimental set-up and formation of brush are shown in Figs. 1 and 2, respectively.

2.3 *Selection of Process Parameters and Their Range*

In the current work, the effect of several input factors such as voltage, tool rotation (RPM), and finishing time on the response, namely surface roughness, has been studied. The levels of the input variables have been chosen based on the past literature, pilot experimentations, and machine capability. The input variables and their levels are shown in Table 2.

2.4 *RSM for Parameter Design*

In current article, response surface methodology's Box–Behnken design has been utilised to carry out the experiments and to develop empirical relationship between input process variables and output. Box and Wilson introduced RSM to study the statistical design in experimentations. In RSM, relationship is established between

Fig. 1 MAF experimental set-up

the independent process parameters and response which is given in Eq. 1.

$$Y_i = f(a_1, a_2, a_3, \ldots a_k) \pm e \tag{1}$$

where Y_i is the output and f is the function of response. a_1, a_2, a_3,…a_k are the input variables, and e is the residual measures experiment error. A second-order regression equation has been used to approximate response function.

The two methods of design in RSM are, namely, central composite design (CCD) and Box–Behnken design (BBD). CCD is used during sequential experimentations; it permits five levels of each parameter containing combinations with high and low level for factors simultaneously. While in BBD, it proposed three levels of parameters only and at same time with a fewer number of experimental runs as compared to CCD for same numbers of input variables. The uses of BBD are confined to a condition when predictions of ultimate response are not required. In the present article, a total of 15 experiments have been performed as per the experimental plan. MINITAB 18 software has been used to generate and analyse the design. The experimental plan with input variables along with output is shown in Table 3.

Fig. 2 Formation of magnetic abrasive brush

Table 2 Process parameters and levels for design of experiments

Symbol	Process parameters	Level		
		−1	0	1
X_1	Voltage	1	3	5
X_2	Tool rotation	100	200	300
X_3	Finishing time	30	45	60

2.5 Process Variables

The difference in surface roughness has been taken as a process response for evaluation. The difference in surface roughness was calculated by the following formula:,

$$\Delta R_a = \text{Initial roughness} - \text{Final roughness}$$

where

Table 3 BBD-based experimental design matrix and response

S. No.	Input parameters			Response
	Voltage (V)	Tool rotation (rpm)	Finishing time (min)	Difference in surface roughness (ΔR_a) (μm)
1	1	100	45	0.3834
2	5	100	45	0.2500
3	1	300	45	0.1867
4	5	300	45	0.1000
5	1	200	30	0.3800
6	5	200	30	0.3200
7	1	200	60	0.2000
8	5	200	60	0.3833
9	3	100	30	0.2534
10	3	300	30	0.4934
11	3	100	60	0.5800
12	3	300	60	0.1533
13	3	200	45	0.2300
14	3	200	45	0.1133
15	3	200	45	0.1833

ΔR_a = difference in surface roughness. The initial and final R_a was measured by Handysurf.

3 Results and Discussions

The empirical model to correlate surface roughness and input variables, i.e. voltage, rotation of electromagnet, and finishing time, is shown in the following regression equation (uncoded units).

$$\begin{aligned}\text{SR diff} = {} & 1.182 - 0.1129\,X_1 + 0.00208\,X_2 - 0.0420\,X_3 \\ & + 0.00066\,X_1X_2 + 0.000005\,X_2^2 + 0.000634\,X_3^2 \\ & + 0.000058\,X_1X_2 + 0.00203X_1X_3 - 0.000111\,X_2X_3\end{aligned}$$

To check whether the model is well fitted with the data or not, ANOVA results are determined. The ANOVA result for surface roughness is shown in Table 4.

Table 4 ANOVA result for surface roughness

Source	DF	Adj SS	Adj MS	*F* value	*P* value
Model	9	0.247129	0.027459	5.25	0.041
Linear	3	0.038855	0.012952	2.47	0.176
Voltage	1	0.001171	0.001171	0.22	0.656
RPM	1	0.035564	0.035564	6.79	0.048
Time	1	0.002119	0.002119	0.40	0.553
Square	3	0.081808	0.027269	5.21	0.054
Voltage * voltage	1	0.000026	0.000026	0.00	0.947
RPM * RPM	1	0.009925	0.009925	1.90	0.227
Time * time	1	0.075130	0.075130	14.35	0.013
Two-way interaction	3	0.126466	0.042155	8.05	0.023
Voltage * RPM	1	0.000545	0.000545	0.10	0.760
Voltage * time	1	0.014799	0.014799	2.83	0.154
RPM * time	1	0.111122	0.111122	21.23	0.006
Error	5	0.026172	0.005234		
Lack-of-fit	3	0.019272	0.006424	1.86	0.368
Pure error	2	0.006900	0.003450		
Total	14	0.273300			

3.1 Model Summary

S	*R*-sq (%)	*R*-sq(adj) (%)	*R*-sq(pred) (%)
0.0723487	90.42	73.19	0.00

From the ANOVA table, it is interpreted that the developed model has *f* value of 5.25 and *p* value of 0.041(less than 0.05) which shows that the model is significant. The adequacy of the model was confirmed further by high value of determination coefficient, i.e. R^2 equals to 90.42%. The *p* value for lack-of-fit is 0.368 which shows that it is insignificant at 95% confidence interval.

3.2 Interactive Effects of Inputs Parameters on Surface Roughness

3D response surface plots were obtained to depict the effect of variation in input parameters on surface roughness. From Fig. 3a, it is concluded that the difference in surface roughness value decreases as tool rotation increases up to a value of 250 rpm,

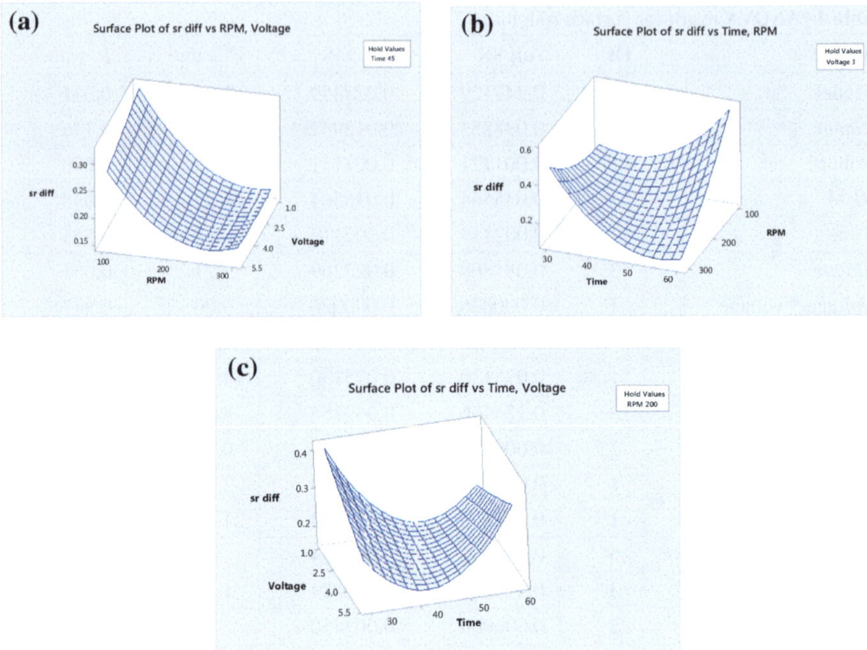

Fig. 3 **a** Effect of tool rotation and voltage on R_a, **b** effect of time and tool rotation on R_a, **c** effect of time and voltage on R_a

and thereafter, it starts increasing. Also, at smaller value of tool rotation, difference in surface roughness increases linearly with increment in voltage, and at the higher value, it increases at a very low rate linearly.

From Fig. 3b, it is inferred that at lower value of tool rotation, difference in surface roughness decreases with increase in finishing time up to a value of 55 min and then starts increasing. Also, at higher value of tool rotation, the difference in surface roughness decreases up to around 40 min and thereafter starts increasing with time. At lower value of finishing time, surface roughness decreases with increase in tool rotation, and at higher value of finishing time, roughness increases linearly with spindle speed. From Fig. 3c, it is concluded that difference in surface roughness firstly decreases with finishing time up to a value of 40 min and then starts increasing, and at lower value of finishing time, surface roughness increases linearly with voltage, and at a higher value of finishing time, it shows slight linear decrement.

3.3 Single Optimisation Through Response Surface Methodology

The range of input parameters, namely voltage, tool rpm, and finishing time and goal of response are shown in Table 5.

The aim of optimisation is to set difference in surface roughness as maximum. The optimisation plots obtained are shown in Fig. 4. Each cell of graph shows how response varies with one of factors while other factors remain fixed. The vertical and horizontal lines show current setting of parameters and current value if response respectively. At the top of column, high low and current setting of parameters within square brackets are given. The composite desirability value shown at left corner is 1. The current setting obtained is voltage of 5 V, tool rpm as 100, and finishing time as 60 min, and the optimised response, i.e. difference in surface roughness, as 0.6268. Once the optimal value of response is obtained, confirmatory experiments are performed for process improvement. The experimental result obtained from confirmatory experiments at current setting of parameters is 0.6542. The experimental error of 4.37% is obtained between predicted and experimental values.

Table 5 Range of parameters and output for desirability

Parameters	Goal	Lower value	Higher value	Weight	Importance
Voltage	In range	1	5	1	1
Tool rpm	In range	100	300	1	1
Finishing time	In range	30	60	1	1
Difference in surface roughness	Maximise	0.1	0.58	1	1

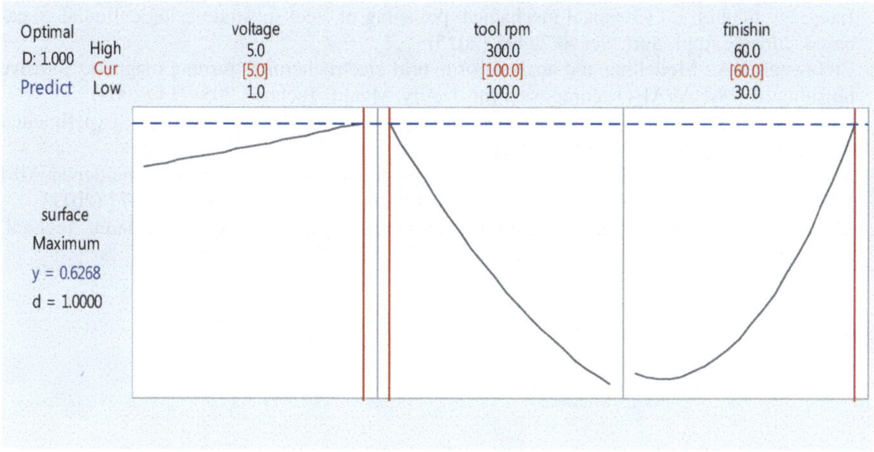

Fig. 4 Optimization plots

4 Conclusion

The predicted values and the experimental values match reasonably well with R^2 of 90.42% for surface roughness which shows that the model is fitted well. For modelling of machining process variables, response surface methodology can be applied successfully to achieve information with a fewer number of experiments. 3D surface plots drawn explain the better understanding of interactions of input factors on the response. The following results were concluded:

1. The difference in surface roughness value shows decrement with an increase in tool rotation with any value of voltage.
2. Surface roughness shows a linear increase with tool rotation at higher finishing time, while at lower finishing time, it shows a slight decrement with tool rotation
3. At any value of voltage, surface roughness firstly decreases up to 40 min and then shows an increasing trend. At a lower value of finishing time and tool rotation, surface roughness increases linearly following a straight line.
4. The optimum value of response, i.e. difference in surface roughness, is obtained as 0.6268 at optimum values of 5 V, 100 rpm tool rotation, and 60 min finishing time with 4.37% error between predicted and experimental value obtained from confirmatory experiments.

References

1. Kumar, H., Singh, S., Kumar, P.: Magnetic abrasive finishing a review. Int. J. Eng. Res. Technol. (IJERT), 1–9 (2013)
2. Shinmura, T., Takazawa, K., Hatano, E.: Study on magnetic abrasive finishing. 325–328 (1990)
3. Jain, V.K.: Advanced Machining Processes. Allied Publisher Pvt. Ltd., New Delhi (2004)
4. Jiang, L., Jianbin, L.: Chemical mechanical polishing of steel substrate using colloidal silica-based slurries. Appl. Surf. Sci. 487–495 (2015)
5. El-Taweel, T.A.: Modelling and analysis of hybrid electrochemical turning magnetic abrasive finishing of 6061 Al/Al_2O_3 composite. Int. J. Adv. Manuf. Technol. 705–714 (2008)
6. Sihag, N., Kala, P., Pandey, P.M.: Chemo assisted magnetic abrasive finishing: experimental investigations. Proced. CIRP 39–43 (2015)
7. Mulik, R.S., Pandey, P.M.: Ultrasonic assisted magnetic abrasive finishing of hardened AISI 52100 steel using unbonded SiC abrasives. Int. J. Refract. Met. Hard Mater. 68–77 (2011)
8. Mulik, R., Pandey, P.M.: Magnetic finishing of hardened AISI 52100 steel. Int. J. Manuf. Technol. 501–515 (2011)

Modeling for Rotary Ultrasonic Drilling of Soda Lime Glass Using Response Surface Methodology

Ranjeet Kumar, Sanjay Mishra and Sanjeev Kumar Singh Yadav

Abstract Rotary ultrasonic drilling (RUD) is an efficient and accurate way to drill holes in hard and brittle materials. Material removal rate and hole circularity are the essential output performance characteristics of a drilled hole in these hard to cut materials. In the present article, an experimental investigation has been performed for RUD in 3 mm thick soda lime glass using response surface methodology. Mild steel tool of diameter 2 mm was used. Effect of ultrasonic power, frequency, spindle speed and abrasive grit number on the material removal rate and hole circularity has been explored. It is observed that with increase in frequency, both MRR and circularity decrease. Ultrasonic power significantly effect hole circularity but has less effect on MRR for different frequency. MRR increases with abrasive grit number but circularity is adversely affected. At a higher value of abrasive grit number, increase in tool rotation increases hole circularity.

Keywords Rotary ultrasonic drilling · Soda lime glass · Response surface methodology · Circularity

1 Introduction

Soda lime glass presents an exclusive application in optical components because of its superior properties like high thermal coefficient of expansion, high wear resistance and low fracture toughness. Because of its high hardness and strength, it is very difficult to machine this material. Aerospace, automobile industries, dental, fluidics and optics are wide areas of application of soda lime glass. Surface cracks and subsurface cracks are major demerits of conventional drilling methods which reduce strength of material and also increase edge chipping, and hence, RUD was introduced [1]. Rotary ultrasonic machining was originated by Mr. Percy Legge (1964) at the

R. Kumar (✉) · S. Mishra
Madan Mohan Malaviya University of Technology, Gorakhpur, Uttar Pradesh 273010, India
e-mail: rjranjeet1220@gmail.com

S. K. S. Yadav
Harcourt Butler Technical University, Kanpur, Uttar Pradesh, India

D. Dutta and B. Mahanty (eds.), *Numerical Optimization in Engineering and Sciences*, Advances in Intelligent Systems and Computing 979,
https://doi.org/10.1007/978-981-15-3215-3_51

United Kingdom Atomic Energy Authority (UKAEA); however, the idea of combining ultrasonic vibrations and drilling was carried out by Brown et al. [2]. RUD is a hybrid non-conventional machining process that incorporates the mechanism of material removal of ultrasonic machining and diamond grinding.

Zhang et al. perfomed RUD on optical K9 glass and observed that there was a significant increase in cutting power, decrease in thrust force and there was small effect on surface roughness. Further, the cutting force increases with increase in feed rate significantly and with spindle speed there is slight decrement. At higher speed and low feed, average surface roughness was reduced [3]. Ding et al. observed that cutting force, surface roughness and torque were reduced by 24%, 23%, 47%, respectively, in RUD of C/SiC as compared to conventional drilling [4]. During RUD of AI7075-T6 alloy, Amini et al. concluded that torque reduces with increase in spindle speed and decreases with reduction in ultrasonic vibrations [5]. Ahmed et al. [6] concluded that RUD produces much smoother surface than conventional drilling in alumina. Azarhoushang et al. [7] concluded that for Inconel 738-LC, increase in vibration amplitude decreases surface roughness. Jiao et al. [8] concluded that for C/SiC composites, feed rate was most significant factor which affects the surface roughness. Wang et al. [9] establish a model for prediction of edge chipping in case of brittle material considering brittle fracture and highest impact force of abrasives due to vibrations. Singh et al. utilized central composite rotatable design based on the response surface methodology for machining of quartz ceramics. The effect of rpm, coolant pressure, ultrasonic power and feed rate on material removal and chipping size was determined. It was found that the feed rate was most significant for any considered response. Chipping size was found to be minimum at low feed and high rpm. High feed and coolant pressure give best MRR [10]. Liu et al. carried out rotary ultrasonic drilling on ceramic materials, and central composite design of response surface methodology has been used to study the influence of design variables. It was found that rotary ultrasonic drilling produces high productivity and hole accuracy than static USM [11].

The literature review indicates that very few experimental investigations of RUD using RSM method has been performed. Therefore, in this study, an attempt has been made to study the effect of ultrasonic power, ultrasonic frequency, spindle speed and abrasive grit number on MRR and hole circularity at the entrance during RUD of 3 mm thick soda lime glass using mild steel tool.

2 Experimental Details

In the current investigation, a soda lime glass sheet with dimension $20 \times 20 \times 3$ mm was chosen as a work material. The chemical composition and mechanical properties of soda lime glass are shown in Tables 1 and 2, respectively.

All the experiments were carried using the RUD setup as shown in Fig. 1. There is a mild steel cutting tool of diameter 2 mm which is axially vibrated, and a constant feed is given to work material toward the tool. Black silicon carbide was used as an

Table 1 Chemical composition of soda lime glass

Element	SiO_2	CaO	Na_2O	MgO	Al_2O_3
Percentage %	70–74	5–14	10–16	0–6	0–3

Table 2 Mechanical properties of soda lime glass

Property	Density	Modulus of elasticity	Poisson ratio	Shear modulus	Knop hardness
Amount	2500 kg/m^3	72,000 N/mm^2	0.23	29,166 N/mm^2	6GPa

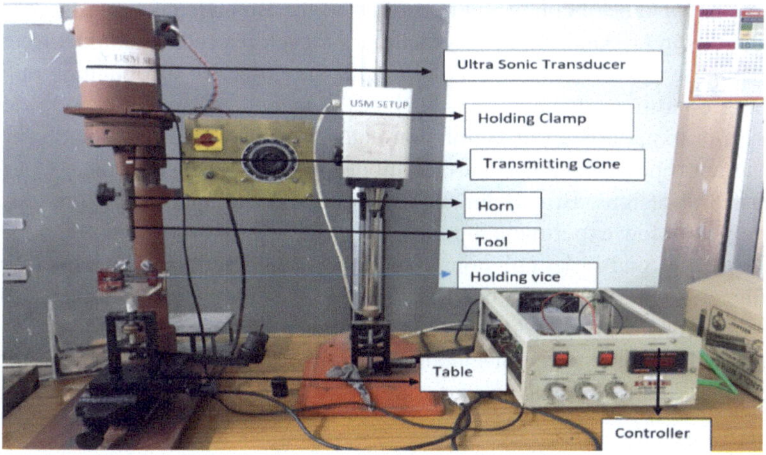

Fig. 1 Rotary ultrasonic drilling setup (HBTU, MED, Kanpur)

abrasive. Water has been used to make slurry of abrasive. A laser tachometer has been used to measure spindle speed and a digital multimeter to measure the frequency of vibration.

2.1 *Selection of Process Parameters and Their Range*

In the current investigation, the effect of several controllable input parameters, i.e., ultrasonic power, frequency, spindle speed and abrasive grit number on the response, namely material removal rate and hole circularity at entrance has been explored. The levels of input parameters have been selected on the basis of pilot experiments and machine capability (Table 3).

Table 3 Input parameters and their values

Variables	Symbol	Level (−1)	Level (0)	Level (1)
Ultrasonic power(%W)	A	70	80	90
Frequency(KHz)	B	21.5	22.5	25.5
Spindle speed(rpm)	C	750	850	950
Abrasive grit number	D	30	46	60

2.2 *RSM for Parameter Design*

Experiments have been conducted according to the Box–Behnken design of RSM. Box and Wilson (1951) utilized this method to develop a relationship between several machining parameters and response with the objective of obtaining optimum machining condition of the system [12]. RSM technique can be used to develop models from experimental or simulation data. It is the process of identifying and fitting an appropriate response surface model from experimental data. Compared to other RSM designs, Box–Behnken design (BBD) can analyze the process more effectively with few experimental points.

In RSM, a second-order polynomial model is considered to explore the influence of parameters on the responses. In BBD, three levels of input parameters are required. Furthermore, it is confined when extreme responses are not required to be predicted. BBD is much cheaper and convenient to run than CCD with the same number of input factors. The RSM analysis was conducted using MINITAB software.

$$y_i = \beta_0 + \sum_{i=1}^{k} \beta_i x_{iu} + \sum_{i=1}^{k} \beta_{ii} x_{iu}^2 + \sum_{i<j}^{k} \beta_{ij} x_{iu} x_{ju} + e_u \tag{1}$$

where y_i is the response, x_{iu} is the coded value of the ith input variable for uth experiment, n is the total number of the input factors, β_0 is a constant, β_i, β_{ii} and β_{ij} are the regression coefficients of second order, and e_u is the experimental error for uth observation [13]. A total of 27 experiments have been done as per the experimental design with four process variables, having three levels each. The BBD-based experimental design matrix and corresponding responses are shown in Table 4.

2.3 *Process Variables*

The weight of the workpiece before and after the machining was measured using scientific weighing balance with minimum readability of 0.0001 g and maximum weight capacity of 220 g and MRR were calculated using Eq. (2)

Table 4 BBD-based experimental design matrix and result

Exp. No.	Process parameters				Response	
	Ultrasonic power % W	Frequency (KHz)	Spindle speed (rpm)	Abrasive Grit No.	MRR (g/min)	Hole circularity at entry
1	70	21.5	850	46	0.0098	0.829
2	90	21.5	850	46	0.0156	0.881
3	70	25.5	850	46	0.0034	0.856
4	90	25.5	850	46	0.0026	0.831
5	80	22.5	750	30	0.0025	0.840
6	80	22.5	950	30	0.0048	0.899
7	80	22.5	750	60	0.0073	0.935
8	80	22.5	950	60	0.0076	0.809
9	70	22.5	850	30	0.0052	0.832
10	90	22.5	850	30	0.0040	0.868
11	70	22.5	850	60	0.0070	0.848
12	90	22.5	850	60	0.0154	0.856
13	80	21.5	750	46	0.0145	0.892
14	80	25.5	750	46	0.0026	0.827
15	80	21.5	950	46	0.0127	0.883
16	80	25.5	950	46	0.0036	0.874
17	70	22.5	750	46	0.0029	0.867
18	90	22.5	750	46	0.0026	0.898
19	70	22.5	950	46	0.0035	0.867
20	90	22.5	950	46	0.0024	0.866
21	80	21.5	850	30	0.0028	0.877
22	80	25.5	850	30	0.0028	0.865
23	80	21.5	850	60	0.0245	0.873
24	80	25.5	850	60	0.0051	0.879
25	80	22.5	850	46	0.0060	0.867
26	80	22.5	850	46	0.0054	0.876
27	80	22.5	850	46	0.0046	0.885

$$\text{MRR} = \frac{\text{initial weight} - \text{final weight}}{\text{time}} \text{gm/min} \tag{2}$$

Hole circularity at entry (C_{entry}) was calculated as the ratio of minimum to maximum diameter of drilled hole at entry side. Since most of the geometrical variation occurs at the entrance, circularity at the entry side will provide a good information about the effect of process parameters on geometrical accuracy. Dino lite's digital microscope was used to visualize the dimension of drilled hole. The geometrically

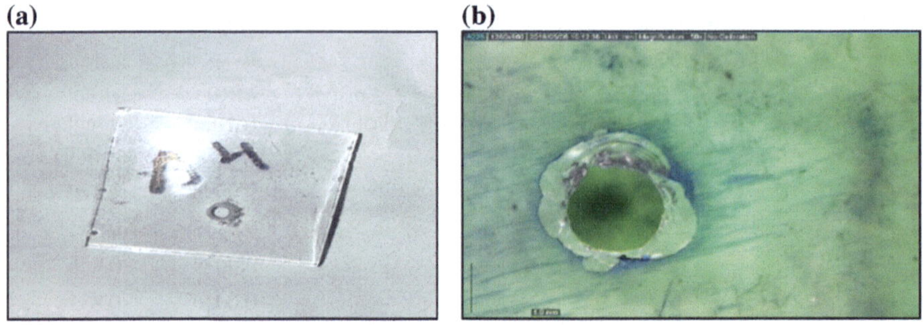

Fig. 2 Microscopic view of drilled hole

circular hole will give the value of hole circularity as one (Fig. 2).

$$C_{entry} = \frac{d_{min}}{d_{max}} \quad (3)$$

3 Results and Discussions

3.1 Mathematical Model

The mathematical model for correlating material removal rate and hole circularity at entry with various machining parameters as stated in experimental design is as follows:

MRR $= 0.804 + 0.00139\ A - 0.0797\ B + 0.000181\ C + 0.00097\ D - 0.000002\ A^2 + 0.001814\ B^2 - 0.000000\ C^2 + 0.000008\ D^2 - 0.000064\ A \times B - 0.000000\ A \times C + 0.000016 A \times D + 0.000002\ B \times C - 0.000105\ B \times D - 0.000000\ C \times D$

Hole circularity at entry $= -1.74 + 0.0538\ A + 0.019\ B - 0.00091\ C + 0.02889\ D - 0.000153\ A^2 - 0.00083\ B^2 + 0.000000\ C^2 - 0.000022\ D2 - 0.000857\ A \times B - 0.000008\ A \times C - 0.000046\ A \times D + 0.000093\ B \times C + 0.000119\ B \times D - 0.000030\ C \times D$

To test whether the model fits the data or not, S values for MRR and hole circularity are obtained as 0.0031888 and 0.0155357, respectively, which are low and values of R2 for both are 84.03% and 84.48%, respectively, which are high. So the data are well fitted in developed model. ANOVA tests have been done to test the adequacy of the developed empirical model for MRR and hole circularity. Tables 5 and 6 show the ANOVA result for MRR and hole circularity, respectively. The *p*-value of the models is much smaller than 0.05 for both responses. Developed regression model for MRR is significant and linear, and square terms are also significant. Regression model for

Table 5 ANOVA table for MRR

Source	Dof	AdjSS	AdjMS	*F*-value	*p*-value	Contribution (%)
Model	14	0.000642	0.000046	4.51	0.006	84.03
Linear	4	0.000341	0.000085	8.38	0.002	46.79
Ultrasonic power (A)	1	0.000001	0.000001	0.12	0.732	1.27
Frequency (B)	1	0.000280	0.000280	27.54	0.000	24.01
Rpm (C)	1	0.000001	0.000001	0.11	0.741	0.05
Abrasive grit no. (D)	1	0.000067	0.000067	6.61	0.025	21.45
Square	4	0.000205	0.000051	5.05	0.013	26.99
A^2	1	0.000000	0.000000	0.03	0.869	1.57
B^2	1	0.000142	0.000142	13.98	0.003	20.61
C^2	1	0.000007	0.000007	0.70	0.418	2.49
D^2	1′	0.000018	0.000018	1.74	0.212	2.32
2 way interaction	6	0.000078	0.000013	1.28	0.335	10.25
AB	1	0.000008	0.000008	0.76	0.401	0.98
AC	1	0.000000	0.000000	0.02	0.902	0.02
AD	1	0.000022	0.000022	2.17	0.166	2.89
BC	1	0.000001	0.000001	0.09	0.770	0.12
BD	1	0.000047	0.000047	4.58	0.054	6.09
CD	1	0.000001	0.000001	0.11	0.743	0.15
Error	12	0.000122	0.000010			15.97
Lack of fit	10	0.000121	0.000012	24.53	0.040	15.84
Pure error	2	0.000001	0.000000			0.13
total	26					100.00

hole circularity is significant, and the interaction of variables is also significant. Lack of fit for MRR and hole circularity have p-values of 0.040 and 0.250, respectively, represents that developed models are adequate at 95% confidence interval.

3.2 *Effect of Process Parameters on MRR and Hole Circularity*

The MRR during RUD decreases with frequency up to 24 kHz but at higher values it begins to increase at constant abrasive grit number. Similarly, the MRR increases with abrasive grit number for constant frequency. The increase of MRR with abrasive grit number is more rapid at lower frequency, whereas at higher frequency the effect of abrasive grit number on MRR decreases as depicted in Fig. 3a, b indicates that the

Table 6 ANOVA table for circularity

Source	Dof	AdjSS	AdjMS	*F*-value	*p*-value	Contribution (%)
Model	14	0.015760	0.001126	4.66	0.006	84.48
Linear	4	0.001146	0.000286	1.19	0.366	11.59
Ultrasonic power (A)	1	0.000084	0.000084	0.35	0.566	4.56
Frequency (B)	1	0.000912	0.000912	3.78	0.076	5.19
Rpm (C)	1	0.000093	0.000093	0.39	0.546	1.66
Abrasive grit no. (D)	1	0.000072	0.000072	0.30	0.595	0.18
Square	4	0.001831	0.000458	1.90	0.176	9.80
A^2	1	0.001250	0.001250	5.18	0.042	7.74
B^2	1	0.000030	0.000030	0.12	0.731	0.17
C^2	1	0.000105	0.000105	0.44	0.522	1.20
D^2	1′	0.000130	0.000130	0.54	0.478	0.70
2 way interaction	6	0.011768	0.001961	8.13	0.001	63.08
AB	1	0.001372	0.001372	5.68	0.034	7.41
AC	1	0.000256	0.000256	1.06	0.323	1.37
AD	1	0.000194	0.000194	0.81	0.387	1.04
BC	1	0.001629	0.001629	6.75	0.023	8.33
BD	1	0.000060	0.000060	0.25	0.628	0.32
CD	1	0.008321	0.008321	34.48	0.000	44.60
Error	12	0.002896	0.000241			15.52
Lack of fit	10	0.002734	0.000273	3.38	0.250	14.66
Pure error	2	0.000162	0.000081			0.87
total	26					100

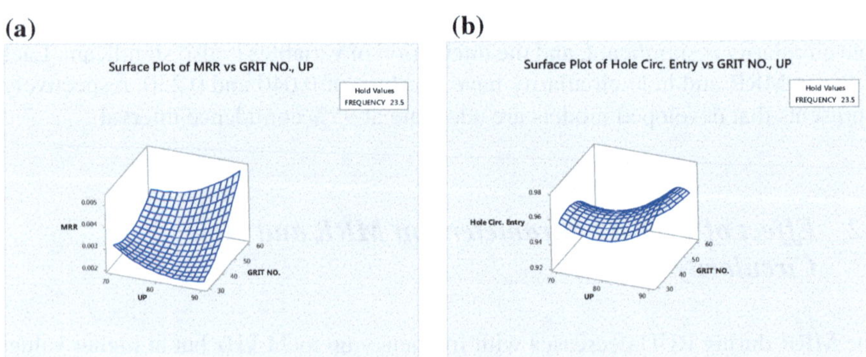

Fig. 3 Effect of frequency and grit number on **a** MRR **b** circularity

hole circularity continuously decreases with frequency for a particular abrasive grit number but the magnitude of effect decreases with increase of abrasive grit number. The increase of abarsive grit number increases the circularity as the finer size of abrasives will lead to less distortions and burrs (Fig. 3b).

The corresponding effect of ultrasonic power and frequency on MRR and circularity has been shown in Fig. 4a, b, respectively. The MRR decreases very rapidly with frequency for all values of ultrasonic power because contact ratio between workpiece and tool is larger. The ultrasonic power increases the MRR very gradually at different frequencies because as the value of ultrasonic power increases, the cutting depth of each abrasive will also increase. The rate of change of circularity with ultrasonic power is very high, and it begins to decrease with ultrasonic power after initial increase at higher frequency because brittle fracture phenemenon will become more dominant than workpicec crushability.

The effect of rotation of the tool and abrasive grit number has been shown in Fig. 5. The rotation of the tool has very minimal effect on MRR for particular abrasive grit number but at certain rotation, MRR increases rapidly with grit number because the mass of particle increases with grit number which produces larger impact over the

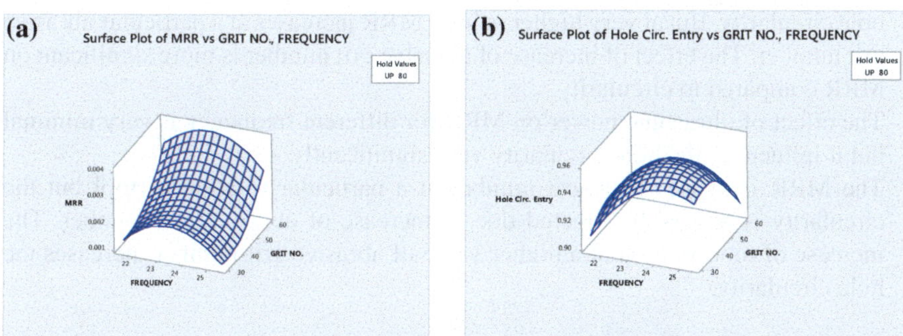

Fig. 4 Effect of frequency and ultrasonic power on **a** MRR **b** circularity

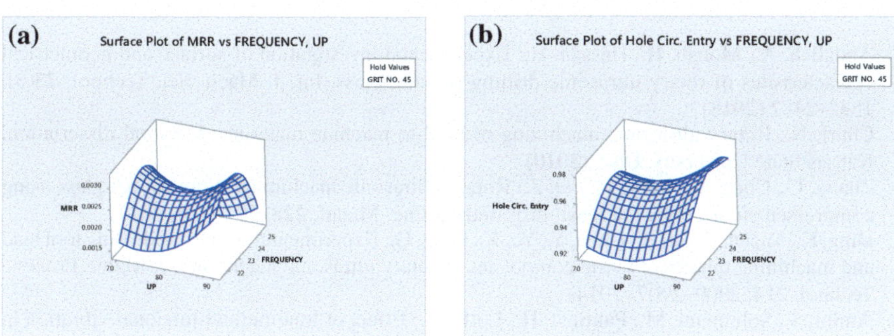

Fig. 5 Effect of grit number and spindle speed on **a** MRR **b** circularity

striking area. The hole circularity shows a reverse trend with respect to these two parameters. The increase of grit number decreases the circularity but the rotation of the tool increases the circularity significantly at higher abrasive grit number because the cutting forces reduced at higher speed.

4 Conclusion

Experimental investigation of rotary ultrasonic drilling was performed with Box–Behnken approach of RSM for creating hole of diameter 2 mm in soda lime glass sheet of thickness 3 mm. Second-order regression model was developed for MRR and hole circularity considering the input process parameters of ultrasonic power, ultrasonic frequency, abrasive grit number and rotation of the mid steel tool. The following conclusions can be drawn for MRR and hole circularity using the response surface plots for different combination of input process parameters.

1. Frequency is found to be the most contributing factor for both the responses for individual effects. The increase of frequency decreases for both the MRR and hole circularity. But at very higher values, MRR increases at a particular abrasive grit number. The effect of increase of abrasive grit number is more significant on MRR compared to circularity.
2. The effect of ultrasonic power on MRR for different frequency is very minimal but it influences the hole circularity very significantly.
3. The MRR increases with grit number at a particular rotation of tool but the circularity is adversely affected due to increase of abrasive grit number. The increase of tool rotation at a higher value of abrasive grit number increases the hole circularity.

References

1. Abdallah, A., Mohab, H., Hassan, H.: Experimental investigation of surface and geometrical characteristics of rotary ultrasonic drilling of soda glass. Int. J. Mach. Sci. Technol. **23**(3), 1532–2482 (2018)
2. Churi, N.: Rotary ultrasonic machining of hard-to-machine materials. Doctoral dissertation, Kansas State University, USA (2010)
3. Zhang, C., Cong, W., Feng, P., Pei, Z.: Rotary Ultrasonic machining of optical K9 glass using compressed air as coolent: a feasibility study. J Eng. Manuf. **228**, 504–514 (2014)
4. Ding, K., Yucan, F., Su, H., Chen, Y., Yu, X., Ding, G.: Experimental studies on drilling tool load and machining quality of C/Sic composites in rotary ultrasonic machining. J. Mater. Process. Technol. **214**, 2900–2907 (2014)
5. Amini, S., Soleimani, M., Paktinat, H., Lotfi, M.: Effect of longitudinal-torsional vibration in ultrasonic assisted drilling. Mater. Manuf. Process **32**, 616–622 (2017)
6. Ahmed, Y., Cong, W.L., Stanco, M.R., Xu, Z.G., Pie, Z.J., Treadwell, C., Zhu, Y.L., Li, Z.C.: Rotary ultrasonic machining of alumina dental ceramics: a preliminary experimental study on surface and subsurface damages. J. Manuf. Sci. Eng. **134**, 064501-1–0645501-5 (2012)

7. Azarhoushang, B., Akbari, J.: Ultrasonic assited drilling of Inconel738-LC. Int. J. Mach. Tools Manuf. **47**, 1027–1033 (2007)
8. Jiao, Y., Hu, P., Pie, Z.J.: Rotary ultrasonic machining of ceramics: design of experiment. Int. J. Manuf. Technol. Manag. **7**, 192–205 (2005)
9. Wang, J., Feng, P., Zhang, J., Zhang, C., Pie, Z.: Modelling of dependency of edge chipping size on the material properties and cutting force for rotary ultrasonic drilling of britte materials. Int. J. Mach. Tools Manuf. 18–27 (2016)
10. Singh, R.P., Singhal. S.: Experimental investigation of machining characteristics in rotary ultrasonic machining of quartz ceramic. J. Mater. Design Appl. 1–20 (2016)
11. Liu, J.W., Baek, D.K., Ko, T.K.: chipping minimization in drilling ceramic materials with rotary ultrasonic machining. Int. J. Adv. Manuf. Technol. **72**, 1527–1535 (2014)
12. Sun, Y., Li, T., Yan, J. Technology optimization for polysaccharides(pop) extraction from the fruiting bodies of pleurotus ostreatus by Box-behnken statistical design. Carbohydr. Polymers 242–247 (2010)
13. Kumar, A., Kumar, V., Kumar, J. Multi-response optimization of process parameters based on response surface methodology for pure titanium using WEDM process. Int. J. Adv. Manuf. Technol. 2645–2668 (2013)

Process Optimization of Digital Conjugate Surfaces: A Review

Pagidi Madhukar, Guru Punugupati, N. Selvaraj and C. S. P. Rao

Abstract Complex surface methods are widely used in various industrial tools such as cutting tools, shell of television, turbine blades etc. Researchers tend to concentrate on conjugate surfaces, which can be produced by digital method to optimize any process. Conjugate surface deals with relative motion. Most of the cases pertain to power transmitting devices such as gears. Providing conjugate surface is very difficult, and to overcome this problem, digital conjugate method has been introduced. It produces event complex conjugate surfaces. In such cases, researchers have focused on Digital Gear Tooth Surface (DGTS). It causes conjugate motion between the gear teeth, which is represented through discrete points, and with these discrete points, the digital surface of conjugate moments is resolved. Computer simulated examples generate and machine the non-standard and complex shapes of digital conjugate tooth surfaces. This technique is not only useful for machining discrete digital gear tooth surfaces and gear tooth surface with complex design, but also other 3D digital surfaces with optimized process.

Keywords Conjugate surface · Digital gear tooth surface · Complex surface

1 Introduction

Technology continues to improve every day according to the demand for modern precision machinery. In such machineries, gear components play a crucial role as power transmitting mechanism. Therefore, to meet the demand and manufacturing of these components requires complicated analysis and treatment of surface engagement. Computer technology helps to develop discrete treatment method.

General purpose CAD/CAM system are relatively mature in modeling various complex surface shapes but they are quite weak in further treatment of complex

P. Madhukar (✉) · G. Punugupati · N. Selvaraj
Department of Mechanical Engineering, National Institute of Technology, Warangal, Telangana 506004, India
e-mail: pmadhu88@gmail.com

C. S. P. Rao
National Institute of Technology, Andhra Pradesh, Tadepalligudem, Andhra Pradesh, India

D. Dutta and B. Mahanty (eds.), *Numerical Optimization in Engineering and Sciences*, Advances in Intelligent Systems and Computing 979,
https://doi.org/10.1007/978-981-15-3215-3_52

surface models. Nowadays, direct machining is being generated using algorithms via two techniques. One of them is Area by Area Machining [1, 2]. According to this technique, typical complex surface assigns are separated from not so complex components and a large or small tool is selected respectively. The second technique involves intersection of planes on the surface to reach the above mentioned objective. For machining of parallel plane strategy, *Z*-map model is used [3]. The intersect of mean least square surface (MLS) is also used to drive planes [4].

Modern engineering technology is encouraging machine-building aligning towards synthesis, digitization and intelligence. This accelerates the research on generating method of digitized surface. The generating digitized surface problem solving depends on the digitized conjugate surface theory [5]. Theory of conjugate surface is a novel technology which examines mutual conversions and the relationships between paired motions and paired geometric graphs under conditions of mechanical transmission and machinery processing. It involves mechanics, mechanism, differential geometry, and can be used for the design of cutting tool contour, cam contour as well as gear tooth face. It can also be used for the synthesis of mechanism, motion analyses, processing simulation and so on.

1.1 Conjugate Surface Concept

To study the theory of conjugate surface, both analytic and digitized surfaces are important and conjugate relationship as well as conditions play a critical role (Fig. 1).

Σ1 and Σ2 are two arbitrary conjugate surfaces, If Σ1 is considered as mother surface,

*S*1 (*O*l *X*1 *Y*l *Z*l) and *S*2 (*O*2 *X*5 *Y*2 *Z*2) are the two coordinates.

Surfaces 1 and 2 are associated at point *M* where *M*1 = M2

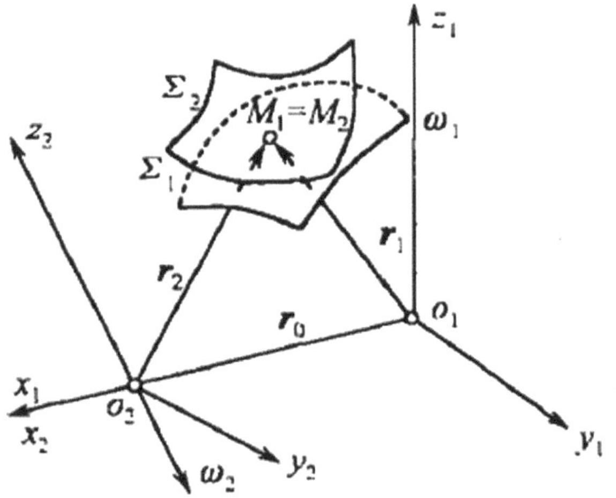

Fig. 1 Two surfaces of conjugate body motion

Position vectors are $r1$ and $r2$

Normal vectors are $n1$ and, $n2$

Surface $S1$ is connected with Σ1, and surface $S2$ with Σ2. Similarly, position vectors $r1$ point to Σ1 and $r2$ point to Σ2. Unit normal vectors $n1$ have point rl on surface Σ1 and $n1$ point $r2$ on surface Σ2 and Σ1 considered as mother surface, the equation may expressed with u & v; then $r1 = r1(u, v)$

Surface Σ1 changes as per law, ɵ is dependent parameters, and it can form a series of surfaces (surface cluster). The equation of these surfaces as followed:

$$\mathrm{r1} = \mathrm{r1}(\mathrm{u\ v}\ \Theta)$$

u & v are the mother surface geometrical parameters, and ɵ is the motion parameter. Figure 2 shows the conjugate profiles and Fig. 3 shown the various applications of conjugate surfaces.

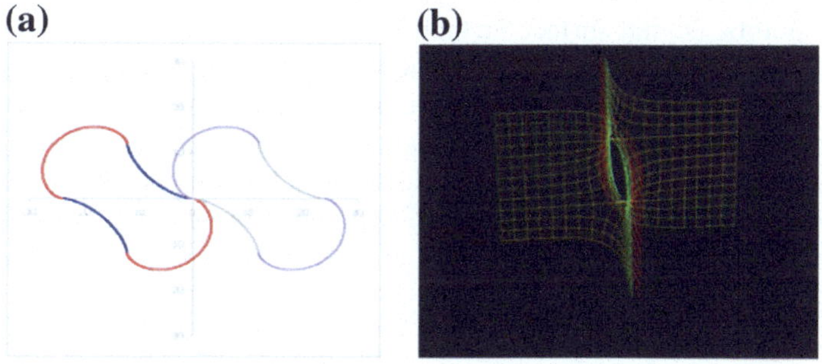

Fig. 2 **a** Conjugate motion of a gear profile, **b** Conjugate profile at contact surfaces

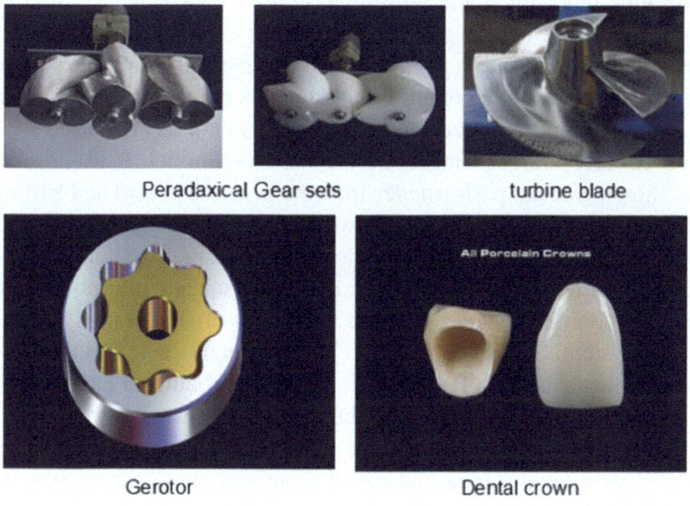

Fig. 3 Various applications of conjugate surfaces

Various conjugate surfaces applications such as Paradaxical gear set, turbine blade, gerotor, dental crowne figures are shown in below.

Researchers have tended to focus on various tooth profiles for improvement of performance over the last decade. Author Litvin et al. [6] investigated and proposed a parabolic tooth profile to minimize transmission errors. Tsay et al. [7, 8] suggested a changed form for helical gears with circular as well as involute tooth. Zhang [9] examined double involute profiles with a transition among them. The scope possibly extended to multiple involute profiles with various pressure angles. Ariga et al. [10] represented a tooth profile with a combination of circular arc and involute. And these combinations are closer to the idea of enforcing splines. Komori [11] developed and proposed a design for a spur gear with zero relative curvature at lots of contact points.

To avoid manufacturing difficulties in conjugate surfaces, digital surface approach was introduced and is widely used. A digital surface implies that, the digitalized points of a adjacent surface can be saved as a three dimensional matrix on a computer. Digital surface is also defined as a point set of Σm which appears like a surface rather than a curve or solid based object. This method describes surfaces in terms of discrete points. Digital surface method not only meets the requirement of complex shape representation as well as geometric design, it also helps in product data exchange and leads to information transmission. Therefore it's widely used in manufacturing and mechanical designs. There are two techniques to find digital surfaces. "Modern Design" is one technique, which includes boundary element method and finite element method. The second one is 3D-digital measurement of manufactured part or real objective models.

2 Literature Review

Luo [12] creatively designed and represented high load capacity cosine gear tooth profiles. It describe the cosine gear generation principle and mathematical model which consist of conjugate tooth profile. Cosine tooth profile and line of action equations are well established based on meshing theory. A new solid model was generated and characterized by computerized simulation such as contact ratio and sliding ratio, bending and contact stresses for this new solid drive model. It was found that the cosine gear showing better performance than the involute gear. The FEM model and stress distribution for each profile is shown in Fig. 4. Ming [13] discussed digital conjugate surfaces for solving computationally intensive problems by dimension reduction interpolation method and also alternative surface interpolation using the interpolation curve to achieve a conjugate surface normal vector as well as relative velocity of the system. According to Lijun [14], conjugate surfaces theoretical analysis is based on the principle of rotation, its beginning being an envelope of derived surface, forming a cylindrical surface of the rotating conjugate surface equation and contact line equation. It explained how the line of contact is drawn by computer and how the round steel leveler rolls and round conjugate motion may be achieved by computer simulation. Gang and Cuo [15] proposed modern computing technology

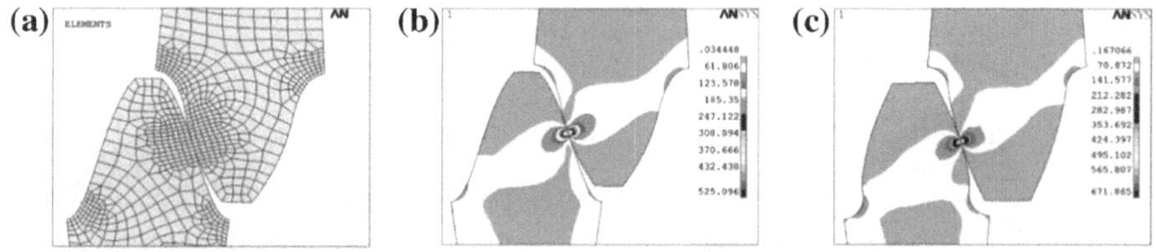

Fig. 4 **a** FEM Model **b** Stress distribution of cosine &**c** Involute profile [12]

and mathematical programming based on the principle of conjugating surfaces. They provided a direct description of the conjugate process with mathematical model and this model is used for both theoretical and digital research. A result obtained by digital simulation method covers the traditional conjugate surface theory of the basic content, adapted to solve interference and non-deformation problems under complex conditions.

Xin et al. [16], proposed a new method of numerical simulation to figure out special problems of conjugate surface in the form of discrete numerical data. The conjugating surface problems are changed into mathematical programs under minimum conditions. In this paper Guo et al. [17] proposed adaptive synthesis to analyze and discretize the conjugate surfaces to design the computer software by taking the conjugation process into account. Ming [18] used MATLAB to solve digitized conjugating surface problems. And he showed that surface interpolation was substituted by curve interpolation, to get the vectors and relative rate of digitized conjugating surfaces. By lower dimension interpolation method, the mathematical model of digitized conjugating surfaces problems were solved and a solution program was developed in MATLAB while the interface was achieved by Visual C++.

Xiao et al. [19], presents a design method for digitized conjugating surface by using MATLAB. The researchers proposed a mathematical model for conjugating surface and solving it according to the conjugating principle and computing feature of MATLAB. They invented a method to extract 1 dimensional arrays from a 3 dimensional array, and its results were optimized by a simple MATLAB function called the design process. Ming [20] has shown how the conjugating surfaces are solved by a mathematical model based on the digital dimension reduction interpolation method. The author used computing software MATLAB for powerful numerical and graphical display capabilities to develop the digital conjugate surface algorithm, with visual C++ to achieve a human-machine interface and processing of data. Masood et al. [21] introduced Direct machining process by providing efficient paths of tool from cloud data, which is saved in STL file format. The primary aim is to attain maximum efficiency in the machining of sculptured surface geometries, accepting complex machining fields. An algorithm was developed to generate various tool paths (Fig. 5a) for direct manufacturing of machining surfaces as well as studied a case study for cloud data (Fig. 5b).

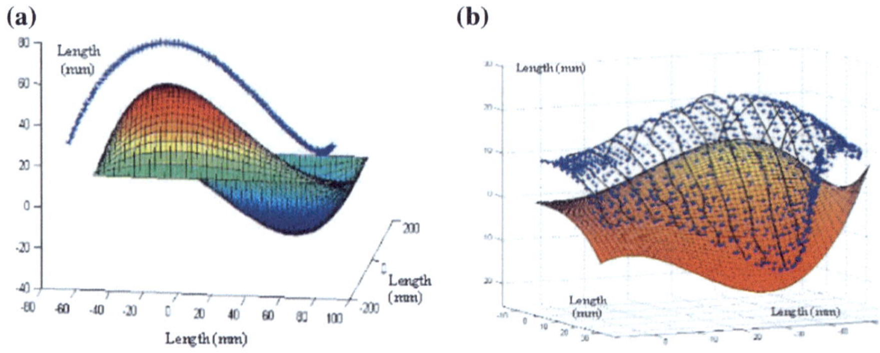

Fig. 5 **a** Tool path **b** Case Study for cloud data on conjugate surface [21]

Wang et al. [22] used machining method for complex surface face gear with hob cutter and CNC machine tools, based on gear hob principle and surface characteristics. Five axis CNC was used for machining complex components. A CNC hob technique process was suggested about the surface orthogonal gear. Yu and Ting [23] presented the first theory as well as exercise of free-form conjugation modeling. They proposed a novel technique of master-slave that extends the conventional computer aided single to dual geometry modeling. They later applied gear geometry to demonstrate the first free-form conjugation modeling method. Litvin et al. [24] developed a simplified approach to determine normal and principal curvatures for complex surfaces where the representation is in three-parameter form. Da Zhun and Liu [25] proposed average comprehensive radius of curvature (ACRC) and a new quality index for characterize the leakage path of the geometrical shape.

Radzevich [26] rectified the problems when machining sculptured surfaces on the multi-axis NC machine. The author developed six requirements and proper considerations for sculptured surface machining (SSM). Chen [27] proposed a pair of line conjugation ruled surfaces which produce instantaneous conjugate movement to one degree of freedom for various succinctness. The base conjugate complex surface is specified to be a minimal surface, a surface with only one parameter for specification of surface curvatures. Wang et al. [28] introduced a generating technique for digital surfaces of gear tooth, and focused on relative conjugate motion between digital gear tooth surfaces and discrete points and also solved non-standard shapes of digital gear tooth surface profiles shapes. As per "pre-calibration error compensation" approach, the machining on final sculpture surface errors are reduced subsequent operations as well as rebuilt the error compensation scenario. These error compensations are represented in Fig. 6 with respect to initial machined results.

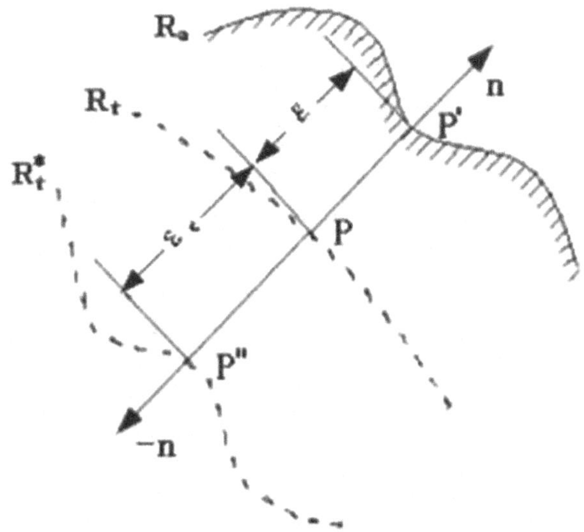

Fig. 6 Error compensations with respect to first machining results [28]

3 Conclusion

The researchers have been trying to solve conjugate surface using digitized surface method, for which one must construct analysis equation for digitized surface. To achieve this, reduced dimension and decomposition techniques and optimization algorithms can highly decelerate the complication of simplifying conjugate surface. Invariant approach, and rigid-body motions are represented by the curvatures of the surface and rotation in terms of a single surface curvature tensor. All the surface quantities are represented by invariants (tensors as well as vectors) and the outcomes of point and line contacts are directly received based on analysis and tensor algebra. This approach will be helpful for theoretical growth in gearing system theory without reference to coordinate systems; the components can be expressed in the form of a matrix, carried out by clearer and easier processing calculations. With this, other complex conjugate shapes can be easily designed, fabricated and simulated with optimized processes.

References

1. Makki, M., Tournier, C., Thiebaut, F., Lartigue, C., Souzani, C.: 5-axis direct machining of rough clouds of points. Comp. Aide. Desi. **7**(4), 591–600 (2010)
2. Makkia, M., Lartiguea, C., Tourniera, C., Thiébauta, F.: Direct duplication of physical models in discrete 5-axis machining. Virt. Phys. Proto. **3**(1), 93–103 (2011)
3. Tchantchane, Z., Bey, M.: Toward the roughing of sculptured surfaces from a regular cloud of points. In: 13th International Research/Expert Conference, Trends in the Development of Machinery and Associated Technology-TMT 2009, pp. 21–24. Hammamet, Tunisia (2009)

4. Zhang, D., Yang, P., Qian, X.: Adaptive NC path generation from massive point data with bounded error. J. Manf. Sci. Eng. ASME. **131** (2009)
5. Lai-yuan, X., Dao-xun, L., Chuan-yun, Y.: Theory of digitized conjugate surface and solution to conjugate surface. Wuh. Univ. J. Nat. Sci. **9**(2), 183–187 (2004)
6. Litvin, F., Fuentes, A.: Gear Geometry and Applied Theory. Cambridge University Press, Cambridge (2004)
7. Tsay, C., Fong, Z.: Tooth contact analysis for helical gears with pinion circular arc teeth and gear involute shaped teeth. ASME J. Mech. Trans. Autom. Des. **111**(2), 278–284 (1989)
8. Tsay, C., Fong, Z.: Computer simulation and stress analysis of helical gears with pinion circular arc teeth and gear involute teeth. Mach. Theory **26**(2), 145–154 (1991)
9. Zhang, G., Xu, H., Long, H.: Double involute gear with ladder shape of tooth. Chin. J. Mech. Eng. **31**, 47–52 (1995)
10. Ariga, Y., Nagata, S.: Load capacity of a new WN gear with basic rack of combined circular and involute profile. ASME J. Mech. Trans. Autom. Des. **107**(4), 565–572 (1985)
11. Komori, T., Ariga, Y., Nagata, S.: A new gears profile having zero relative curvature at many contact points (LogiX tooth profile). J. Mech. Des. **112**, 430–436 (1990)
12. Luo, S., Wu, Y., Wang, J.: The generation principle and mathematical models of a novel cosine gear drive. Mech. Mach. Theory **43**(12), 1543–1556 (2008)
13. Ming, Q.: Digitized conjugating surfaces dimension reduced solution. Nanj. Univ. Mech. Eng. Chinese J. TH123
14. Lijun, D.: Conjugate Surface theoretical analysis and computer drawing. Chinese J. TH126.2
15. Gang, Y., Cao, L.: Conjugate Surfaces digital simulation principle. Mech. Eng. Dal. Univ. Tech. China TH132.4
16. Xin, L. Jian, L., Delun, W.: Study of principle of numerical simulation about conjugating surface. Dept. Mech. Eng. DUT, Beijing Chemical Engineering College
17. Guo, W.Z., Zou, H., Wang, S.: Discretization and analysis principle of conjugating surfaces for adaptive synthesis. Mech. Eng. **02** (2000)
18. Ming, Q.: Study on digitized conjugating surfaces based on MATLAB. Coll. Mech. Eng. Nanj. Univ. Sci. Tech. Nanjing, Jiangsu 210094, China
19. Xiao, L., Liao, D., Yi, C.: Digitized conjugating surface design and simulation using MATLAB. Huazhong Univ. Sci. Tech. Wuhan 430074
20. Ming, Q.: Conjugating surface based on MATLAB digital. Nanj. Univ. Sci. Mech. Eng. China, TP391.7
21. Masood, A., Siddiqui, R., Pinto, M., Rehman, H.: Maqsood: tool path generation, for complex surface machining, using point cloud data. Procedia CIRP **26**, 397–402 (2015)
22. Wang, Y.Z., Hou, L.W., Liu, Y.Q., Shen, R., Wu, C.H., Ge, H.X., Zhang, Z.Z., Zou, W. and Li, Y.: Research on machining method of complex surface face gears. Appli. Mech. Mate. **163**, 233–237 (2012)
23. Ting, K., Yu, B.: Free-form conjugation modeling and gear tooth profile design. J. Mech. Robot. ASME **5**, 011001–011010 (2013)
24. Litvin, F.L., Chen, N.X., Chen, J.S.: Computerized determination of curvature relations and contact ellipse for conjugate surfaces. Comput. Methods Appl. Mech. Engrg. **125**, 151–170 (1995)
25. Xiao, D.Z., Liu, D.T.: Computer aided curvature analysis along the contact line of conjugate helical surfaces. J. Mate. Proc. Tech. **61**, 67–71 (1996)
26. Radzevich, S.P.: Conditions of proper sculptured surface machining. Comput. Aided Design **34**, 727–740 (2002)
27. Chen, C.-H.: Line conjugation ruled surfaces for representation of instantaneous conjugate motion of one DOF. Mech. Machine Theory. **37**, 1105–1116 (2002)
28. Wang, F., Yi, C., Wang, T., Yang, S., Zhao, G.: A generating method for digital gear tooth surfaces. Int. J. Adv. Manuf. Technol. **28**, 474–485 (2006)

Optimization of Wear Parameters of AA7150-TiC Nanocomposites by Taguchi Technique

Pagidi Madhukar, N. Selvaraj, Vipin Mishra and C. S. P. Rao

Abstract Wear optimization of titanium carbide reinforced nanocomposite fabricated through the vortex, double stir casting technique was carried out through L25 orthogonal array of Taguchi method. The experimental work was executed on pin on disc tribometer with different input parameters such as weight percentage (0, 0.5, 1.0, 1.5, 2.0%) of titanium carbide, sliding distance (500, 1000, 1500, 2000, 2500 m), sliding speed (0.5, 1.0, 1.5, 2.0, 2.5 m/s), and applied load (5, 10, 15, 20, 25 N) under dry sliding conditions. The analysis of individual parameter influence on responses was investigated through Analysis of Variance (ANOVA) and regression analysis was used to correlate the fit of the model with experimental data of wear.

Keywords AA7150 · TiC · Two-step stir casting · Taguchi technique · Wear

1 Introduction

This aerospace and automobile sectors are always on the lookout for advanced materials with higher grade of mechanical and tribological properties which combine easy of fabrication with the ability to vary composition of alloy to have a range of desired properties. But virtue of being lightweight and because of high strength and stiffness Al as matrix material is prefered [1–3]. Aluminium metal composite is a suitable candidate for aerospace applications, manufacturing of automobile components, and gas turbine engine. Al–Mg–Cu–Zn alloys are mostly used in aerospace and automobile sectors because of highly corrosion resistant and having high strength. AA7150 finds application in upper and lower wings of aeroplane, upper wings of stringers, floor beams, and seat rails. The strength of AMMNC depends on the size of ceramic particles, interparticle spacing, and nature of matrix and reinforcement

P. Madhukar (✉) · N. Selvaraj · V. Mishra
Department of Mechanical Engineering, National Institute of Technology, Warangal, Telangana, India
e-mail: pmadhu88@gmail.com

C. S. P. Rao
National Institute of Technology, Andhra Pradesh, Tadepalligudem, Andhra Pradesh, India

D. Dutta and B. Mahanty (eds.), *Numerical Optimization in Engineering and Sciences*, Advances in Intelligent Systems and Computing 979,
https://doi.org/10.1007/978-981-15-3215-3_53

interface. Therefore, reinforcement of nano-size particles was chosen due to its ability to change morphological and interfacial characteristics of the nanocomposite. Incorporating nanoparticles is a critical issue in high-density liquid due to floating of particles at the liquid metal surface. To prevent the floating, vortex method is introduced, which redirects ceramic particles into the molten liquid. Two-step stir casting process is chosen to fabricate the nanocomposites with uniform distribution of ceramic particles.

Wear is found to frequently appear at mating parts which leads to frequent replacement of components, because of abrasion. Abrasive wear takes place when hard ceramic particles penetrate a soft surface and displace material in the form of elongated chips and slivers [4]. The tribological behaviour of metal matrix composites depends on the interface between the particles and the matrix, composition of the matrix, and particle distribution. These conditions depend in turn on the type of environment, contact area, counter surface, applied load, sliding speed, and geometry [5]. The principle tribological parameters, such as applied load [6–8], sliding speed [9, 10], and percentage of reinforcing particles, control friction and wear performance. To know the parameter influence and its optimization, Taguchi technique was adopted for TiC reinforced nanocomposites. Taguchi method is a useful tool for improving the performance, process, product design, system, and reduction in experimental cost and time [11]. Abbas et al. [12] established the Taguchi technique to deal with output responses influenced by multi-variables. Therefore, this technique can optimize process parameters and optimal combinations through Analysis of Variance (ANOVA) approach [13].

The present study is intended to look into the effects of sliding distance, sliding velocity, load, and wt% of ceramic reinforcement on responses such as wear loss. Titanium carbide (TiC) nanoparticulate reinforced nanocomposites are produced by two-step stir casting process. Wear test is performed with pin on disc apparatus to the most effective control parameter that is identified on the reference variable by using Taguchi method. L25 orthogonal array was used based on parameters and levels. ANOVA was used to investigate design parameters and its effect on wear behaviour of the composites.

2 Materials and Methods

In this work, commercially available AA7150 Ingots were used as matrix material and titanium carbide which were used as reinforcements; titanium carbide particles had >99.9% purity and 30–50 nm average particle size. A number of methods are available to produce aluminium metal matrix nanocomposite, and out of these methods, two methods are most extensively used for fabrication. One is a solid-state method. This method is expensive and time-consuming, and large production of AMMNC is difficult to achieve through powder metallurgy method. Liquid-state method (i.e., stir casting) is preferred for fabrication because it is easy, enables mass production, and can generate complex shapes and produce uniform dispersion of reinforcement particle with better particle matrix, easier control of matrix structure, and low cost of processing which are other reasons for choosing liquid-state method.

But the problem of stir casting method with nano-size particles is that when it is added to the molten material, poor wettability and non-uniform dispersion occur because of very high specific area of nanoparticles. To avoid these problems, a new technique called the vertex method is used and this uses a mechanical rotary stirrer. Line diagram of stir casting machine is represented in Fig. 1. Preheated nano ceramic reinforcements and flux are incorporated slowly into the vortex of molten metal alloy to build good wetting between ceramic reinforcement and aluminium alloy melt pool. This melt was poured into die and specimens were machined as per ASTM G99 standard.

Samples were made of cylindrical shape with a length of 30 mm and diameter of 8 mm and used to conduct wear test on pin on disc tribometer by varying applied loads (5, 10, 15, 20, 25 N), sliding velocity (0.5, 1.0, 1.5, 2.0, 2.5 m/s), sliding distance (500, 1000, 1500, 2000, 2500 m), and weight percentage (0.5, 1.0, 1.5, 2.0%) at dry sliding conditions. Figure 2 shows the experimental set-up.

The experimental design was developed by considering four independent parameters and five levels through Taguchi technique. The factorial designs are very effective when a large number of factors are studied, and it is very hard to conduct all test combinations [14]. L25 orthogonal array was used for four factors of the five levels considered, and their interactions were tabulated by Taguchi method. The output

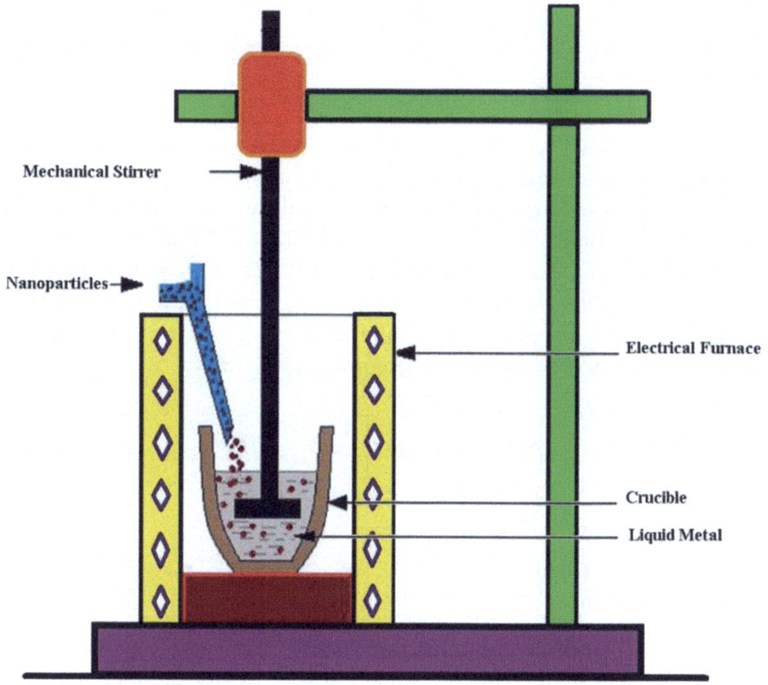

Fig. 1 Stir casting line diagram

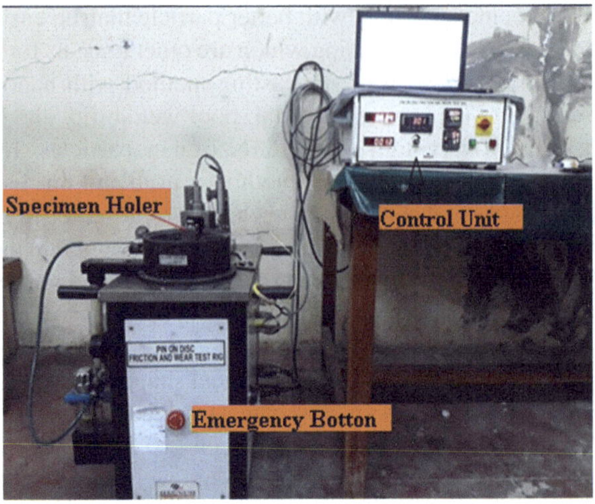

Fig. 2 Pin on disc tribometer

Table 1 Levels and factors in the experimental plan

S. No	Process parameters	Units	Symbol	Levels				
				L1	L2	L3	L4	L5
1	Load	N	L	5	10	15	20	25
2	Sliding distance	m	SD	500	1000	1500	2000	2500
3	Sliding velocity	m/s	SV	0.5	1	1.5	2	2.5
4	Weight % of TiC	wt%	wt%	0	0.5	1	1.5	2

response that was created for the model is wear loss. A total of 25 experiments runs were executed based on the runs created by Taguchi model (Table 1).

The analysis part was carried out for experimental data using MiniTab17 software. The results were tabulated into a signal to noise ratio. The main objective of the exercise was to find processes insensitive to noise factors as well as performance. It was applied to calculate quality deviation from the targets [14]. Wear loss from the S/N ratio was determined using "smaller is better", and it can be measured by logarithmic loss function as shown in the equation below:

$$\text{S/N} = -10\log\left[\frac{1}{n}\left(\sum y^2\right)\right]$$

where

y = experimental data (responses)
n = number of observations

The major objective of the model is to minimize the wear loss of the composite material. ANOVA was then favoured to find out statistically significant parameters.

3 Results and Discussions

The wear investigations were carried out as per Taguchi L25 orthogonal array, and weight loss values were tabulated for different combinations of process parameters as shown in Table 2. The experimental values for the wear loss were transferred to S/N ratios to evaluate the quality characteristics, and the results generated are shown in Table 2. The individual parameter influence on output responses as wear loss has been measured using S/N ratio. The individual parameter and its ranking for wear for different levels are tabulated in Table 4.

Figure 3 shows the main effects plot to wear loss of titanium carbide reinforced composites. From the graph, it is confirmed that the wear loss decreases with increase of weight percentages [15] due to hard phase ceramic particles, which promote the

Table 2 Experimental results of L25 orthogonal array for wear loss

No.	wt%	SD	SV	Load	Wear Loss	S/N ratio
1	0	500	0.5	10	0.0009	60.9151
2	0	1000	1	20	0.0052	45.6799
3	0	1500	1.5	30	0.0082	41.7237
4	0	2000	2	40	0.0143	36.8933
5	0	2500	2.5	50	0.0291	30.7221
6	0.5	500	1	30	0.0098	40.1755
7	0.5	1000	1.5	40	0.014	37.0774
8	0.5	1500	2	50	0.0265	31.5351
9	0.5	2000	2.5	10	0.0221	33.1122
10	0.5	2500	0.5	20	0.0103	39.7433
11	1	500	1.5	50	0.0131	37.6546
12	1	1000	2	10	0.0066	43.6091
13	1	1500	2.5	20	0.0115	38.7860
14	1	2000	0.5	30	0.0129	37.7882
15	1	2500	1	40	0.0133	37.5230
16	1.5	500	2	20	0.0004	67.9588
17	1.5	1000	2.5	30	0.0073	42.7335
18	1.5	1500	0.5	40	0.0054	45.3521
19	1.5	2000	1	50	0.0042	47.5350
20	1.5	2500	1.5	10	0.0076	42.3837
21	2	500	2.5	40	0.009	40.9151
22	2	1000	0.5	50	0.0041	47.7443
23	2	1500	1	10	0.0033	49.6297
24	2	2000	1.5	20	0.0011	59.1721
25	2	2500	2	30	0.0018	54.8945

Table 3 Analysis of Variance for wear loss

Source	DF	Seq SS	Adj SS	Adj MS	F	P	%P
wt%	4	0.000549	0.000549	0.000137	8.25	0.006	40.52
SD	4	0.000125	0.000125	0.000031	1.88	0.208	9.22
SV	4	0.000267	0.000267	0.000067	4.02	0.045	19.70
Load	4	0.000281	0.000281	0.000070	4.23	0.040	20.74
Resid Error	8	0.000133	0.000133	0.000017			9.81
Total	24	0.001355					

Table 4 Response table for means

Level	wt%	SD	SV	Load
1	0.011540	0.006640	0.006720	0.008100
2	0.016540	0.007440	0.007160	0.005700
3	0.011480	0.010980	0.008800	0.008000
4	0.004980	0.010920	0.009920	0.011200
5	0.003860	0.012420	0.015800	0.015400
Delta	0.012680	0.005780	0.009080	0.009700
Rank	1	4	3	2

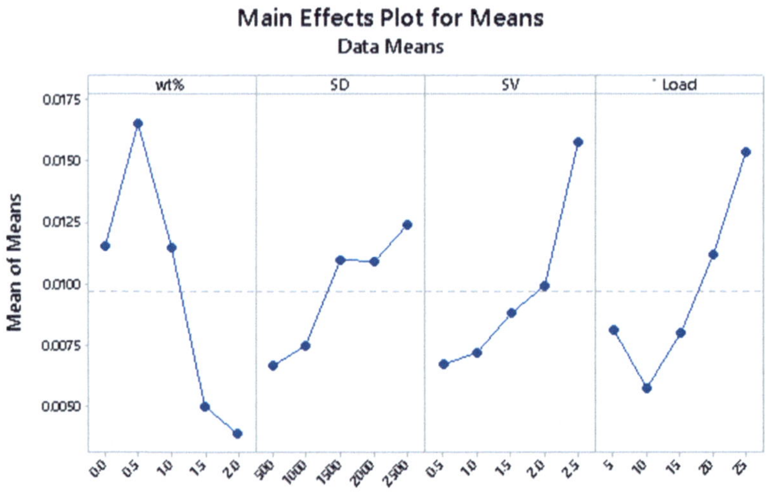

Fig. 3 Main effective plots for wear loss

hardness in the composite. It is also observed that the sliding distance, velocity, and load follow the same trend in the graph.

Wear loss increases with the increase of sliding distance, velocity, and applied load due to high frictional heat at the surface of the couple which leads to softening material, resulting in more wear loss [16]. It was also observed that the effect of the applied load becomes very critical when the sliding distance becomes high. 2% wt. of TiC, 500 m of SD, 0.5 m/s of SV, and 10 N of load are the optimal values to minimize wear loss.

Table 4 shows the response table for wear loss of AA7150-TiC nanocomposites, and it is observed that the weight percentage has the highest role on wear loss and then applied a load, sliding velocity, and distances. This observation is made by given raking Table 4. The responses are based on delta value, and it is obtained by the difference between highest to the lowest value of the column.

The regression equation for wear loss

$$\begin{aligned}\text{Wear Loss} = & -0.00175 - 0.00538\,\text{wt\%} + 0.000003\,\text{SD} \\ & + 0.00418\,\text{SV} + 0.000402\,\text{Load}\end{aligned} \quad (1)$$

Linear regression was done for wear loss, and it is represented in Eq. (1) where it predicts the correlation with 90.2% adjacent $R2$ value to confirm the goodness of fit for experimental and modelled data. Therefore, 90.0.2% ensures that the model is a good correlation between input parameters to responses.

The strong influence of individual parameters was obtained by different maximum and minimum of mean S/N ratio values. The main effect plots for wear loss are represented graphically in Fig. 3. From the results, analysis was done for the main effects of wear loss, and this gave the optimal parameter combinations for minimum wear loss as 2% of TiC, 500 m of SD, 0.5 m/s of SV, and 10 N of the load.

Analysis of Variance was used to find the most significant influencing parameter for the output response. Table 3 shows the ANOVA values for wear. This analysis is at a level of 95% confidence. The percentage contribution (%P) of individual parameters for the final response can be measured for all variables as the ratio of the individual sum of squares of parameters to the total sum of squares. This can be observed for each parameter as shown in Table 3 with the load having the most impact with high statistical value (40.52%) on wear loss followed by the load (20.74%), SV (19.7%), and SD% (9.22%). Other combined interactions are neglected because their contribution is negligible.

4 Conclusion

Wear optimization of AA7150-TiC nanocomposites was investigated through dry sliding wear test using L25 orthogonal array Taguchi method. Linear regression model was developed and confirmed with adjacent $R2$ 90.2% of goodness fit. It was found that 2% of TiC, 500 m of SD, 0.5 m/s of SV, and 10 N of load were the optimal input parameter values to achieve minimum wear loss. From ANOVA analysis, weight percentage was seen as the parameter that most influences wear loss with a percentage of contribution (40.52%) on AA7150-TiC nanocomposites, followed by the load (20.74%), SV (19.7%), and SD% (9.22%), respectively.

Acknowledgements "We acknowledge Dr. M. Raja Vishwanathan, Head of Humanities & Social Science Department, National Institute of Technology, Warangal, Telangana, INDIA, for proof reading of the paper and useful suggestions."

References

1. Seah, K.H.W., Sharma, S.C., Rao, P.R., Girish, B.M.: Mechanical properties of as-cast and heat-treated ZA-27/silicon carbide particulate composites. J. Mater. Desig. **16**, 277–281 (1995)
2. Shehadeh, Lina M., Jalham, Issam S.: The effect of adding different percentages of manganese (Mn) and copper (Cu) on the mechanical behavior of aluminum. Jord. J. Mech. Indu. Eng. **10**, 19–26 (2016)
3. Prasad, S.V., Asthana, R.: Aluminium-metal matrix composites for automotive applications: tribological considerations. Tribol **17**, 445–453 (2004)
4. Sahin, Y., Özdin, K.: A model for the abrasive wear behavior of aluminium based composites. Matri. Desig. **29**, 728–733 (2008)
5. Rohatgi, P.K., Schultz, B.F., Daoud, A., Zhang, W.W.: Tribological performance of A206 aluminum alloy containing silica sand particles. Trib Intr **43**, 455–466 (2010)
6. Prasad, K., Prasad, S.V., Das, A.A.: Abrasion-induced microstructural changes and material removal mechanisms in squeeze-cast aluminium alloy-silicon carbide composites. J. Matri. Sci. **27**, 4489–4494 (1992)
7. Bai, N.P., Ramasesh, B.S., Surappa, M.K.: Dry sliding wear of A356-Al-SiCp composites. Wear **157**, 295–304 (1992)
8. Bhansali, K.J., Mehrabian, R.: Abrasive wear of aluminum matrix composites. J. Metals **34**, 30–34 (1982)
9. Lee, S.Kim, Han, K.S., Lim, T.: Wear behavior of aluminium matrix composite materials. J. Matri. Sci. **27**, 793–800 (1992)
10. Sannino, P., Rack, H.J.: Dry sliding wear of discontinuously reinforced aluminum composites: review and discussion. Wear **189**, 1–19 (1995)
11. Montgomery, D.C.: Design and Analysis of Experiments, 4th edn. Wiley Inc., NY (2006)
12. Al-Refaie, Abbas, Tai-Hsi, Wu, Li, Ming-Hsien: An effective approach for solving the multi-response problem in Taguchi method. Jor. J. Mech. Indus. Eng. **4**, 314–323 (2010)
13. Basavarajappa, S., Chandramohan, G.: Wear studies on metal matrix composites: a Taguchi approach. J. Mater. Sci. Techn. **21**, 845–850 (2005)
14. Roy, R.K.: Design of Experiments Using Taguchi Approach. Wiley, New York
15. Tyagi, R.: Synthesis and tribological characterization of in situ cast Al–TiC composites. Wear **259**, 569–576 (2005)
16. Baradeswaran, A., Elaya Perumal, A.: Wear and mechanical characteristics of Al 7075/graphite composites. Compos. B **56**, 472–476 (2014)

Influence of Pulse GMA Process Variables on Penetration Shape Factor of AISI 304L Welds

Rati Saluja and K. M. Moeed

Abstract Weld geometrical parameters are imperative as these affect weld performance and soundness. A weld penetration shape factor is also indexed as 'form factor' that is ratio between weld width (*w*) and weld penetration (*p*). The ratio of 3:2 (*w*:*p*) is optimum in gas metal arc welding intended for quality welds. Increase in heat input increases penetration shape factor that results in weld shape imperfections at weld root. A larger penetration shape factor is susceptible to surface cracking whether a smaller value is prone to centreline cracking in Grade 304L welds. Hence, efforts have been made to understand direct and interactive effects of heat input on penetration shape factor by implementation of 5-level 3-factors central composite rotatable design using response surface methodology. A predictive model is crafted with the measured response as a function of welding current, arc voltage and shielding gas flow rate. Analysis of variance and confirmatory tests are attempted to ensure the accuracy of mathematical model within permissible limit.

Keywords ANOVA · Austenitic stainless steel · Gas metal arc welding · Shape factor

1 Introduction

Semi-automatic pulsed gas metal arc welding machines (p-GMAW) are widely opted for joining of AISI 304L austenitic stainless steels (ASSs) for industrial applications. Grade 304L steels are worldwide utilized in manufacturing and fabrication industries owing to its superior qualities at elevated and cryogenic temperatures. Grade 304L could be welded easily but extremely difficult to maintain flat due to its high coefficient of thermal expansion [1]. The primarily solution to minimize the weld defects in 304L welds is either steel must be welded in lesser time or heat input should be minimized. Heat input is a function of range of process variables as welding current, arc voltage, electrode efficiency and welding speed. Moreover, the kind and flow rate of shielding gas also aid variation in heat input. Deeper penetrations are difficult to

R. Saluja (✉) · K. M. Moeed
Department of Mechanical Engineering, Integral University, Lucknow, UP 226028, India
e-mail: ratisaluja@gmail.com

D. Dutta and B. Mahanty (eds.), *Numerical Optimization in Engineering and Sciences*, Advances in Intelligent Systems and Computing 979,
https://doi.org/10.1007/978-981-15-3215-3_54

achieve at low heat input whether the application of high heat increases the deposition rate that affects weld geometrical characteristics. Wider weld width (w) introduces a wider HAZ whether excess root penetration (p) results in poor weld fit-up and surplus post-mechanical operations [2]. Hence, it becomes mandatory to keep a balance among input process parameters and weld geometrical characteristics. Various researches have been conducted in the past few years to determine the influence of welding parameters on geometrical characteristics. To attain evenness between both important geometrical characteristics and to understand penetration pattern, an index termed as weld penetration shape factor (w/p) has been introduced. Omajene et al. had investigated the influence of process variables on bead geometrical parameters for underwater welding. Their research summarized that bead shape primarily depends upon the applied heat input, cooling rate, speed of welding, weld design and the water atmosphere for wet underwater welds [3]. Venkatesan et al. developed four-factor, five-level central composite rotatable designs (CCRDs) for determination of angular distortion, bowing distortion and bead geometry on ferritic steel on flux shielded arc (FCA) welds. Research concluded that increased bead width increases angular distortion whether increased penetration decreases bowing distortion [4]. Prabhu et al. developed a predictive model for quantification of convexity index (CI) for pulsed gas metal arc welding parameters using five-level CCRD for IS 2062 structural steel welds. They concluded that CI increases with increase in current whether reduces with increase in welding speed [1]. Later Om et al. had also explained the influence of submerged arc process (SAW) variables on weld penetration shape factor (WPSF) for austenitic steel surfacing. They summarized that WPSF reduces with increase in wire feed rate and welding speed and enhances if voltage and preheat current are increased [5]. Sharma et al. had also assessed bead geometrical features as a function of pulse GMA process variables. They stated that peak current, pulse width and welding speed affect geometrical features in comparison with shielding gas flow rate or frequency [6]. Sudhakran et al. had also investigated the effect of tungsten inert gas variables on weld depth to width ratio (p/w). They explicated that increase in current and shielding gas flow rate enhances p/w ratio whether an increase in rest two parameters reduces the measured response [2]. Sooner Aghakhani et al. studied effect of GMA process variables (wire feed speed, voltage, NTPD and welding speed) on weld dilution for ST-37 plates. Wire feed rate had been found as the most dominating variable among preferred variables [7].

Most of the literature had focused for prediction of bead geometrical parameters, but lesser attention is paid to weld geometrical characteristics. Bead geometry analysis is favourable during surfacing, but conclusive results of weld geometrical analysis could be effortlessly employed for determination of geometrical responses during manufacturing as well as in repair. Several efforts had been made by researchers to increase the depth of penetration and weld width, without noticing the correlation between WPSF concerning parent metal plate thickness. An optimum WPSF maintains equilibrium among weld width and penetration and controls weld volume and properties. Literature review concluded that despite the availability of numerous mathematical predictive models, rare work has been reported to determine weld penetration shape factor (WPSF) for Grade 304L austenitic stainless pulse welds.

Researchers had focused to craft predictive model for FCAW, SAW, GMAW or automatic pulsed GMAW, but there is not a single research which has been accomplished on semi-automatic pulsed GMAW equipment. These machines are equipped with software that allows the user to initiate pulse mode of metal transfer during welding. Pulse metal transfer is obtained using the current value above transition current, which is above 220 A for stainless steel, for voltage range (18 V, −34 V) using a mixture of 98% Argon with 2% oxygen [8]. Therefore, pulsed GMAW is preferred for joining of SA 240 Grade 304L ASSs 6-mm-thick plates. Occurrence of Si intensifies the existence of N_2 that facilitates retained ferrite in weld, and hence, AWS A 5.9 ER-308LSi (1.2 mm diameter) has been preferred over ER-308L as filler material. Execution of design of matrix (DOE) is channelized the experimental investigations in systematic fashion and makes easier to link relation among input process variables and response. Hence, central composite rotatable design (CCRD) is applied to ensure accuracy and precision during development of mathematical model. Analysis of variance (ANOVA) and confirmatory test also signify that developed predictive model could be implemented for determination of WPSF for Grade 304L stainless steel pulsed welds.

2 Experimental Methodology

ESAB Aristo MIG/MAG U5000I advance multiprocess synergic pulse welding equipment has been used for welding of SA240 AISI 304L austenitic base metal plates of dimension (300 mm × 150 mm × 6 mm). This is a three-phase, 50 Hz, 500 A, 100% duty cycle welding machine made in USA Integrated weld-cloud 2.0 software which allows alternation in preferred process variables along with selection of additional variables. Variation in heat input could be obtained with variation in welding current (*I*), arc voltage (*V*) and shielding gas flow rate (*Q*). At one side, '*I*' and '*V*' enhance the heat input, and on the other side, '*Q*' reduces the same. Hence, all three process variables are preferred as independent process variables. Since '*I*' and '*V*' have opted as independent variables, welding speed has to be varied to fill the molten filler metal in the weld groove, and hence, welding speed is not selected as an independent variable. Alteration in variables amends the geometrical features of weld and thus varies the WPSF.

The element compositions of the base and filler metal are detailed in Table 1. The details of independent process parameters and additional investigational variables for fabrication of 304L plates are detailed in Tables 2 and 3.

2.1 *Fixing the Range of Independent Process Variables*

The functioning range of the preferred variables was quantified with trial runs by fixing two of the parameters w.r.t. the parameter, whose limits were to be examined. The

Table 1 Units, notation and levels of welding parameters

Process parameters	Units	Notation	Levels				
			−1.682	−1	0	1	1.682
Shielding gas flow rate	Litres/min	*Q*	10	12	14	16	18
Welding current	Amperage	*I*	220	226	235	244	250
Voltage	Volt	*V*	25	26	27	28	29

three-factor five-level CCRD is executed with full replication for preferred process parameters for the implementation of design of experiments (DOE) for measurement of WPSF. DOE consists of eight factorial for possible combinations of $(\pm 1, \pm 1, \pm 1)$, six centre (0,0,0) and six axial $\{(\pm 1.682,0,0), (0, \pm 1.682, 0), (0, 0, \pm 1.682)\}$ points. Overall 20 experimental runs of CCRD evaluate the direct and interactive impact of the input variables on the WPSF. Levels of DOE are quantified with the help of Eq. (1) as mentioned in Table 1. Here, X_i = required coded value of variable, X = general value of the variable within lowest limit ($X_{\min}$) and the highest limit ($X_{\max}$) of variable [2].

$$X_i = 1.682\left[\frac{2X - (X_{\max} + X_{\min})}{(X_{\max} - X_{\min})}\right] \quad (1)$$

2.2 *Measurement of Weld Geometrical Features*

Weld width was measured with digital vernier calliper (Make: Mitutoyo, model no: 500-196-30, least count: 0.02 mm), and weld penetration was measured using vernier height gauge (Make: Kristeel Shinwa Industries Ltd., model no: 2918, least count: 0.01 mm). Five readings on mid-points are taken for weld width and two readings at front depth and back depth for measurement of weld penetration were taken and WPSF was quantified with average readings of every investigation. The measured values of weld width, penetration, WPSF, predicted WPSF, residuals (e_u), sum of square (*SS*) and standard deviation (σ) are tabulated in Table 2. e_u interpret divergence among observed and predicted WPSF of conducted experimental runs whether *SS* refers to total variability of the WPSF in DOE [9].

3 Application of Analysis of Variance (ANOVA)

The WPSF of AISI 304L could be stated as a function of variables as WPSF = *f* (*I*, *V*, *Q*). The predictive mathematical model is structured with RSM-based second-order regression technique as crafted below in Eq. (2). Here, b_o = independent term,

Table 2 Coded design matrix with observed and predicted WPSF, e_u, SSE and σ for 304L welds

S. No.	Run order	Coded parameters			Observed WPSF	Predicted WPSF	Residual (e_u)	SSE	SD(σ)
		I	*V*	*Q*					
1	2	−1	−1	−1	1.500	1.33	0.0072	0.00939	0.007205
2	10	1	−1	−1	1.498	1.50	−0.0022	0.02061	0.002195
3	13	−1	1	−1	1.482	1.42	−0.0005	0.00139	0.000495
4	9	1	1	−1	1.410	1.42	−0.0092	0.01539	0.009195
5	15	−1	−1	1	1.405	1.40	0.0076	0.01539	0.007605
6	3	1	−1	1	1.365	1.42	−0.0011	0.01739	0.001095
7	16	−1	1	1	1.593	1.42	−0.0043	0.01261	0.004295
8	18	1	1	1	1.511	1.59	−0.0010	0.01139	0.000995
9	7	−1.682	0	0	1.403	1.51	0.0033	0.01861	0.003305
10	19	1.682	0	0	1.341	1.50	−0.0006	0.00261	0.000595
11	12	0	−1.682	0	1.413	1.56	−0.0026	0.00111	0.002595
12	11	0	1.682	0	1.517	1.58	0.0048	0.02339	0.004805
13	17	0	0	−1.682	1.580	1.41	−0.0011	0.00861	0.001095
14	4	0	0	1.682	1.561	1.42	−0.0176	0.03739	0.017595
15	20	0	0	0	1.413	1.42	0.0224	0.01539	0.022405
16	6	0	0	0	1.416	1.41	−0.0049	0.02461	0.004895
17	14	0	0	0	1.400	1.42	0.0007	0.00061	0.000705
18	1	0	0	0	1.417	1.37	−0.0042	0.03061	0.004195
19	8	0	0	0	1.418	1.51	0.0006	0.01061	0.000605
20	5	0	0	0	1.440	1.48	0.0026	0.00661	0.002605

Table 3 Significance testing (ANOVA) of developed model

Parameter	First-order term		Second-order term		Third-order term		Error term		F-value	p-value
	SS	DOF	SS	DOF	SS	DOF	SS	DOF		
WPSF	0.0185	3	0.0261	3	0.0495	3	0.0008	5	95.06	0.0001

the coefficient b_i = linear term for X_i, b_{ii} = coefficient of quadratic terms, b_{ij} = coefficient of interaction terms $X_i\ X_j$ and ε_i = cumulative error of predictive model [9].

$$Y = b_o + \sum_{i=1}^{k} b_i X_i + \sum_{i=1}^{k} b_{ii} X_i^2 + \sum_{i=1}^{k} b_{ij} X_i X_j + \varepsilon_i \tag{2}$$

Regression coefficients signify control of input variables over WPSF. A '+' regression coefficient points affirmative association among variable and response whether '−' regression coefficient shows the converse correlation between variable and response [9]. The F and p test has been conducted, to check accuracy of predictive mathematical model with analysis of variance technique as displayed in Table 3. Fisher's value (=95.06) implies towards significance of developed model for 5% level of significance. P-values less than 0.05 indicate model terms which are significant. The Lack of Fit (F-value = 0.3494) indicates insignificancy due to this case I, V, IV, VQ, Q^2, I^2 and V^2 which are significant model terms whether Q and IQ are insignificant. The final predictive model was constructed with significant regression coefficients as displayed in Eq. (3) with the help of Design Expert 11.0 software [10].

$$\begin{aligned} \text{WTSF} = 1.417 - 0.022(I) + 0.029(V) - 0.0163(I^2) + 0.0165(V^2) \\ + 0.0538(Q^2) + 0.054(VQ) - 0.014(IV) + \varepsilon \end{aligned} \tag{3}$$

4 Validation of Results

The coefficient of determination of adjusted regression model (R_a^2) is evaluated against coefficient of determination of full regression model (R_f^2). The coefficient of determination of predicted model ($R_p^2 = 0.9635$) is in reasonable agreement with the R_a^2 (0.9780); that indicates the divergence is less than 0.0145.

Calculated S/N ratio (34.80) also points a reasonable indication as adequate precision larger than 4 is desirable. Confirmatory tests as displayed in Table 4 also point

Table 4 Results of conformity test for WPSF for Grade 304L welds

S. No.	Q (LPM)	I (A)	V (V)	Observed WPSF	Predicted WPSF	Error
1	11	220	27	1.48	1.50	0.84
2	11	220	28	1.47	1.46	0.22
3	16	231	27	1.44	1.43	0.39
4	15	236	25	1.35	1.39	2.76
5	17	238	25	1.39	1.38	0.97

towards accuracy of developed model, as error between predicted and observed WPSF is obtained below 2.76%.

5 Effect of Process Variables on WPSF

The developed mathematical model has been pertained for understanding the direct and quadratic effects of preferred process parameters on WPSF in 304L welds by substituting the numeral values of the *Q*, *I* and *V* for conducted set of investigations.

5.1 Direct Effect of Shielding Gas Flow Rate on WPSF

From Fig. 1, it is inferred that WPSF suppresses as *Q* is increased from 10 to 14 L/min, afterwards it increases as *Q* approaches to its maximal value (18 L/min). A surplus *Q* compels the molten metal to weld groove, and hence, the penetration increases that

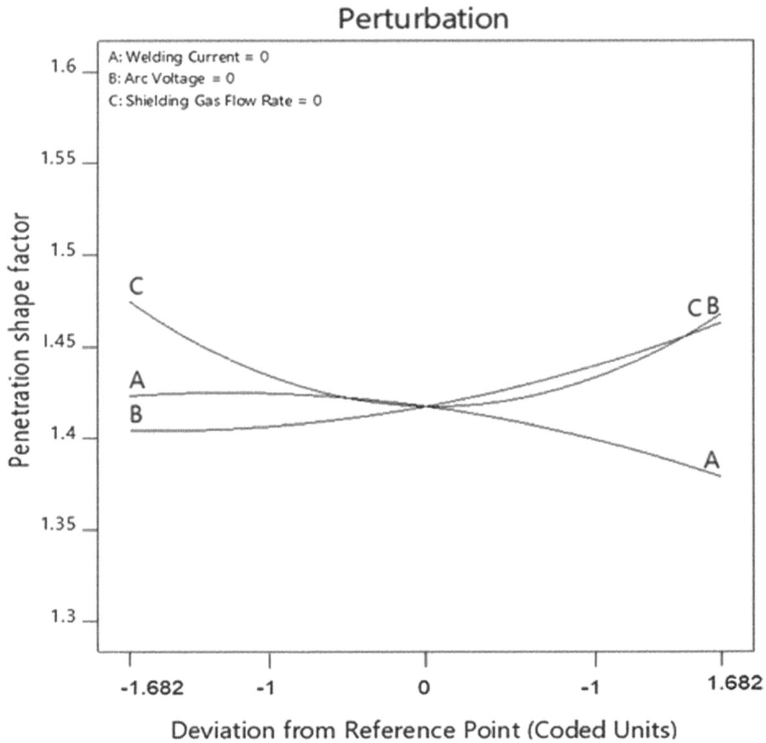

Fig. 1 Perturbation plot displaying the effect of process variables on WPSF

results in decrease in WPSF. As Q crosses its centre value, WPSF starts increasing. This increase in WPSF could be attributable to reduction in heat input caused by the additional increase in Q, as high Q provides an aided cooling effect to Grade 304L welds. Due to enhanced cooling, the layers of weld root solidify, prior to layers of weld width that causes reduction in penetration in comparison with weld width. Thus, it could be concluded that WPSF initially reduces and soon after increases when shielding gas flow rate passes its mean value.

5.2 Direct Effect of Welding Current on WPSF

Figure 1 infers that WPSF decreases with increase in welding current. Increase in the welding current enhances the metal deposition rate that improves penetration depth in comparison with weld width. WPSF is inversely proportional to penetration depth, and hence with increase in current and penetration depth, WPSF decreases.

5.3 Direct Effect of Arc Voltage on WPSF

From Fig. 1, it can be inferred that WPSF increases, as the arc voltage is increased from 25 to 29 V. It can also be interpreted as, at higher voltage, arc column widens that increases the weld width of Grade 304L weld. Because WPSF is directly proportional to weld width, WPSF increases with increase in arc voltage.

5.4 Interactive Effects Among Welding Current and Voltage on WPSF

For estimating the interaction effects of 'I–V' shielding gas flow rate is set to centre value of CCRD (14 L/min). For maximal value of 'I' and intermediary values of 'V', minimum value of WPSF (1.35) has been obtained as displayed in Fig. 2. For higher voltage values (>28 V) and minimum to intermediary values of welding current (220–240 A), optimal values of WPSF (1.5–1.53) could be obtained. Moderate variations in WPSF are seen with alteration in current at low voltage whether higher variations are noticed in WPSF if voltage is increased for a fixed value of current. Drastic reduction in WPSF has been analysed if current is increased beyond 242 A.

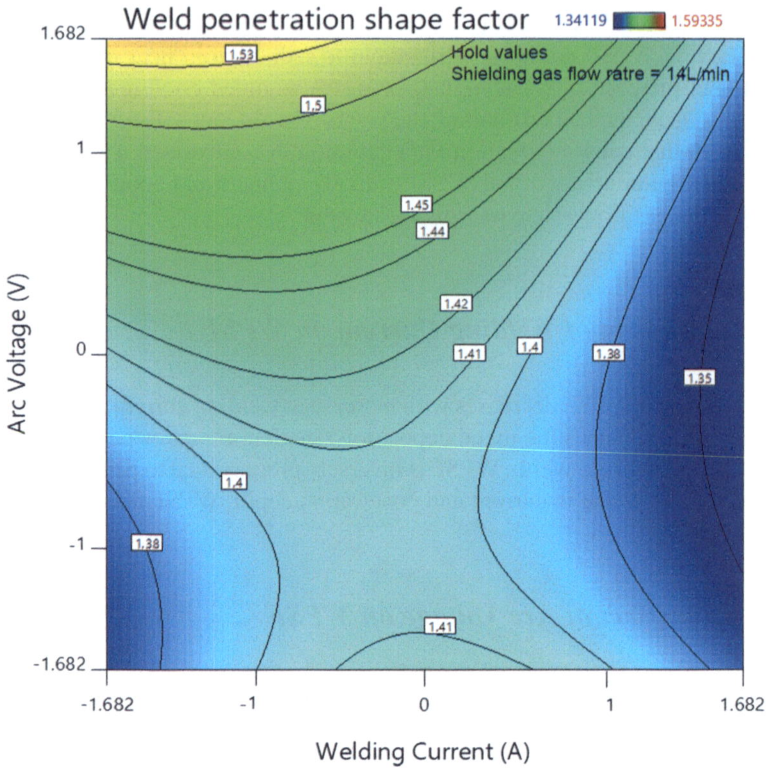

Fig. 2 Contour plot displaying interaction effects of '*I*–*V*' on WPSF

5.5 Interactive Effects of Shielding Gas Flow Rate and Welding Current on WPSF

Response surface plot demonstrated in Fig. 3 illustrates that maximal WPSF (1.55) is obtained for the combination of either minimal value of welding current or gas flow rate or maximum value of both the factors. For all intermediary values of gas flow rate ($14 < Q < 16$), moderate values of WPSF (1.36 < WPSF < 1.5) are obtained for any value of welding current. That indicates nearly equal WPSF could be obtained for a fixed value of '*I*' if '*Q*' is varied within its intermediary values. Optimal values of WPSF (1.5) could be obtained for gas flow rate either at 16.6 L/min or 10–11 L/min for current range between 220 and 245 A.

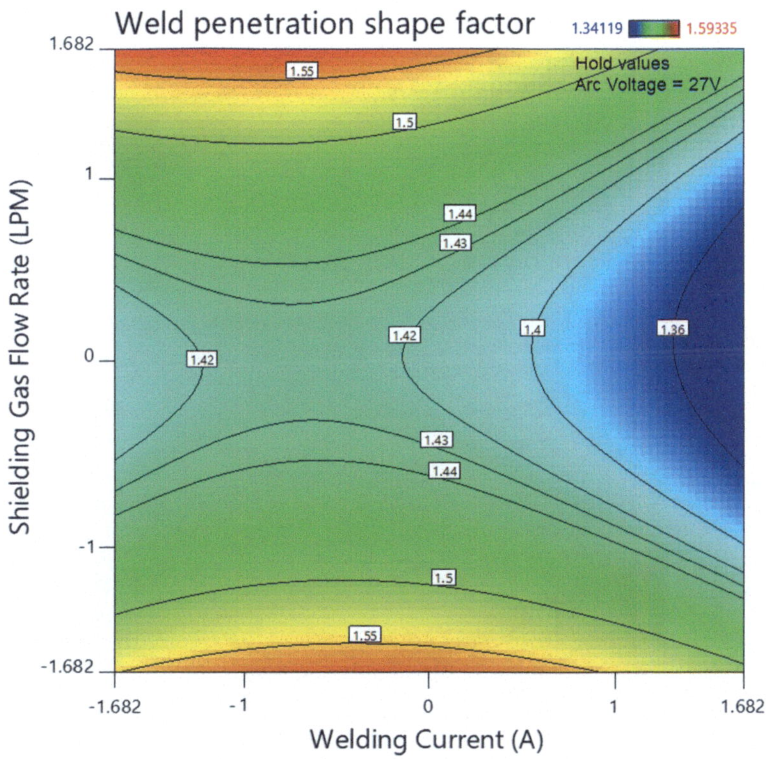

Fig. 3 Contour plot displaying interaction effects of '*Q*–*I*' on WPSF

5.6 Interactive Effects of Arc Voltage and Shielding Gas Flow Rate on WPSF

Figure 4 illustrates that maximal WPSF (1.70–1.75) is obtained for combination of either minimal value of '*V*' and '*Q*' or maximum value of both the factors. For increasing '*Q*' within 10–18 L/min for a fixed value of '*V*', a drastic reduction in WPSF is achieved for a voltage range from 25 to 26.5 V. Sooner this pattern diminishes and WPSF increases by minor alteration in any/both of the variables.

6 Conclusion

The experimental investigation is accomplished for prediction of WPSF for Grade 304L pulse GMA welds. Various concluding elucidation summarized that developed mathematical model for prediction of WPSF as a function of *I*, *V* and *Q* could be practised for the preferred range of process parameters. Experimental investigation

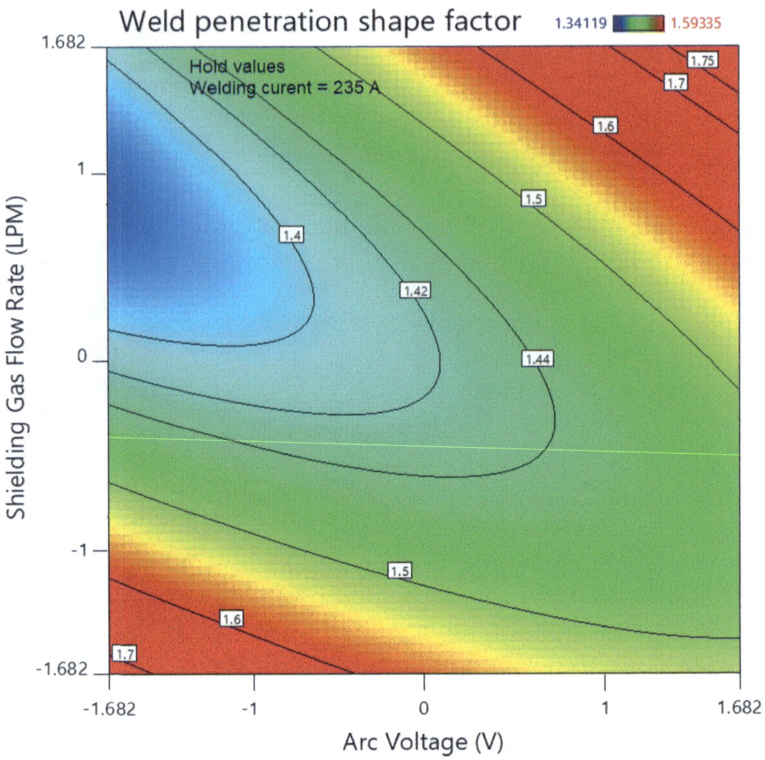

Fig. 4 Contour plot displaying interaction effects of '*Q*–*V*' on WPSF

summarizes that welding current has negative, and arc voltage has a positive influence on WPSF for 304L welds. Effect of shielding gas was found initially negative than it became positive as gas flow rate was increased beyond its centre value. The effect of arc voltage was more pronounced than other two preferred parameters. The minimum WPSF (1.341) was attained for combination of ($I = 250$ A, $V = 27$ V, $Q = 14$ L/min), and maximum WPSF was attained at ($I = 226$A, $V = 28$ V, $Q = 16$ L/min). Optimal WPSF (1.50) had been achieved for centre values ($I = 235$ A, $V = 27$ V, $Q = 14$ L/min) of all three process parameters. The WPSF of 304 welds was attained in acceptable range for weld geometrical features for 304L arc welds, and thus, it can be concluded that developed mathematical model could be easily employed for prediction of weld penetration shape factor for pulsed gas metal arc welds.

References

1. Prabhu, R., Alwarsamy, T., Ramakrishnan, R., Gargyan, P.S.: Development of mathematical model for prediction of convexity Index in cladding by pulse MIG welding process. Pro. Mat. Sci. **5**, 663–668 (2014)
2. Sudhakaran, R., VeL Murugan, V., Senthilkumar, K.M.: Effect of welding process parameters on weld bead geometry and optimization of process parameters to maximize depth to width ratio for stainless steel gas tungsten arc welded plates using genetic algorithm. Eur. J. Sci. Res. **62**, 76–94 (2011)
3. Omajene, J.E., Martikainen, J., Kah, P.: Effect of welding parameters on weld bead shape for welds done underwater. Int. J. Mech. Eng. App. **2**(6), 128–134 (2014)
4. Venkatesan, M.V., Murugan, N.: Role of FCA welding process parameters, on bead profile, angular and bowing distortion of ferritic stainless steel sheets. J. Eng. Sci. Tech **9**(1), 107–122 (2014)
5. Om, H., Pandey, S.: Effect of process parameters on weld penetration shape factor in ASAW based surfacing. Int. J. Inn. Res. Sci. Eng. Techn. **3**(6), 13590–13596 (2014)
6. Sharma, A., Verma, D.K., Arora, N.: A scheme of comprehensive assessment of weld bead geometry. Int. J. Ad. Manuf. Tech. 1–22 (2015)
7. Aghakhani, M., Mehrdad, E., Hayati, E.: Parametric optimization of gas metal arc welding process by Taguchi method on weld dilution. Int. J. Model. Optim. **1**(3), 216–220 (2011)
8. Ghosh, P.K.: Concept of pulse current gas metal arc welding process, Pulse current gas metal arc welding characteristics, control and applications, Mat. Form. Mach Tribo. 31–45 (2017)
9. Montgomery, D.C., Peck, E.F.: Introduction to Linear Regression Analysis. Wiley, New York (2005)
10. Design Expert Trial Version 11.0. Stat-ease Inc. (2018)

Numerical Optimization of Trench Film Cooling Parameters Using Response Surface Approach

V. G. Krishna Anand and K. M. Parammasivam

Abstract Film cooling is a widely used cooling method to safeguard the surface of gas turbine components from adverse impact of hot mainstream gases. The trench film cooling is a type of film cooling where the film cooling holes are placed within the cut slots designed on the film cooling surface. Numerical optimization is performed on trench film parameters using response surface-based approach. The trench parameters used in this study are trench length, trench depth, trench film hole compound angle, blowing ratio and area averaged film effectiveness is used as the response factor. Response surface contours were generated, and analysis of variance was performed. The results of the study show that the trench parameters, viz. trench depth, blowing ratio have higher significance on response factor. The optimized trench parameters were trench length 2D, trench height 0.83D, trench film hole compound angle 83.12° and blowing ratio of 1.96.

Keywords Trench parameters · Film cooling · Response surface approach

1 Introduction

Gas turbines are widely used in electric power generation, aircraft propulsion units and marine propulsion systems. Most of the components of gas turbine may require extensive cooling, in order to safeguard itself from hot combustion gases. The gas turbine operates above a temperature range of 1500 °C, and most of the materials used on turbine components may melt at these higher temperature environments. In order to safeguard the gas turbine components and for its continuous operation, the cooling modes are applied on both internal and external surfaces of turbine components. The internal cooling is applied through impingement jets, and film cooling is used for external surface cooling. The low temperature air required for film cooling is obtained from compressor stages, and this air is directed through ducts and is finally ejected through film cooling holes available on the cooling surface. In traditional film cooling, the film cooling holes are directly placed on the surface that require protection from hot stream gases; however, this method can lead to decreased effects on film cooling

V. G. K. Anand (✉) · K. M. Parammasivam
Department of Aerospace Engineering, MIT Campus, Anna University, Chennai 600 044, India

D. Dutta and B. Mahanty (eds.), *Numerical Optimization in Engineering and Sciences*, Advances in Intelligent Systems and Computing 979,
https://doi.org/10.1007/978-981-15-3215-3_55

with increase in coolant ejection rates. In trench film cooling, the film cooling holes are embedding within cut slots, and cooling air is ejected from film holes are passed through the trenched slots. Lu et al. [1] performed experimental studies to understand the effects of trench geometry on film cooling performance. The results of their study showed that the height of the trench has a significant influence on trench cooling performance and a trench length of two times the film hole diameter delivered higher cooling effectiveness. Lee and Kim [2] performed numerical studies with the goal to identify the effective trench dimensions. In comparison with other trench geometry, a trench length of two times the film hole diameter and trench depth of one times the film hole diameter delivered higher cooling performance. Lakzian [3] carried out numerical studies to determine the optimum parameters on film cooling. The optimized film parameters determined from the study are blowing, density ratio and turbulence intensity. Alshehaby et al. [4] performed numerical studies to optimize the shaped film holes parameters. The results of their study determined the optimized size of slot film hole. Elsayed et al. [5] used CFD and simplex method to optimize the film cooling parameters that can deliver maximum film cooling effectiveness. The results of the study show that decrease in stream angle of film hole leads to higher film effectiveness at higher blowing rates. Lee et al. [6] used advanced algorithm with surrogate method to optimize the parameters of laid-shaped film holes with the aim to obtain higher FCE and lower aerodynamic losses. Savastano et al. [7] performed numerical studies and optimized the parameters of effusion plates. Wang et al. [8] used NN-GA method and optimized important parameters of fan film holes that can deliver enhanced film cooling effectiveness. Oguntade et al. [9] performed numerical investigations and showed that modifications on trench outlets delivered improved film effectiveness on cooling surface. It is observed from the literatures that very few or no studies have been carried out on the optimization of trench film cooling parameters. The aim of the present study is to perform numerical optimization of trench film cooling parameters with the goal to maximize the trench film effectiveness using response surface approach. The trench FC parameters used for this study are trench length (w), trench height (d), trench hole compound angle (β) and blowing rate (M).

2 Response Surface Approaches

The response surface approach is a statistical method to develop approximate quadratic model based on input parameters and response factor. The range of trench parameters used in this study is trench length (2–3 D), trench height (0.5–0.9 D), trench hole compound angle (45^0–135^0), blowing ratio (0.4–2.0), and D is the diameter of the film hole (4 mm). The AA film effectiveness is chosen as a response factor. The BBD is used and the chosen range of trench parameters and arranged into an experimental matrix system (EMS). The modeled EMS consists of 29 cases of different combinations of trench parameters. Computational investigations were carried out to determine the response factor for each case of EMS.

2.1 Numerical and Experimental Details

The computational domain is scaled down based on experimental test section geometry. The domain consists of mainstream section, trench section and film cooling hole. The critical regions of computational domain with higher thermal and pressure gradients are regions close to the wall surface, trench region and film hole system. A higher dense mesh system is used to cover the critical regions of the computational domain. The details of computational model used in the study are shown in Fig. 1. The *x*-axis and *z*-axis represent the longitudinal and lateral extent of the computational domain, and *y*-axis represents the vertical extent of the domain.

Structured cluster mesh system is used for mainstream with trench region and unstructured mesh model is used for film hole. Grid independence test is carried out with three different meshes, viz. 0.85, 2.1 and 3.8 million cells and with span averaged FCE used as assessment parameter for grid independence tests. The R-k-ε model with EWT is used for turbulence modeling. The vertical sides of the computational domain are specified with periodic BCs. The mainstream flow is maintained at a velocity of 13 m/s. The velocity of the coolant stream is varied as per the requirement of blowing ratio.

The experiments were carried out on facility developed at the laboratory where the free stream flow is heated with a matrix of heater assemblies. The test section is a square duct made of plexiglass sheet assembled at the center of the test facility. The

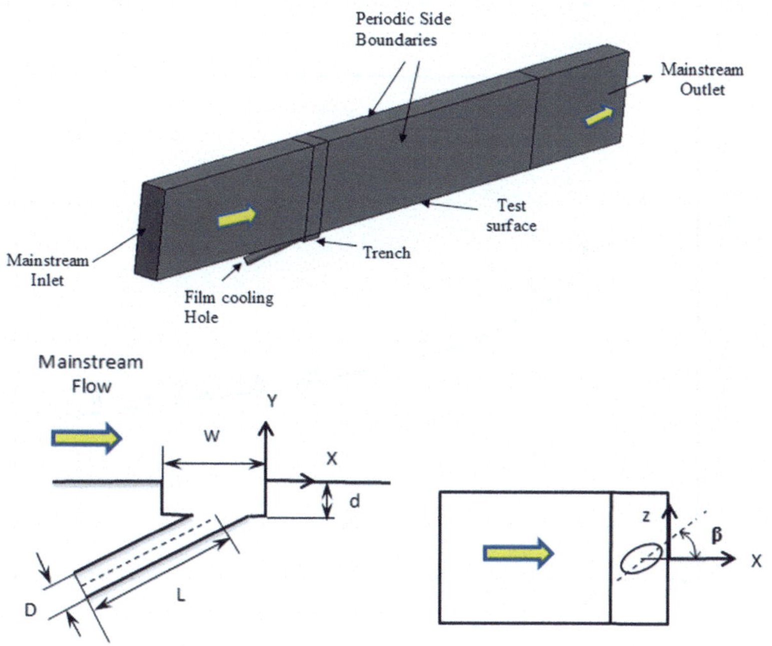

Fig. 1 Computational details and trench film paramters

unit other than test section is fabricated out of MS sheet of thickness 3.5 mm. The film cooling air is supplied at room temperature from storage tanks. The test section is fabricated from low thermal conductivity material to make it to satisfy the adiabatic wall boundary condition. *K*-type thermocouples were used for the measurements of temperatures along the test plates.

3 Results and Discussions

The computational results need to be validated, and this is carried out with a single case identified from EMS of response surface methodology. Fig. 2 shows the plot for the comparison of numerical results with the experimental data for an *M* of 1.2. The computational results are mostly lower than experimental results all along the test surface.

The response surface plots generated were used to study the pattern of variation of trench parameters, viz. trench length, trench height, film hole compound angle and blowing rate against the area averaged film cooling effectiveness of test surface. The analysis of variance was performed to identify the influential parameters that can affect the film cooling effectiveness on test surface. The parameter with the *P* value < 0.05 is considered to be significant for film effectiveness. Here, the terms trench height and blowing ratio are more influential parameter followed by film hole compound angle. Fig. 3 shows the variation of increase of trench height with

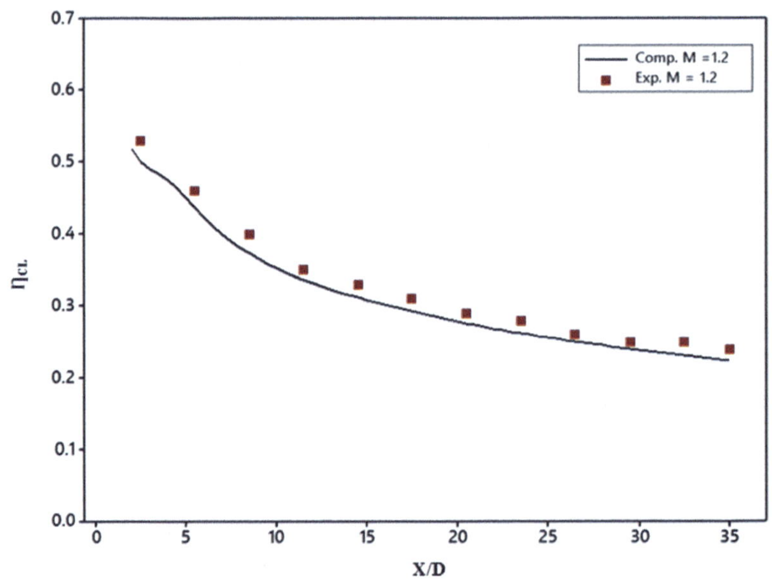

Fig. 2 Comparison of computational and experimental results of at *M* of 1.2

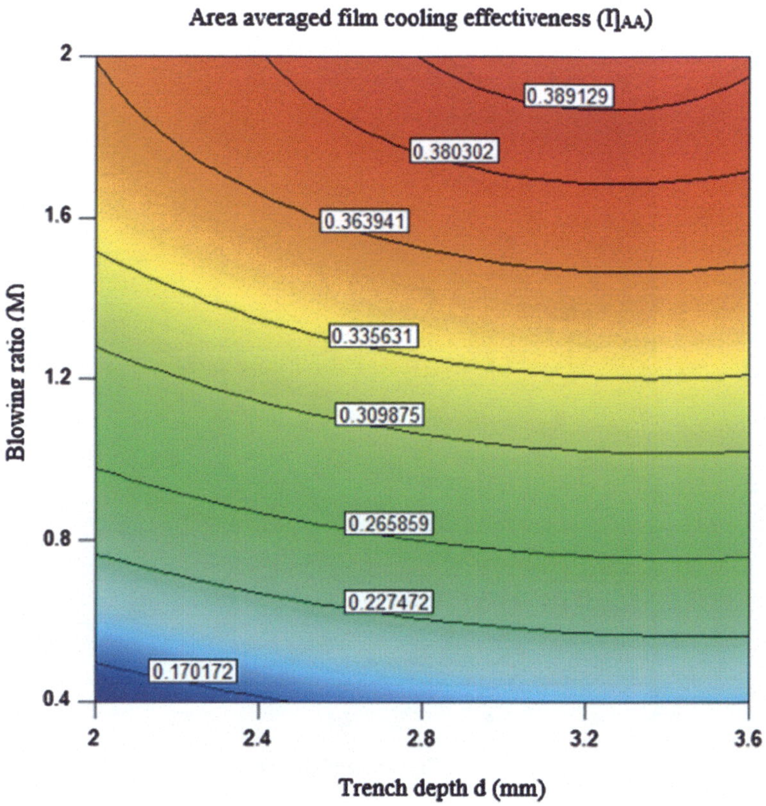

Fig. 3 Variation of trench height and blowing ratio with AA film effectiveness

blowing ratio against the response parameter. The higher trenches in the range of 2.8–3.6 mm have areas of higher film effectiveness. The variation of trench height and compound angle of film hole against the area averaged film effectiveness is shown in Fig. 4. The trench height above 0.7D and film compound angle around 90° have regions of higher film effectiveness, and the numerical optimization is performed on trench parameters by setting it to be in the range of selected trench parameters and with a goal to maximize the response factor. The optimization is performed based on desirability function of response surface approach. The optimization is performed on trench parameters to satisfy a desirability value of 1. The optimized trench parameters are trench length 2D, trench height (or) depth 0.83D, trench film hole compound angle 83.12° and blowing ratio of 1.96.

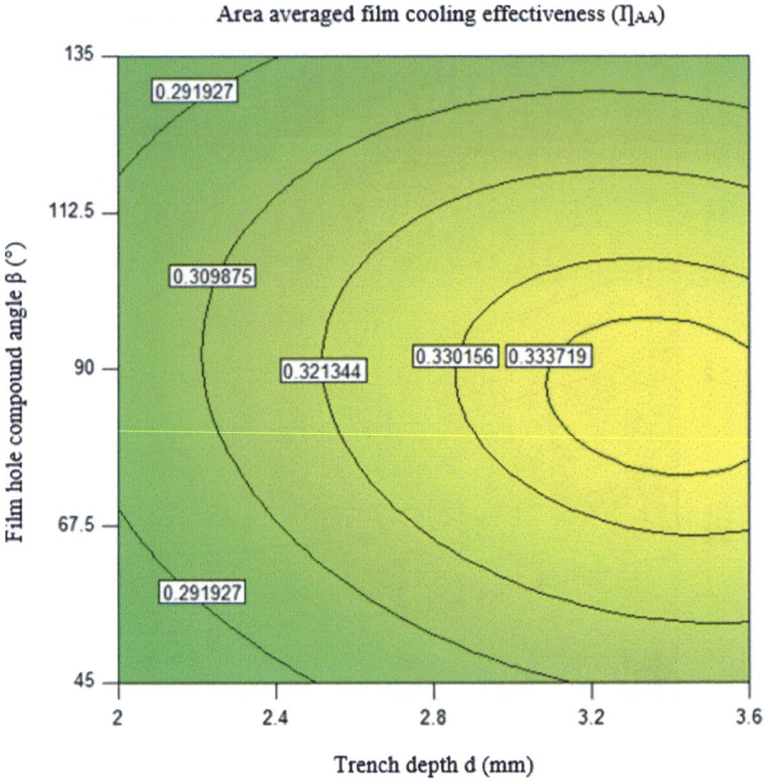

Fig. 4 Variation of trench height and FHCA with AA film effectiveness

4 Conclusion

The numerical optimization of trench film cooling parameters, viz. trench length, trench depth, film hole compound angle and blowing ratio were performed through response surface approach. The AA film effectiveness was used a response parameter. Experimental matrix system based on BBD was generated based on the selected range of trench parameters. Computational study was used to determine the response factor of EMS. Experimental results performed for validation show close agreement with numerical data. The optimized trench parameters were trench length 2D, trench height 0.83D, trench film hole compound angle 83.12° and blowing ratio of 1.96.

Acknowledgements The first author wishes to acknowledge Department of Science and Technology, Govt. of India for award of INSPIRE Fellowship.

References

1. Lu, Y., Dhungel, A., Ekkad, S.V., Bunker, R.S.: Effect of trench width and depth on film cooling from cylindrical holes embedded in trenches, ASME J. Turbomach. **131**(1), 011003-1-13 (2009)
2. Lee, K.D., Kim, K.Y.: Film cooling performance of cylindrical holes embedded in a transverse trench. Numer. Heat Transfer **65**(2), 127–143 (2014)
3. Kelishami, M.K., Lakzian, E.: Optimization of blowing ratio for film cooling on a flat plate. Int. J Num. Methods Heat Fluid Flow **27**(1), 104–119 (2017)
4. Alshehaby, M.M., Ragab, K.E., El-Gabry, L.: Numerical optimization of geometric parameters for shaped film cooling holes. In: Proceedings of ASME Turbo Expo 2017: turbomachinery Technical Conference and Exposition, GT-2017-65063, V05AT12A017, June 26–30. ASME, North Carolina (2017)
5. Elsayed, A.M., Owis, F.M., Rahman, M.M.A.: Film cooling optimization using numerical computation of the compressible viscous flow equations and simplex algorithm, Int. J Aerospace Eng. **2013**, Article ID 859465, 1–24 (2013)
6. Lee, K.D., Kim, S.M., Kim, K.Y.: Multi-objective optimization of a row of film cooling holes using an evolutionary algorithm and surrogate modeling. Numeri. Heat Transfer **63**(8), 623–641 (2013)
7. Savastano, W., Pranzitelli, A., Andrews, G.E., Biancolini, M.E., Ingham, D.B., Pourkashanian, M.: Goal driven shape optimization for conjugate heat transfer in an effusion plate. In: Proceedings of ASME Turbo Expo 2015: Turbine Technical Conference and Exposition, GT-2015-42251, V05AT10A003, June 15–19. ASME, Canada (2015)
8. Wang, C., Zhang, J., Zhou, J.: Optimization of a fan shaped hole to improve film cooling performance by RBF neural network and genetic algorithm. Aerospace Sci. Tech. **58**(1), 18–25 (2016)
9. Oguntade, H.I., Andrews, G.E., Ad, B., Ingham, D.B., Pourkashanian, M.: Improved trench film cooling with shaped trench outlets. ASME J Turbomach **135**(2), 1–15 (2013)

Analysis of Low Molecular Proteins Obtained from Human Placental Extract Considered as New Strategic Biomaterial for Pulp-Dentinal Regeneration

Ashmitha K. Shetty, Swaroop Hegde, Anitha Murali, Ashish J. Rai and Qhuba Nasreen

Abstract Dental pulp infections and trauma to pulpal tissue could apparently cause inflammation and eventually cause regeneration of pulp in cellular level as process of physiological events. Pulp response to infection is similar to any other tissue response to infection in the body. New methodologies explored in the field of regenerative dentistry involve grouping of growth factors, scaffolds and mesenchymal stem cells (MSCs). The generation of functional dentine by effective stimulation of odontoblast has not yet been achieved though there are numerous strategies have been employed in regenerative dentistry. Need for creating bioactive material enables the pulpal tissue to heal itself instead of permanently removing the injured tissue and placing prostheses. Human placental extract (HPE) was obtained by mechanical method from tissue of full-term individual placenta. The pH of the solution was estimated to be favourable for osteogenesis. The protein depletion technique was employed in order to obtain low molecular weight proteins may contribute to the variation in obtaining potential proteins. HPE obtained was characterized after in-solution digestion and mass spectrometry. Ten proteins were identified out of which multiple inositol polyphosphate phosphatase 1, DACT2, proliferation marker protein Ki-67 and zinc finger BED domain-containing protein 1 were of interest. The proteins of

A. K. Shetty (✉)
Department of Pedodontics and Preventive Dentistry, Faculty of Dental Sciences, M.S. Ramaiah University of Applied Sciences, Bangalore 560054, India

S. Hegde
Department of Conservative and Endodontics, Faculty of Dental Sciences, M.S. Ramaiah University of Applied Sciences, Bangalore 560054, India

A. Murali
Department of Pharmacology, Faculty of Pharmacy, M.S. Ramaiah University of Applied Sciences, Bangalore 560 054, India

A. J. Rai
Department of Oral and Maxillofacial Surgery, AB Shetty College of Dental Sciences, NITTE University, Mangalore 575018, India

Q. Nasreen
Department of Pharmacology. M.S. Ramaiah, University of Applied Sciences, Bangalore 560 054, India

D. Dutta and B. Mahanty (eds.), *Numerical Optimization in Engineering and Sciences*, Advances in Intelligent Systems and Computing 979,
https://doi.org/10.1007/978-981-15-3215-3_56

interest were either involved in cell proliferation directly or indirectly. Considering this information, use of this extract on human dental pulp cell can be functional.

Keywords Human placental extract · Pulp therapy · Mass spectrometry

Abbreviations

MALDI-TOF/TOF	Matrix-assisted laser desorption/ionization—Time-of-flight mass spectrometry
ESI-Q-TOF-MS/MS	Electrospray ionization quadrupole time-of-flight mass spectrometry

1 Introduction

Endodontic therapy for mangement of pulp involvement in case of deep dentinal caries currently involves removal of necrotic pulp tissue and replacing it with inert materials [1]. The common disadvantage with these materials will be loss of sensitivity and vitality and sometime result in secondary infection or post-operative fracture [2].

New methodologies have been explored in the field of regenerative dentistry involving grouping of growth factors, scaffolds and mesenchymal stem cells (MSCs) [3]. The generation of functional dentine by effective stimulation of odontoblast has not yet been achieved though there are numerous strategies have been employed in regenerative dentistry [3–5]. Need for creating bioactive material enables the pulpal tissue to heal itself instead of permanently removing the injured tissue and placing prostheses.

The human placenta ensures mutual exchange of nutrients of mother and foetus, determining growth and development of the foetus thus making placenta an exceptional temporary organ [6]. Placenta is believed to be an allogenic in nature, and its genotype could be entirely foreign [7].

The human placental organ possesses an extensive collection of bioactive proteins which are necessary for growth and development of foetus. The studies in the past have suggested the placental extracts comprise uracil, tyrosine, tryptophan and phenylalanine exhibiting strong anti-inflammatory and anti-oxidant activities [8–11]. However, at the same time, a large amount of research was little systemized and not always correlated with conventional pharmaceutical and other methods of treatment. Recent developments of cell therapy approaches along with opportunities for auto-banking significantly increased the interest in the placenta as a source of biological material.

The major aim of this study was to extract proteins effectively and characterization of this extract in most promising way for best clinical use in regeneration of dental pulp cells.

The present objective is critical as extraction of human placental extract has numerous proteins, lipids and glycoproteins. Thus, method of extract is important because maximum proteins of clinical relevance should be obtained. To confirm this, mass spectrometry is done for characterization of proteins in the extract.

2 Experimental Procedure

Ethical committee and associated hospital of Ramaiah University of Applied Sciences, Bangalore, provided approval for the study. Placenta was collected immediately following C-section deliveries ($n = 1$). Fresh full-term human placenta (40 weeks of gestation) weighing between 400 and 600 g within 30 mins of delivery was obtained. Excess blood was drained, and placenta was taken for further preparation. Placenta of the patient with the history of growth restriction, intrauterine, foetal abnormalities, hypertension, anaemia, diabetes, use of drugs or tobacco, premature labour/induced, multiple pregnancy, other medical or obstetric complications (HIV and Hepatitis) were excluded. A portion of placental tissue was minced using sterile steel blade. The minced tissue was processed in food processor placed in ice bath before operating at 2000 rpm for 8–10 mins. This forms a paste-like homogenate. A broad-spectrum proteolytic enzyme inhibitor cocktail (Sigma Aldrich, catalogue number P9599) was added before processing. To the homogenate 2:1(v/v) acetonitrile was added and vortex briefly, this was allowed to stand for 30 mins to precipitate high molecular weight proteins. This mixture was centrifuged at 14,000 rpm (13107xg) for 10 mins at 4 °C to get supernatant (~550 μL). This was separated and transferred to a new eppendorf tube and stored at −20 °C.

About 200 μl of final extract was subjected to mass spectrometric analysis for characterization of proteins. UltrafleXtreme MALDI-TOF/TOF (Bruker Daltonics) and Bruker Daltonics ESI-Q-TOF-(Maxis Impact) mass spectroscopy analysis was performed at Molecular Biophysics Unit, Indian Institute of Sciences, Bangalore. In-solution digestion was achieved by addition of 1 μL Trypsin, and reaction is incubated overnight at 30 °C.

Peptide sequencing was identified MASCOT and SwissProt software. Peptide mass fingerprinting (PMF) searches in MASCOT software were identified at 0.2 Da peptide tolerance. Identification was based on four peptides matching as significant. Randomized searches after the obtained peak list are entered in the query. MS/MS analysis was for peaks obtained by ESI-Q-TOF-MS/MS (Bruker Daltonics) were listed in the query for MS/MS search using software MASCOT sequence query. Search was optimized by fixing the parameters. The data were accepted for identification of peptide at 56 or more and significant at $p < 0.05$. SwissProt database was used for protein sequencing and protein identification. Parameter was fixed for taxa as homo sapiens, missed cleavages were set 1; ±0.1 Da for peptide mass tolerance;

carbamidomethyl (C) as fixed modification and oxidized methionine for variable modification.

3 Results and Discussions

Human placental extract obtained from 1 gm of homogenate into approximate 30 ml of aqueous extract solution. The final lyophilized 0.028 gm (28 mg) powder was obtained from 6 ml of extract and stored at 2–3 °C (Fig 1). About ten proteins were identified by MALDI-TOF/TOF (Table 1) and six proteins by ESI-Q-TOF (Table 2) mass spectroscopy analysis.

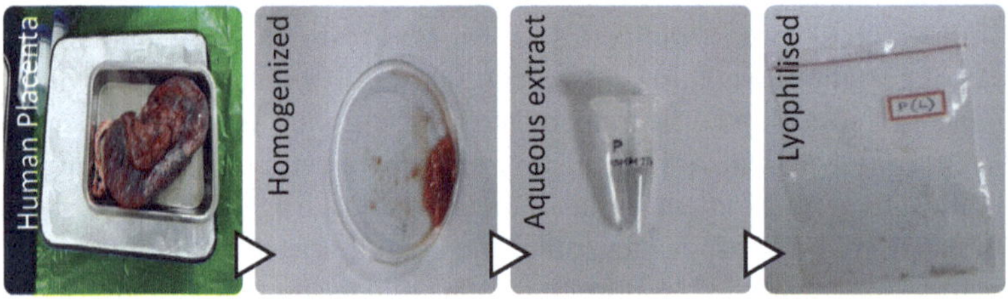

Fig. 1 Lyophilized extract obtained from placental homogenate

Table 1 Proteins identified using MALDI-TOF-MS/MS

Protein	Acc. No	Mw (kDa)	Score	Matched peaks
ADP-ribosylation factor-like protein 17	Q8IVW1	19.6	20	2
Protein CASC10	Q5T4H9	15.1	25	3
Zinc finger BED domain-containing protein 1	O96006	78.1	32	7
Keratin-associated protein 15-1	Q3L176	14.9	20	2
Putative uncharacterized protein encoded by ZNF503-AS2	A6NEH8	20.8	36	5
Multiple inositol polyphosphate phosphatase 1	Q9UNW1	55.6	31	8
CAP-Gly domain-containing linker protein 2	Q9UDT6	116.2	42	16
Proliferation marker protein Ki-67	P46013	358.4	46	32
39S ribosomal protein L40, mitochondrial	Q9NQ50	24.4	24	5
Haemoglobin subunit beta	P68871	16.1		

Table 2 Proteins identified by ESI-Q-TOF/TOF

Mass	Gene	Protein	Sequence
16104	HBB_HUMAN	Haemoglobin subunit beta	K.VNVDEVGGEALGR.L
71352	ALBU_HUMAN	Serum albumin	K.LVAASQAALGL. K.DVFLGMFLYEYAR.R K.DVFLGMFLYEYAR.R K.EFNAETFTFHADICTLSEK.E
15306	HBA_HUMAN	Haemoglobin subunit alpha	K.VGAHAGEYGAEALER.M
521940	STAR9_HUMAN	StAR-related lipid transfer protein 9	R.RQVGEAAAGR.G
83287	DACT2_HUMAN	Dapper homologue 2	R.AGGPLAR.G
68726	EWS_HUMAN	RNA-binding protein EWS	R.GGPGGPGGPGGPMGRMGGR.G

3.1 Determination of pH of the Solution

The pH of human placental extract was estimated at 7.44 as per pH digital metre. The pH of HPE was found to be alkaline and mimics the physiological pH $\approx$ 7.4. The pH of human placental extract has not been mentioned in any of the literature. Thus, an attempt to determine the pH of the extract would be useful to determine bioactivity when it is used in therapy. Pulp-capping material like Biodentine shows pronounced anti-bacterial property because of its increased pH to 12.5. Disinfection of dentine and growth inhibition of bacteria was observed due to increased pH as calcium was released during the setting phase [12]. The pH of the biomaterial plays a vital role as activity alkaline phosphatase is highest at a pH of 7.37. Mineralization is sensitive to pH variation. Though a higher and an alkaline pH is required for osteoinduction, but the activity diminishes above 7.37 level. Moreover, at 8.0 pH, mineralization diminishes in in -vitro and in -vivo. Thus, pH variation is crucial at pulp and pulp-capping intersurfaces [13].

3.2 Determination of Protein Concentration by Bradford Assay

Total protein concentration estimated was 175 μg/ml. Whereas in extract prepared by water-soluble method, Bradford method showed 1 ml contained 50 μg protein [14]. Thus, protein extraction method in the present study employed acetonitrile depletion which removed high molecular weight protein leaving behind low molecular weight protein. Study on consequence of porcine placental extract on proliferation of human fibroblasts showed that protein concentration was 100–200μg/ml was adequate for proliferation of somatic cells originating from mesenchymal stem cells in -vitro [15] (Table 3).

3.3 *Identification of Potential Proteins of Interest (Table 3)*

3.3.1 MINP1_HUMAN Multiple Inositol Polyphosphate Phosphatase 1

Multiple inositol polyphosphate phosphatase (MIPP-1) acts as an enzyme involved in inositol phosphate metabolism. It catalyses higher inositol polyphosphates to inositol (1,4,5) trisphosphate-3 (Ins(1,4,5)P3) which in turn is a second messenger which consecutively regulates calcium internal storage in the cell. (Ins(1,4,5)P3) plays a crucial role cellular processes like metabolism, contraction, proliferation and neural activity [16]. For calcium mobilization from Ins (1, 4, 5) P3 sensitive calcium pools which are linked to calcium channels, Ins (1, 4, 5)P3 should bind to receptors associated with it. Calcium release mediated by Ins (1, 4, 5) P3 can be regulated by various drugs and physiological processes [17]. It is observed that MINPP1 in -vivo participated in homeostatic regulation of and inositol pentakisphosphate InsP5 and inositol hexakisphosphate InsP6 which are concerned cell growth and differentiation of chondrocytes [18]. MIPP1 is highly expressed in liver, placenta and kidney. It is observed that human MIPP1 is clinical significant as an increased transformation of inositol phosphate is important in ossification of bone in normal as well as pathological situation such as osteoarthritis and healing of fracture [19].

3.3.2 DACT2_HUMAN

Dapper homologue (DACT) 2 belongs to DACT gene family. DACT gene involves in Wnt signalling pathway [20]. The two signalling systems in animals are Wnt signalling and transforming growth factor-beta (TGF-β) signalling. They control various processes required for embryonic development, cell differentiation, proliferation and migration. Its role is also significant in cancer, tissue regeneration and stem cell behaviour. DACT 2 is specific in modulating TGF β-dependent wound healing [21].

3.3.3 Proliferation Marker Protein Ki-67

Ki-67 involves in wound healing in early stages of cell proliferation by association with nuclear antigen synthesis in cell cycle phases. Considered as a potential marker for keratinocyte proliferation and remodelling of epithelium [22]. During mitosis is seen that the proteins are relocated to the surface, but Ki-67 protein is present during

Table 3 Biological functions of identified protein according to UniProt

Mass	Gene	Protein	Biological pathways UniProt
19607	ARL17_HUMAN	ADP-ribosylation factor-like protein 17	They are Arf GTPase-activating proteins (GAPs) mainly involved in cell migration Helps in protein movement It amends formations of vesicles contained in Golgi apparatus
78106	ZBED1_HUMAN	Zinc finger BED domain-containing protein 1	Promotes cell proliferation by binding to DNA element of the promoter region of the gene Initiates transcription by binding to histone H1 promoter
14968	KR151_HUMAN	Keratin-associated protein 15-1	It is part of essential hair keratin-associated proteins (KRTAP), by creating cross-linking bond with hair keratin residues like cysteine forming to rigid and resistant hair shaft Keratin protein matrix with high-sulphur and high-glycine-tyrosine keratins
55654	MINP1_HUMAN	Multiple inositol polyphosphate phosphatase 1	Induces transition of chondrocytes to proliferate or causes of hypertrophy Involves in endochondral ossification in bone development
116231	CLIP2_HUMAN	CAP-Gly domain-containing linker protein 2	Involves in control of organelle translocations specific to brain Helps to link dentritic lamellar body the neuronic gap junction to microtubules
358474	KI67_HUMAN	Proliferation marker protein Ki-67	It covers considerable portion of chromosome surface, thus maintaining mitotic chromosomes dispersed in the cytoplasm. Due to its high net electric charge behaves as a surfactant on the chromosome dispersing and allowing chromosomal motility independently

(continued)

Table 3 (continued)

Mass	Gene	Protein	Biological pathways UniProt
521940	STAR9_HUMAN	StAR-related lipid transfer protein 9	Plays role in mitosis by stabilizing the pericentriolar material
83287	DACT2_HUMAN	Dapper homologue 2	It involves in regulating signalling pathways intracellular during formation of the foetus It reduces TGF-beta signalling inhibits re-epithelization of skin wounds
68726	EWS_HUMAN	RNA-binding protein EWS	May cause unusual activation of target genes of fusion proteins But generally functions as a transcriptional repressor These proteins may be involved during the progression of tumour

all the phase of cell division that is G1, S, G2 and mitosis, and it is not present at the resting phase G0. Thus, Ki-67 makes an excellent cell proliferative maker [23].

3.3.4 Zinc Finger BED Domain-Containing Protein 1(ZBED 1)

ZBED genes control range of functions in vertebrate tissues. Specifically, ZBED1 is associated with cell proliferation. It controls ribosomal protein gene essential for cell proliferation. ZBED1 is also identified as human-Ac, hDREF and TRAMP [24]. hDREF behaves as a transcriptional factor in the process of cell proliferation. It is linked with up regulation of ribosomal protein (RP) genes. Consequently, up-regulated RP genes modulate the transition of G1-S phase of cell proliferation. The expression of hDREF reaches peak during S phase in normal human fibroblast. Thus, hDERF plays a significant task in cell cycle progression [25].

4 Conclusion

HPE was obtained by mechanical method from tissue of full-term individual placenta. The pH of the solution was estimated to be favourable for osteogenesis. The protein depletion technique was employed may contribute to variation in obtaining potential low molecular weight proteins. Ten proteins were identified by mass spectrometry, out of which multiple inositol polyphosphate phosphatase 1, DACT2, proliferation marker protein Ki-67 and zinc finger BED domain-containing protein 1 were of interest. These proteins either involved in cell proliferation directly or indirectly. Since placental extract have been used in an extensive range of therapies, the proteins obtained in present study correlated with the data available on confirmed in -vivo and in -vitro experiments. In view of this information, the use of this extract on human dental pulp can be purposeful.

Acknowledgements This study was executed in Faculty of Pharmacy at M.S. Ramaiah University of Applied Sciences, Bangalore. We acknowledge Molecular Biophysics Unit (MBU), Indian Institute of Science (IISc), Bangalore, for mass spectroscopic analysis.

References

1. Yu, C., Abbott, P.V.: An overview of the dental pulp: its functions and responses to injury. Aust. Dent. J. **52**, S4–S6 (2007)
2. Cordeiro, M.M., Dong, Z., Kaneko, T., Zhang, Z., Miyazawa, M., Shi, S., Smith, A.J., Nör, J.E.: Dental pulp tissue engineering with stem cells from exfoliated deciduous teeth. J. Endod. **34**(8), 962–969 (2008)
3. Albuquerque, M.T., Ryan, S.J., Münchow, E.A., Kamocka, M.M., Gregory, R.L., Valera, M.C., Bottino, M.C.: Antimicrobial effects of novel triple antibiotic paste–mimic scaffolds on Actinomyces Naeslundii biofilm. J. Endod. **41**(8), 1337–1343 (2015)
4. Galler, K.M., D'Souza, R.N., Federlin, M., Cavender, A.C., Hartgerink, J.D., Hecker, S., Schmalz, G.: Dentin conditioning codetermines cell fate in regenerative endodontics. J. Endod. **37**(11), 1536–1541 (2011)
5. Yang, J., Yuan, G., Chen, Z.: Pulp regeneration: current approaches and future challenges. Frontiers Physiol. **7**, 58 (2016)
6. Huppertz, B.: Placental origins of preeclampsia: challenging the current hypothesis. Hypertension **51**(4), 970–975 (2008)
7. Veenstra van Nieuwenhoven, A.L., Heineman, M.J., Faas, M.M.: The immunology of successful pregnancy. Hum. Reprod. Update **9**(4), 347–357 (2003)
8. Togashi, S., Takahashi, N., Iwama, M., Watanabe, S., Tamagawa, K., Fukui, T.: Antioxidative collagen-derived peptides in human-placenta extract. Placenta **23**(6), 497–502 (2002)
9. Park, S.Y., Phark, S., Lee, M., Lim, J.Y., Sul, D.: Anti-oxidative and anti-inflammatory activities of placental extracts in benzo [a] pyrene-exposed rats. Placenta **31**(10), 873–879 (2010)
10. Sur, T.K., Biswas, T.K., Ali, L., Mukherjee, B.: Anti-inflammatory and anti-platelet aggregation activity of human placental extract. Acta Pharmacol. Sinica **24**(2), 187–192 (2003)
11. Banerjee, K.K., Bishayee, A., Chatterjee, M.: Anti-inflammatory effect of human placental extract: a biochemical mechanistic approach. Rivista europea per le scienze mediche e farmacologiche=European review for medical and pharmacological sciences=Revue europeenne pour les sciences medicales et pharmacologiques **14**(6), 361–366 (1992)
12. Singh, H., Kaur, M., Markan, S., Kapoor, P.: Biodentine: a promising dentin substitute. J. Interdiscipl. Med. Dent. Sci. **2**(140), 2 (2014)
13. Song, M., Bo, Y., Kim, S., Hayashi, M., Smith, C., Sohn, S., Kim, E., Lim, J., Stevenson, R.G., Kim, R.H.: Clinical and molecular perspectives of reparative dentin formation: lessons learned from pulp-capping materials and the emerging roles of calcium. Dent. Clin. **61**(1), 93–110 (2017)
14. Samiei, F., Jamshidzadeh, A., Noorafshan, A., Ghaderi, A.: Human placental extract ameliorates structural lung changes Iinduced by Amiodarone in rats. Iranian J. Pharm. Res. IJPR **15**(Suppl), 75 (2016)

15. Yoshizawa, R.S.: Public perspectives on the utilization of human placentas in scientific research and medicine. Placenta **34**(1), 9–13 (2013)
16. Yu, J., Leibiger, B., Yang, S.N., Caffery, J.J., Shears, S.B., Leibiger, I.B., Barker, C.J., Berggren, P.O.: Cytosolic multiple inositol polyphosphate phosphatase in the regulation of cytoplasmic free Ca2 + concentration. J. Biol. Chem. **278**(47), 46210–46218 (2003)
17. Berridge, M.J., Irvine, R.F.: Inositol phosphates and cell signalling. Nature **341**(6239), 197 (1989)
18. Kilaparty, S.P., Agarwal, R., Singh, P., Kannan, K., Ali, N.: Endoplasmic reticulum stress-induced apoptosis accompanies enhanced expression of multiple inositol polyphosphate phosphatase 1 (Minpp1): a possible role for Minpp1 in cellular stress response. Cell Stress Chaperones **21**(4), 593–608 (2016)
19. Caffrey, J.J., Hidaka, K., Matsuda, M., Hirata, M., Shears, S.B.: The human and rat forms of multiple inositol polyphosphate phosphatase: functional homology with a histidine acid phosphatase up-regulated during endochondral ossification. FEBS Lett. **442**(1), 99–104 (1999)
20. Wang, S., Dong, Y., Zhang, Y., Wang, X., Xu, L., Yang, S., Li, X., Dong, H., Su, L., Ng, S.S.M., Chang, Z.: DACT2 is a functional tumor suppressor through inhibiting Wnt/β-catenin pathway and associated with poor survival in colon cancer. Oncogene **34**(20), 2575 (2015)
21. Schubert, F.R., Sobreira, D.R., Janousek, R.G., Alvares, L.E., Dietrich, S.: Dact genes are chordate specific regulators at the intersection of Wnt and Tgf-β signaling pathways. BMC Evol. Biol. **14**(1), 157 (2014)
22. Loeffelbein, D.J., Rohleder, N.H., Eddicks, M., Baumann, C.M., Stoeckelhuber, M., Wolff, K.D., Drecoll, E., Steinstraesser, L., Hennerbichler, S, Kesting, M.R.: Evaluation of human amniotic membrane as a wound dressing for split-thickness skin-graft donor sites. BioMed Res. Int. (2014)
23. Scholzen, T., Gerdes, J.: The Ki-67 protein: from the known and the unknown. J. Cell. Physiol. **182**(3), 311–322 (2000)
24. Hayward, A., Ghazal, A., Andersson, G., Andersson, L., Jern, P.: ZBED evolution: repeated utilization of DNA transposons as regulators of diverse host functions. PLoS ONE **8**(3), e59940 (2013)
25. Yamashita, D., Sano, Y., Adachi, Y., Okamoto, Y., Osada, H., Takahashi, T., Yamaguchi, T., Osumi, T., Hirose, F.: hDREF regulates cell proliferation and expression of ribosomal protein genes. Mol. Cell. Biol. **27**(6), 2003–2013 (2007)

Predictive Data Optimization of Doppler Collision Events for NavIC System

P. Sathish and D. Krishna Reddy

Abstract Navigation with Indian Constellation (NavIC) is satellite-based navigation system developed by Indian Space Research Organization (ISRO), India. It consists of seven satellites, among them, three are geostationary (GEO) satellites, and the rest are geosynchronous satellites. There are several factors that effect the positional accuracy of the NavIC system, and among them, one of the important parameter is Doppler collision (DC). The occurrence of the DC depends upon the usage of geostationary (GEO) satellites in position estimation. The DC may occur when the relative Doppler shift between two satellites is less than the bandwidth of receiver code tracking loop. To analyze the DC events, the required navigational data are collected from IRKS-GPS-SBAS (IGS) receiver located at low latitude station. There are variations in the values of Doppler shift of GEO satellites and have a significant impact on the Doppler collision prediction. To predict the DC accurately, the data optimization is very much essential for precise positioning applications. This paper presents the predictive data optimization of DC events by using various filtering methods such as moving average, Savitzky–Golay and median filter. Among three methods, Savitzky–Golay method provides better data optimization to predict short-term DC events as compared with other methods.

Keywords Doppler collision · NavIC · Data optimization · GEO satellites

1 Introduction

NavIC system gives position, navigation and timing (PNT) services over the Indian landmass extended up to 1500 km beyond its political boundaries [1]. Doppler collision is not a major constraint in GPS or satellite systems that use frequency division

P. Sathish (✉) · D. Krishna Reddy
Department of ECE, Chaitanya Bharathi Institute of Technology, Hyderabad, Telangana, India

D. Krishna Reddy
e-mail: dkreddy@cbit.ac.in

D. Dutta and B. Mahanty (eds.), *Numerical Optimization in Engineering and Sciences*, Advances in Intelligent Systems and Computing 979,
https://doi.org/10.1007/978-981-15-3215-3_57

multiple access (GLONASS). The prediction of DC events is useful in RTK applications in order to decrease the pseudorange error [2, 3]. Prediction of DC events in NavIC system with data optimization gives better positional accuracy [4, 5].

2 DC Predictive Analysis Methods for NavIC

The Doppler shift (DS) of GEO satellites is calculated from RINEX data as shown in Fig. 1. By using different filtering techniques namely moving average, Savitzky–Golay and median filtering, Doppler shift fluctuations of GEO satellites are smoothed. The time occurrence of the DC for different days is calculated using simple arithmetic progression, DC is predicted for forthcoming days (Figs. 2 and 3).

2.1 Moving Average Filter Method

Moving average method is the simplest technique for smoothening of signals consisting of equidistant points. The smoothed point $(y_k)_s$ is the average of an odd numbers of consecutive $2n + 1$ $(n = 1, 2, 3 \ldots)$ points of the raw data [6].

$$(Y_k)_s = \sum_{i=-n}^{n} \frac{Y_{k+i}}{(2n + 1)} \tag{1}$$

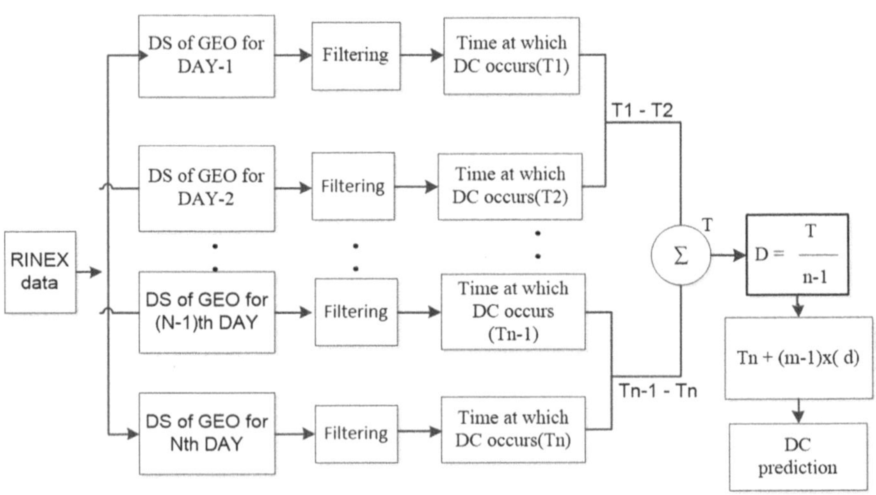

Fig. 1 Block diagram for Doppler collision prediction in NavIC

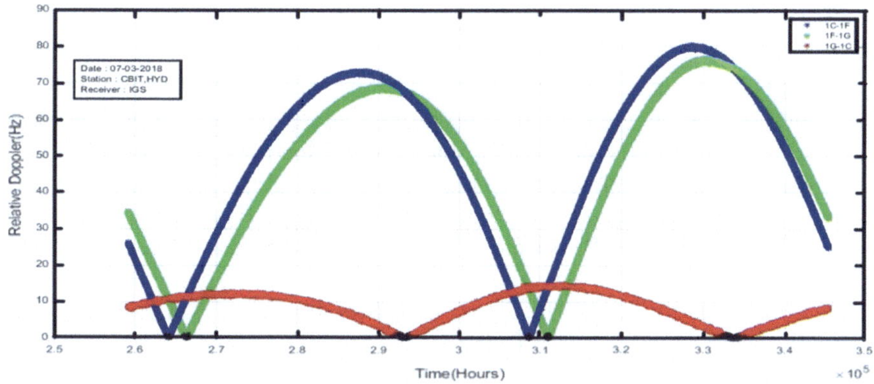

Fig. 2 Doppler collision for IRNSS GEO satellites

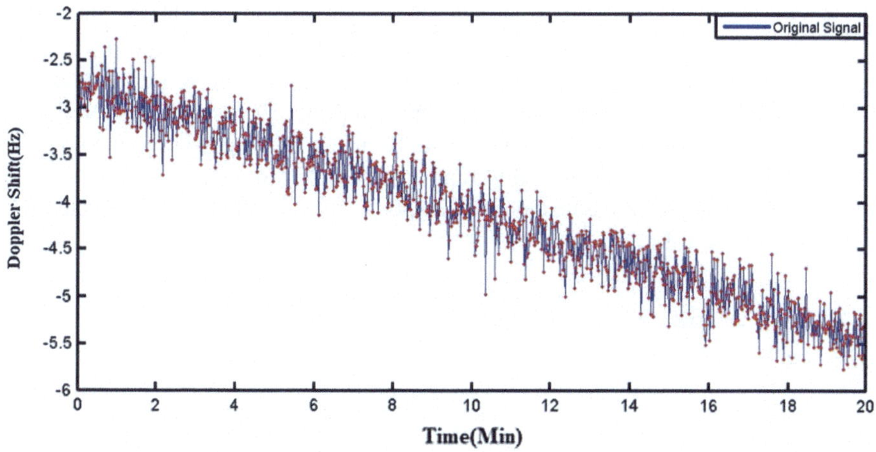

Fig. 3 Doppler shift fluctuations of IRNSS-1C satellite

Savitzky and Golay filter method is a digital filtering method, to increase the precision of the data points without distorting the signal tendency using convolution [7].

$$Y_j = \sum_{i=-m}^{i=m} \frac{C_i Y_{j+i}}{N} \quad (2)$$

Median filter method is a nonlinear digital filtering method, often used to eliminate fluctuations from signal. It is pre-processing step which provides better removal of the noise of the signal [8, 9].

3 Results and Discussions

The relative Doppler shift for the three GEO satellite pairs 1C–1F, 1C–1G and 1F–1G represented in blue, green and red color, respectively, as shown Fig. 5. The black color portion in the graph represents the duration for which Doppler collision occurs on that particular day which means the relative Doppler frequencies between the two satellites are less than the code lock loop bandwidth, i.e., 1 Hz (Table 1).

For IRNSS 1C–1F and 1F–1G, the Doppler collision duration is less than 17 min, whereas the duration of Doppler collision for IRNSS 1C–1G is quite long. First DC event was started at 14:38:17 IST and long lasting for 38 min and 26 s, and the second event was occurring at 1:39:33 IST long lasting for 1 h 04 min 31 s respectively as shown in Table 2.

In order to predict the Doppler collision events more accurately and to minimize the error of estimation, the fluctuations in the Doppler frequencies of the satellites as shown in Fig. 6 should be normalized using different filtering methods with minimum data samples.

The analysis of DC prediction is carried out using three different methods namely moving average, Savitzky–Golay and median filtering. Further, these methods are also evaluated for data optimization. The predicted DC events are compared with observed DC events using IGS receiver data. By using moving average method, the calculated smoothed signal with different data sample values is shown in Fig. 4. When the data samples are considered as 20 per one frame length, the predicted DC error is minimum value.

Table 1 DC events starting time and duration of IRNSS GEO satellites

GEO Satellite pairs	DC EVENT 1		DC EVENT 2	
	Start time (IST)	Duration (IST)	Start time (IST)	Duration (IST)
1C–1F	6:49:09	7:34	19:10:24	7:43
1F–1G	7:49:08	8:31	19:49:25	7:41
1C-1G	**14:38:17**	**38:26**	**1:39:33**	**1:04:31**

[a]*IST* Indian Standard Time

Table 2 Comparison of various prediction DC methods for data optimization

Prediction method	DC events	Predicted		Observed		DC error
		Start time	Duration	Start time	Duration	
Moving average	Event 1	14:40:40	38:26	14:29:42	43:37	5:11
	Event 2	1:55:07	39:25	1:53:18	38:48	0:47
Savitzky–Golay	**Event 1**	**14:40:33**	**31:35**	**14:42:48**	**31:38**	**0:03**
	Event 2	**1:55:11**	**39:30**	**1:53:18**	**38:48**	**0:42**
Median filter	Event 1	14:38:39	39:01	14:38:17	38:26	0:35
	Event 2	1:45:28	54:30	1:39:33	1:04:31	10:01

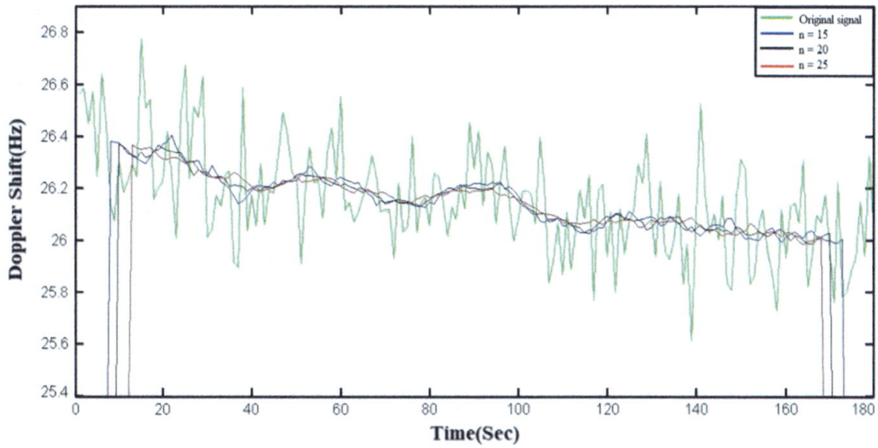

Fig. 4 Doppler shift estimation using moving average method

Similarly, by using Savitzky–Golay and median filtering methods, the calculated Doppler shift values with different data samples are shown in Figs. 5 and 6. It is observed that for short-term prediction, the Savitzky–Golay method gives less DC error as shown in Fig. 5, whereas for long-term prediction, median filter gives less DC error.

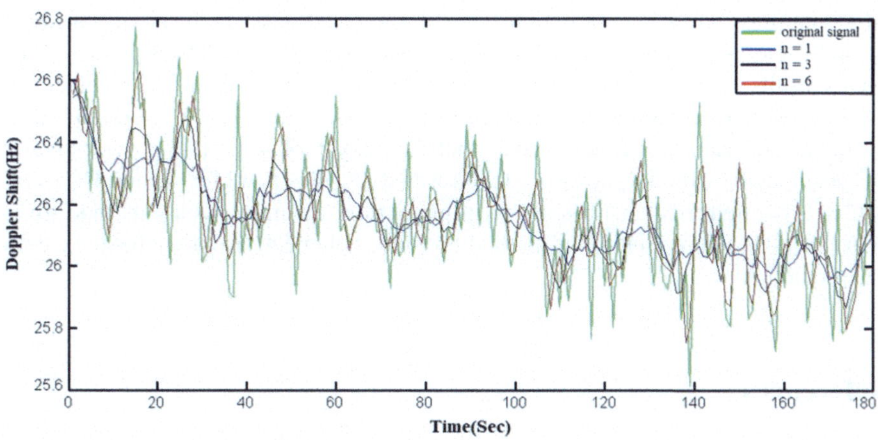

Fig. 5 Doppler shift estimation using Savitzky–Golay method

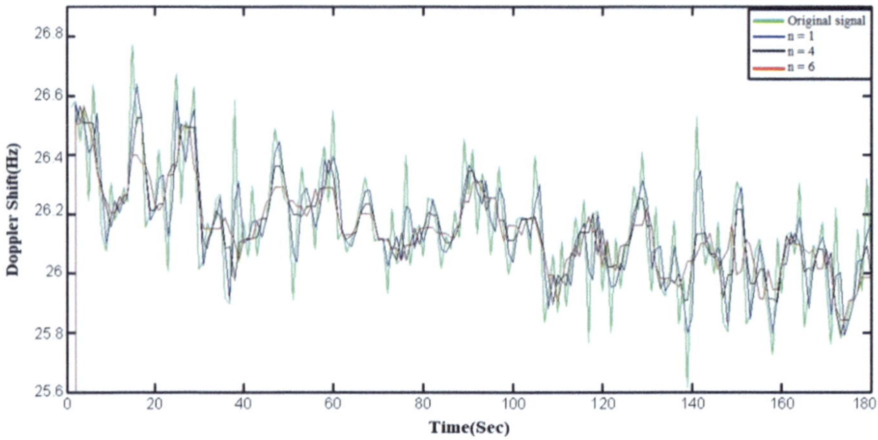

Fig. 6 Doppler shift estimation using median filter method

4 Conclusion

The time duration for Doppler collision events for IRNSS 1C–1G satellite pair is more, i.e., 1 h 43 min as compared with other two satellites pairs (IRNSS 1C–1F, IRNSS 1F–1G). An efficient algorithm has developed to estimate and predict Doppler collision in NavIC system. To remove the fluctuations in Doppler shift of GEO satellites and for DC prediction, three filtering methods namely moving average, Savitzky–Golay and median filter are implemented successfully. These predicted DC events results are compared with observed DC events obtained from IGS receiver. All the three methods are implemented for different data samples in order to optimize data for better DC accuracy. It is observed that Savitzky–Golay method provides better DC accuracy, i.e., the DC error is 0.03 s for first DC event and for second DC event 0.42 s. Also noticed that for short-term prediction, Savitzky–Golay method gives better results, and median filter method provides better DC accuracy for long-term DC prediction.

References

1. Murthynjaya, L., Ramasubramanian, R.: IRNSS Signal in Space Interface Control Document for Standard Positioning Service, Version 1.1. ISRO, India (2017, August)
2. Bhandari, V., O'Keefe, K.: What is Doppler collision and is it a problem in GNSS?, Inside GNSS Article **11**(1), 26–36 (2016). USA
3. Spilker, J.J., Axelrad, P., Parkinson, B.W., Enge, P. (eds.): Global Positioning System: theory and Applications, vol. I. American Institute of Aeronautics and Astronautics, pp. 547–566 (1996)
4. Bhandari, V., O'Keefe, K.: Characterization of Doppler collision and its impact on carrier phase ambiguity resolution using geostationary satellites. GPS Solutions **21**(4), 1707–1720 (2017). Springer, Heidelberg

5. Cox, D.T., Shallberg, K.W., Manz, A.: Definition and analysis of WAAS receiver multipath error envelopes. J. Navig. **46**(4), 271–282 (1999)
6. Ekhosuehi, V., Dickson, E., Omorogbe, A.: On forecast performance using a class of weighted moving average processes for time Series. J. Nat. Sci. Res. **6**(13), 87–92 (2016)
7. Bian, J.H., Li, A., Song, M., Ma, L., Jiang, J.: Reconstruction of NDVI time-series datasets of MODIS based on Savitzky-Golay filter. J. Remote Sens. (2010). 1007–4619 04-725-17
8. Misra, P., Enge, P.: Global Positioning System: signals, Measurements, and Performance, 2nd edn.,, pp. 147–193. Ganaga-Jamuna Press, Lincoln (2006)
9. Garin, L., Rousseau, J.-M.: Enhanced strobe correlator multipath rejection for code & carrier. In: Proceedings of ION GPS 1997, 16–19 September, pp. 559–568. Kansas City, USA (1997)

The manufacturer's authorised representative in the EU is Springer Nature Customer Service Centre GmbH, Europaplatz 3, 69115 Heidelberg, Germany. If you have any concerns regarding our products, please contact ProductSafety@springernature.com

Printed and bound by CPI Group (UK) Ltd, Croydon, CR0 4YY
30/09/2026
02225556-0017